AF872763

Algebraic Biology, Matrix Genetics, and Genetic Intelligence

Other Titles by the Authors

Symmetrical Analysis Techniques for Genetic Systems and Bioinformatics: Advanced Patterns and Applications

Mathematics of Bioinformatics: Theory, Practice, and Applications

Algebraic Biology, Matrix Genetics, and Genetic Intelligence

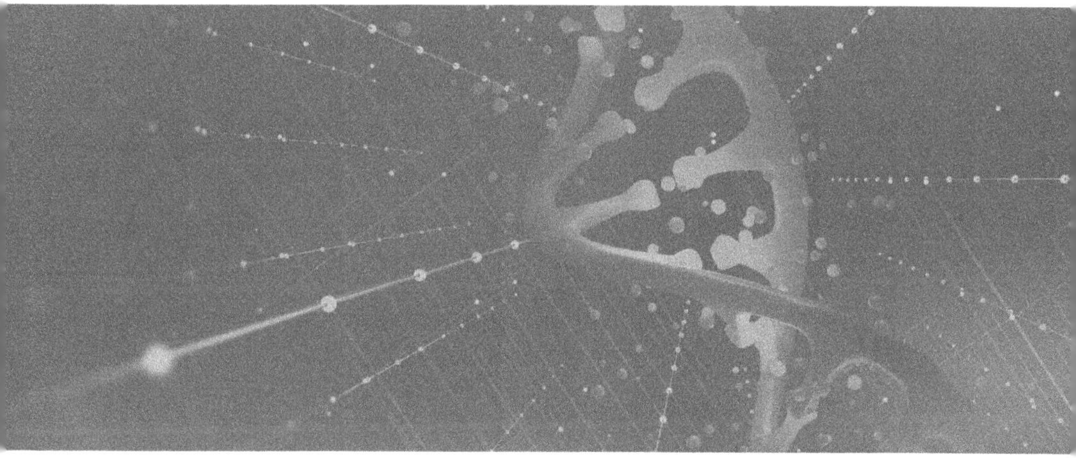

Sergey Petoukhov
Russian Academy of Sciences, Russia

Matthew He
Nova Southeastern University, USA

World Scientific

NEW JERSEY • LONDON • SINGAPORE • BEIJING • SHANGHAI • HONG KONG • TAIPEI • CHENNAI • TOKYO

Published by

World Scientific Publishing Co. Pte. Ltd.
5 Toh Tuck Link, Singapore 596224
USA office: 27 Warren Street, Suite 401-402, Hackensack, NJ 07601
UK office: 57 Shelton Street, Covent Garden, London WC2H 9HE

Library of Congress Control Number: 2023018062

British Library Cataloguing-in-Publication Data
A catalogue record for this book is available from the British Library.

ALGEBRAIC BIOLOGY, MATRIX GENETICS, AND GENETIC INTELLIGENCE

Copyright © 2024 by World Scientific Publishing Co. Pte. Ltd.

All rights reserved. This book, or parts thereof, may not be reproduced in any form or by any means, electronic or mechanical, including photocopying, recording or any information storage and retrieval system now known or to be invented, without written permission from the publisher.

For photocopying of material in this volume, please pay a copying fee through the Copyright Clearance Center, Inc., 222 Rosewood Drive, Danvers, MA 01923, USA. In this case permission to photocopy is not required from the publisher.

ISBN 978-981-127-825-9 (hardcover)
ISBN 978-981-127-826-6 (ebook for institutions)
ISBN 978-981-127-827-3 (ebook for individuals)

For any available supplementary material, please visit
https://www.worldscientific.com/worldscibooks/10.1142/13468#t=suppl

Desk Editor: Shaun Tan Yi Jie

Typeset by Stallion Press
Email: enquiries@stallionpress.com

Preface

Many sciences turn to biology in connection with the anthropomorphic slogans "man is the measure of all things" and "man is the cosmos" in the hope of finding support for solving their own problems. This point of view can be seen, for example, in the book about the human phenomenon [1], whose author believed the following. To find out how the world was formed and what its further fate is, one should "decipher" the person; because of this, future synthetic science will take man as a basis; this will be a new era in science, in which there will be a complete understanding that man as a "subject of knowledge" is the key to the whole science of nature. The book [2] also supports that the theory of living systems — but not physics — is the most appropriate center for the vision of the world. Here one can recall the statement of Descartes in his book *Description of the Human Body* about the importance of personal cognition: "*There is no more fruitful occupation than to try to know itself*".

Poincaré stated that the true, only goal of science is the discovery not of a mechanism, but of unity [3]. It is the hope of the authors of this book that its presented results will further this goal by revealing the hidden relationships between the genetic coding system and various fields of science and culture.

Today, the sequence of nitrogenous bases in DNA is recognized as the main source of biological information. Therefore, in many sciences and arts — linguistics, music, psychology, etc. — attempts are made to establish structural links with the genetic coding system, to take its features as the basis for the classification and comprehension systems in their field. Our book, using extensive phenomenological materials, convincingly shows that the general biological structures associated with the genetic inheritance of organisms are related to algebraic structures known in mathematics, physics and computer science. In other words, there is a vast intersection between the world of biological phenomena and the world of algebraic abstractions. One of the presented examples of the study of this intersection is the discovery of universal rules for the stochastic organization of DNA information sequences in the genomes of higher and lower organisms. These phenomenological rules of genomes have the form of non-trivial algebraic invariants that remain unchanged over millions of years of biological evolution, during which millions of species of organisms die off and new species arise (although locally genomic sequences are modified by mutations, natural selection mechanisms, etc.).

The development of modern mathematical natural sciences is based on the use of certain mathematical tools. Mathematical tools of theoretical research can be compared with glasses for a visually impaired person: adequate glasses provide a person with a clear and beautiful picture of reality, which he had previously seen as blurred and hidden by fog. Darwin once wrote: "*I have deeply regretted that I did not proceed far enough at least to understand something of the great leading principles of mathematics; for men thus endowed seem to have an extra sense.*" [4].

It is known about "*the unreasonable effectiveness of mathematics in the natural sciences*" [5]. However, in biological research, mathematics is traditionally used as an auxiliary tool for processing experimental data since biological bodies, in terms of their atomic and molecular composition, represent an extremely complex and heterogeneous system. But is mathematics, which ensured the progress of modern physics, computer science, and other sciences about inanimate objects, not able to generate a similar progress in the field of knowledge of living matter by revealing its secrets as a mathematically organized entity? The content of this book supports the idea that mathematics can help in the progress in knowledge of living matter.

In biology itself today, from a physical and mathematical point of view, there are very few things which are well formalized, and which have a level of general biological significance. The genetic code in this sense occupies a very special place. Its molecular biological foundations have been largely investigated, formalized and are the same for all living things. In connection with the identified key value of genetic informatics, a new definition of life itself has appeared: "*Life is a partnership between genes and mathematics.*" [6]. But what kinds of mathematics are a partner of genes? The content of our book attracts attention to matrix-tensor mathematics as an effective tool for revealing of hidden regularities of the genetic coding system. In the book, this tool allows effective studying of emergent algebraic properties of the complex structured system of DNA and RNA alphabets jointly with regular features of the genetic coding.

In particular, it led to the doctrine of energy-informational evolution based on bio-antenna arrays. This doctrine considers the evolution of biological systems for the first time in connection with the amazing emergent properties of antenna arrays, which are widely used in many areas of technology. The unique properties of antenna arrays in engineering provide noise-resistant multi-channel operation with weak signals against a background of strong noise. Accordingly, the named doctrine explains the similar abilities of living organisms and Mendel's law of independent noise-resistant inheritance of traits.

The book contains examples of biological phenomena confirming that biological evolution has long used the amazing emergent properties of antenna arrays and their wave functioning: complex faceted eyes of insects and invertebrates, inherited phenomena of biological echolocation and electrolocation, and so on. It is assumed that the fundamental problem of biological dissymmetry and molecular chirality, considered since the time of L. Pasteur, is associated with the polarization of electromagnetic waves of bio-antenna arrays. This doctrine allows a new understanding of tissue regeneration and other biological phenomena, as well as the development of new technologies. In the whole, the presented doctrine states that bio-antenna arrays with their coordinated waves orchestrate life, that is, have a key meaning. Correspondingly, one can suppose that the secret of the structural organization and origin of the genetic code, as well as the origin of organisms, must be sought not in the random combination of the molecular elements of the genetic code, but in the emergent properties of self-organizing systems of bio-antenna arrays with their wave energy activity.

The operation of antenna arrays is based on resonant mechanisms, which use the fundamental physical principle of energy minimization: in resonant combining of parts into a single unit, each member of the ensemble requires less energy for performing its own work than when working individually. The book devotes much attention to resonant mechanisms in biological phenomena.

Recent technological progress in the determination of genomic sequences yields many millions of gene sequences now. But what do these sequences tell us and what generalities and rules govern them? The modern situation in the theoretic field of genetic informatics can be characterized by the following citation: "*What will we have when these genomic sequences are determined? What do we have now in the 10 million nucleotides of sequence data determined to date? We are in the position of Johann Kepler when he first began looking for patterns in the volumes of data that Tycho Brahe had spent his life accumulating. We have the program that runs the cellular machinery, but we know very little about how to read it.*" [7]. In our view, the discovery of hidden regularities in the genetic code system, which are described in the book, can help in understanding this problem.

From an information viewpoint, biological organisms are informational essences. They obtain genetic information from their ancestors and transmit it to descendants. In the biological literature, it is possible quite often to come across the statement that living organisms are the texts starting from a molecular level of their organization. Just from the information-hereditary point of view all living organisms are unified: all of them have identical bases of system of genetic coding. A conception of informational nature of living organisms is reflected in the words by Chernavskiy: "*Notions of "information" or "valuable information" are not utilized in physics of non-biological nature because they are not needed there. On the contrary, in biology notions of "information" and especially "valuable information" are main ones; understanding and description of phenomena in biological nature are impossible without these notions. A specificity of "living matter" lies in these notions."* [8].

The data of the book are connected with the genetic intelligence, that is, the part of the inherited potentialities of living organisms which solves the following task: how is it possible, using information from genetic molecules, to build, for example, from a single fertilized cell a complex organism with trillions of cells in such a way that it possesses parental traits, which are reproduced in a noise-resistant multi-channel way, despite strong noises and constant changes of nutritional conditions and external influences. The properties of the genetic intelligence can serve as a prototype for creation of artificial intelligence systems of the genomorphic type.

Organization of the Book

The book is organized into 14 chapters. A brief description of each chapter follows.

Chapter 1 is devoted to main information on structures of the genetic code system, its symmetries and binary-oppositional features. The representation of DNA alphabets of n-plets in the form of genetic matrices of a single tensor family is explained. Special attention is paid to the Vertebrate Mitochondrial Code, which is the most symmetrical and ancient in the set of the known dialects of the genetic code. A connection of the genetic matrices with Rademacher functions is noted. Dyadic-shift decompositions of the genetic matrices are used, which reveal a connection of the genetic matrices with sets of orthogonal matrices and algebras of hypercomplex numbers. Some algebraic features of the genetic matrices are described. It is shown that the study of the emergent properties of the genetic system of multiple binary oppositions reveals its algebraic connection with the ancient principle "like begets like". This principle, known in genetics in connection with the complementary replication of DNA strands, turns out to be important for understanding the structural properties of the entire genetic coding system.

Chapter 2 continues the theme of a close connection of the genetic system with binary opposition principles, dyadic groups of binary numbers, and some algebras of hypercomplex numeric systems. Special attention is devoted to the well-known method of algebraic bit-reversible holography, which is used in processing of digital information and based on dyadic groups of binary numbers. An application of its algorithms to analysis of matrix presentations of DNA alphabets reveals unexpected relations of the genetic system with the algebra-holographic concept. Relations between the bit-reversible holography, the "like begets like" principle, and binary-complementarities in matrix genetics are analyzed. Connections of the genetic matrices with Hadamard matrices and Walsh functions, which are widely used in noise immunity coding, are shown. Some applications of Hamiltonian quaternions and genetic octetons for modeling biological phenomena are shown. The problem of genetic intelligence and inherited intellectual abilities of organisms is considered.

Chapter 3 is devoted to using the theory of resonances and its algebra-matrix tools for modeling features of the genetic coding systems. The concept of multi-resonance genetics is presented. The analogy is shown between Punnett squares of Mendelian genetics and tables of the tensor inheritance of eigenvalues of matrices from the theory of resonances. The postulate is formulated that the resonances and the corresponding oscillatory energies are key factors for genetic encoding. Using this postulate and the notion of dominant and recessive eigenvalues of resonances, the known segregation inside genetic alphabets is explained. Also, the theory of resonances of oscillatory systems with many degrees of freedom is used for modeling algebraic features of the genetic code system, the basic psychophysical law of Weber-Fechner and phyllotaxis laws. Thoughts on the morpho-resonance field, which is related to morphogenesis and metric tensors, are described. The usefulness of using the achievements and phenomena of vibrational mechanics for understanding many biomechanical phenomena is noted. In addition, the application of stochastic resonance mechanisms to the modeling and understanding of some biological phenomena is discussed.

Chapter 4 presents a family of universal rules of the stochastic organization of DNA sequences in genomes of higher and lower organisms. These rules generalize Chargaff's second rule. They were discovered due to the described method of oligomer sums. These rules, which are related to the system of binary oppositions in molecular genetics, hold not only complete genomic DNAs but also in so-called sparse DNA epi-chains. Some of these rules are connected with algebraic projectors. Using the genetic matrices, graphical representations of the rules are proposed. The described universal rules of genomic DNAs indicate the existence of non-trivial algebraic invariants of a globally genomic nature, which remained unchanged over million years of biological evolution, during which millions of species of organisms die off and new ones appear (although locally genomic sequences are modified by mutations, natural selection mechanisms, etc.). This seems to be connected with the question of the origin of life since these patterns are presented also in archaea and bacteria.

Chapter 5 connects the topic of universal rules of the stochastic-deterministic organization of genomic DNAs with a general topic of the interrelationship of randomness and determinism in biological bodies. Analysis and modeling of this interrelationship by formalisms of matrix algebra and matrix genetics seem to be important for receiving new understanding of the role of randomness and determinism in biological self-organization and also for producing new approaches to build bridges between biology, informatics, and mathematical natural science. The biological dualism "stochastics-determinism" is considered in connections with phenomena of Gestalt biology and bio-symmetries. Gestalt biology includes inherited biological phenomena — morphogenetic, homeostatic, senso-motoric, etc., which are akin to those in Gestalt psychology. Considering the biological importance of the principle "like begets like", special attention is paid to bio-symmetries, based on the groups of similarity and local-similarity (that is, conformal or Möbius) transformations.

Chapter 6 is devoted to the doctrine of energy-informational evolution on bio-antenna arrays and their wave functions. The importance of this doctrine is connected with amazing emergent properties of digital antenna arrays (frequently called Intelligence Antenna), which are widely used in a great number of technologies. Examples of biological phenomena are shown, which confirm that biological evolution has long used the amazing emergent properties of antenna arrays and their wave functioning: complex faceted eyes of insects and invertebrates, inherited phenomena of biological echolocation and electrolocation, and so on. This doctrine was initially generated by analogies between genomic percentage matrices and the matrices in the known tensor-matrix theory of digital antenna arrays. Some applications of this doctrine to problems of genetic coding, morphogenesis, and regeneration phenomena are considered. The presented data allow the thinking that it is not genes that are the dictators of all life activity, since they themselves are built into the information-energy coherences of bio-antenna arrays and are produced by them. Correspondingly, the secret of the structural organization and origin of the genetic code, as well as the origin of organisms, must be sought not in the random combination of the molecular elements of the genetic code, but in the emergent properties of self-organizing systems of bio-antenna arrays with their wave energy activity.

Chapter 7 considers applications of mathematical formalisms of quantum biology and quantum informatics for modeling features of the genetic code system. The author of the first work on quantum biology P. Jordan claimed that life's missing laws were the rules of chance and probability of the quantum world that were somehow scaled up inside living organisms. This chapter notes that the above-described results of studying probabilities in genomic DNAs confirm his prediction. To model the universal rules of stochastic organization of genomic DNAs, connected with the evolutionary bio-antenna doctrine, the notion of "2*n*-qubits" systems is used. The chapter describes the theme of the affinity between this doctrine and the formalisms of quantum informatics and quantum mechanics. Briefly speaking, the doctrine links together quantum informatics and antenna array theory regarding genetic and inherited physiological phenomena. Many authors have long suggested the quantum-informational nature of living bodies. In this book, the analysis of living bodies at the deepest — genomic — level leads to more convincing and fruitful pieces of evidence about organisms as quantum information entities (or as quantum-like entities). In this approach, the described universal stochastic features of genomic DNAs reveal that each of these DNAs is a regular multi-level web of 2-qubit states. Graphical representations of the stochastic organization of genomic DNAs are proposed. A connection between the described universal rules of genomic DNAs and Frohlich's theory about collective quantum effects in biological systems is discussed.

Chapter 8 considers questions of mathematical variety of algebraic tools and notions, which are related to non-trivial structures of molecular ensembles of the genetic coding system and are used to reveal the emergent properties of the genetic system. It presents algebraic connections between the numeric properties of nucleobase alphabets and the following mathematical objects: 1) metric tensors, the notion of which is useful to simulate morphogenetic phenomena and to develop the proposed concept of the morpho-resonance field; 2) formalisms of the theory of resonances, which was long ago used in modeling biological processes; 3) special numeric values — the golden ratio and the ring ratio; 4) doubly stochastic matrices used long ago in different scientific directions and are connected with unitary matrices on the basis of a known theorem; 5) hyperbolic numbers and their algebraic extensions. It is noted that the presented results equally apply to the alphabets of DNA and RNA.

Chapter 9 presents the mathematical operations that allow algebraic modeling of interrelations between stochastic molecular processes and inherited deterministic structures in biological bodies. Features of the described universal rules of stochastic organization of genomic DNAs are connected with algebraic operations called "tensor-unitary transformations". These operations allow algebraic modeling of the biological dualism "stochastics-determinism", physiological Gestalt phenomena, and some other biological structures. Algebraic properties of tensor-unitary transformations and their possible profits for applications in algebraic biology and understanding of genetic intelligence are presented. It is supposed that they are useful for the development of quantum biomechanics to model biosimilar quantum systems capable of tensor self-development with the inheritance of memory of past states. Arguments for developing the theory of stochastic determinism as an antipode to the known theory of deterministic chaos are formulated.

Chapter 10 represents the results of studying hidden algebraic rules regarding quantitative characteristics of n-plets groupings in information structures of long DNAs. This study uses the oligomer sums method. Numeric data of DNA sequences of many genomes of eukaryotes and prokaryotes are presented. Also, analysis of some long genes and giant viruses by oligomer sums method is described. Additional evidence is provided that the nucleotide genomic sequences have universal algebraic invariants in the course of biological evolution despite mutations, the pressure of natural selection, and other evolutionary factors. New mathematical tools and approaches for an in-depth study of the biological world and its evolution appear. In particular, it is shown that the oligomer sums method can be used for the analysis of amino acid primary sequences in long proteins (the example of the protein titin is presented). Also, analysis of some long genes and giant viruses by this method is described. It is emphasized that the genomic hyperbolic rules are cardinally different from the well-known hyperbolic Zipf's law. The presented results are interesting also for discussions concerning various well-known theories of biological evolution: Darwinism, nomogenesis, orthogenesis, etc. The described study is a continuation of research on symmetries in biological objects. This study further illustrates the effectiveness of symmetry analysis in natural systems.

Chapter 11 describes results testifying that already at the deepest level of biological organization — at the level of genomic DNAs associated with quantum informatics — fractal phenomena work, which totally permeate the overlying inherited biostructures, and which are also connected to stochastic mechanisms of Gestalt biology. In inherited biological structures, many fractal-like patterns are known, the ensembles of which gradually arise and become more complex during the ontogenetic development of organisms from a unicellular to a multicellular state. In developmental biology, the formation of inherited fractal-like patterns in the course of ontogenesis can be considered as one of the options for implementing the ancient principle "like gives rise to like" connected with the genetic coding system (as is considered in Chapter 1). Many authors devoted their publications to biological fractals. The existence of fractal globules in the three-dimensional architecture of whole genomes for providing maximally dense packing of chromosomes is known. By contrast to this known fact, our book presents quite another thing: the fractal-like organization of long DNA texts, in particular, in connection with epi-chains of genomic DNAs and fractal genetic nets or trees, which are related to tetra-group symmetries of these texts. Fragments of this fractal-like organization of genomic DNAs are also presented in Chapters 3, 4, 6, 7, and 10. Also, the chapter considers a connection between fractal features of genomic DNAs and the phenomenon of vegetative propagations, which accompanies ontogenetic development of many organisms and is one of the manifestations of the principle "like begets like".

Chapter 12 is devoted to the problem of studying structural relationships between biomechanical and genetic systems. This area of research is conditionally called "genetic biomechanics". All biomechanical systems of the body must be structurally consistent with the genetic coding system to be transmitted in the encoded form to the next generations. Correspondingly, their structural features can bear the imprints of genetic coding structures. The main attention of this chapter is paid to the algebraic modeling of the kinematics of multi-parametric or multi-block biomechanical bodies such as a human body having arms, legs, fingers, head, torso, etc. A person can move these parts of the body in concert or relatively independently of each other. To model the biomechanical kinematics of multi-block bodies in vector parameter spaces, it is important to find appropriate matrix operators. Given the increase in the dimension of the configuration space of body parameters during its ontogenetic development, the desired types of matrix operators should have an additional property: when they are tensorial expanded, they must retain the ability to similarly model changes in the tensorial expanded configuration space. It turns out that the system of structured DNA alphabets in their matrix form of presentation is associated with matrix operators of this type; in other words, the genetic structures themselves suggest the way to the construction of such operators and their tensor extensions. These operators are related to the concept of multi-dimensional and multi-block united-hypercomplex numeric systems. The description of their properties and applications is presented in this chapter. The proposed systems of united-hypercomplex numbers (briefly, of U-numbers) present a new class of mathematical models for biology. They principally differ from hypercomplex numbers widely used in modern mathematical natural sciences.

Chapter 13 describes interrelations between the holistic system of parameters of molecular genetics and ratios of the musical harmony of Pythagorean musical tunings, which have hidden connections with the golden ratio and musical Fibonacci-stage tunings. These connections were discovered due to the study of the emergent properties of the holistic system of binary oppositions and hydrogen bonds in the DNA parameters using its matrix representation. Music is a game with acoustic resonances, to which people are remarkably predisposed. For tens of thousands of years, people have created musical instruments. The same musical tunings were used in different nations and epochs without knowledge of mathematics. It is one of the pieces of evidence about inherited feelings of musical harmony and its hidden connection with the genetic code system. Leibniz declared that music is the arithmetic of the soul. The presented results additionally testify in favor that molecular-genetic systems are organized on ensembles of resonances and that binary sets of resonances are the basis of internal binary genetic computers. They give materials for a new slogan: a living body is a musical instrument (a synthesizer with an abundance of rearrangements of resonant modes). These results are essential for the problem of genetic bases of aesthetics and inborn feeling of harmony, and they can be used for development of a cultural direction of "genetic art". They are used in the Moscow State Tchaikovsky Conservatory to create so-called "genetic music". These results can be applied for tasks of music therapy, which is very popular around the world. The discovery of the connections between the genetic code system in its matrix representation and the golden ratio testifies in favor of the molecular-genetic base of many known facts about the aesthetic meaning of the golden section.

Chapter 14 presents a perspective approach to study holographic-like properties of organisms by using the known notion of logic holography based on Walsh functions and binary oppositions. Physical waves used in physical holography can be modeled digitally. Correspondingly noise immunity and other properties of optical and acoustical holography can be modeled digitally and electrically, in particular, by using Walsh functions and logic operations concerning dyadic groups of binary numbers. The pioneering work about holography by Walsh waves is [9]. In this digital Walsh holography, objects, whose digital holograms should be made, are represented in forms of 2^n-dimensional vectors. Due to materials of our book, one can believe that principles of logical holography and appropriate logic operations are one of the bases for the genetic system. Correspondingly, the theory of genetic logical holography is developed, where mathematics of logical holography and logical operations are used for modeling genetic phenomena. This mathematics includes lattice functions, logical operations with them, dyadic spaces, dyadic groups of binary numbers, logic modulo-2 addition, dyadic convolution, dyadic derivatives as well as isomorphic and homomorphic transformations of Walsh related to peculiarities of molecular-genetic systems. This algebraic modeling of genetic phenomena gives possibilities of new heuristic associations and new understanding of natural phenomena. The chapter contains a few examples of algebraic models and approaches in this new field of genetic research. Considering the presented results, one can believe that genetic sequences are algebra-logic holograms, and to read the objective image hidden in them, it is necessary to subject them to a restoring transformation. From the standpoint of the theory of genetic logic holography, life seems to be a complex system of genetic Boolean functions and their spectral logic by using Walsh-Hadamard transformations and genetic logical holograms. Briefly speaking, life is a system of genetic logic holograms.

Sergey Petoukhov
Matthew He

About the Authors

Dr. Sergey Petoukhov is Full Professor and Chief of the Laboratory of Biomechanical Systems, Mechanical Engineering Research Institute, Russian Academy of Sciences, Russia. He is also Editor-in-Chief of the *International Journal of Mathematical Sciences and Computing*, and co-editor of two Springer book series *Advances in Intelligent Systems and Computing* and *Lecture Notes on Data Engineering and Communications Technologies*. He has authored 7 books and published more than 200 research papers in the areas of theoretical and mathematical biology (in particular, in problems of genetic coding and bioinformatics), biomechanics, crystallography, theories of symmetry and self-organization. He holds a few leadership positions, including Advisory Board Chairman of the International Symmetry Association (ISA), Vice-Chair of the International Advisory Board of Directors of the Research Association of Modern Education and Computer Science, and Vice-President of the International Society of Natural Medicine.

Dr. Matthew He is Full Professor of Mathematics at Nova Southeastern University (NSU), USA. He is also Editor-in-Chief of the *International Journal of Information Technology and Computer Science* and serves as Editor for a few other journals. He has authored/edited over 10 books and published more than 100 research papers in the areas of approximation theory, computational mathematics, information theory, bioinformatics and cognitive informatics. He received the World Academy of Sciences Achievement Award in 2003 and 2010 in recognition of his research contributions to the field of computing. He received the Professor of the Year Award in Excellence of Teaching and Research in 2002 at NSU. He holds a Ph.D. in Mathematics from University of South Florida, USA.

Contents

Chapter 1

Genetic Code Systems, Emergence Features, and Analysis by Algebraic Matrices

Modern science connects many basic secrets of living matter with the genetic code. Molecular genetics has discovered that all organisms are identical to each other by their basic molecular-genetic structures, where many molecular elements are combined into a single system possessing effective emergent properties to provide inheritance phenomena. Amazingly, a basic structural system of genetic coding turns out to be strikingly simple. Its simplicities and orderliness throw down a challenge to specialists from many scientific fields. One of the most prospective ways to study the structural organization of the genetic system and its emergent properties is proposed by matrix analysis, which has a great number of applications in various scientific fields, including classical and quantum mechanics, informatics, and so on. Matrix genetics combines many achievements of applications of matrix analysis in studying genetics and inherited physiological phenomena. This chapter introduces this multi-aspect topic.

1.1. Basic elements of genetic code systems

Due to the remarkable works of many scientists, modern science knows basic phenomenological data about molecular structures of the genetic code including the genetic alphabet of 4 nucleobases in DNA and RNA, 64 triplets coding 20 amino acids, etc. All living organisms are unified. All of them have identical molecular bases of the genetic coding system. These bases are amazingly simple. For realization of the genetic messages, which encode sequences of amino acids in proteins, all kinds of organisms utilize

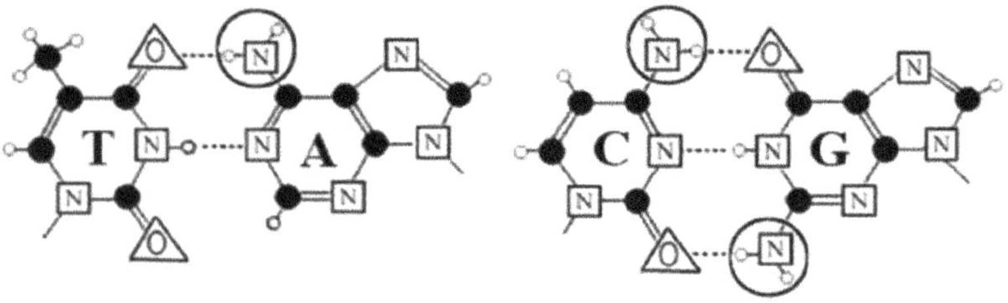

Fig. 1.1.1. The complementary pairs of the four nitrogenous bases in DNA: A-T (adenine and thymine), C-G (cytosine and guanine). Hydrogen bonds in these pairs are shown by dotted lines. Black circles are atoms of carbon; small white circles are atoms of hydrogen; squares with the letter N are atoms of nitrogen; triangles with the letter O are atoms of oxygen. Amides (or amino groups) NH_2 are marked by big circles.

in their molecules of heredity the sets of only four "letters" or nitrogenous bases: adenine (A), cytosine (C), guanine (G), and thymine (T) in DNA or uracil (U) in RNA (Fig. 1.1.1). Linear sequences of these four letters contain the genetic information for protein synthesis in all living bodies.

The given sets of the four letters are usually considered as the elementary alphabets of a genetic code. In this book, we will usually appeal to the elementary alphabet of DNA — A, C, G, T — though many of the described algebraic aspects are related to the elementary alphabet of RNA — A, C, G, U — as well. These four letters form the complementary pairs C-G and A-T, because they stand opposite each other in the double helix of DNA. C and G are connected by three hydrogen bonds, A and T (or A and U) by two (Fig. 1.1.1).

Genetic information, which is transferred by molecules of heredity, defines the primary structure of proteins of biological organisms. Proteins are the main dense component of biological organisms. Many thousand kinds of proteins exist. Each of them possesses its own individual function. In particular, all biological enzymes, which provide phenomenal speeds of many biochemical reactions in organisms, are proteins. Enzymes speed up chemical reactions by 10^{10}–10^{14} times [10]. The whole harmonic system of metabolism depends on proteins. Each coded protein exists in the form of a chain of 20 kinds of inherited amino acids. All amino acids in proteins are connected by the same type of chemical bond, which is named peptide bond. 20 amino acids, which are encoded genetically, have the following traditional abbreviations, which are used in our book: Ala – alanine, Arg – arginine, Asn – asparagine, Asp – aspartic acid, Cys – cysteine, Gln – glutamine, Glu – glutamic acid, Gly – glycine, His – histidine, Ile – isoleucine, Leu – leucine, Lys – lysine, Met – methionine, Phe – phenylalanine, Pro – proline, Ser – serine, Thr – threonine, Trp – tryptophan, Tyr – tyrosine, Val – valine.

A sequence of amino acids in a protein chain is defined by genetic triplets. A triplet (or a codon) is a block of three neighbor nitrogenous bases along a filament of DNA or RNA. The number of triplets which can be constructed from the four-letter alphabet is equal to $4^3 = 64$. Each triplet has its coded meaning: it encodes one of 20 kinds of amino acids or plays the role of a stop signal or start-signal for protein synthesis. Each codon has its anti-codon, which consists of the corresponding complementary letters: for example, the triplet CTG has the anti-codon GAC.

The genetic code is named "the degeneracy code" because its 64 triplets encode 20 amino acids and different amino acids are encoded by different quantities of triplets. Hypotheses about a connection between this degeneracy and the noise immunity of genetic information have existed since the time of the discovery of the genetic code. Symmetries in the structures of degeneracy of the genetic code are described below. Many dialects of the genetic code exist in biological organisms and their subsystems, which differ from one another by some differences in correspondences between triplets and elements encoded by them (see details: http://www.ncbi.nlm.nih.gov/Taxonomy/Utils/wprintgc.cgi).

Modern science knows a great number of DNA sequences in many genes and genomes of different species. Many of the sequences are located in the GenBank (https://www.ncbi.nlm.nih.gov/genbank/).

1.2. Symmetries and binary principles in the molecular genetic system

Science does not know why the nucleotide alphabet of DNA has been created by nature from just four letters (A, T, C, G), and why just these very simple molecules were chosen for the DNA alphabet (out of millions of possible molecules). But science knows [11–14] that these four molecules are interrelated by means of their symmetrical peculiarities in the united molecular ensemble, having three pairs of binary-oppositional traits or indicators (Table 1.2.1):

1) Two letters are purines (A and G), and the other two are pyrimidines (C and T). From the standpoint of these binary-oppositional traits one can denote C = T = 0, A = G = 1. From the standpoint of these traits, any of the DNA sequences is represented by a corresponding binary sequence. For example, GCATGAAGT is represented by 101011110;
2) Two letters are amino-molecules (A and C) and the other two are keto-molecules (G and T). From the standpoint of these traits, one can designate A = C = 0, G = T = 1. Correspondingly, the same sequence as above, GCATGAAGT, is represented by another binary sequence, 100110011;
3) The pairs of complementary letters, A-T and C-G, are linked by two and three hydrogen bonds, respectively. From the standpoint of these traits, one can designate C = G = 0, A = T = 1. Correspondingly, the same sequence, GCATGAAGT, is read as 001101101.

Accordingly, each DNA sequence is the carrier of three parallel messages on three different binary languages. At the same time, these three types

Table 1.2.1. Three kinds of binary-oppositional indicators of the nitrogenous bases of DNA and their corresponding binary sub-alphabets.

№	Binary symbols	C	A	G	T/U
1	0_1 — pyrimidines 1_1 — purines	0_1	1_1	1_1	0_1
2	0_2 — amino 1_2 — keto	0_2	0_2	1_2	1_2
3	0_3 — three hydrogen bonds 1_3 — two hydrogen bonds	0_3	1_3	0_3	1_3

of binary representations form a common logic set on the basis of logic operation of modulo-2 addition denoted by the symbol $\oplus$: modulo-2 addition of any two such binary representations of the DNA sequence gives a sum which is equal to the third binary representation of the same DNA sequence: for example, $101011110 \oplus 100110011 = 001101101$. Recall the rules of the bit-wise modulo-2 addition: $0 \oplus 0 = 0$; $0 \oplus 1 = 1$; $1 \oplus 0 = 1$; $1 \oplus 1 = 0$. (The logic operation of modulo-2 addition is actively used in quantum informatics [15]).

Considering the phenomenological fact that each of the DNA letters is uniquely defined by any two kinds of binary-oppositional indicators (Table 1.2.1), these genetic letters can be represented as corresponding pairs of binary symbols, for example, from the standpoint of two first binary-oppositional indicators. It is convenient for us for further description to show at the first position of each of the letters its binary symbol from the second pair of binary-oppositional indicators (the indicator "amino or keto": C = A = 0, T = G = 1) and at the second positions of each of the letters its binary symbol from the first pair of binary-oppositional indicators (the indicator "pyrimidine or purine": C = T = 0, A = G = 1). In this case the letter C is represented by the binary symbol $0_2 0_1$ (that is, as 2-bit binary number), A by $0_2 1_1$, T by $1_2 0_1$, and G by $1_2 1_1$. Using these representations of separate letters, each of 16 duplets is represented as a concatenation of the binary symbols of its letters (that is, as 4-bit binary number): for example, the duplet CC is represented as 4-bit binary number $0_2 0_1 0_2 0_1$, CA as $0_2 0_1 0_2 1_1$, etc. By extension, each of the 64 triplets is represented as a concatenation of the binary symbols of its letters (that is, as 6-bit binary number): for example, the triplet CCC is represented as 6-bit binary number $0_2 0_1 0_2 0_1 0_2 0_1$, CCA as $0_2 0_1 0_2 0_1 0_2 1_1$, etc. In general, each n-plet is represented as a concatenation of the binary symbols of its letters (below we will not show these indexes 2 and 1 of separate letters in binary representations of n-plets but will remember that each of the positions corresponds to its own kind of indicators from the first or second set of indicators in Table 1.2.1).

It is convenient to represent DNA alphabets of 4 nucleotides, 16 duplets, 64 triplets, ... 4^n n-plets in the form of square tables (Fig. 1.2.1),

	0	1
0	C 00	A 01
1	T 10	G 11

	00	01	10	11
00	CC 0000	CA 0001	AC 0010	AA 0011
01	CT 0100	CG 0101	AT 0110	AG 0111
10	TC 1000	TA 1001	GC 1010	GA 1011
11	TT 1100	TG 1101	GT 1110	GG 1111

	000	001	010	011	100	101	110	111
000	CCC 000000	CCA 000001	CAC 000010	CAA 000011	ACC 000100	ACA 000101	AAC 000110	AAA 000111
001	CCT 001000	CCG 001001	CAT 001010	CAG 001011	ACT 001100	ACG 001101	AAT 001110	AAG 001111
010	CTC 010000	CTA 010001	CGC 010010	CGA 010011	ATC 010100	ATA 010101	AGC 010110	AGA 010111
011	CTT 011000	CTG 011001	CGT 011010	CGG 011011	ATT 011100	ATG 011101	AGT 011110	AGG 011111
100	TCC 100000	TCA 100001	TAC 100010	TAA 100011	GCC 100100	GCA 100101	GAC 100110	GAA 100111
101	TCT 101000	TCG 101001	TAT 101010	TAG 101011	GCT 101100	GCG 101101	GAT 101110	GAG 101111
110	TTC 110000	TTA 110001	TGC 110010	TGA 110011	GTC 110100	GTA 110101	GGC 110110	GGA 110111
111	TTT 111000	TTG 111001	TGT 111010	TGG 111011	GTT 111100	GTG 111101	GGT 111110	GGG 111111

Fig. 1.2.1. The square tables of DNA alphabets of 4 nucleotides, 16 duplets and 64 triplets with a strict arrangement of all components. Each of the tables is constructed in line with the principle of binary numeration of its column and rows (see explanations in text).

whose rows and columns are numerated by binary symbols in line with the following principle: entries of each column are numerated by binary indicators "pyrimidine or purine" (C = T = 0, A = G = 1) in line with the first set of binary indicators in Table 1.2.1 (for example, the triplet CAG and all other triplets in the same column are the combination "pyrimidine-purine-purine" and so this column is correspondingly numerated 011). By contrast, entries of each row are numerated by binary indicators "amino or keto"

(C = A = 0, T = G = 1) in line with the second set of indicators in Table 1.2.1 (for example, the same triplet CAG and all other triplets in the same row are the combination "amino-amino-keto" and so this row is correspondingly numerated 001; correspondingly, the triplet CAG is enumerated by 001011). In such tables (Fig. 1.2.1), each of 4 letters, 16 duplets, 64 triplets, etc., takes automatically its own individual place and all components of the alphabets are arranged in a strict order. This strict ordering of the relative positions for all members of the DNA alphabets proves useful in revealing hidden regularities and rules in the genetic coding system.

One can see in Fig. 1.2.1 that all complementary n-plets are located inverse-symmetrically with respect to the center of the appropriate matrix. Correspondingly, $2n$-bit binary numbering of each n-plet is transformed into $2n$-bit numbering of its complementary n-plet (that is, n-plet of the opposite strand of DNA) by the mutual interchanging or inversion of digits $0 \leftrightarrow 1$ in it. For example, by this complementary operation, the numbering 001010 of the triplet CAT becomes numbering 110101 of its complementary triplet GTA. Recall that this inversion $0 \leftrightarrow 1$ of all bits in binary numbers is well known in mathematics and computing for generation of complementary (or complement) binary numbers (https://en.wikipedia.org/wiki/Method_of_complements). In the so-called ones' complement binary system, the complement of a binary number is created by the inversion $0 \leftrightarrow 1$ on each of its bits [16]. In such a system, a number is negated (converted from positive to negative or vice versa) by computing its complement. The complement of a positive binary number is interpreted as negative binary one. From the point of view of this binary system, each of the pairs of complementary n-plets in the matrices of DNA alphabets in Fig. 1.2.1 is represented by one positive and one negative binary number. For example, in the matrix [C, A; T, G], adenine A is represented by binary number 01 and its complementary guanine G is represented by binary-complement number 10 that can be interpreted as a negative number regarding number 01. In other words, in their binary representations, the algebraic relation between A and T can be conditionally denoted as A = −T, their binary-complement interrelation characterized by opposite signs. In the same way, C and G are represented in Fig. 1.2.1 by complement binary numbers 00 and 11, which can be interpreted as interrelated positive and negative numbers. Correspondingly, one can also use the similar conditional denotation C = –G. In matrix genetics, this mutual inversion $0 \leftrightarrow 1$ is called the binary-complementary operation and is actively used in §§ 1.7 and 1.8 and also in subsequent chapters on the realization of the ancient principle "like begets like" in matrix genetics.

We should emphasize that three separate genetic tables (Fig. 1.2.1) form the joint tensor family of matrices since they are interrelated by the known operation of the tensor (or Kronecker) product of matrices. It should be especially explained. By definition, under tensor (or Kronecker) multiplication of two matrices, each entry of the first matrix is linked with the whole second matrix; correspondingly, the tensor product (whose operation is denoted by the symbol ⊗) of two square matrices V and W of the orders m and n is the matrix $Q = V \otimes W = ||v_{ij} \times W||$ with the order $m \times n$ [17]. Most frequently, arithmetic multiplication of numbers is used as such a linking operation. But besides arithmetic multiplication, other operations also can be used as a linking operation in tensor multiplication of matrices [18]. With regard to the symbolic genetic matrices in Fig. 1.2.2,

$$\begin{bmatrix} C & A \\ T & G \end{bmatrix} \otimes \begin{bmatrix} C & A \\ T & G \end{bmatrix} = \begin{bmatrix} C^* \begin{bmatrix} C & A \\ T & G \end{bmatrix}, & A^* \begin{bmatrix} C & A \\ T & G \end{bmatrix} \\ T^* \begin{bmatrix} C & A \\ T & G \end{bmatrix}, & G^* \begin{bmatrix} C & A \\ T & G \end{bmatrix} \end{bmatrix} =$$

CC	CA	AC	AA
CT	CG	AT	AG
TC	TA	GC	GA
TT	TG	GT	GG

$$\begin{bmatrix} C & A \\ T & G \end{bmatrix} \otimes \begin{bmatrix} C & A \\ T & G \end{bmatrix} \otimes \begin{bmatrix} C & A \\ T & G \end{bmatrix} =$$

CCC	CCA	CAC	CAA	ACC	ACA	AAC	AAA
CCT	CCG	CAT	CAG	ACT	ACG	AAT	AAG
CTC	CTA	CGC	CGA	ATC	ATA	AGC	AGA
CTT	CTG	CGT	CGG	ATT	ATG	AGT	AGG
TCC	TCA	TAC	TAA	GCC	GCA	GAC	GAA
TCT	TCG	TAT	TAG	GCT	GCG	GAT	GAG
TTC	TTA	TGC	TGA	GTC	GTA	GGC	GGA
TTT	TTG	TGT	TGG	GTT	GTG	GGT	GGG

Fig. 1.2.2. The tensor family of genetic matrices $[C, A; T, G]^{(n)}$ (here the tensor power $n = 2, 3$) of DNA alphabets of 4 nucleotides, 16 duplets and 64 triplets. The symbol ⊗ means the tensor product.

the well-known concatenation operation serves as such a linking operation. Concatenation is well known in computer programming and formal language theory as the operation for joining individual string elements into a single chain [19]. Fig. 1.2.2 shows examples where the second tensor power of the symbolic (2 × 2)-matrix [C, A; T, G] of 4 DNA monoplets automatically gives the symbolic matrix of 16 duplets; the third tensor power of the same matrix of 4 DNA letters gives the symbolic matrix of 64 triplets with the strict arrangement of entries. In some other parts of this book, arithmetic multiplication of numbers also is used in the tensor families of numeric matrices as the linking operation under the tensor product.

One can see from a comparison of Fig. 1.2.2 with Fig. 1.2.1 that the tensor powers of the genetic matrix of 4 letters [C, A; T, G] produce — in a new way — the same square tables of the DNA alphabets with the same strict arrangements of alphabetic components without any mention about binary-oppositional traits of genetic letters C, A, T and G, which were used to construct the square tables in Fig. 1.2.1.

So, the structural organization of the system of DNA alphabets relates to the algebraic operation of the tensor product. It is important since the operation of the tensor product is well known in mathematics, physics, and informatics, where it provides a way of putting vector spaces together to form larger vector spaces. *This construction is crucial to understanding the quantum mechanics of multiparticle systems* [15]. Besides, the tensor product is one of the basic instruments in quantum informatics.

The tensor (or Kronecker) product of matrices arises in a natural way in the problem of searching a matrix. The eigenvalues of matrix A ⊗ B are equal to a product of $c_i \times d_j$, where c_i and d_j are eigenvalues of the matrices A and B [17]. This property is used in subsequent chapters about multi-resonance genetics and the inheritance of traits in Mendelian genetics. The tensor product is also connected with generations of fractal structures [18] and is used widely in many fields of mathematics, theoretical physics, informatics, etc.

The presentation of ensembles of elements of the genetic coding system in the form of tensor families of genetic matrices (or "genomatrices" briefly) has appeared as a useful tool to investigate structures of the genetic code from the viewpoint of their analogy with the theory of discrete signals processing, noise immunity coding, quantum informatics, etc. The scientific direction, which deals with such matrix presentation of the ensembles of genetic elements and their numeric parameters, is named "matrix genetics" [12, 13].

1.3. Symmetrical degeneracy matrix of the Vertebrate Mitochondrial Code

64 triplets encode 20 amino acids that make up the proteins of living bodies, as well as stop signals. How are the 20 amino acids and stop signals of protein synthesis located in the matrix of 64 triplets (Figs. 1.2.1, 1.2.2)? This arrangement cannot be predicted since amino acids and nucleotides are completely different in structure. The number of variants for the arrangement of amino acids with some repetitions to fill the entire (8×8)-matrix is immense: $>> 10^{100}$ (for comparison, in physics, the lifetime of the universe is estimated at 10^{17} seconds). Will this arrangement of amino acids be chaotic, or will it suddenly turn out to be regularly symmetrical?

It turns out that from an ocean of possibilities, nature has chosen an algebraically regular version of the repetition and arrangement of amino acids and stop signals in this matrix of 64 triplets, shown in Fig. 1.3.1 for the case of the Vertebrate Mitochondrial Genetic Code (this dialect of the genetic code is considered the most symmetrical and ancient [20]); in the theory of symmetry, it is customary to start the analysis with symmetric patterns, then move on to study cases of violation. The shown matrix of encoded amino acids and stop signals consists of pairs of adjacent rows, identical in the composition of amino acids and stop codons.

This tensor-matrix regularity is one of the pieces of evidence for the deep connection of genetic coding with the formalisms of quantum informatics. The work [21] shows that the strict ordering of triplets in their algebraic

CCC Pro	CCA Pro	CAC His	CAA Gln	ACC Thr	ACA Thr	AAC Asn	AAA Lys
CCT Pro	CCG Pro	CAT His	CAG Gln	ACT Thr	ACG Thr	AAT Asn	AAG Lys
CTC Leu	CTA Leu	CGC Arg	CGA Arg	ATC Ile	ATA Met	AGC Ser	AGA Stop
CTT Leu	CTG Leu	CGT Arg	CGG Arg	ATT Ile	ATG Met	AGT Ser	AGG Stop
TCC Ser	TCA Ser	TAC Tyr	TAA Stop	GCC Ala	GCA Ala	GAC Asp	GAA Asp
TCT Ser	TCG Ser	TAT Tyr	TAG Stop	GCT Ala	GCG Ala	GAT Asp	GAG Asp
TTC Phe	TTA Leu	TGC Cys	TGA Trp	GTC Val	GTA Val	GGC Gly	GGA Gly
TTT Phe	TTG Leu	TGT Cys	TGG Trp	GTT Val	GTG Val	GGT Gly	GGG Gly

Fig. 1.3.1. The symmetrical arrangement of 20 amino acids and stop codons in the matrix of 64 triplets. A case of the Vertebrate Mitochondrial Genetic Code is shown. The generally accepted abbreviations for these amino acids and stop codons (Stop) are used. Pairs of adjacent rows, identical in the composition of amino acids and stop codons, are highlighted in the same color.

matrix (Fig. 1.2.1) allows discovery of important hidden regularities in the set of 20 proteinogenic amino acids.

1.4. The phenomenon of binary separation of DNA alphabets into equal halves

Let us discuss one more phenomenological symmetry connected with the known binary-oppositional separation of the DNA alphabet of 64 triplets — according to their code properties — into two equal sub-alphabets illustrated by Fig. 1.4.1: 32 triplets with strong roots (i.e., triplets starting with 8 strong duplets CC, CT, CG, AC, TC, GC, GT, GG) and 32 triplets with weak roots (i.e., triplets starting with the other 8 duplets) [22, 23]. Coding value of triplets with strong roots is independent of the letter on their third position. For example, the four triplets with the same strong root CGC, CGA, CGT, CGC encode the same amino acid Arg, though they have different letters on their third position. On the contrary, the coding value of triplets with

THE STANDARD CODE	
8 subfamilies of the triplets with strong roots and the amino acids which are encoded by them	8 subfamilies of the triplets with weak roots and the amino acids which are encoded by them
CCC, CCT, CCA, CCG → Pro	CAC, CAT, CAA, CAG → His, His, Gln, Gln
CTC, CTT, CTA, CTG → Leu	AAC, AAT, AAA, AAG → Asn, Asn, Lys, Lys
CGC, CGT, CGA, CGG → Arg	ATC, ATT, ATA, ATG → Ile, Ile, Ile, Met
ACC, ACT, ACA, ACG → Thr	AGC, AGT, AGA, AGG → Ser, Ser, Arg, Arg
TCC, TCT, TCA, TCG → Ser	TAC, TAT, TAA, TAG → Tyr, Tyr, Stop, Stop
GCC, GCT, GCA, GCG → Ala	TTC, TTT, TTA, TTG → Phe, Phe, Leu, Leu
GTC, GTT, GTA, GTG → Val	TGC, TGT, TGA, TGG → Cys, Cys, Stop, Trp
GGC, GGT, GGA, GGG → Gly	GAC, GAT, GAA, GAG → Asp, Asp, Glu, Glu
THE VERTEBRATE MITOCHONDRIAL CODE	
CCC, CCT, CCA, CCG → Pro	CAC, CAT, CAA, CAG → His, His, Gln, Gln
CTC, CTT, CTA, CCG → Leu	AAC, AAT, AAA, AAG → Asn, Asn, Lys, Lys
CGC, CGT, CGA, CGG → Arg	ATC, ATT, ATA, ATG → Ile, Ile, Met, Met
ACC, ACT, ACA, ACG → Thr	AGC, AGT, AGA, AGG → Ser, Ser, Stop, Stop
TCC, TCT, TCA, TCG → Ser	TAC, TAT, TAA, TAG → Tyr, Tyr, Stop, Stop
GCC, GCT, GCA, GCG → Ala	TTC, TTT, TTA, TTG → Phe, Phe, Leu, Leu
GTC, GTT, GTA, GTG → Val	TGC, TGT, TGA, TGG → Cys, Cys, Trp, Trp
GGC, GGT, GGA, GGG → Gly	GAC, GAT, GAA, GAG → Asp, Asp, Glu, Glu

Fig. 1.4.1. Two examples of the basic scheme of the genetic code degeneracy with 32 triplets with strong roots (denoted in black) and 32 triplets with weak roots. Top: the case of the Standard Code. Bottom: the case of the Vertebrate Mitochondrial Code. Triplets which changed their code meanings in relation to the Standard Code are highlighted in yellow. All initial data are taken from http://www.ncbi.nlm.nih.gov/Taxonomy/Utils/wprintgc.cgi.

weak roots depends on the letter on their third position. For example, in the grouping of the four triplets with the same weak root CAC, CAT, CAA and CAG, two triplets (CAC, CAT) encode the amino acid His and the other two (CAA, CAG) encode another amino acid Gln. All triplets with strong roots are marked in black in Fig. 1.4.1 in contrast to triplets with weak roots. This figure shows the correspondence between the set of 64 triplets and the corresponding set of 20 amino acids with stop signals (Stop) of protein synthesis in the Standard Code and the Vertebrate Mitochondrial Code.

In the constructed tensor family of matrices $[C, A; T, G]^{(n)}$ (Fig. 1.2.2) for DNA alphabets of 16 duplets, 64 triplets, and 256 tetraplets, we mark n-plets with strong roots in black (Fig. 1.4.2).

Fig. 1.4.2 shows that the noted phenomenological binary-oppositional separations of DNA alphabets of n-plets ($n = 2, 3, 4, \ldots$) produce very symmetric features of black-and-white mosaics in each of the shown genetic matrices:

1) Left and right matrix halves are mirror-anti-symmetric to each other in their colors since any pair of cells, disposed of by the mirror-symmetric manner in these halves, has opposite colors.
2) Diagonal quadrants of the matrix are identical to each other from the viewpoint of their mosaic.
3) The adjacent rows 1–2, 3–4, etc., are identical to each other from the viewpoint of the mosaic.
4) A sequence of black and white cells in each row has a meander-like character: black and white fragments have identical lengths. Such mosaic of each row corresponds to a meander-like form of one of Rademacher functions that take only two values «+1» and «−1» and whose examples are shown in Fig. 1.4.2. Rademacher functions are connected with the theory of orthogonal series and theory of probabilities. For example, every statement about the Rademacher functions can be interpreted from the point of view of the theory of probability (see details in

	00	01	10	11
00	CC	CA	AC	AA
01	CT	CG	AT	AG
10	TC	TA	GC	GA
11	TT	TG	GT	GG

	000	001	010	011	100	101	110	111
000	CCC	CCA	CAC	CAA	ACC	ACA	AAC	AAA
001	CCT	CCG	CAT	CAG	ACT	ACG	AAT	AAG
010	CTC	CTA	CGC	CGA	ATC	ATA	AGC	AGA
011	CTT	CTG	CGT	CGG	ATT	ATG	AGT	AGG
100	TCC	TCA	TAC	TAA	GCC	GCA	GAC	GAA
101	TCT	TCG	TAT	TAG	GCT	GCG	GAT	GAG
110	TTC	TTA	TGC	TGA	GTC	GTA	GGC	GGA
111	TTT	TTG	TGT	TGG	GTT	GTG	GGT	GGG

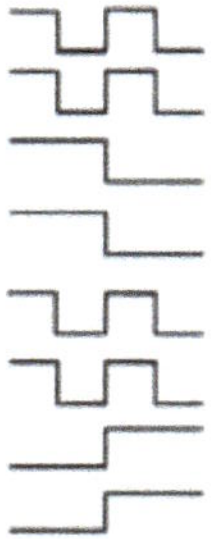

	0000	0001	0010	0011	0100	0101	0110	0111	1000	1001	1010	1011	1100	1101	1110	1111
0000	CCCC	CCCA	CCAC	CCAA	CACC	CACA	CAAC	CAAA	ACCC	ACCA	ACAC	ACAA	AACC	AACA	AAAC	AAAA
0001	CCCT	CCCG	CCAT	CCAG	CACT	CACG	CAAT	CAAG	ACCT	ACCG	ACAT	ACAG	AACT	AACG	AAAT	AAAG
0010	CCTC	CCTA	CCGC	CCGA	CATC	CATA	CAGC	CAGA	ACTC	ACTA	ACGC	ACGA	AATC	AATA	AAGC	AAGA
0011	CCTT	CCTG	CCGT	CCGG	CATT	CATG	CAGT	CAGG	ACTT	ACTG	ACGT	ACGG	AATT	AATG	AAGT	AAGG
0100	CTCC	CTCA	CTAC	CTAA	CGCC	CGCA	CGAC	CGAA	ATCC	ATCA	ATAC	ATAA	AGCC	AGCA	AGAC	AGAA
0101	CTCT	CTCG	CTAT	CTAG	CGCT	CGCG	CGAT	CGAG	ATCT	ATCG	ATAT	ATAG	AGCT	AGCG	AGAT	AGAG
0110	CTTC	CTTA	CTGC	CTGA	CGTC	CGTA	CGGC	CGGA	ATTC	ATTA	ATGC	ATGA	AGTC	AGTA	AGGC	AGGA
0111	CTTT	CTTG	CTGT	CTGG	CGTT	CGTG	CGGT	CGGG	ATTT	ATTG	ATGT	ATGG	AGTT	AGTG	AGGT	AGGG
1000	TCCC	TCCA	TCAC	TCAA	TACC	TACA	TAAC	TAAA	GCCC	GCCA	GCAC	GCAA	GACC	GACA	GAAC	GAAA
1001	TCCT	TCCG	TCAT	TCAG	TACT	TACG	TAAT	TAAG	GCCT	GCCG	GCAT	GCAG	GACT	GACG	GAAT	GAAG
1010	TCTC	TCTA	TCGC	TCGA	TATC	TATA	TAGC	TAGA	GCTC	GCTA	GCGC	GCGA	GATC	GATA	GAGC	GAGA
1011	TCTT	TCTG	TCGT	TCGG	TATT	TATG	TAGT	TAGG	GCTT	GCTG	GCGT	GCGG	GATT	GATG	GAGT	GAGG
1100	TTCC	TTCA	TTAC	TTAA	TGCC	TGCA	TGAC	TGAA	GTCC	GTCA	GTAC	GTAA	GGCC	GGCA	GGAC	GGAA
1101	TTCT	TTCG	TTAT	TTAG	TGCT	TGCG	TGAT	TGAG	GTCT	GTCG	GTAT	GTAG	GGCT	GGCG	GGAT	GGAG
1110	TTTC	TTTA	TTGC	TTGA	TGTC	TGTA	TGGC	TGGA	GTTC	GTTA	GTGC	GTGA	GGTC	GGTA	GGGC	GGGA
1111	TTTT	TTTG	TTGT	TTGG	TGTT	TGTG	TGGT	TGGG	GTTT	GTTG	GTGT	GTGG	GGTT	GGTG	GGGT	GGGG

Fig. 1.4.2. Binary-oppositional separations of DNA alphabets of n-plets into equal sub-alphabets of n-plets with strong roots (denoted in black) and weak roots (explanation in text). At the right of the matrix of 64 triplets, Rademacher functions illustrate its meander-like mosaic.

[24, 25]). It is interesting since probabilities in biological phenomena play important roles and a biological dualism “stochastics-determinism” exists, which is considered in subsequent chapters. Also, we will return to Rademacher functions later when considering genetic Hadamard matrices and their Walsh functions playing important roles in noise immunity communication, quantum mechanics and other fields.

These mosaic genetic matrices are also interesting not only by their connection with binary-oppositional principles but also by their connection to a method of algebraic holography considered in the next chapter.

Black and white cells of the symbolic matrices in Fig. 1.4.2 reflect the opposition of two sub-alphabets of n-plets with strong and weak roots and therefore can be represented by elements +1 and −1 in them. In this representation, the numeric matrices in Fig. 1.4.3 appear.

Since the numerical matrices in Fig. 1.4.3 are closely related to the Rademacher functions, they are conventionally called Rademacher genetic

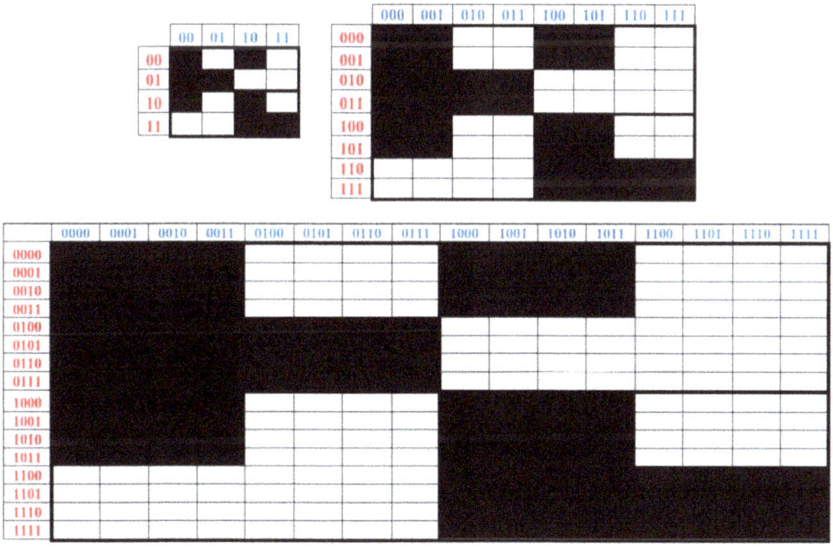

Fig. 1.4.3. Numerical representation of mosaic matrices for DNA alphabets of 16 duplets, 64 triplets, and 256 tetraplets from Fig. 1.4.2 is shown. The black and white cells of the matrices represent elements +1 and −1, respectively.

matrices. Do these Rademacher genetic matrices have any essential algebraic meaning? Yes, they do. The next section proves it by decomposition of the genetic matrices.

1.5. Dyadic-shift decompositions into unitary matrices and hypercomplex numbers

For special decompositions of genetic matrices in Fig. 1.4.3, we use structures of matrices of dyadic shifts long known in theory of discrete signal processing, sequency theory by Harmuth, etc. [26, 27]. These matrices are constructed on the basis of mathematical operation of modulo-2 addition for binary numbers met already in § 1.2. Modulo-2 addition is utilized broadly in the theory of discrete signal processing as a fundamental operation for binary variables. For example, modulo-2 addition of two binary numbers 110 and 101, which are equal to 6 and 5 respectively in decimal notation, gives the result $110 \oplus 101 = 011$, which is equal to 3 in decimal notation. The expression (1.1) shows an example of the series of binary numbers:

$$000, 001, 010, 011, 100, 101, 110, 111 \tag{1.1}$$

The series (1.1) forms a dyadic group, in which modulo-2 addition serves as the group operation [28]. The distance in this symmetry group is known as the Hamming distance. Since the Hamming distance satisfies the conditions of a metric group, the dyadic group is a metric group. The modulo-2 addition of any two binary numbers from (1.1) always results in a new number from the same series. The number 000 serves as the unit element of this group: for example, $010 \oplus 000 = 010$. The reverse element for any number in this group is the number itself: for example, $010 \oplus 010 = 000$. One can see in Fig. 1.2.1 that dyadic groups play an important role in the tensor family of constructed genetic matrices of DNA alphabets, where ordered sets of columns, rows, and n-plets are enumerated on basis-complete dyadic groups of binary numbers. Dyadic groups are one of the main personages in this book.

Above in § 1.2, the notion "binary-complementary operation" was introduced regarding interchange 0 ↔ 1 in binary numbers. Two binary numbers that are converted into each other under interchange 0 ↔ 1 are complementary to each another. For example, in the dyadic group (1.1), the pairs of complementary numbers are the following: 000-111, 001-110, 010-101, 011-100 (in the decimal system, they correspond to pairs of numbers 0–7, 1–6, 2–5, 3–4). In a pair of complementary numbers, one of them is always even and the other is odd, that is, any pair of complementary numbers is a pair of even and odd numbers. Accordingly, those columns (rows) that are enumerated by complementary binary numbers are complementary to each another. In the genetic matrices in Figs. 1.4.2 and 1.4.3, complementary columns are located mirror-symmetric in the left and right halves of the matrices, and complementary rows are located mirror-symmetric in the upper and lower halves. In different parts of this book, it will be shown that structural organization of the genetic system relates to complementary binary numbers and complementary columns (and rows) in genetic matrices; for example, it concerns structured sets of DNA alphabets and also universal stochastic rules of DNA in genomes of higher and lower organisms.

The series (1.1) can be transformed by modulo-2 addition with the binary number 001 into a new series of the same numbers:

$$001, 000, 011, 010, 101, 100, 111, 110 \qquad (1.2)$$

Such changes in the initial binary sequence, produced by modulo-2 addition of its members with any binary number from the same dyadic group, are termed dyadic shifts [26–28]. If any system of elements demonstrates its connection with dyadic shifts, it indicates that the structural organization of its system is related to the logic of modulo-2 addition. In particular, our book shows that the structural organization of genetic systems is related to the logic of modulo-2 addition.

Fig. 1.5.1 shows an example of such matrices of dyadic shifts from [26]. In such matrices, each row and column are numerated by means of binary numbers from an appropriate dyadic group. Each matrix cell has its binary

	000	001	010	011	100	101	110	111
000	000	001	010	011	100	101	110	111
001	001	000	011	010	101	100	111	110
010	010	011	000	001	110	111	100	101
011	011	010	001	000	111	110	101	100
100	100	101	110	111	000	001	010	011
101	101	100	111	110	001	000	011	010
110	110	111	100	101	010	011	000	001
111	111	110	101	100	011	010	001	000

Fig. 1.5.1. An (8×8)-matrix of dyadic shifts with binary numerations of its rows and columns.

numeration from the same dyadic group. This numeration of any cell is the result of modulo-2 addition of binary numerations of the column and row of this cell. For example, the cell from the column 110 and the row 101 obtains the binary numeration 011 by means of such addition.

In each of such dyadic-shift $(2^n \times 2^n)$-matrices, the whole set of numerated cells is separated into 2^n subsets, each of which includes only 2^n cells enumerated by the same kind of binary numbers. For example, in Fig. 1.5.1, the (8×8)-matrix contains 8 subsets of cells, where each subset contains only 8 cells enumerated by the same number from the dyadic group (1.1).

Each of the Rademacher genetic $(2^n \times 2^n)$-matrices in Fig. 1.4.3, having black-and-white mosaics of distributions of numbers +1 and −1 in them, can be dyadic-shift decomposed into sums of 2^n sparse matrices. For example, the appropriate decomposition of the Rademacher (4×4)-matrix of 16 duplets into the sum of 4 sparse matrices $e_0 + e_1 + e_2 + e_3$ is shown in Fig. 1.5.2.

Each of these 4 sparse matrices turns out to satisfy the condition of an orthogonal matrix $j_k \times j_k^T = E$, where $k = 0, 1, 2, 3$, the symbol T denotes the transposition of a matrix, and E is the identity matrix. It means that all these 4 matrices are orthogonal matrices. An orthogonal matrix presents a linear transformation, which preserves the inner product of vectors, and therefore acts as an isometry of Euclidean space, such as rotation, reflection or roto-reflection preserving vector lengths; in other words, it is a unitary transformation.

$$\begin{array}{|c|c|c|c|}\hline 1 & -1 & 1 & -1 \\ \hline 1 & 1 & -1 & -1 \\ \hline 1 & -1 & 1 & -1 \\ \hline -1 & -1 & 1 & 1 \\ \hline \end{array} = \begin{vmatrix} 1,0,0,0 \\ 0,1,0,0 \\ 0,0,1,0 \\ 0,0,0,1 \end{vmatrix} + \begin{vmatrix} 0,-1,0,0 \\ 1,0,0,0 \\ 0,0,0,-1 \\ 0,0,1,0 \end{vmatrix} + \begin{vmatrix} 0,0,0,-1 \\ 0,0,-1,0 \\ 0,-1,0,0 \\ -1,0,0,0 \end{vmatrix} + \begin{vmatrix} 0,0,1,0 \\ 0,0,0,-1 \\ 1,0,0,0 \\ 0,-1,0,0 \end{vmatrix} = e_0+e_1+e_2+e_3$$

*	e_0	e_1	e_2	e_3
e_0	e_0	e_1	e_2	e_3
e_1	e_1	$-e_0$	e_3	$-e_2$
e_2	e_2	$-e_3$	e_0	$-e_1$
e_3	e_3	e_2	e_1	e_0

Fig. 1.5.2. The dyadic-shift decomposition of the Rademacher (4 × 4)-matrix R_4 of the DNA alphabet of 16 duplets (from Fig. 1.4.3) into 4 sparse orthogonal matrices e_0, e_1, e_2, e_3, whose sets are closed relative to multiplication; e_0 is the identity matrix. The multiplication table for this set is shown, which matches with the multiplication table of the Cockle split-quaternion algebra used in the Poincare conformal disk model of hyperbolic geometry. The symbol for this model is presented from https://commons.wikimedia.org/wiki/Category:Poincar%C3%A9_disk_models.

It turns out that the set of these 4 sparse matrices (Fig. 1.5.2) is closed under multiplication and matches the corresponding multiplication table, known as the multiplication table of basic elements in the four-dimensional algebra of split-quaternions of Cockle, known since 1849 (Fig. 1.5.2). The obtained result shows that the genetic numeric (4×4)-matrix R_4 of 16 duplets (Fig. 1.4.3) can be interpreted as split-quaternions with unit coordinates.

Split-quaternions are used in the Poincaré conformal disk model to describe hyperbolic motions in Lobachevsky's hyperbolic geometry [29; https://en.wikipedia.org/wiki/Split-quaternion]. The connection between the genetic coding system and the Poincaré conformal disk model of the Lobachevski hyperbolic geometry, as described, reveals the genetic basis for the known relationship between inherited physiological phenomena and hyperbolic geometry. Let us recall some. Articles on non-Euclidean geometry of the space of visual perception have been published in different countries beginning with the pioneering work [30]. The author of the work [31], who carried out particularly careful experiments concerning this visual perception with the participation of 200 subjects and using 1,300 visual patterns, summarized: "*Poincaré's model of hyperbolic space ... shows fairly good agreement with the experimental results.*"

The work [32] describes many data about conformal bio-symmetries in inherited morphogenetic configurations. [33] states that the spatio-temporal organization of locomotion is associated with hyperbolic turns and the geometry of Minkowski. In this regard, its author put forward the "locomotor theory of relativity" and wrote about the relativistic brain and relativistic biomechanics. A study of the regular rearrangements of phyllotactic lattices during the growth of organisms led to the assertion that living matter is structurally linked to the geometry of Minkowski [34, 35].

A similar dyadic-shift decomposition of the Rademacher genetic (8 × 8)-matrix R_8, which presents the structural alphabet of 64 triplets (Fig. 1.4.3), gives a sum of 8 sparse matrices r_0, r_1, r_2, r_3, r_4, r_5, r_6, r_7, which are shown in Fig. 1.5.3.

Each of these 8 sparse matrices also satisfies the condition of an orthogonal matrix $r_k \times r_k^T = E$, where $k = 0, 1, 2, \ldots, 7$, the symbol T denotes the matrix transposition, and E is the identity matrix. The set of these 8 matrices is closed relative to multiplication and matches a multiplication table of basic elements for an appropriate algebra of eight-dimensional numeric system. The obtained result shows that the genetic numeric (8 × 8)-matrix of 64 duplets (Fig. 1.4.3) can be interpreted as an appropriate eight-dimensional number with unit coordinates.

The multiplication table (Fig. 1.5.3) shows as well that there exist special triads of matrices from the matrix set r_0, r_1, r_2, r_3, r_4, r_5, r_6, r_7. In each of these triads, the product of its two members gives its third member (with the sign "+" or "−"), that is, members of any of such triads form a closed grouping under multiplication. Moreover, in each of the triads, indexes of its members are strictly interconnected via their binary presentations: in such triads, modulo-2 addition of binary representations for indexes of any two members always gives the binary presentation of the index of the third member. For example, one of the triads is a subset r_1, r_2, r_3 with the following interconnections among its members: $r_1 \times r_2 = r_3$, $r_1 \times r_3 = r_2$, $r_2 \times r_1 = r_3$, $r_2 \times r_3 = -r_1$, $r_3 \times r_1 = r_2$, $r_3 \times r_2 = -r_1$. The decimal indexes 1, 2, and 3 of members

1	1	-1	-1	1	1	-1	-1		1,0,0,0,0,0,0,0		0,1,0,0,0,0,0,0		0,0,-1,0,0,0,0,0	
1	1	-1	-1	1	1	-1	-1		0,1,0,0,0,0,0,0		1,0,0,0,0,0,0,0		0,0,0,-1,0,0,0,0	
1	1	1	1	-1	-1	-1	-1		0,0,1,0,0,0,0,0		0,0,0,1,0,0,0,0		1,0,0, 0,0,0,0,0	
1	1	1	1	-1	-1	-1	-1	=	0,0,0,1,0,0,0,0	+	0,0,1,0,0,0,0,0	+	0,1,0, 0,0,0,0,0	+
1	1	-1	-1	1	1	-1	-1		0,0,0,0,1,0,0,0		0,0,0,0,0,1,0,0		0,0,0,0,0,0,-1,0	
1	1	-1	-1	1	1	-1	-1		0,0,0,0,0,1,0,0		0,0,0,0,1,0,0,0		0,0,0,0,0,0,0,-1	
-1	-1	-1	-1	1	1	1	1		0,0,0,0,0,0,1,0		0,0,0,0,0,0,0,1		0,0,0,0,1,0,0, 0	
-1	-1	-1	-1	1	1	1	1		0,0,0,0,0,0,0,1		0,0,0,0,0,0,1,0		0,0,0,0,0,1,0, 0	

0,0,0,-1,0,0,0,0		0,0,0,0,1,0, 0,0		0,0,0,0,0,1, 0,0		0,0,0,0,0,0,-1,0		0,0,0,0,0,0,0,-1
0,0,-1,0,0,0,0,0		0,0,0,0,0,1, 0,0		0,0,0,0,1,0, 0,0		0,0,0,0,0,0,0,-1		0,0,0,0,0,0,-1,0
0,1, 0,0,0,0,0,0		0,0,0,0,0,0,-1,0		0,0,0,0,0,0,0,-1		0,0,0,0,-1,0,0,0		0,0,0,0,0,-1,0,0
1,0, 0,0,0,0,0,0	+	0,0,0,0,0,0,0,-1	+	0,0,0,0,0,0,-1,0	+	0,0,0,0,0,-1,0,0	+	0,0,0,0,-1,0,0,0
0,0,0,0,0,0,0,-1		1,0,0,0,0,0,0, 0		0,1,0,0,0,0, 0,0		0,0,-1,0,0,0,0,0		0,0,0,-1,0,0,0,0
0,0,0,0,0,0,-1,0		0,1,0,0,0,0,0, 0		1,0,0,0,0,0, 0,0		0,0,0,-1,0,0,0,0		0,0,-1,0,0,0,0,0
0,0,0,0,0,1, 0,0		0,0,-1,0,0,0,0,0		0,0,0,-1,0,0,0,0		-1,0,0,0,0,0,0,0		0,-1,0,0,0,0,0,0
0,0,0,0,1,0, 0,0		0,0,0,-1,0,0,0,0		0,0,-1,0,0,0,0,0		0,-1,0,0,0,0,0,0		-1,0,0,0,0,0,0,0

*	r_0	r_1	r_2	r_3	r_4	r_5	r_6	r_7
r_0	r_0	r_1	r_2	r_3	r_4	r_5	r_6	r_7
r_1	r_1	r_0	r_3	r_2	r_5	r_4	r_7	r_6
r_2	r_2	r_3	$-r_0$	$-r_1$	$-r_6$	$-r_7$	r_4	r_5
r_3	r_3	r_2	$-r_1$	$-r_0$	$-r_7$	$-r_6$	r_5	r_4
r_4	r_4	r_5	r_6	r_7	r_0	r_1	r_2	r_3
r_5	r_5	r_4	r_7	r_6	r_1	r_0	r_3	r_2
r_6	r_6	r_7	$-r_4$	$-r_5$	$-r_2$	$-r_3$	r_0	r_1
r_7	r_7	r_6	$-r_5$	$-r_4$	$-r_3$	$-r_2$	r_1	r_0

Fig. 1.5.3. The dyadic-shift decomposition of the Rademacher genetic (8×8)-matrix R_8 of the DNA alphabet of 64 triplets (from Fig. 1.4.3) into 8 sparse orthogonal matrices r_0, r_1, r_2, r_3, r_4, r_5, r_6, r_7, where r_0 is the identity matrix. For this set, closed relative to multiplication, its multiplication table is shown at the bottom.

in this triad have binary presentations 001, 010, 011. The modulo-2 addition of any two of these binary indexes gives the third binary index: $001 \oplus 010 = 011$, $010 \oplus 011 = 001$, $011 \oplus 001 = 010$. Table 1.5.1 shows 7 triads, which possess analogical features, as the multiplication table (Fig. 1.5.3) confirms.

These seven triads together with the identity matrix r_0 define seven algebras of four-dimensional hypercomplex numbers with their multiplication tables presented in Fig. 1.5.4. One can see that the first three tables define commutative algebras because they are symmetrical; the other four multiplication tables define non-commutative algebras because they

Table 1.5.1. Seven special triads of elements from the set of 8 basis elements of the algebra of eight-dimensional numbers (Fig. 1.5.3). Each of these triads is closed relative to multiplication of its members with each other, as the multiplication table in Fig. 1.5.3 confirms. Interconnections of indexes of members in each of the triads via the modulo-2 addition are shown.

Triads of basis elements	Modulo-2 addition of binary indexes	Decimal notations of these binary indexes
r_1, r_2, r_3	$001 \oplus 010 = 011; 010 \oplus 01 = 001; 011 \oplus 001 = 010$	$001 = 1; 010 = 2; 011 = 3$
r_1, r_4, r_5	$001 \oplus 100 = 101; 100 \oplus 101 = 001; 101 \oplus 001 = 100$	$001 = 1; 100 = 4; 101 = 5$
r_1, r_6, r_7	$001 \oplus 110 = 111; 110 \oplus 111 = 001; 111 \oplus 001 = 110$	$001 = 1; 110 = 6; 111 = 7$
r_2, r_4, r_6	$010 \oplus 100 = 110; 100 \oplus 110 = 010; 110 \oplus 010 = 100$	$010 = 2; 100 = 4; 110 = 6$
r_2, r_5, r_7	$010 \oplus 101 = 111; 101 \oplus 111 = 010; 111 \oplus 010 = 101$	$010 = 2; 101 = 5; 111 = 7$
r_3, r_4, r_7	$011 \oplus 100 = 111; 100 \oplus 111 = 011; 011 \oplus 111 = 100$	$011 = 3; 100 = 4; 111 = 7$
r_3, r_5, r_6	$011 \oplus 101 = 110; 101 \oplus 110 = 011; 011 \oplus 110 = 101$	$011 = 3; 101 = 5; 110 = 6$

	1	r_1	r_2	r_3
1	1	r_1	r_2	r_3
r_1	r_1	1	r_3	r_2
r_2	r_2	r_3	-1	$-r_1$
r_3	r_3	r_2	$-r_1$	-1

	1	r_1	r_4	r_5
1	1	r_1	r_4	r_5
r_1	r_1	1	r_5	r_4
r_4	r_4	r_5	1	r_1
r_5	r_5	r_4	r_1	1

	1	r_1	r_6	r_7
1	1	r_1	r_6	r_7
r_1	r_1	1	r_7	r_6
r_6	r_6	r_7	1	r_1
r_7	r_7	r_6	r_1	1

	1	r_2	r_4	r_6
1	1	r_2	r_4	r_6
r_2	r_2	-1	$-r_6$	r_4
r_4	r_4	r_6	1	r_2
r_6	r_6	$-r_4$	$-r_2$	1

	1	r_2	r_5	r_7
1	1	r_2	r_5	r_7
r_2	r_2	-1	$-r_7$	r_5
r_5	r_5	r_7	1	r_2
r_7	r_7	$-r_5$	$-r_2$	1

	1	r_3	r_4	r_7
1	1	r_3	r_4	r_7
r_3	r_3	-1	$-r_7$	r_4
r_4	r_4	r_7	1	r_3
r_7	r_7	$-r_4$	$-r_3$	1

	1	r_3	r_5	r_6
1	1	r_3	r_5	r_6
r_3	r_3	-1	$-r_6$	r_5
r_5	r_5	r_6	1	r_3
r_6	r_6	$-r_5$	$-r_3$	1

Fig. 1.5.4. Multiplication tables of four-dimensional algebras of hypercomplex numbers, whose basis elements belong to matrix triads in Table 1.5.1. The disposition of indexes of elements in each of the tables corresponds to the principle of dyadic shifts. The digit 1 refers to the identity matrix r_0.

are asymmetrical. The last four multiplication tables coincide — up to a permutation of columns and rows — with the multiplication table of basic elements for the algebra of split-quaternions by Cockle in Fig. 1.5.2.

A similar dyadic-shift decomposition of the numeric (16×16)-matrix, which presents the structural alphabet of 256 tetraplets (Fig. 1.4.3), gives a sum of 16 sparse orthogonal matrices, whose set is also closed under multiplication and determines a multiplication table of basis elements for an appropriate algebra of a 16-dimensional numeric system.

All these results about connections of matrix presentations in structured alphabets of the genetic code system with described algebras and algebraic methods give pieces of evidence that genetic structures are not accidental but are complex algebraic entities. Some interesting results were obtained in the field of matrix genetics about algebraic features of the evolution of genetic code dialects [36]. We continue descriptions of algebraic features of the genetic code systems in later chapters.

This chapter presented initial data about a close connection of the genetic system with binary-oppositional principles and binary numbers, which are used in classical and quantum computers. Many authors posit that living beings are molecular computers and that natural laws are embedded in molecular texts of DNA for molecular controlling systems (see, for example, [37, 38]). Binary-oppositional features are typical not only for the molecular-genetic system but also for genetically inherited physiological systems. For example, a physiological "all-or-none" law for excitable tissues exists: a nerve cell or a muscle fiber gives only the answer "yes" or "no" under the action of various stimuli [39].

1.6. Regarding permutations of positions inside *n*-plets in mosaic matrices $[C\ A; T\ G]^{(n)}$

The theory of discrete signal processing pays special attention to permutations of information elements. Cyclic permutations in triplets are specially studied in connection with the phenomenology of genetic sequences and the problem of point mutations (see, for example, [40–45]). All the possible simultaneous permutations of positions inside all the triplets lead to new genetic mosaic matrices, whose mosaic rows again correspond to Rademacher functions and whose dyadic-shift decompositions define the multiplication table of the same algebra of eight-dimensional hypercomplex numbers (Fig. 1.5.3).

It is obvious that a simultaneous permutation of positions in triplets transforms most of the triplets in cells of the initial genomatrix $[C\ A; T\ G]^{(3)}$. For example, in the cyclic transformation of the order 1-2-3 of positions into the order 2-3-1, the triplet CAG is transformed into the triplet AGC, etc. Because each of the triplets is connected with the binary numeration of its column and row, these binary numerations are also transformed correspondingly; for example, the binary numeration 011 is transformed into 110.

Six variants of the order of positions inside triplets are possible: 1-2-3, 2-3-1, 3-1-2, 3-2-1, 2-1-3, 1-3-2. The initial matrix $[C\ A; T\ G]_{123}^{(3)}$ is related with the first of these orders (Figs. 1.4.2, 1.4.3). The other five genetic matrices $[C\ A; T\ G]_{231}^{(3)}$, $[C\ A; T\ G]_{312}^{(3)}$, $[C\ A; T\ G]_{321}^{(3)}$, $[C\ A; T\ G]_{213}^{(3)}$, $[C\ A; T\ G]_{132}^{(3)}$, which correspond to five other orders, are shown in Fig. 1.6.1 (subscripts indicate the order of positions in triplets). In these matrices, black-and-white mosaic of each row corresponds again to one of the Rademacher functions. The replacement of all the triplets with strong and weak roots by entries "+1" and "−1" correspondingly transforms these genomatrices into their Rademacher forms R_{231}, R_{312}, R_{321}, R_{213}, R_{132} (Fig. 1.6.2).

Black and white cells of the symbolic matrices in Fig. 1.6.1 reflect the opposition of two sub-alphabets of triplets with strong and weak roots and therefore can be represented by elements +1 and −1 in them. In this representation, the numeric matrices in Fig. 1.6.2 appear.

The whole set of these 6 matrices R_{123}, R_{231}, R_{312}, R_{321}, R_{213}, R_{132}, which arise under cyclic permutations of positions of nucleobases in triplets, can be presented in the form of a cyclic diagram in Fig. 1.6.3.

$[C\ A;\ T\ G]_{231}^{(3)} =$

	000 (0)	010 (2)	100 (4)	110 (6)	001 (1)	011 (3)	101 (5)	111 (7)
000 (0)	CCC 000 (0)	CAC 010 (2)	ACC 100 (4)	AAC 110 (6)	CCA 001 (1)	CAA 011 (3)	ACA 101 (5)	AAA 111 (7)
010 (2)	CTC 010 (2)	CGC 000 (0)	ATC 110 (6)	AGC 100 (4)	CTA 011 (3)	CGA 001 (1)	ATA 111 (7)	AGA 101 (5)
100 (4)	TCC 100 (4)	TAC 110 (6)	GCC 000 (0)	GAC 010 (2)	TCA 101 (5)	TAA 111 (7)	GCA 001 (1)	GAA 011 (3)
110 (6)	TTC 110 (6)	TGC 100 (4)	GTC 010 (2)	GGC 000 (0)	TTA 111 (7)	TGA 101 (5)	GTA 011 (3)	GGC 001 (1)
001 (1)	CCT 001 (1)	CAT 011 (3)	ACT 101 (5)	AAT 111 (7)	CCG 000 (0)	CAT 010 (2)	ACG 100 (4)	AAG 110 (6)
011 (3)	CTT 011 (3)	CGT 001 (1)	ATT 111 (7)	AGT 101 (5)	CTG 010 (2)	CGG 000 (0)	ATG 110 (6)	AGG 100 (4)
101 (5)	TCT 101 (5)	TAT 111 (7)	GCT 001 (1)	GAT 011 (3)	TCG 100 (4)	TAG 110 (6)	GCG 000 (0)	GAG 010 (2)
111 (7)	TTT 111 (7)	TGT 101 (5)	GTT 011 (3)	GGT 001 (1)	TTG 110 (6)	TGG 100 (4)	GTG 010 (2)	GGG 000 (0)

$[C\ A;\ T\ G]_{312}^{(3)} =$

	000 (0)	100 (4)	001 (1)	101 (5)	010 (2)	110 (6)	011 (3)	111 (7)
000 (0)	CCC 000 (0)	ACC 100 (4)	CCA 001 (1)	ACA 101 (5)	CAC 010 (2)	AAC 110 (6)	CAA 011 (3)	AAA 111 (7)
100 (4)	TCC 100 (4)	GCC 000 (0)	TCA 101 (5)	GCA 001 (1)	TAC 110 (6)	GAC 010 (2)	TAA 111 (7)	GAA 011 (3)
001 (1)	CCT 001 (1)	ACT 101 (5)	CCG 000 (0)	ACG 100 (4)	CAT 011 (3)	AAT 111 (7)	CAT 010 (2)	AAG 110 (6)
101 (5)	TCT 101 (5)	GCT 001 (1)	TCG 100 (4)	GCG 000 (0)	TAT 111 (7)	GAT 011 (3)	TAG 110 (6)	GAG 010 (2)
010 (6)	CTC 010 (2)	ATC 110 (6)	CTA 011 (3)	ATA 111 (7)	CGC 000 (0)	AGC 100 (4)	CGA 001 (1)	AGA 101 (5)
011 (3)	TTC 110 (6)	GTC 010 (2)	TTA 111 (7)	GTA 011 (3)	TGC 100 (4)	GGC 000 (0)	TGA 101 (5)	GGC 001 (1)
001 (3)	CTT 011 (3)	ATT 111 (7)	CTG 010 (2)	ATG 110 (6)	CGT 001 (1)	AGT 101 (5)	CGG 000 (0)	AGG 100 (4)
111 (7)	TTT 111 (7)	GTT 011 (3)	TTG 110 (6)	GTG 010 (2)	TGT 101 (5)	GGT 001 (1)	TGG 100 (4)	GGG 000 (0)

$[C\ A;\ T\ G]_{321}^{(3)} =$

	000 (0)	100 (4)	010 (2)	110 (6)	001 (1)	101 (5)	011 (3)	111 (7)
000 (0)	CCC 000 (0)	ACC 100 (4)	CAC 010 (2)	AAC 110 (6)	CCA 001 (1)	ACA 101 (5)	CAA 011 (3)	AAA 111 (7)
100 (4)	TCC 100 (4)	GCC 000 (0)	TAC 110 (6)	GAC 010 (2)	TCA 101 (5)	GCA 001 (1)	TAA 111 (7)	GAA 011 (3)
010 (2)	CTC 010 (2)	ATC 110 (6)	CGC 000 (0)	AGC 100 (4)	CTA 011 (3)	ATA 111 (7)	CGA 001 (1)	AGA 101 (5)
110 (6)	TTC 110 (6)	GTC 010 (2)	TGC 100 (4)	GGC 000 (0)	TTA 111 (7)	GTA 011 (3)	TGA 101 (5)	GGC 001 (1)
001 (1)	CCT 001 (1)	ACT 101 (5)	CAT 011 (3)	AAT 111 (7)	CCG 000 (0)	ACG 100 (4)	CAT 010 (2)	AAG 110 (6)
101 (5)	TCT 101 (5)	GCT 001 (1)	TAT 111 (7)	GAT 011 (3)	TCG 100 (4)	GCG 000 (0)	TAG 110 (6)	GAG 010 (2)
011 (3)	CTT 011 (3)	ATT 111 (7)	CGT 001 (1)	AGT 101 (5)	CTG 010 (2)	ATG 110 (6)	CGG 000 (0)	AGG 100 (4)
111 (7)	TTT 111 (7)	GTT 011 (3)	TGT 101 (5)	GGT 001 (1)	TTG 110 (6)	GTG 010 (2)	TGG 100 (4)	GGG 000 (0)

Fig. 1.6.1. Five genetic matrices $[C\ A;\ T\ G]_{231}^{(3)}$, $[C\ A;\ T\ G]_{312}^{(3)}$, $[C\ A;\ T\ G]_{321}^{(3)}$, $[C\ A;\ T\ G]_{213}^{(3)}$, $[C\ A;\ T\ G]_{132}^{(3)}$, which correspond to orders of positions in triplets 2-3-1, 3-1-2, 3-2-1, 2-1-3, 1-3-2 in relation to the initial matrix $[C\ A;\ T\ G]_{123}^{(3)}$ in Fig. 1.4.2. Black and white cells contain triplets with strong and weak roots, respectively. Binary numerations of columns and rows are shown with their decimal equivalents in brackets.

$[C\ A;\ T\ G]_{213}^{(3)} =$

	000 (0)	001 (1)	100 (4)	101 (5)	010 (2)	011 (3)	110 (6)	111 (7)
000 (0)	CCC 000 (0)	CCA 001 (1)	ACC 100 (4)	ACA 101 (5)	CAC 010 (2)	CAA 011 (3)	AAC 110 (6)	AAA 111 (7)
001 (1)	CCT 001 (1)	CCG 000 (0)	ACT 101 (5)	ACG 100 (4)	CAT 011 (3)	CAT 010 (2)	AAT 111 (7)	AAG 110 (6)
100 (4)	TCC 100 (4)	TCA 101 (5)	GCC 000 (0)	GCA 001 (1)	TAC 110 (6)	TAA 111 (7)	GAC 010 (2)	GAA 011 (3)
101 (5)	TCT 101 (5)	TCG 100 (4)	GCT 001 (1)	GCG 000 (0)	TAT 111 (7)	TAG 110 (6)	GAT 011 (3)	GAG 010 (2)
010 (2)	CTC 010 (2)	CTA 011 (3)	ATC 110 (6)	ATA 111 (7)	CGC 000 (0)	CGA 001 (1)	AGC 100 (4)	AGA 101 (5)
011 (3)	CTT 011 (3)	CTG 010 (2)	ATT 111 (7)	ATG 110 (6)	CGT 001 (1)	CGG 000 (0)	AGT 101 (5)	AGG 100 (4)
110 (6)	TTC 110 (6)	TTA 111 (7)	GTC 010 (2)	GTA 011 (3)	TGC 100 (4)	TGA 101 (5)	GGC 000 (0)	GGC 001 (1)
111 (7)	TTT 111 (7)	TTG 110 (6)	GTT 011 (3)	GTG 010 (2)	TGT 101 (5)	TGG 100 (4)	GGT 001 (1)	GGG 000 (0)

$[C\ A;\ T\ G]_{132}^{(3)} =$

	000 (0)	010 (2)	001 (1)	011 (3)	100 (4)	110 (6)	101 (5)	111 (7)
000 (0)	CCC 000 (0)	CAC 010 (2)	CCA 001 (1)	CAA 011 (3)	ACC 100 (4)	AAC 110 (6)	ACA 101 (5)	AAA 111 (7)
010 (2)	CTC 010 (2)	CGC 000 (0)	CTA 011 (3)	CGA 001 (1)	ATC 110 (6)	AGC 100 (4)	ATA 111 (7)	AGA 101 (5)
001 (1)	CCT 001 (1)	CAT 011 (3)	CCG 000 (0)	CAG 010 (2)	ACT 101 (5)	AAT 111 (7)	ACG 100 (4)	AAG 110 (6)
011 (3)	CTT 011 (3)	CGT 001 (1)	CTG 010 (2)	CGG 000 (0)	ATT 111 (7)	AGT 101 (5)	ATG 110 (6)	AGG 100 (4)
100 (4)	TCC 100 (4)	TAC 110 (6)	TCA 101 (5)	TAA 111 (7)	GCC 000 (0)	GAC 010 (2)	GCA 001 (1)	GAA 011 (3)
110 (6)	TTC 110 (6)	TGC 100 (4)	TTA 111 (7)	TGA 101 (5)	GTC 010 (2)	GGC 000 (0)	GTA 011 (3)	GGA 001 (1)
101 (5)	TCT 101 (5)	TAT 111 (7)	TCG 100 (4)	TAG 110 (6)	GCT 001 (1)	GAT 011 (3)	GCG 000 (0)	GAG 010 (2)
111 (7)	TTT 111 (7)	TGT 101 (5)	TTG 110 (6)	TGG 100 (4)	GTT 011 (3)	GGT 001 (1)	GTG 010 (2)	GGG 000 (0)

Fig. 1.6.1. (*Continued*)

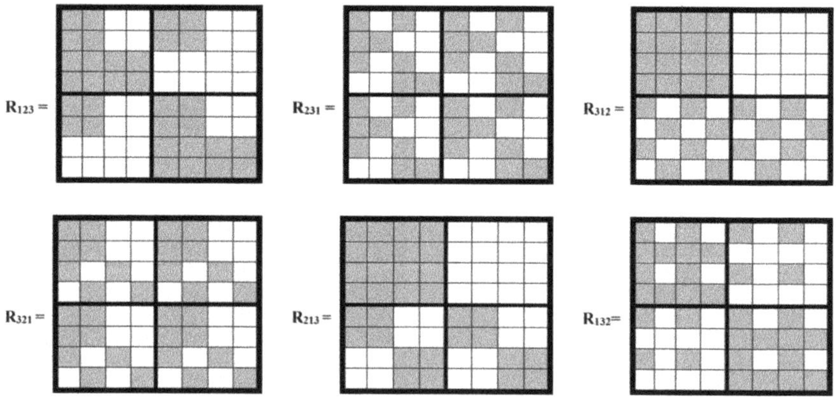

Fig. 1.6.2. Numeric matrices R_{123}, R_{231}, R_{312}, R_{321}, R_{213}, R_{132} represent symbolic genetic matrices from Fig. 1.6.1. Each black (white) cell contains the entry "+1" (−1).

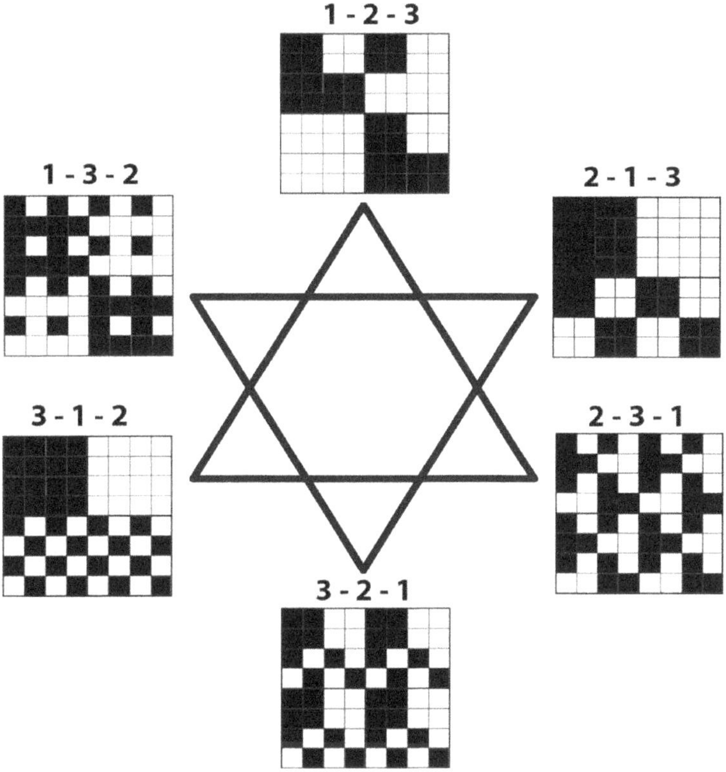

Fig. 1.6.3. The diagram representation for the whole set of matrices of 64 triplets with strong roots (black) and weak roots (white) for all possible variants of simultaneous permutations of positions in all triplets. The numbers above the matrices show the order of nucleobases when rearranging positions in triplets.

Each of the numeric matrices R_{231}, R_{312}, R_{321}, R_{213}, R_{132} can be also decomposed into the sum of 8 sparse matrices r_0, r_1, r_2, r_3, r_4, r_5, r_6, r_7 in accordance with dyadic numerations of its cells in Fig. 1.6.1. More precisely, any sparse matrix r_k ($k = 0, 1, \ldots, 7$) of such decomposition contains entries "+1" or "−1" from the appropriate matrix in Fig. 1.6.1 only in those cells, which correspond to cells with the same dyadic numeration "k"; all the other cells of the matrix r_k contain zero. Fig. 1.6.4 shows all the sets of sparse matrices r_k of such dyadic-shift decompositions.

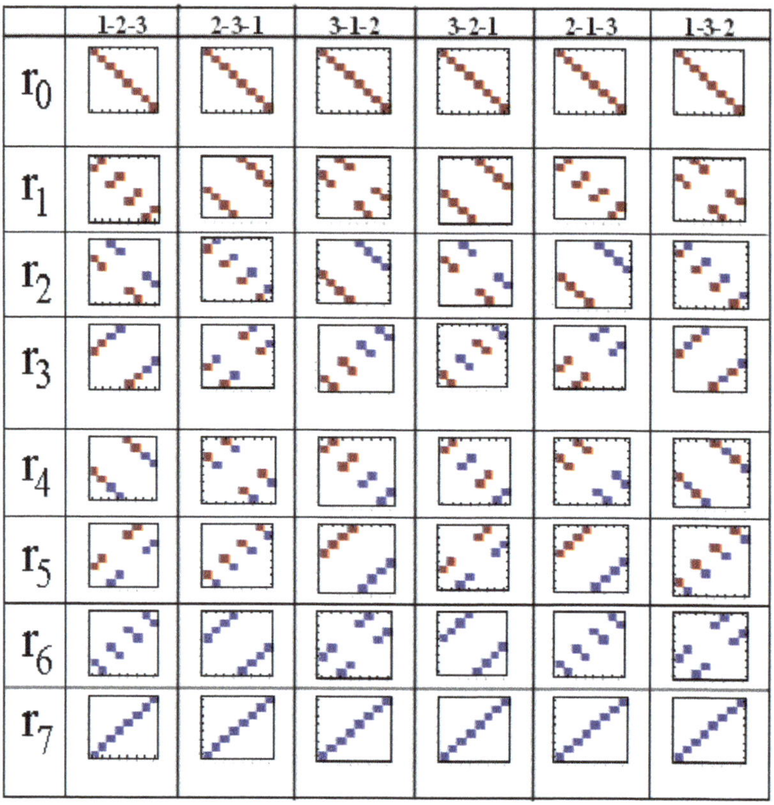

Fig. 1.6.4. The sets of sparse (8 × 8)-matrices r_0, r_1, r_2, r_3, r_4, r_5, r_6, r_7 for the dyadic-shift decomposition of the numeric matrices R_{123}, R_{231}, R_{312}, R_{321}, R_{213}, R_{132} from Figs. 1.6.2 and 1.6.3. Red and blue cells contain entries "+1" and "−1", respectively. White cells contain zero. Each of the columns of the table is denoted by the index of the corresponding matrix.

One can see from Fig. 1.6.3 that each of the 6 sets with eight sparse matrices r_0, r_1, r_2, r_3, r_4, r_5, r_6, r_7 is unique and different from other sets (r_0 is the identity matrix in all the sets). Unexpected facts are that, firstly, each of these sparse matrices is orthogonal; secondly, each of these 6 sets is closed in relation to multiplication; thirdly, each of these sets corresponds to the same multiplication table from Fig. 1.5.3 (with accuracy up to a permutation of columns and rows in tables, that is, up to a choice of names for sparse matrices r_1, r_2, r_3, r_4, r_5, r_6, r_7). It means that this genetic algebra of eight-dimensional hypercomplex numbers possesses at least 6 forms of its matrix representation. The matrices R_{123}, R_{231}, R_{312}, R_{321}, R_{213}, R_{132} are different forms of matrix representation of the same eight-dimensional hypercomplex number having unit coordinates. Our results demonstrate that the DNA alphabet of 64 triplets and the degeneracy of the genetic code are structured by Nature not arbitrarily but in such a regular way that all the possible variants of positional permutations in triplets preserve the described connection of this structured alphabet with the algebra of eight-dimensional numbers.

In the next chapter, we return to these genetic matrices R_{123}, R_{231}, R_{312}, R_{321}, R_{213}, R_{132} under their algorithmic transformations into Hadamard matrices on the basis of an additional binary-opposition feature of the genetic code system.

1.7. The principle "like begets like" in molecular and matrix genetics

1.7.1. *Complementary replication as a key principle in molecular genetics*

The DNA double helix model created by J.D. Watson and F. Crick in 1953 gave a powerful impetus to the development of genetic research. It showed to the world a recursive algorithm for the complementary replication of DNA strands, which ensures the replication of the genetic information recorded on these strands. In this process, DNA is separated into two complementary strands. Each strand of the original DNA molecule serves as a template for the production of its new complementary counterpart. This seminal work by Watson and Crick was perceived as the discovery of a key secret of life, corresponding to the ancient notion that "like begets like". Scientists were struck by how simple and beautiful this explanation of the replication and preservation of genetic information based on the mechanism of complementarity turned out to be. It was emphasized that it is this complementarity that provides the most important properties of DNA as a carrier of hereditary information [46].

The complementary replication of DNA in all living organisms is essential for cell division during growth and repair of damaged tissues, while it also ensures that each of the new cells receives its own copy of the DNA. The cell possesses the distinctive property of division, which makes complementary replication of DNA essential. Complementary replication of DNA strands occurs at an astonishing rate. For example, the well-known bacteria *E. coli* has a speed of replication of over 1,000 bases per second [47].

The complementary replication of DNA strands is based on the oppositions of complementary nucleotides A-T and C-G. But these oppositions are just particular cases of a wide set of mutually related binary oppositions in the integral system of genetic coding. Other types of such oppositions are: amino (A and C) vs. keto (G and T); three hydrogen bonds vs. two hydrogen bonds in complementary pairs of nucleotides; purines (A and G) vs. pyrimidines (C and T); n-plets with strong roots vs. n-plets with weak roots (Fig. 1.4.1). Usually, in genetics, each type of binary opposition is considered on its own out of their compete ensemble. But the organism is one, and this holistic system of binary oppositions may have important emergent properties that its individual parts do not have and that need to be studied.

The set of the named types of binary oppositions is comfortably represented in the described tensor family of matrices for DNA alphabets of n-plets with the numbering of their components based on dyadic groups (Fig. 1.2.1). This tensor family of genetic matrices makes it possible to algebraically study the emergent properties of alphabets of DNA n-plets and the genetic coding. The analysis of Rademacher genetic matrices (Fig. 1.4.1) reveals that it is built in accordance with the principle "like begets like" and the algebraic notion of the binary-complementary replication in structures of genetic $(2^n \times 2^n)$-matrices representing phenomenological properties of the genetic system in their connection with dyadic groups of binary numbers [48]. Let us explain this.

The new notion of the binary-complementary replication is used, which is connected with the binary operation of interchange $0 \leftrightarrow 1$ introduced above in connection with the complementary columns or rows (see §§ 1.2 and 1.5).

The complementary columns are located mirror-symmetric in the left and right halves of the considered matrix. The complementary rows are located in its upper and lower halves. Each pair of complementary columns contains one column enumerated by even numbers and one column enumerated by odd numbers (the same holds for each pair of complementary rows). All n-plets in one of the columns are complementary to n-plets in the second column (the same holds for n-plets in each pair of complementary rows).

The purpose of this section is to describe results of an in-depth analysis of the principle "like begets like" in the genetic coding system. These results show that the molecular complementary replication of DNA strands is accompanied by the presence of an algebraic version of the principle "like begets like" in the matrix representations of DNA alphabets connected with dyadic groups of binary numbers. This algebraic version is based on binary-oppositional structures in the genetic molecular system represented by the corresponding sets of binary numbers and matrices of DNA alphabets. The obtained results allow thinking that the principle "like begets like" (or a complementary replication in a wide sense) is systemic in the genetic organization and is connected with algebraic features of biological organization. Correspondingly, this biological principle can be additionally modeled by algebraic-matrix methods and approaches. Such algebraic-matrix modeling of the genetic coding system gives new ways for studying and understanding the key role of the principle "like begets like" in genetic and other inherited physiological complexes. The considered algebraic binary-complementary replication in phenomenological structures of genetic $(2^n \times 2^n)$-matrices is connected with the above introduced notion of pairs of complementary columns or rows.

1.7.2. *Complementary-replicated genetic matrices and an even-odd column decomposition of the matrix of 64 triplets*

Let us demonstrate the first example of the binary-complementary replication principle in the Rademacher genetic matrix of 64 triplets in

	000 (0)	001 (1)	010 (2)	011 (3)	100 (4)	101 (5)	110 (6)	111 (7)
000 (0)	+1	+1	−1	−1	+1	+1	−1	−1
001 (1)	+1	+1	−1	−1	+1	+1	−1	−1
010 (2)	+1	+1	+1	+1	−1	−1	−1	−1
011 (3)	+1	+1	+1	+1	−1	−1	−1	−1
100 (4)	+1	+1	−1	−1	+1	+1	−1	−1
101 (5)	+1	+1	−1	−1	+1	+1	−1	−1
110 (6)	−1	−1	−1	−1	+1	+1	+1	+1
111 (7)	−1	−1	−1	−1	+1	+1	+1	+1

=

	(0)	(1)	(2)	(3)	(4)	(5)	(6)	(7)
(0)	+1		−1		+1		−1	
(1)	+1		−1		+1		−1	
(2)	+1		+1		−1		−1	
(3)	+1		+1		−1		−1	
(4)	+1		−1		+1		−1	
(5)	+1		−1		+1		−1	
(6)	−1		−1		+1		+1	
(7)	−1		−1		+1		+1	

+

	(0)	(1)	(2)	(3)	(4)	(5)	(6)	(7)
(0)		+1		−1		+1		−1
(1)		+1		−1		+1		−1
(2)		+1		+1		−1		−1
(3)		+1		+1		−1		−1
(4)		+1		−1		+1		−1
(5)		+1		−1		+1		−1
(6)		−1		−1		+1		+1
(7)		−1		−1		+1		+1

Fig. 1.7.1. The even-odd decomposition of the Rademacher genetic matrix of 64 triplets (from Fig. 1.4.3) as the sum of two sparse complementary matrices: left, the even-columns matrix containing only non-zero columns having even numberings; right, the odd-columns matrix containing only non-zero columns having odd numberings. Empty cells contain zero entries. Numbers in brackets are decimal values of binary numberings.

Fig. 1.4.3 having symmetric and mirror-antisymmetric features described above. Fig. 1.7.1 shows that the numeric presentation of this matrix, containing entries +1 and −1 (whose locations correspond to triplets with strong and weak roots), is the sum of two sparse matrices, one of which contains only non-zero columns enumerated by even numbers and the other contains only non-zero columns enumerated by odd numbers. Each pair of complementary columns is separated among these two matrices. Correspondingly, the sparse matrix with even-numerated columns is conditionally called the even-columns matrix; all its non-zero columns correspond to triplets, which contain pyrimidines C or T at their ends (by this reason, this sparse matrix can be also called the pyrimidine-columns matrix). The sparse matrix with odd-numerated columns is called the odd-columns matrix; all its non-zero columns correspond to triplets, which contain purines A or G at their ends (by this reason, this matrix can be called the purine-columns matrix).

The even-columns (8 × 8)-matrix in Fig. 1.7.1 is the sum of 4 sparse (8 × 8)-matrices $s_0 + s_1 + s_2 + s_3$ shown in Fig. 1.7.2 (such decomposition is called the column dyadic-tensor-shift decomposition by reasons explained below). The set of these 4 matrices s_0, s_1, s_2, s_3 is closed relative to multiplication and corresponds to a multiplication table in Fig. 1.7.2, right. It is interesting that this multiplication table has already appeared above (Fig. 1.5.2) in a quite different case: the dyadic-shift decomposition of the mosaic (4 × 4)-matrix of 16 duplets. This table matches the multiplication table of the Cockle split-quaternions algebra used in the Poincare conformal disk model of hyperbolic geometry.

Analogically, the odd-columns matrix (Fig. 1.7.1, right) is the sum of 4 sparse matrices $p_0 + p_1 + p_2 + p_3$ shown in Fig. 1.7.3. The set of these 4 matrices p_0, p_1, p_2, p_3 is closed regarding multiplication and defines the

s_0 =

1	0	0	0	0	0	0	0
1	0	0	0	0	0	0	0
0	0	1	0	0	0	0	0
0	0	1	0	0	0	0	0
0	0	0	0	1	0	0	0
0	0	0	0	1	0	0	0
0	0	0	0	0	0	1	0
0	0	0	0	0	0	1	0

s_1 =

0	0	−1	0	0	0	0	0
0	0	−1	0	0	0	0	0
1	0	0	0	0	0	0	0
1	0	0	0	0	0	0	0
0	0	0	0	0	0	−1	0
0	0	0	0	0	0	−1	0
0	0	0	0	1	0	0	0
0	0	0	0	1	0	0	0

s_2 =

0	0	0	0	0	0	−1	0
0	0	0	0	0	0	−1	0
0	0	0	0	−1	0	0	0
0	0	0	0	−1	0	0	0
0	0	−1	0	0	0	0	0
0	0	−1	0	0	0	0	0
−1	0	0	0	0	0	0	0
−1	0	0	0	0	0	0	0

s_3 =

0	0	0	0	1	0	0	0
0	0	0	0	1	0	0	0
0	0	0	0	0	0	−1	0
0	0	0	0	0	0	−1	0
1	0	0	0	0	0	0	0
1	0	0	0	0	0	0	0
0	0	−1	0	0	0	0	0
0	0	−1	0	0	0	0	0

×	s_0	s_1	s_2	s_3
s_0	s_0	s_1	s_2	s_3
s_1	s_1	$-s_0$	s_3	$-s_2$
s_2	s_2	$-s_3$	s_0	$-s_1$
s_3	s_3	s_2	s_1	s_0

Fig. 1.7.2. The column dyadic-tensor-shift decomposition of the even-columns matrix (from Fig. 1.7.1, left) into 4 sparse matrices s_0, s_1, s_2, s_3, whose set is closed relative to multiplication; s_0 plays the role of the identity matrix in this set. The multiplication table for this set is shown on the right, which matches with the multiplication table of the four-dimensional algebra of Cockle split-quaternions used in the Poincare conformal disk model of hyperbolic geometry. The symbol of this model is presented.

p_0

0	1	0	0	0	0	0	0
0	1	0	0	0	0	0	0
0	0	0	1	0	0	0	0
0	0	0	1	0	0	0	0
0	0	0	0	0	1	0	0
0	0	0	0	0	1	0	0
0	0	0	0	0	0	0	1
0	0	0	0	0	0	0	1

p_1

0	0	0	−1	0	0	0	0
0	0	0	−1	0	0	0	0
0	1	0	0	0	0	0	0
0	1	0	0	0	0	0	0
0	0	0	0	0	0	0	−1
0	0	0	0	0	0	0	−1
0	0	0	0	0	1	0	0
0	0	0	0	0	1	0	0

p_2

0	0	0	0	0	0	0	−1
0	0	0	0	0	0	0	−1
0	0	0	0	0	−1	0	0
0	0	0	0	0	−1	0	0
0	0	0	−1	0	0	0	0
0	0	0	−1	0	0	0	0
0	−1	0	0	0	0	0	0
0	−1	0	0	0	0	0	0

p_3

0	0	0	0	0	1	0	0
0	0	0	0	0	1	0	0
0	0	0	0	0	0	0	−1
0	0	0	0	0	0	0	−1
0	1	0	0	0	0	0	0
0	1	0	0	0	0	0	0
0	0	0	−1	0	0	0	0
0	0	0	−1	0	0	0	0

×	p_0	p_1	p_2	p_3
p_0	p_0	p_1	p_2	p_3
p_1	p_1	$-p_0$	p_3	$-p_2$
p_2	p_2	$-p_3$	p_0	$-p_1$
p_3	p_3	p_2	p_1	p_0

Fig. 1.7.3. The column dyadic-tensor-shift decomposition of the odd-columns matrix (from Fig. 1.7.1, right) into 4 sparse matrices p_0, p_1, p_2, p_3, whose set is closed relative to multiplication; p_0 plays the role of the identity matrix in this set. The multiplication table for this set is shown, which matches the multiplication table of the Cockle split-quaternions algebra used in the Poincare conformal disk model of hyperbolic geometry. The symbol of this model is presented.

multiplication table in Fig. 1.7.3, right. This multiplication table coincides with the multiplication table of the four-dimensional algebra, which was received above for the even-columns matrix (Fig. 1.7.2). Both the even-columns matrix and the odd-columns matrix present Cockle's split-quaternions with unit coordinates (these split-quaternions have different forms of their matrix representations, with which these even-columns and odd-columns genetic matrices turn out to be associated). Correspondingly, both these genetic matrices are connected with the Poincare conformal disk model of hyperbolic geometry.

Now let us show that the summation of the even-columns and odd-columns matrices, which are binary-complementary to each other and connected with the four-dimensional algebra, gives the summary matrix W as a new algebraic entity, which is connected already with the eight-dimensional algebra. This summary Rademacher matrix (Fig. 1.7.1, top and Fig. 1.7.5, left) — under its column dyadic-tensor-shift decomposition — is the sum of 8 sparse matrices $v_0 + v_1 + v_2 + v_3 + v_4 + v_5 + v_6 + v_7$ shown in Fig. 1.7.4. The set of these matrices v_0, v_1, v_2, v_3, v_4, v_5, v_6, v_7 is closed relative to multiplication and matches the multiplication table (Fig. 1.7.4, bottom) of an eight-dimensional algebra. This algebra has interesting properties, which

$$v_0 = \begin{bmatrix} 1&0&0&0&0&0&0&0\\ 1&0&0&0&0&0&0&0\\ 0&0&1&0&0&0&0&0\\ 0&0&1&0&0&0&0&0\\ 0&0&0&0&1&0&0&0\\ 0&0&0&0&1&0&0&0\\ 0&0&0&0&0&0&1&0\\ 0&0&0&0&0&0&1&0 \end{bmatrix} \quad v_1 = \begin{bmatrix} 0&1&0&0&0&0&0&0\\ 0&1&0&0&0&0&0&0\\ 0&0&0&1&0&0&0&0\\ 0&0&0&1&0&0&0&0\\ 0&0&0&0&0&1&0&0\\ 0&0&0&0&0&1&0&0\\ 0&0&0&0&0&0&0&1\\ 0&0&0&0&0&0&0&1 \end{bmatrix} \quad v_2 = \begin{bmatrix} 0&0&-1&0&0&0&0&0\\ 0&0&-1&0&0&0&0&0\\ 1&0&0&0&0&0&0&0\\ 1&0&0&0&0&0&0&0\\ 0&0&0&0&0&0&-1&0\\ 0&0&0&0&0&0&-1&0\\ 0&0&0&0&1&0&0&0\\ 0&0&0&0&1&0&0&0 \end{bmatrix} \quad v_3 = \begin{bmatrix} 0&0&0&-1&0&0&0&0\\ 0&0&0&-1&0&0&0&0\\ 0&1&0&0&0&0&0&0\\ 0&1&0&0&0&0&0&0\\ 0&0&0&0&0&0&0&-1\\ 0&0&0&0&0&0&0&-1\\ 0&0&0&0&0&1&0&0\\ 0&0&0&0&0&1&0&0 \end{bmatrix}$$

$$v_4 = \begin{bmatrix} 0&0&0&0&1&0&0&0\\ 0&0&0&0&1&0&0&0\\ 0&0&0&0&0&0&-1&0\\ 0&0&0&0&0&0&-1&0\\ 1&0&0&0&0&0&0&0\\ 1&0&0&0&0&0&0&0\\ 0&0&-1&0&0&0&0&0\\ 0&0&-1&0&0&0&0&0 \end{bmatrix} \quad v_5 = \begin{bmatrix} 0&0&0&0&0&1&0&0\\ 0&0&0&0&0&1&0&0\\ 0&0&0&0&0&0&0&-1\\ 0&0&0&0&0&0&0&-1\\ 0&1&0&0&0&0&0&0\\ 0&1&0&0&0&0&0&0\\ 0&0&0&-1&0&0&0&0\\ 0&0&0&-1&0&0&0&0 \end{bmatrix} \quad v_6 = \begin{bmatrix} 0&0&0&0&0&0&-1&0\\ 0&0&0&0&0&0&-1&0\\ 0&0&0&0&-1&0&0&0\\ 0&0&0&0&-1&0&0&0\\ 0&0&-1&0&0&0&0&0\\ 0&0&-1&0&0&0&0&0\\ -1&0&0&0&0&0&0&0\\ -1&0&0&0&0&0&0&0 \end{bmatrix} \quad v_7 = \begin{bmatrix} 0&0&0&0&0&0&0&-1\\ 0&0&0&0&0&0&0&-1\\ 0&0&0&0&0&-1&0&0\\ 0&0&0&0&0&-1&0&0\\ 0&0&0&-1&0&0&0&0\\ 0&0&0&-1&0&0&0&0\\ 0&-1&0&0&0&0&0&0\\ 0&-1&0&0&0&0&0&0 \end{bmatrix}$$

×	v0	v1	v2	v3	v4	v5	v6	v7
v0	v0	v1	v2	v3	v4	v5	v6	v7
v1	v0	v1	v2	v3	v4	v5	v6	v7
v2	v2	v3	-v0	-v1	-v6	-v7	v4	v5
v3	v2	v3	-v0	-v1	-v6	-v7	v4	v5
v4	v4	v5	v6	v7	v0	v1	v2	v3
v5	v4	v5	v6	v7	v0	v1	v2	v3
v6	v6	v7	-v4	-v5	-v2	-v3	v0	v1
v7	v6	v7	-v4	-v5	-v2	-v3	v0	v1

Fig. 1.7.4. The column dyadic-tensor-shift decomposition of the sum of the even-columns matrix and the odd-columns matrix (this summary Rademacher matrix W is shown in Fig. 1.7.1, top) into 8 sparse matrices v_0, v_1, v_2, v_3, v_4, v_5, v_6, v_7, whose set is closed relative to multiplication. The multiplication table for this set is shown at the bottom.

were described in previous publications without a connection to the topic of complementary replications presented here [12, 13, 49, 50].

This summary matrix W generates algorithmically its complementary-replicated analogue W_R by means of the interchange of numbers $0 \leftrightarrow 1$ in the binary numerating of its columns with the corresponding rearrangement

W =

	0	1	2	3	4	5	6	7
0	1	1	−1	−1	1	1	−1	−1
1	1	1	−1	−1	1	1	−1	−1
2	1	1	1	1	−1	−1	−1	−1
3	1	1	1	1	−1	−1	−1	−1
4	1	1	−1	−1	1	1	−1	−1
5	1	1	−1	−1	1	1	−1	−1
6	−1	−1	−1	−1	1	1	1	1
7	−1	−1	−1	−1	1	1	1	1

↔

W_R =

	7	6	5	4	3	2	1	0
0	−1	−1	1	1	−1	−1	1	1
1	−1	−1	1	1	−1	−1	1	1
2	−1	−1	−1	−1	1	1	1	1
3	−1	−1	−1	−1	1	1	1	1
4	−1	−1	1	1	−1	−1	1	1
5	−1	−1	1	1	−1	−1	1	1
6	1	1	1	1	−1	−1	−1	−1
7	1	1	1	1	−1	−1	−1	−1

Fig. 1.7.5. The Rademacher genetic matrix W of 64 triplets (from Fig. 1.4.2) and its complementary-replicated matrix W_R, which are transformed to each other by the algorithm based on binary-oppositions in the DNA nucleobases alphabet (the purine-pyrimidine transformation, see explanations in the text). Cells containing entries +1 correspond to locations of triplets with strong roots. The numbering of columns and rows is shown in the decimal system.

of the columns (that is, rearrangements of columns located mirror-symmetrically in the left and right halves of the matrix). This interchanging algorithm 0 ↔ 1 in binary numbers provides interchange in any pair of complementary columns that differ from each other in the content of triplets with purine and pyrimidine endings, in some analogy with the complementarity of purines and pyrimidines in DNA double strands. For example, the column with number 110 (which corresponds to the nucleotide order "purine-purine-pyrimidine" in all its triplets) takes the place of the column with number 001 (which corresponds to the order "pyrimidine-pyrimidine-purine" in all its triplets). Fig. 1.7.5 shows the summary matrix W and its complementary-replicated analogue W_R, which is generated by this algorithm based on binary-oppositions in the DNA nucleobases alphabet and which is also connected by its meander-like mosaic with meander-like Rademacher functions.

Applying this complementary-replicating algorithm to the complementary-replicated matrix W_R generates the original Rademacher matrix W, that is, W and W_R are mutual complementary-replicated matrices resembling two complementary strings of DNA. This algorithm is recursive and its application allows generation of such pairs of complementary-replicated matrices again and again. So, the ancient notion that "like begets like" surprisingly turns out to be realized in genetics not only for complementary strings of DNA but also for the numeric presentation of the genetic matrix of 64 triplets, which is based on the binary-oppositional properties of their DNA alphabet and of the degeneracy of the genetic coding. In other words, molecular complementary-replicated properties of DNA strings exist jointly with algebraic binary-complementary replication properties of the considered alphabetical matrix of the genetic code. Both of these properties are parts of genetics of the whole organism and so are interrelated. These algebraic complementary-replicated properties of genetic matrices allow the application of effective algebraic methods for further study of genetics to include it in the field of modern mathematical natural sciences in connection with multi-dimensional algebras, hyperbolic geometry, theory of resonances, etc. Some results of such algebra-genetic studies are presented in this book.

The complementary-replicated matrix W_R — under its column dyadic-tensor-shift decomposition — is the sum of 8 sparse matrices $q_0 + q_1 + q_2 + q_3 + q_4 + q_5 + q_6 + q_7$ shown in Fig. 1.7.6. The set of these matrices q_0, q_1, q_2, q_3, q_4, q_5, q_6, q_7 is closed relative to multiplication and matches the multiplication table (Fig. 1.7.6, bottom) of an eight-dimensional algebra. This new multiplication table turns out to be a complementary-replicated analogue of the multiplication table shown for the similar decompositions of the matrix W in Fig. 1.7.4: in these multiplication tables, each value of the product $q_i \times q_k$ is equal to the value $v_i \times v_k$ but with an opposite sign (here indexes i, k = 0, 1, 2, 3, 4, 5, 6, 7).

The action of complementary-replicated (8 × 8)-matrices W and W_R on an arbitrary eight-dimensional vector $\overline{X}$ generates two new vectors that

$$q_0 = \begin{bmatrix} -1&0&0&0&0&0&0&0\\ -1&0&0&0&0&0&0&0\\ 0&0&-1&0&0&0&0&0\\ 0&0&-1&0&0&0&0&0\\ 0&0&0&0&-1&0&0&0\\ 0&0&0&0&-1&0&0&0\\ 0&0&0&0&0&0&-1&0\\ 0&0&0&0&0&0&-1&0 \end{bmatrix} \quad q_1 = \begin{bmatrix} 0&-1&0&0&0&0&0&0\\ 0&-1&0&0&0&0&0&0\\ 0&0&0&-1&0&0&0&0\\ 0&0&0&-1&0&0&0&0\\ 0&0&0&0&0&-1&0&0\\ 0&0&0&0&0&-1&0&0\\ 0&0&0&0&0&0&0&-1\\ 0&0&0&0&0&0&0&-1 \end{bmatrix}$$

$$q_2 = \begin{bmatrix} 0&0&1&0&0&0&0&0\\ 0&0&1&0&0&0&0&0\\ -1&0&0&0&0&0&0&0\\ -1&0&0&0&0&0&0&0\\ 0&0&0&0&0&0&1&0\\ 0&0&0&0&0&0&1&0\\ 0&0&0&0&-1&0&0&0\\ 0&0&0&0&-1&0&0&0 \end{bmatrix} \quad q_3 = \begin{bmatrix} 0&0&0&1&0&0&0&0\\ 0&0&0&1&0&0&0&0\\ 0&-1&0&0&0&0&0&0\\ 0&-1&0&0&0&0&0&0\\ 0&0&0&0&0&0&0&1\\ 0&0&0&0&0&0&0&1\\ 0&0&0&0&0&-1&0&0\\ 0&0&0&0&0&-1&0&0 \end{bmatrix}$$

$$q_4 = \begin{bmatrix} 0&0&0&0&-1&0&0&0\\ 0&0&0&0&-1&0&0&0\\ 0&0&0&0&0&0&1&0\\ 0&0&0&0&0&0&1&0\\ -1&0&0&0&0&0&0&0\\ -1&0&0&0&0&0&0&0\\ 0&0&1&0&0&0&0&0\\ 0&0&1&0&0&0&0&0 \end{bmatrix} \quad q_5 = \begin{bmatrix} 0&0&0&0&0&-1&0&0\\ 0&0&0&0&0&-1&0&0\\ 0&0&0&0&0&0&0&1\\ 0&0&0&0&0&0&0&1\\ 0&-1&0&0&0&0&0&0\\ 0&-1&0&0&0&0&0&0\\ 0&0&0&1&0&0&0&0\\ 0&0&0&1&0&0&0&0 \end{bmatrix}$$

$$q_6 = \begin{bmatrix} 0&0&0&0&0&0&1&0\\ 0&0&0&0&0&0&1&0\\ 0&0&0&0&1&0&0&0\\ 0&0&0&0&1&0&0&0\\ 0&0&1&0&0&0&0&0\\ 0&0&1&0&0&0&0&0\\ 1&0&0&0&0&0&0&0\\ 1&0&0&0&0&0&0&0 \end{bmatrix} \quad q_7 = \begin{bmatrix} 0&0&0&0&0&0&0&1\\ 0&0&0&0&0&0&0&1\\ 0&0&0&0&0&1&0&0\\ 0&0&0&0&0&1&0&0\\ 0&0&0&1&0&0&0&0\\ 0&0&0&1&0&0&0&0\\ 0&1&0&0&0&0&0&0\\ 0&1&0&0&0&0&0&0 \end{bmatrix}$$

×	q_0	q_1	q_2	q_3	q_4	q_5	q_6	q_7
q_0	$-q_0$	$-q_1$	$-q_2$	$-q_3$	$-q_4$	$-q_5$	$-q_6$	$-q_7$
q_1	$-q_0$	$-q_1$	$-q_2$	$-q_3$	$-q_4$	$-q_5$	$-q_6$	$-q_7$
q_2	$-q_2$	$-q_3$	q_0	q_1	q_6	q_7	$-q_4$	$-q_5$
q_3	$-q_2$	$-q_3$	q_0	q_1	q_6	q_7	$-q_4$	$-q_5$
q_4	$-q_4$	$-q_5$	$-q_6$	$-q_7$	$-q_0$	$-q_1$	$-q_2$	$-q_3$
q_5	$-q_4$	$-q_5$	$-q_6$	$-q_7$	$-q_0$	$-q_1$	$-q_2$	$-q_3$
q_6	$-q_6$	$-q_7$	q_4	q_5	q_2	q_3	$-q_0$	$-q_1$
q_7	$-q_6$	$-q_7$	q_4	q_5	q_2	q_3	$-q_0$	$-q_1$

Fig. 1.7.6. The column dyadic-tensor-shift decomposition of the matrix W_R (Fig. 1.7.5) into 8 sparse matrices q_0, q_1, q_2, q_3, q_4, q, q_6, q_7, whose set is closed relative to multiplication. The multiplication table for this set is shown at the bottom.

are complementary to each other: the corresponding coordinates of both generated vectors are the same in their absolute values, but have opposite signs. A numerical example of this with a voluntary vector $\bar{X} = [1, 2, 3, 4, 5, 6, 7, 8]$ is shown by expression (1.3):

$$\begin{aligned} \bar{X} \times W &= [1, 2, 3, 4, 5, 6, 7, 8] \times W = [6, 6, -22, -22, 22, 22, -6, -6] = \bar{Y} \\ \bar{X} \times W_R &= [1, 2, 3, 4, 5, 6, 7, 8] \times W_R = [-6, -6, 22, 22, -22, -22, 6, 6] = -\bar{Y} \end{aligned} \tag{1.3}$$

Another interesting property of the Rademacher genetic complementary-replicated matrices W and W_R is that — by their repeated action on the emerging vectors (1.3) — one can generate as many complementary-replicated vectors as desired. In this case, the quadrupling of coordinate values in the vectors occurs, reminiscent of the quadrupling of genetic information during the meiosis division of germ cells, under which one cell generates 4 similar cells with a complete set of DNAs in each. The following example (1.4), using the notations from (1.3), illustrates this quadrupling of coordinate values with a regular changing of signs "+" and "–":

$$\bar{Y} \times W_R = -4 \times \bar{Y}; \quad \bar{Y} \times W_R^{\,2} = 4^2 \times \bar{Y}; \quad \bar{Y} \times W_R^{\,3} = -4^3 \times \bar{Y};$$
$$\bar{Y} \times W_R^{\,4} = 4^4 \times \bar{Y}; \quad \text{etc.} \tag{1.4}$$

The expression (1.5) shows one more property of the Rademacher complementary-replicated matrices W and W_R:

$$W \times W_R = W_R \times W = -4W_R \tag{1.5}$$

The matrix W/4 is an oblique projector since $(W/4)^2 = W/4$. In contrast, the matrix W_R corresponds to another condition: $(W_R/4)^2 = -W_R/4$.

Each of the resulting vectors $\bar{X} \times W$ and $\bar{X} \times W_R$ is always a complementary palindrome: the sequence of its coordinates, which is read in forward order, coincides with the sequence, which is read in reverse order and having coordinates with the opposite sign (see the example (1.3)). This algebraic feature of the action of complementary-replicated matrices is interesting, since in molecular genetics the problem of complementary palindromes has long been known. Here one should recall the difference between an ordinary palindrome and a complementary palindrome. By definition, an ordinary palindrome is a string that reads the same from beginning and from the end. By contrast, a complementary palindrome in molecular genetics is a fragment of a chain of DNA or RNA which becomes an ordinary palindrome if each symbol in one half of the fragment is replaced by its complementary symbol ($A \leftrightarrow T$, $C \leftrightarrow G$) [51]. For instance, AGCTCGCGAGCT is a complementary palindrome. In nucleotide sequences of DNA and RNA, a great number of complementary palindromes and ordinary palindromes exists [51, 52]. For instance, families of repetitive sequences occupy about one-third of the human genome. The importance of the problem of repeats in genetic sequences is reflected in the fact that 6,000 articles were published on this subject during the 20 years before 1991 [53].

One should add that the theme of the complementary columns (and rows) in the described genetic matrices is also essential in connection with the universal rules of stochastic organization of DNA in genomes of higher and lower organisms. These rules will be presented later in this book. They include the rule of approximate equality for the sums of probabilities of n-plets belonging to the even column and the odd column of each pair of complementary columns (the same is true for each pair of complementary rows).

1.7.3. *Complementary-replicated genetic matrices and an even-odd rows decomposition of the matrix of 64 triplets*

Let us show that similar algebraic results arise in the case of the so-called "rows dyadic-tensor-shift decomposition" of the same mosaic matrix of 64 triplets from Fig. 1.4.3. This matrix has pairs of complementary rows, which are located mirror-symmetric in its top and bottom halves; as it was noted above in § 1.5, each pair contains one row with even numbers and one row with odd numbers. Fig. 1.7.7 shows that the numeric presentation of this matrix, containing entries +1 and −1 (whose locations correspond to triplets with strong and weak roots, respectively), is the sum of two sparse matrices, one of which contains only non-zero rows enumerated by even numbers and the other contains only non-zero rows enumerated by odd numbers. Each pair of complementary rows is separated among these two matrices. Correspondingly, the sparse matrix with even-numerated rows is conditionally called the even-rows matrix of the row type; all its non-zero rows correspond to triplets, which contain amino-molecules A or C at their ends (by this reason, this sparse matrix can be also called the amino-rows matrix). The sparse matrix with odd-numerated rows is called the odd-rows matrix; all its non-zero rows correspond to triplets, which contain keto-molecules G or T at their ends (by this reason, this sparse matrix can be also called the keto-rows matrix).

The even-rows (8 × 8)-matrix in Fig. 1.7.7 is the sum of 4 sparse (8 × 8)-matrices $u_0 + u_1 + u_2 + u_3$ shown in Fig. 1.7.8. The set of these 4 matrices u_0, u_1, u_2, u_3 is closed relative to multiplication and corresponds to the multiplication table in Fig. 1.7.8, right. This table is again the multiplication table of the four-dimensional algebra of Cockle split-quaternions, which

	000 (0)	001 (1)	010 (2)	011 (3)	100 (4)	101 (5)	110 (6)	111 (7)
000 (0)	+1	+1	−1	−1	+1	+1	−1	−1
001 (1)	+1	+1	−1	−1	+1	+1	−1	−1
010 (2)	+1	+1	+1	+1	−1	−1	−1	−1
011 (3)	+1	+1	+1	+1	−1	−1	−1	−1
100 (4)	+1	+1	−1	−1	+1	+1	−1	−1
101 (5)	+1	+1	−1	−1	+1	+1	−1	−1
110 (6)	−1	−1	−1	−1	+1	+1	+1	+1
111 (7)	−1	−1	−1	−1	+1	+1	+1	+1

=

	(0)	(1)	(2)	(3)	(4)	(5)	(6)	(7)
(0)	+1	+1	−1	−1	+1	+1	−1	−1
(1)								
(2)	+1	+1	+1	+1	−1	−1	−1	−1
(3)								
(4)	+1	+1	−1	−1	+1	+1	−1	−1
(5)								
(6)	−1	−1	−1	−1	+1	+1	+1	+1
(7)								

+

	(0)	(1)	(2)	(3)	(4)	(5)	(6)	(7)
(0)								
(1)	+1	+1	−1	−1	+1	+1	−1	−1
(2)								
(3)	+1	+1	+1	+1	−1	−1	−1	−1
(4)								
(5)	+1	+1	−1	−1	+1	+1	−1	−1
(6)								
(7)	−1	−1	−1	−1	+1	+1	+1	+1

Fig. 1.7.7. The even-odd presentation of the mosaic matrix of 64 triplets (from Fig. 1.4.3) as the sum of two sparse complementary matrices: the left matrix, called the even-rows matrix, contains only non-zero rows having even numberings; the matrix at right, called the odd-rows matrix, contains only non-zero rows having odd numberings. Empty cells contain zero entries. Numbers in brackets are decimal values of binary numberings.

u_0

1	1	0	0	0	0	0	0
0	0	0	0	0	0	0	0
0	0	1	1	0	0	0	0
0	0	0	0	0	0	0	0
0	0	0	0	1	1	0	0
0	0	0	0	0	0	0	0
0	0	0	0	0	0	1	1
0	0	0	0	0	0	0	0

u_1

0	0	−1	−1	0	0	0	0
0	0	0	0	0	0	0	0
1	1	0	0	0	0	0	0
0	0	0	0	0	0	0	0
0	0	0	0	0	0	−1	−1
0	0	0	0	0	0	0	0
0	0	0	0	1	1	0	0
0	0	0	0	0	0	0	0

u_2

0	0	0	0	1	1	0	0
0	0	0	0	0	0	0	0
0	0	0	0	0	0	−1	−1
0	0	0	0	0	0	0	0
1	1	0	0	0	0	0	0
0	0	0	0	0	0	0	0
0	0	−1	−1	0	0	0	0
0	0	0	0	0	0	0	0

u_3

0	0	0	0	0	0	−1	−1
0	0	0	0	0	0	0	0
0	0	0	0	−1	−1	0	0
0	0	0	0	0	0	0	0
0	0	−1	−1	0	0	0	0
0	0	0	0	0	0	0	0
−1	−1	0	0	0	0	0	0
0	0	0	0	0	0	0	0

×	u_0	u_1	u_2	u_3
u_0	u_0	u_1	u_2	u_3
u_1	u_1	$-u_0$	$-u_3$	u_2
u_2	u_2	u_3	u_0	u_1
u_3	u_3	$-u_2$	$-u_1$	u_0

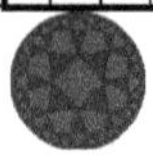

Fig. 1.7.8. The row dyadic-tensor-shift decomposition of the even-row matrix (from Fig. 1.7.7, left) into 4 sparse matrices u_0, u_1, u_2, u_3, whose set is closed relative to multiplication; u_0 plays the role of the identity matrix in this set. The multiplication table for this set is shown on the right, which matches the multiplication table of the four-dimensional algebra of Cockle split-quaternions used in the Poincare conformal disk model of hyperbolic geometry. The symbol of this model is presented.

a_0

0	0	0	0	0	0	0	0
1	1	0	0	0	0	0	0
0	0	0	0	0	0	0	0
0	0	1	1	0	0	0	0
0	0	0	0	0	0	0	0
0	0	0	0	1	1	0	0
0	0	0	0	0	0	0	0
0	0	0	0	0	0	1	1

a_1

0	0	0	0	0	0	0	0
0	0	-1	-1	0	0	0	0
0	0	0	0	0	0	0	0
1	1	0	0	0	0	0	0
0	0	0	0	0	0	0	0
0	0	0	0	0	0	-1	-1
0	0	0	0	0	0	0	0
0	0	0	0	1	1	0	0

a_2

0	0	0	0	0	0	0	0
0	0	0	0	1	1	0	0
0	0	0	0	0	0	0	0
0	0	0	0	0	0	-1	-1
0	0	0	0	0	0	0	0
1	1	0	0	0	0	0	0
0	0	0	0	0	0	0	0
0	0	-1	-1	0	0	0	0

a_3

0	0	0	0	0	0	0	0
0	0	0	0	0	0	-1	-1
0	0	0	0	0	0	0	0
0	0	0	0	-1	-1	0	0
0	0	0	0	0	0	0	0
0	0	-1	-1	0	0	0	0
0	0	0	0	0	0	0	0
-1	-1	0	0	0	0	0	0

*	a_0	a_1	a_2	a_3
a_0	a_0	a_1	a_2	a_3
a_1	a_1	$-a_0$	$-a_3$	a_2
a_2	a_2	a_3	a_0	a_1
a_3	a_3	$-a_2$	$-a_1$	a_0

Fig. 1.7.9. The row dyadic-tensor-shift decomposition of the even-row matrix (from Fig. 1.7.7, right) into 4 sparse matrices a_0, a_1, a_2, a_3, whose set is closed relative to multiplication; a_0 plays the role of the identity matrix in this set. The multiplication table for this set is shown on the right, which matches the multiplication table of the four-dimensional algebra of Cockle split-quaternions used in the Poincare conformal disk model of hyperbolic geometry. The symbol of this model is presented.

we met above in Figs. 1.5.2, 1.7.2, 1.7.3 and which is used in the Poincare conformal disk model of hyperbolic geometry.

Analogically, the odd-rows matrix (Fig. 1.7.7 at right) is the sum of 4 sparse matrices $a_0 + a_1 + a_2 + a_3$ shown in Fig. 1.7.9. The set of these 4 matrices a_0, a_1, a_2, a_3 is closed regarding multiplication and defines the multiplication table in Fig. 1.7.9, right. This multiplication table coincides with the multiplication table of the four-dimensional algebra, which was received above for the even-rows matrix (Fig. 1.7.8) and for even-columns and odd-columns matrices (Figs. 1.7.2, 1.7.3). Both the even-rows matrix and the odd-rows matrix present Cockle's split-quaternions with unit coordinates, which are connected with the Poincare conformal disk model of hyperbolic geometry.

The sum of the even-rows matrix and the odd-rows matrix gives the genetic matrix W in Fig. 1.7.1 at the top, which was above analyzed jointly with its complementary-replicated analogue W_R (Figs. 1.7.4–1.7.6).

Similar approaches using even-odd structures and dyadic-tensor-shift decompositions are also appropriate to analyze binary-complementary replicated properties of Rademacher genetic matrices of higher orders, for example, the (16×16)-matrix of 256 tetraplets in Fig. 1.4.2.

Different forms of implementation of the fundamental biological principle "like begets like" (or a complementary replication in a wide sense) can be seen at different levels of inherited biological organization. For example, in the brain of humans and animals, which has mirror-complementary hemispheres (left and right), mirror neurons are known. A mirror neuron is a neuron that fires both when an animal acts and when the animal observes the same action performed by another. Thus, the neuron "mirrors" the behavior of the other, as though the observer were itself acting.

The theme of mirror neurons, whose functioning is based on one of the forms of the complementary replication principle, provokes wide scientific researches and debates since it concerns cognitive functions, the origin of language, learning facilitation, automatic imitation, motor mimicry, autism, human capacity of emotions such as empathy, and many other problems (see, for example, [54, 55]). In 2014, *Philosophical Transactions of the Royal Society B* published a special issue entirely devoted to mirror neuron research [56]. One of the questions that arose is the following: where do mirror neurons come from? [57].

The above-described results of our studies in the field of matrix genetics give pieces of evidence that the system of mirror neurons and the system of DNA complementary replication are not isolated parts of the organism, but are particular parts of a bio-algebraic complex realizing the phenomenon "like begets like". Our body structure with its left and right halves, having left and right sensory motor systems, also can be considered as one of the manifestations of this complementary-replicating complex. Another example is given by our visual perception, whose optical system of the eye provides the transmission of the external image to the retina in complementary inverted and reduced forms. Although the image on the retina is inverted, we can see objects in the direct form by some complementary-replicating action of our brain.

Correspondingly, complementary replication is a systemic phenomenon in genetic organization. It is not that the molecules of two strands of DNA randomly docked, formed a complementary pair and began to repeat the process of complementary replication at breakneck speed. Another point of view is proposed: the DNA replication phenomenon is part of a holistic bio-algebraic genetic complex of complementary replication, parts of which manifest themselves at different levels of organization of the being, up to the functioning of the brain with its mirror neurons and the ability to empathize and imitate external events. Algebraic formalisms, described in this book, provide possibilities to model features of this complex and to develop algebraic biology.

1.7.4. *Dyadic-tensor-shift decompositions of genetic matrices*

Let us return to the dyadic-shift decomposition of the (4×4)-matrix shown in Fig. 1.4.2 into 4 sparse matrices e_0, e_1, e_2, e_3, the set of which is closed under multiplication and defines the multiplication table of the Cockle split-quaternion algebra. Tensorial multiplying each of these (4×4)-matrices by the (2×2)-matrix $[1, 1; 1, 1]$ gives 8 sparse matrices of the 8th order, coinciding with the above-described matrices v_0, v_1, v_2, v_3, v_4, v_5, v_6, v_7 in Fig. 1.7.4. The sum of these 8 matrices gives the above-described matrix

W in Fig. 1.7.5, which is an example of a dyadic-tensor-shift matrix. Its decomposition into 8 mentioned sparse matrices v_0, v_1, …, v_7 is called dyadic-tensor-shift decomposition, related to the property of the complementary replication of genetic matrices as it turns out.

Let us emphasize that the dyadic-tensor-shift decomposition of the Rademacher genetic matrix of 64 triplets (Fig. 1.4.3) relates to the following: the set of 64 triplets is divided into two equal subsets not only on the basis of strong and weak roots of the triplets, but also on the basis of whether the triplets end in pyrimidines (C and T) or purines (A and G). Accounting for this factor has above led to new interesting results on the algebraic features of the genetic coding system, associated with the criterion of even and odd numberings of columns and rows in the Rademacher mosaic matrix of 64 triplets. In this matrix, all columns with even numberings (0, 2, 4, 6) contain only triplets ending in pyrimidines C and T; in contrast, all columns with odd numberings (1, 3, 5, 7) contain only triplets ending in purines A and G. In the same matrix, all rows with even numberings contain only triplets ending in amino-molecules C and A; in contrast, all rows with odd numberings contain only triplets ending in keto-molecules T and G.

1.8. Complementary replications and the matrix of 64 triplets under its twice-complementary transformation

This section continues research in the field of matrix genetics and phenomena of algebraic-biological binary oppositions, aimed at demonstrating the key role of the principle "like begets like" and complementary replications in the genetic coding system including tensor families of genetic matrices.

For this additional research, let us consider a transformation of the Rademacher genetic matrix of 64 triplets from Fig. 1.4.3 under the simultaneous interchange of numbers $0 \leftrightarrow 1$ in the binary numerating of its columns and rows with appropriate rearrangements of the columns and the rows. These rearrangements of numberings correspond to simultaneous molecular interchanges inside all pairs of complementary columns and also all pairs of complementary rows defined above by indicators of purine-pyrimidine endings and amino-keto endings in triplets. Fig. 1.8.1 shows a new Rademacher matrix B, which arises under such twice-complementary transformation and which is conditionally called the twice-complementary matrix. This twice-complementary matrix B satisfies the following conditions: $B^2 = 4B$, $(B/2)^2 = B/2$, that is, the asymmetrical matrix B/2 is an oblique projector. The matrix B is the sum of two matrices: the odd-columns matrix and the even-columns matrix, which are shown in Fig. 1.8.1, at the bottom.

	111 (7)	110 (6)	101 (5)	100 (4)	011 (3)	010 (2)	001 (1)	000 (0)
111 (7)	1	1	1	1	-1	-1	-1	-1
110 (6)	1	1	1	1	-1	-1	-1	-1
101 (5)	-1	-1	1	1	-1	-1	1	1
100 (4)	-1	-1	1	1	-1	-1	1	1
011 (3)	-1	-1	-1	-1	1	1	1	1
010 (2)	-1	-1	-1	-1	1	1	1	1
001 (1)	-1	-1	1	1	-1	-1	1	1
000 (0)	-1	-1	1	1	-1	-1	1	1

=

	7	6	5	4	3	2	1	0
7	1		1		-1		-1	
6	1		1		-1		-1	
5	-1		1		-1		1	
4	-1		1		-1		1	
3	-1		-1		1		1	
2	-1		-1		1		1	
1	-1		1		-1		1	
0	-1		1		-1		1	

+

	7	6	5	4	3	2	1	0
7		1		1		-1		-1
6		1		1		-1		-1
5		-1		1		-1		1
4		-1		1		-1		1
3		-1		-1		1		1
2		-1		-1		1		1
1		-1		1		-1		1
0		-1		1		-1		1

Fig. 1.8.1. The twice-complementary matrix B (top), which is generated by the twice-complementary transformation from the Rademacher matrix of 64 triplets (from Fig. 1.4.3); its black cells correspond to triplets with strong roots. This matrix B is the sum of the odd-columns matrix BOC and the even-columns matrix BEC, whose numberings of columns and rows are shown in decimal notations (bottom). Empty cells contain zeros.

It turns out that the odd-columns matrix BOC is the sum of 4 new sparse matrices $b_0 + b_1 + b_2 + b_3$, whose set is closed relative to multiplication and defines a multiplication table of the algebra of split-quaternions of Cockle as shown in Fig. 1.8.2 (compare with Figs. 1.7.2 and 1.7.3). This algebra is used in the Poincare conformal disk model of hyperbolic geometry.

It turns out also that the even-columns matrix BEC (from Fig. 1.8.1, bottom right) is the sum of 4 new sparse matrices $k_0 + k_1 + k_2 + k_3$, whose set is closed relative to multiplication and defines as well a multiplication table of the algebra of split-quaternions of Cockle as shown in Fig. 1.8.3 (compare with the same multiplication table in Figs. 1.7.2, 1.7.3, and 1.8.2).

Twice-complementary matrix B (Fig. 1.8.1, top) can be also presented as the sum of the odd-rows matrix BOR and the even-rows matrix BER as shown in Fig. 1.8.4.

The odd-rows matrix BOR (Fig. 1.8.4) is the sum of 4 sparse matrices $h_0 + h_1 + h_2 + h_3$, whose set is closed relative to multiplication and defines a multiplication table of the same algebra of split-quaternions of Cockle as shown in Fig. 1.8.5 (compare with the same multiplication table in Figs. 1.7.2, 1.7.3, 1.8.2, and 1.8.3).

b_0 =

1	0	0	0	0	0	0	0
1	0	0	0	0	0	0	0
0	0	1	0	0	0	0	0
0	0	1	0	0	0	0	0
0	0	0	0	1	0	0	0
0	0	0	0	1	0	0	0
0	0	0	0	0	0	1	0
0	0	0	0	0	0	1	0

b_1 =

0	0	1	0	0	0	0	0
0	0	1	0	0	0	0	0
−1	0	0	0	0	0	0	0
−1	0	0	0	0	0	0	0
0	0	0	0	0	0	1	0
0	0	0	0	0	0	1	0
0	0	0	0	−1	0	0	0
0	0	0	0	−1	0	0	0

b_2 =

0	0	0	0	−1	0	0	0
0	0	0	0	−1	0	0	0
0	0	0	0	0	0	1	0
0	0	0	0	0	0	1	0
−1	0	0	0	0	0	0	0
−1	0	0	0	0	0	0	0
0	0	1	0	0	0	0	0
0	0	1	0	0	0	0	0

b_3 =

0	0	0	0	0	0	−1	0
0	0	0	0	0	0	−1	0
0	0	0	0	−1	0	0	0
0	0	0	0	−1	0	0	0
0	0	−1	0	0	0	0	0
0	0	−1	0	0	0	0	0
−1	0	0	0	0	0	0	0
−1	0	0	0	0	0	0	0

×	b_0	b_1	b_2	b_3
b_0	b_0	b_1	b_2	b_3
b_1	b_1	$-b_0$	$-b_3$	b_2
b_2	b_2	b_3	b_0	b_1
b_3	b_3	$-b_2$	$-b_1$	b_0

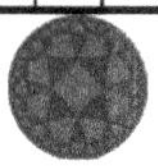

Fig. 1.8.2. The dyadic-tensor-shift decomposition of the odd-columns matrix BOC (from Fig. 1.8.1, bottom left) into 4 sparse matrices b_0, b_1, b_2, b_3, whose set is closed relative to multiplication and defines a multiplication table of the algebra of split-quaternions of Cockle (compare with Figs. 1.7.2 and 1.7.3). This split-quaternions algebra is used in the Poincare conformal disk model of hyperbolic geometry whose symbol is again presented.

k_0 =

0	1	0	0	0	0	0	0
0	1	0	0	0	0	0	0
0	0	0	1	0	0	0	0
0	0	0	1	0	0	0	0
0	0	0	0	0	1	0	0
0	0	0	0	0	1	0	0
0	0	0	0	0	0	0	1
0	0	0	0	0	0	0	1

k_1 =

0	0	0	1	0	0	0	0
0	0	0	1	0	0	0	0
0	−1	0	0	0	0	0	0
0	−1	0	0	0	0	0	0
0	0	0	0	0	0	0	1
0	0	0	0	0	0	0	1
0	0	0	0	0	−1	0	0
0	0	0	0	0	−1	0	0

k_2 =

0	0	0	0	0	−1	0	0
0	0	0	0	0	−1	0	0
0	0	0	0	0	0	0	1
0	0	0	0	0	0	0	1
0	−1	0	0	0	0	0	0
0	−1	0	0	0	0	0	0
0	0	0	1	0	0	0	0
0	0	0	1	0	0	0	0

k_3 =

0	0	0	0	0	0	0	−1
0	0	0	0	0	0	0	−1
0	0	0	0	0	−1	0	0
0	0	0	0	0	−1	0	0
0	0	0	−1	0	0	0	0
0	0	0	−1	0	0	0	0
0	−1	0	0	0	0	0	0
0	−1	0	0	0	0	0	0

×	k_0	k_1	k_2	k_3
k_0	k_0	k_1	k_2	k_3
k_1	k_1	$-k_0$	$-k_3$	k_2
k_2	k_2	k_3	k_0	k_1
k_3	k_3	$-k_2$	$-k_1$	k_0

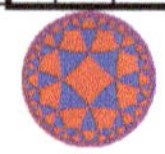

Fig. 1.8.3. The dyadic-tensor-shift decomposition of the even-columns matrix BEC (from Fig. 1.8.1) into 4 sparse matrices k_0, k_1, k_2, k_3, whose set is closed relative to multiplication and defines again a multiplication table of the algebra of split-quaternions of Cockle. The symbol of the Poincare conformal disk model of hyperbolic geometry is shown.

	111 (7)	110 (6)	101 (5)	100 (4)	011 (3)	010 (2)	001 (1)	000 (0)
111 (7)	1	1	1	1	−1	−1	−1	−1
110 (6)	1	1	1	1	−1	−1	−1	−1
101 (5)	−1	−1	1	1	−1	−1	1	1
100 (4)	−1	−1	1	1	−1	−1	1	1
011 (3)	−1	−1	−1	−1	1	1	1	1
010 (2)	−1	−1	−1	−1	1	1	1	1
001 (1)	−1	−1	1	1	−1	−1	1	1
000 (0)	−1	−1	1	1	−1	−1	1	1

=

	7	6	5	4	3	2	1	0
7	1	1	1	1	−1	−1	−1	−1
6								
5	−1	−1	1	1	−1	−1	1	1
4								
3	−1	−1	−1	−1	1	1	1	1
2								
1	−1	−1	1	1	−1	−1	1	1
0								

+

	7	6	5	4	3	2	1	0
7								
6	1	1	1	1	−1	−1	−1	−1
5								
4	−1	−1	1	1	−1	−1	1	1
3								
2	−1	−1	−1	−1	1	1	1	1
1								
0	−1	−1	1	1	−1	−1	1	1

Fig. 1.8.4. The twice-complementary matrix B (top) is the sum of the odd-rows matrix BOR (bottom left) and the even-rows matrix BER (bottom right), whose numberings of columns and rows are shown in decimal notations (bottom). Empty cells contain zeros. Black cells correspond to triplets with strong roots.

h_0 =

1	1	0	0	0	0	0	0
0	0	0	0	0	0	0	0
0	0	1	1	0	0	0	0
0	0	0	0	0	0	0	0
0	0	0	0	1	1	0	0
0	0	0	0	0	0	0	0
0	0	0	0	0	0	1	1
0	0	0	0	0	0	0	0

h_1 =

0	0	1	1	0	0	0	0
0	0	0	0	0	0	0	0
-1	-1	0	0	0	0	0	0
0	0	0	0	0	0	0	0
0	0	0	0	0	0	1	1
0	0	0	0	0	0	0	0
0	0	0	0	-1	-1	0	0
0	0	0	0	0	0	0	0

h_2 =

0	0	0	0	-1	-1	0	0
0	0	0	0	0	0	0	0
0	0	0	0	0	0	1	1
0	0	0	0	0	0	0	0
-1	-1	0	0	0	0	0	0
0	0	0	0	0	0	0	0
0	0	1	1	0	0	0	0
0	0	0	0	0	0	0	0

h_3 =

0	0	0	0	0	0	-1	-1
0	0	0	0	0	0	0	0
0	0	0	0	-1	-1	0	0
0	0	0	0	0	0	0	0
0	0	-1	-1	0	0	0	0
0	0	0	0	0	0	0	0
-1	-1	0	0	0	0	0	0
0	0	0	0	0	0	0	0

*	h_0	h_1	h_2	h_3
h_0	h_0	h_1	h_2	h_3
h_1	h_1	$-h_0$	$-h_3$	h_2
h_2	h_2	h_3	h_0	h_1
h_3	h_3	$-h_2$	$-h_1$	h_0

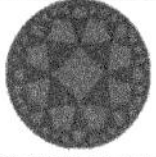

Fig. 1.8.5. The dyadic-tensor-shift decomposition of the odd-rows matrix BOR (from Fig. 1.8.4) into 4 sparse matrices h_0, h_1, h_2, h_3, whose set is closed relative to multiplication and defines a multiplication table of the algebra of split-quaternions of Cockle. The symbol of the Poincare conformal disk model of hyperbolic geometry, which is connected with this algebra, is shown.

y_0 =

0	0	0	0	0	0	0	0
1	1	0	0	0	0	0	0
0	0	0	0	0	0	0	0
0	0	1	1	0	0	0	0
0	0	0	0	0	0	0	0
0	0	0	0	1	1	0	0
0	0	0	0	0	0	0	0
0	0	0	0	0	0	1	1

y_1 =

0	0	0	0	0	0	0	0
0	0	1	1	0	0	0	0
0	0	0	0	0	0	0	0
-1	-1	0	0	0	0	0	0
0	0	0	0	0	0	0	0
0	0	0	0	0	0	1	1
0	0	0	0	0	0	0	0
0	0	0	0	-1	-1	0	0

y_2 =

0	0	0	0	0	0	0	0
0	0	0	0	-1	-1	0	0
0	0	0	0	0	0	0	0
0	0	0	0	0	0	1	1
0	0	0	0	0	0	0	0
-1	-1	0	0	0	0	0	0
0	0	0	0	0	0	0	0
0	0	1	1	0	0	0	0

y_3 =

0	0	0	0	0	0	0	0
0	0	0	0	0	0	-1	-1
0	0	0	0	0	0	0	0
0	0	0	0	-1	-1	0	0
0	0	0	0	0	0	0	0
0	0	-1	-1	0	0	0	0
0	0	0	0	0	0	0	0
-1	-1	0	0	0	0	0	0

*	y_0	y_1	y_2	y_3
y_0	y_0	y_1	y_2	y_3
y_1	y_1	$-y_0$	$-y_3$	y_2
y_2	y_2	y_3	y_0	y_1
y_3	y_3	$-y_2$	$-y_1$	y_0

Fig. 1.8.6. The dyadic-tensor-shift decomposition of the even-rows matrix BER (from Fig. 1.8.4) into 4 sparse matrices y_0, y_1, y_2, y_3, whose set is closed relative to multiplication and defines a multiplication table of the algebra of split-quaternions of Cockle. The symbol of the Poincare conformal disk model of hyperbolic geometry, which is connected with this algebra, is shown.

The even-rows matrix BER (Fig. 1.8.4) is the sum of 4 sparse matrices $y_0 + y_1 + y_2 + y_3$, whose set is closed relative to multiplication and defines a multiplication table of the same algebra of split-quaternions of Cockle as shown in Fig. 1.8.6 (compare with the same multiplication table in Figs. 1.7.2, 1.7.3, 1.8.2, 1.8.3, and 1.8.5).

1.9. Genetic matrices, root-complementarity, and hyperbolic numbers

The materials of §§ 1.7 and 1.8 do not at all exhaust the vast topic of the implementation of the principle "like begets like" in matrix genetics based on binary oppositions in the structure of the genetic system, including complementary replications in the regular structuring of genetic matrices. This topic should be intensively and systematically studied in future. Here let us show only some additional examples of complementary interrelations in genetic matrices (that is, matrices built on the basis of binary-oppositional properties of the molecular ensembles of the genetic system).

It turns out that introduction of an additional notion of "root-complementarity" is useful for understanding and modeling the non-trivial structurization of a molecular genetic system. By definition, two n-bit binary numbers ($n = 3, 4, 5, \ldots$) form a root-complementary pair if they transfer into each other under the interchange of numbers $0 \leftrightarrow 1$ only in their roots. For example, such interchanging transforms number 000 into number 110, that is, binary numbers 000 and 110 form a root-complementary pair. Correspondingly, in the genetic matrices, two columns (or rows) form a root-complementary pair if their binary numberings form a root-complementary pair. Below we show that this topic of root-complementarity pairs is connected with the algebra of two-dimensional hyperbolic (double) numbers.

Let us return to the even-columns matrix in Fig. 1.7.1, bottom left. It is the sum of two matrices shown in Fig. 1.9.1, bottom. The first matrix contains only two non-zero columns, which are numerated by root-complementary binary numbers 000 and 110 (that is, numbers 0 and 6 in decimal notation) and is denoted C_{06}. The second matrix contains only two non-zero columns, which are numerated by root-complementary binary numbers 010 and 100 (that is, numbers 2 and 4 in decimal notation) and is denoted C_{24}.

Fig. 1.9.2 shows that the first matrix C_{06} is decomposed into two matrices e_0 and e_1, whose set is closed relative to multiplication and defines the multiplication table of two-dimensional algebra of hyperbolic (or double) numbers z = x + yj, where x and y are real numbers, $j^2 = 1$, and $j \neq \pm 1$ [58]. The multiplication table of basis elements of this algebra is shown in Fig. 1.9.2, right.

Fig. 1.9.3 shows that the second matrix C_{24} (from Fig. 1.9.1) is decomposed into the sum of two matrices g_0 and g_1, whose set is also closed

	(0)	(1)	(2)	(3)	(4)	(5)	(6)	(7)
(0)	+1		−1		+1		−1	
(1)	+1		−1		+1		−1	
(2)	+1		+1		−1		−1	
(3)	+1		+1		−1		−1	
(4)	+1		−1		+1		−1	
(5)	+1		−1		+1		−1	
(6)	−1		−1		+1		+1	
(7)	−1		−1		+1		+1	

=

(0)	(1)	(2)	(3)	(4)	(5)	(6)	(7)
+1						−1	
+1						−1	
+1						−1	
+1						−1	
+1						−1	
+1						−1	
−1						+1	
−1						+1	

+

(0)	(1)	(2)	(3)	(4)	(5)	(6)	(7)
		−1		+1			
		−1		+1			
		+1		−1			
		+1		−1			
		−1		+1			
		−1		+1			
		−1		+1			
		−1		+1			

Fig. 1.9.1. The decomposition of the even-columns matrix from Fig. 1.7.1 (bottom left) into the sum of matrices C_{06} (at left) and C_{24} (at right). Empty cells contain zeros.

e_0 =

1	0	0	0	0	0	0	0
1	0	0	0	0	0	0	0
1	0	0	0	0	0	0	0
1	0	0	0	0	0	0	0
0	0	0	0	0	0	−1	0
0	0	0	0	0	0	−1	0
0	0	0	0	0	0	1	0
0	0	0	0	0	0	1	0

e_1 =

0	0	0	0	0	0	−1	0
0	0	0	0	0	0	−1	0
0	0	0	0	0	0	−1	0
0	0	0	0	0	0	−1	0
1	0	0	0	0	0	0	0
1	0	0	0	0	0	0	0
−1	0	0	0	0	0	0	0
−1	0	0	0	0	0	0	0

×	e_0	e_1
e_0	e_0	e_1
e_1	e_1	e_0

Fig. 1.9.2. The decomposition of the matrix C_{06} (from Fig. 1.9.1) into the sum of two matrices $e_0 + e_1$, whose set is closed relative to multiplication. Their multiplication table is shown on the right.

relative to multiplication and defines the same multiplication table of two-dimensional algebra of hyperbolic numbers.

Now let us turn to the odd-columns matrix in Fig. 1.7.1, bottom right. It is the sum of two matrices shown in Fig. 1.9.4, bottom. The first matrix contains only two non-zero columns, which are numerated by root-

$g_0 =$

0	0	−1	0	0	0	0	0
0	0	−1	0	0	0	0	0
0	0	1	0	0	0	0	0
0	0	1	0	0	0	0	0
0	0	0	0	1	0	0	0
0	0	0	0	1	0	0	0
0	0	0	0	1	0	0	0
0	0	0	0	1	0	0	0

$g_1 =$

0	0	0	0	1	0	0	0
0	0	0	0	1	0	0	0
0	0	0	0	−1	0	0	0
0	0	0	0	−1	0	0	0
0	0	−1	0	0	0	0	0
0	0	−1	0	0	0	0	0
0	0	−1	0	0	0	0	0
0	0	−1	0	0	0	0	0

×	g_0	g_1
g_0	g_0	g_1
g_1	g_1	g_0

Fig. 1.9.3. The decomposition of the matrix C_{24} (from Fig. 1.9.1) into the sum of two matrices $g_0 + g_1$, whose set is closed relative to multiplication. Their multiplication table, shown on the right, corresponds to the algebra of hyperbolic numbers.

	(0)	(1)	(2)	(3)	(4)	(5)	(6)	(7)
(0)		+1		−1		+1		−1
(1)		+1		−1		+1		−1
(2)		+1		+1		−1		−1
(3)		+1		+1		−1		−1
(4)		+1		−1		+1		−1
(5)		+1		−1		+1		−1
(6)		−1		−1		+1		+1
(7)		−1		−1		+1		+1

=

(0)	(1)	(2)	(3)	(4)	(5)	(6)	(7)
	+1						−1
	+1						−1
	+1						−1
	+1						−1
	+1						−1
	+1						−1
	−1						+1
	−1						+1

+

(0)	(1)	(2)	(3)	(4)	(5)	(6)	(7)
			−1		+1		
			−1		+1		
			+1		−1		
			+1		−1		
			−1		+1		
			−1		+1		
			−1		+1		
			−1		+1		

Fig. 1.9.4. The decomposition of the odd-columns matrix from Fig. 1.7.1 (bottom right) into the sum of matrices K_{17} (left) and K_{35} (right). Empty cells contain zeros.

complementary binary numbers 001 and 111 (that is, numbers 1 and 7 in decimal notation) and is denoted K_{17}. The second matrix contains only two non-zero columns, which are numerated by root-complementary binary numbers 011 and 101 (that is, numbers 3 and 5 in decimal notation) and is denoted K_{35}.

Fig. 1.9.5 shows that the first matrix K_{17} is decomposed into the sum of two matrices q_0 and q_1, whose set is closed relative to multiplication and defines the multiplication table of two-dimensional algebra of hyperbolic numbers.

Fig. 1.9.6 shows that the second matrix K_{35} (from Fig. 1.9.4) is decomposed into the sum of two matrices h_0 and h_1, whose set is closed

q_0 =

0	1	0	0	0	0	0	0
0	1	0	0	0	0	0	0
0	1	0	0	0	0	0	0
0	1	0	0	0	0	0	0
0	0	0	0	0	0	−1	0
0	0	0	0	0	0	−1	0
0	0	0	0	0	0	1	0
0	0	0	0	0	0	1	0

q_1 =

0	0	0	0	0	0	−1	0
0	0	0	0	0	0	−1	0
0	0	0	0	0	0	−1	0
0	0	0	0	0	0	−1	0
0	1	0	0	0	0	0	0
0	1	0	0	0	0	0	0
0	−1	0	0	0	0	0	0
0	−1	0	0	0	0	0	0

×	q_0	q_1
q_0	q_0	q_1
q_1	q_1	q_0

Fig. 1.9.5. The decomposition of the matrix K_{17} (from Fig. 1.9.4) into the sum of two matrices $q_0 + q_1$, whose set is closed relative to multiplication. Their multiplication table, shown on the right, corresponds to algebra of hyperbolic numbers.

h_0 =

0	0	0	−1	0	0	0	0
0	0	0	−1	0	0	0	0
0	0	0	1	0	0	0	0
0	0	0	1	0	0	0	0
0	0	0	0	0	1	0	0
0	0	0	0	0	1	0	0
0	0	0	0	0	1	0	0
0	0	0	0	0	1	0	0

h_1 =

0	0	0	0	0	1	0	0
0	0	0	0	0	1	0	0
0	0	0	0	0	−1	0	0
0	0	0	0	0	−1	0	0
0	0	0	−1	0	0	0	0
0	0	0	−1	0	0	0	0
0	0	0	−1	0	0	0	0
0	0	0	−1	0	0	0	0

×	h_0	h_1
h_0	h_0	h_1
h_1	h_1	h_0

Fig. 1.9.6. The decomposition of the matrix K_{35} (from Fig. 1.9.4) into the sum of two matrices $h_0 + h_1$, whose set is closed relative to multiplication. Their multiplication table, shown on the right, corresponds to algebra of hyperbolic numbers.

relative to multiplication and defines the same multiplication table of two-dimensional algebra of hyperbolic numbers.

1.10. Some concluding remarks

This chapter has presented data about an existing complex of algebraic binary-complementary replications in the multicomponent binary organization of the genetic coding system, which is realized along with molecular complementary replication of DNA strands. This complex can be used to model the molecular replication on the language of algebraic matrices and dyadic groups of binary numbers.

This chapter has also shown connections of the ancient principle "like begets like" with holistic families of structural molecular ensembles of the genetic coding system in their algebraic-matrix presentations (in addition to the known connection between this principle and double-stranded DNAs). We believe that this principle plays a key role in genetics, and therefore, within the framework of algebraic biology, many features of inherited biological structures can and should be studied precisely in connection with this principle. These include, for example, the following: morphogenetic symmetries; biological fractal-like patterns; conformal-geometric features of the space of visual perception according to the mentioned works [30, 31]; and the golden ratio, which since the Renaissance is regarded as a mathematical symbol of self-reproduction. Further study of the algebraic relationships between the genetic system and inherited physiological complexes will increasingly reveal the key biological role of the ancient principle "like begets like" at different levels of biological organization. Accordingly, in the following chapters of this book, we will return to this fundamental principle and its modified versions, for example:

— in Chapter 3, the resonant integration of parts into a whole based on energy minimization can be briefly called the principle "like attracts like". In addition, the well-known Huygens phenomenon of synchronization of many oscillating pendulums having the same natural frequencies and standing on a common vibration platform can be briefly called "like synchronizes like";

— in Chapter 6, on the doctrine of evolution based on antenna arrays, one can note that the operation of highly efficient phased antenna arrays is based on phase-shifted waves of the same frequency, which can be briefly characterized by the words "like uses like" but with a shift in time.

Chapter 2 Genetic Code Systems, Algebraic Holography, and Noise Immunity

This chapter continues the theme of a close connection between the genetic system and binary opposition principles, dyadic groups of binary numbers, and some algebras of hypercomplex numeric systems. Special attention is devoted to the well-known method of the algebraic bit-reversible holography, which is used in processing of digital information and based on dyadic groups of binary numbers. Its application to analysis of structures of DNA alphabets reveals unexpected relations of the genetic system with the algebra-holographic concept. The problem of genetic intelligence and inherited intellectual abilities of organisms is considered.

2.1. Algebraic bit-reversible holography and structured alphabets of DNA

The theme of holographic principles is very popular in physiology and cognitive research, as well as in modern physics with its concepts of the holographic universe. Physical holography, discovered by D. Gabor, provided science and technology with surprisingly effective means of storing and processing information based on its following properties:

- Holographic memory is distributed over the entire volume (area) of the carrier;
- Reducing the volume of the medium only worsens the detail of the image, but does not destroy its parts;

- The possibility of many high-quality images on one medium by changing the angle of the laser beam during their recording and recreation (in contrast to ordinary photography, in which shooting different images at the same frame of film spoils each of them);
- All information is recorded on a flat two-dimensional plate; however, a three-dimensional volumetric image is recreated on its basis.

These amazing properties of physical holography have given rise to many new approaches in biology and physics, many of which are associated with them. For example, in biology, the concept of the connection of brain functions with the properties of holography is widely known, presented in K. Pribram's book *Languages of the Brain* [59]. This book emphasizes that holographic description is unmatched for explaining problems of perception, especially problems of image formation and recognition. There is a lot of physiological data to support this concept. For example, Lashley [60] defined the following: no matter which part of the rat's brain we remove, we cannot destroy the memories of how to perform the complex actions that the rat learned before surgery. In embryology, even 100 years ago, G. Driesch showed that by separating the blastomeres of sea urchin eggs from each other, it is possible to grow from one embryonic cell (blastomere) normal (albeit reduced) larvae with all their organs [61–63]. Such "holographic" phenomena of distributed information have been repeatedly confirmed in many taxonomic groups of multicellular organisms — from sponges to mammals. Many physiological phenomena, the properties of which resemble holographic ones, are described, for example, in the book *The Holographic Universe* [64].

But the brain and the nervous system have appeared at a relatively late stage of biological evolution. A great number of species lived perfectly in the past and are now living without neuronal networks. It is also evident that the mentioned holographic-like phenomena of embryology cannot be reduced to the work of the nervous system. But all morphogenetic, brain, and other physiological structures are genetically inherited, and therefore it is natural to look for the origins of the similarity between holography and non-local informatics of living organisms at the level of the genetic system.

It should be emphasized that algebraic holography is based entirely on numerical algebraic methods and does not require special physical devices such as photosensitive plates and laser emitters, in contrast to physical holography. One of the methods of algebraic holography is known in the technology of digital informatics under the term "bit-reversible holography". This kind of algebraic holography is connected, in particular, with noise immunity coding and algorithms of fast Fourier transform [65–68].

Let us describe a consistency of DNA informatics with the bit-reversible holography method [69–71]. Bit reversion means reading binary numbers in the opposite direction: for example, 001 becomes 100 (which correspond to decimal numbers 1 and 5, respectively).

Fig. 2.1.1 shows a 512 × 512-pixel matrix, in which all columns and rows are sequentially numbered with binary numbers and in which the letter A is drawn. Reading each of these binary numbers in reverse order leads to renumbering of columns and rows and to their new placement with the transformation of the entire image (the second frame in Fig. 2.1.1). If now part of this frame is painted over or deleted, then reapplying the reverse reading of the binary numbers with the permutation will restore the original image.

2.1.1. *Genetic bit-reversed matrices and their algebraic properties*

Now let us return to genetic mosaic matrices from the tensor family [C, A; T, G]$^{(n)}$ (Fig. 1.4.2) to study transformations of the matrix mosaics under the bit-reversion in binary numberings of columns and rows of these

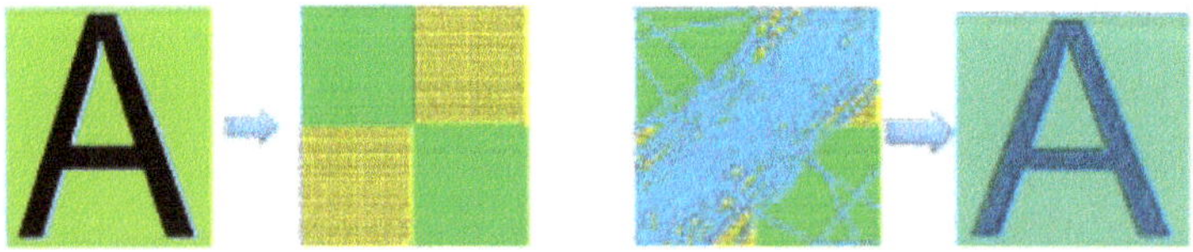

Fig. 2.1.1. An example of the bit-reversible holography (from https://habrahabr.ru/post/155471/).

	00	10	01	11
00	CC	AC	CA	AA
10	TC	GC	TA	GA
01	CT	AT	CG	AG
11	TT	GT	TG	GG

	000	100	010	110	001	101	011	111
000	CCC	ACC	CAC	AAC	CCA	ACA	CAA	AAA
100	GCC	TCC	GAC	TAC	GCA	TCA	GAA	TAA
010	CGC	AGC	CTC	ATC	CGA	AGA	CTA	ATA
110	GGC	TGC	GTC	TTC	GGA	TGA	GTA	TTA
001	CCG	ACG	CAG	AAG	CCT	ACT	CAT	AAT
101	GCG	TCG	GAG	TAG	GCT	TCT	GAT	TAT
011	CGG	AGG	CTG	ATG	CGT	AGT	CTT	ATT
111	GGG	TGG	GTG	TTG	GGT	TGT	GTT	TTT

	0000	1000	0100	1100	0010	1010	0110	1110	0001	1001	0101	1101	0011	1011	0111	1111
0000	CCCC	ACCC	CACC	AACC	CCAC	ACAC	CAAC	AAAC	CCCA	ACCA	CACA	AACA	CCAA	ACAA	CAAA	AAAA
1000	TCCC	GCCC	TACC	GACC	TCAC	GCAC	TAAC	GAAC	TCCA	GCCA	TACA	GACA	TCAA	GCAA	TAAA	GAAA
0100	CTCC	ATCC	CGCC	AGCC	CTAC	ATAC	CGAC	AGAC	CTCA	ATCA	CGCA	AGCA	CTAA	ATAA	CGAA	AGAA
1100	TTCC	GTCC	TGCC	GGCC	TTAC	GTAC	TGAC	GGAC	TTCA	GTCA	TGCA	GGCA	TTAA	GTAA	TGAA	GGAA
0010	CCTC	ACTC	CATC	AATC	CCGC	ACGC	CAGC	AAGC	CCTA	ACTA	CATA	AATA	CCGA	ACGA	CAGA	AAGA
1010	TCTC	GCTC	TATC	GATC	TCGC	GCGC	TAGC	GAGC	TCTA	GCTA	TATA	GATA	TCGA	GCGA	TAGA	GAGA
0110	CTTC	ATTC	CGTC	AGTC	CTGC	ATGC	CGGC	AGGC	CTTA	ATTA	CGTA	AGTA	CTGA	ATGA	CGGA	AGGA
1110	TTTC	GTTC	TGTC	GGTC	TTGC	GTGC	TGGC	GGGC	TTTA	GTTA	TGTA	GGTA	TTGA	GTGA	TGGA	GGGA
0001	CCCT	ACCTT	CACT	AACT	CCAT	ACAT	CAAT	AAAT	CCCG	ACCG	CACG	AACG	CCAG	ACAG	CAAG	AAAG
1001	TCCT	GCCT	TACT	GACT	TCAT	GCAT	TAAT	GAAT	TCCG	GCCG	TACG	GACG	TCAG	GCAG	TAAG	GAAG
0101	CTCT	ATCT	CGCT	AGCT	CTAT	ATAT	CGAT	AGAT	CTCG	ATCG	CGCG	AGCG	CTAG	ATAG	CGAG	AGAG
1101	TTCT	GTCT	TGCT	GGCT	TTAT	GTAT	TGAT	GGAT	TTCG	GTCG	TGCG	GGCG	TTAG	GTAG	TGAG	GGAG
0011	CCTT	ACTT	CATT	AATT	CCGT	ACGT	CAGT	AAGT	CCTG	ACTG	CATG	AATG	CCGG	ACGG	CAGG	AAGG
1011	TCTT	GCTT	TATT	GATT	TCGT	GCGT	TAGT	GAGT	TCTG	GCTG	TATG	GATG	TCGG	GCGG	TAGG	GAGG
0111	CTTT	ATTT	CGTT	AGTT	CTGT	ATGT	CGGT	AGGT	CTTG	ATTG	CGTG	AGTG	CTGG	ATGG	CGGG	AGGG
1111	TTTT	GTTT	TGTT	GGTT	TTGT	GTGT	TGGT	GGGT	TTTG	GTTG	TGTG	GGTG	TTGG	GTGG	TGGG	GGGG

Fig. 2.1.2. The result of the bit-reversible rearrangement in binary numberings of columns and rows of mosaic matrices of DNA alphabets from Fig. 1.4.2.

matrices. This bit-reverse rearrangement of these binary numberings in the matrices in Fig. 1.4.2 generates a family of multi-block matrices whose mosaics consist of repetitions of the mosaic (4 × 4)-matrices of 16 duplets (Fig. 2.1.2). These repetitions can be considered as one more connection with the ancient principle "like begets like", which is linked to the genetic matrices as presented in Chapter 1.

Black and white cells of the matrices in Fig. 2.1.2 reflect the opposition of two sub-alphabets of n-plets with strong and weak roots and therefore can be represented by elements +1 and −1 in them. The mosaic of each row in these matrices consists of equal-length sections with entries +1 and −1, which corresponds to one or another Rademacher function. One can see that these multi-block mosaics of all emerging numerical matrices of DNA alphabets have the form of "matrix crystals" since they consist of repetitions of the mosaic for the (4 × 4)-matrix of 16 duplets, which we conditionally term α-matrix (Fig. 2.1.3). Correspondingly, these multi-block matrices we term α-matrix crystals (or matrix α-crystals); below their binary-complementary analogue in the form of β-matrix crystals are also considered. Does this "universal" repeating block — α-matrix — have a meaningful algebraic meaning? Yes, it has.

Fig. 2.1.4 shows that this block is the sum of 4 sparse matrices, the set of which is closed under multiplication and determines the corresponding

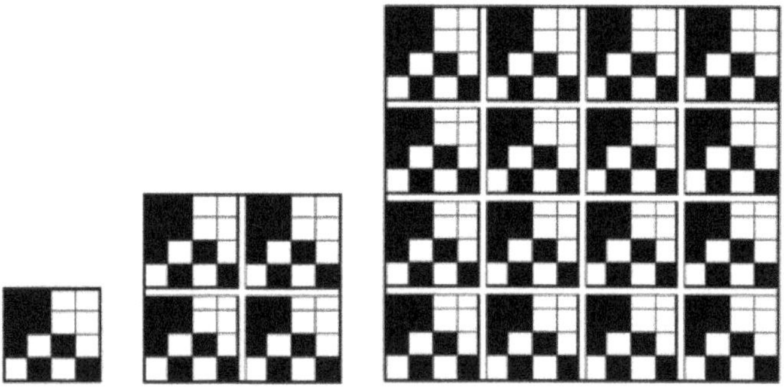

Fig. 2.1.3. Numerical representation of mosaic matrices of DNA alphabets from Fig. 2.1.2 is shown. The black and white cells of the matrices represent elements +1 and −1, respectively.

$$\begin{bmatrix} 1 & 1 & -1 & -1 \\ 1 & 1 & -1 & -1 \\ 1 & -1 & 1 & -1 \\ -1 & 1 & -1 & 1 \end{bmatrix} = \begin{vmatrix} 1,0,0,0 \\ 0,1,0,0 \\ 0,0,1,0 \\ 0,0,0,1 \end{vmatrix} + \begin{vmatrix} 0,0,-1,0 \\ 0,0,0,-1 \\ 1,0,0,0 \\ 0,1,0,0 \end{vmatrix} + \begin{vmatrix} 0,0,0,-1 \\ 0,0,-1,0 \\ 0,-1,0,0 \\ -1,0,0,0 \end{vmatrix} + \begin{vmatrix} 0,1,0,0 \\ 1,0,0,0 \\ 0,0,0,-1 \\ 0,0,-1,0 \end{vmatrix} = j_0+j_1+j_2+j_3$$

*	j_0	j_1	j_2	j_3
j_0	j_0	j_1	j_2	j_3
j_1	j_1	$-j_0$	j_3	$-j_2$
j_2	j_2	$-j_3$	j_0	$-j_1$
j_3	j_3	j_2	j_1	j_0

Fig. 2.1.4. The dyadic-shift decomposition of the α-matrix of the DNA alphabet of 16 duplets (from Fig. 2.1.3) into 4 sparse matrices j_0, j_1, j_2, j_3, whose set is closed concerning multiplication. The multiplication table for this set is shown, which matches the multiplication table of the Cockle split-quaternion algebra used in the Poincaré conformal disk model of hyperbolic geometry. The symbol for this model is shown from https://commons.wikimedia.org/wiki/Category:Poincar%C3%A9_disk_models.

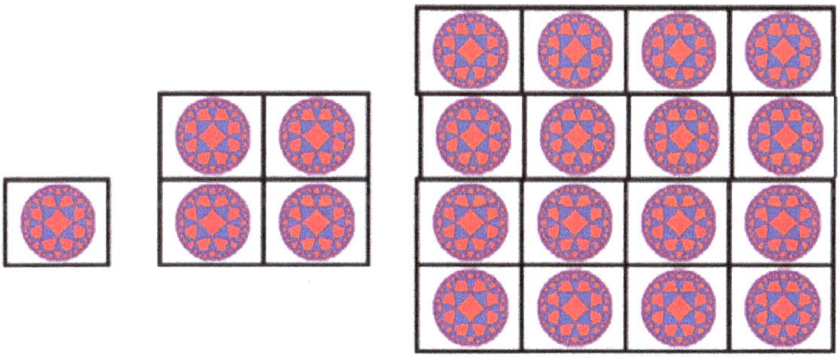

Fig. 2.1.5. "Crystals" of genetic matrices from Fig. 2.1.3 are presented in an artistic image.

multiplication table; this table is known as the multiplication table of the algebra of four-dimensional split-quaternions of Cockle, which was already met in Chapter 1 (Fig. 1.5.2). Split-quaternions are used in the Poincaré conformal disk model to describe hyperbolic motions in Lobachevsky's hyperbolic geometry [29; https://en.wikipedia.org/wiki/Split-quaternion].

Using the symbol of the Poincaré conformal disk model of hyperbolic geometry from Fig. 2.1.4, it is possible to present the obtained block mosaic matrices of structured DNA alphabets in an artistic form to facilitate heuristic associations (Fig. 2.1.5).

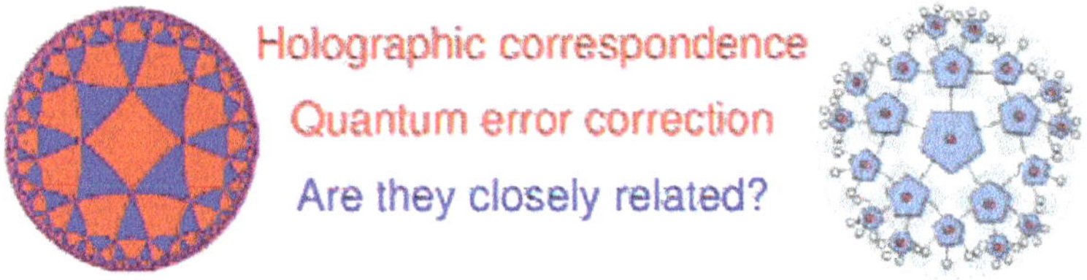

Fig. 2.1.6. The illustration of the theme of holographic quantum error-correcting codes from [72]. The illustration is reproduced with the kind permission of Prof. J. Preskill.

The family of block-unified matrices of DNA alphabets (discovered by the method of bit-reversible holography), whose blocks are conjugated with the conformal disk model of hyperbolic geometry, unexpectedly echoes the theme of "*holographic quantum error-correcting codes*", which is developed at the California Institute of Technology in the USA in connection with the same Poincaré conformal disk model and which is presented in [72] (Fig. 2.1.6).

The theme also includes the consideration of space-time as a quantum error-correcting code [73]. The following quotes from these works, which describe quantum error-correcting codes with a tensor network structure, clarify these parallels in more detail. "*The tensor network is supported on a uniform tiling of hyperbolic space, known as a hyperbolic tessellation. … We shall focus on examples based on tilings of two-dimensional hyperbolic space, which are specific realizations of uniform hyperbolic tilings known as hyperbolic tessellations*" [73]. "*Holographic quantum error-correcting codes are constructed by contracting perfect tensors according to a tilling of hyperbolic space by polygons*" [72].

The bit-reversed analog of the initial matrix of 64 triplets from Fig. 1.3.1 also contains an impressive symmetrical arrangement of all the amino acids and stop codons, which are encoded by the triplets, as Fig. 2.1.7 shows: the bit-reversed matrix has identical upper and lower halves in composition and arrangement of amino acids and stop codons. So, genetic informatics is closely related to the principles of algebraic holography (see additional details in the preprint [70]). It correlates with known physiological works about "holographic brain" (see, for example, [74]).

	000	100	010	110	001	101	011	111
000	CCC Pro	ACC Thr	CAC His	AAC Asn	CCA Pro	ACA Thr	CAA Gln	AAA Lys
100	TCC Ser	GCC Ala	TAC Tyr	GAC Asp	TCA Ser	GCA Ala	TAA Stop	GAA Glu
010	CTC Leu	ATC Ile	CGC Arg	AGC Ser	CTA Leu	ATA Met	CGA Arg	AGA Stop
110	TTC Phe	GTC Val	TGC Cys	GGC Gly	TTA Leu	GTA Val	TGA Trp	GGA Gly
001	CCT Pro	ACT Thr	CAT His	AAT Asn	CCG Pro	ACG Thr	CAG Gln	AAG Lys
101	TCT Ser	GCT Ala	TAT Tyr	GAT Asp	TCG Ser	GCG Ala	TAG Stop	GAG Glu
011	CTT Leu	ATT Ile	CGT Arg	AGT Ser	CTG Leu	ATG Met	CGG Arg	AGG Stop
111	TTT Phe	GTT Val	TGT Cys	GGT Gly	TTG Leu	GTG Val	TGG Trp	GGG Gly

Fig. 2.1.7. The symmetrical arrangement of 20 amino acids and stop codons in the bit-reversed matrix of 64 triplets. A case of the Vertebrate Mitochondrial Genetic Code is shown for comparison with Fig. 1.3.1. Identical colors mark the pairs of rows with identical compositions of amino acids and stop codons.

Holographic principles have long been used for the physical understanding of the world. The famous physicist David Bohm believed that human consciousness is part of the universal hologram of the entire human race and that the entire Universe has a holographic structure. Bohm worked with K. Pribram [59, 74] on the theory that the brain works like a hologram following quantum mathematical principles and characteristics of wave patterns. Nobel Laureate in Physics Gerard t'Hooft put forward the principle of holography in the structure of the world, which is actively developed in modern physics by many authors (see, for example, [75]). Revealing the described connections of the genetic coding system with the Poincaré conformal disk model of the Lobachevski hyperbolic geometry indicates the genetic basis of the known facts of the relationship of physiological phenomena with hyperbolic geometry (see Chapter 1).

It should be emphasized that the bit-reversed transformations in dyadic groups of binary numbers are important not only for algebraic holography but also for fast Fourier transforms, which are widely applied in engineering, music, science, and mathematics. A fast Fourier transform is an algorithm that computes the discrete Fourier transform of a sequence, or its inverse. In 1994, G. Strang described the fast Fourier transform as "*the most important numerical algorithm of our lifetime*" [76]. The book [26, Chapter 4] gives an example of the important role of the method of data permutation with binary inversion for one of the variants of the algorithm of the fast Fourier transformation. In this example, a sequence of members of the dyadic group 000, 001, 010, 011, 100, 101, 110, 111 is reformed into the bit-reversed sequence of binary numbers 000, 100, 010, 110, 001, 101, 011, 111, which was used above in the described analysis of binary structures of the interconnected genetic alphabets by the method of the bit-reversible holography (Figs. 2.1.2, 2.1.7). These and other facts allow thinking that the genetic system has a connection with a fast Fourier transformation (or with a fast Hadamard transformation).

One can add that each of the numeric matrices in Fig. 2.1.3 can be constructed from the α-matrix by the algorithm (2.1) based on its tensor product with the matrix [1, 1; 1, 1]:

$$[1, 1; 1, 1]^{(n)} \otimes [1, 1, -1, -1; 1, 1, -1, -1; 1, -1, 1, -1; -1, 1, -1, 1] \quad (2.1)$$

The matrix [1, 1; 1, 1] used in (2.1) is the matrix representation of a two-dimensional hyperbolic number with unit coordinates. Deep structural connections of genetic informatics with 2^n-dimensional hyperbolic numbers are described below in this book and in [77–82].

2.1.2. *The bit-reversible holography and the "like begets like" principle in matrix genetics*

Let us return to the α-matrix of 16 duplets (from Fig. 2.1.2), whose dyadic-shift decomposition revealed its connection with the multiplication table of the Cockle split-quaternion algebra (Fig. 2.1.4). Now we create a new matrix

	11	01	10	00
11	1	-1	1	-1
01	-1	1	-1	1
10	-1	-1	1	1
11	-1	-1	1	1

	11	01	10	00
11	GG	TG	GT	TT
01	AG	CG	AT	CT
10	GA	TA	GC	TC
00	AA	CA	AC	CC

Fig. 2.1.8. The β-matrix (left), which is a binary-complementary analogue of the α-matrix (from Figs. 2.1.2–2.1.4). The symbolic matrix of 16 duplets (right), which is a prototype of the numeric β-matrix.

Fig. 2.1.9. Examples of β-matrix crystals; black and white cells contain numbers +1 and −1 correspondingly (see explanation in the text).

called β-matrix (Fig. 2.1.8), which is a binary-complementary analogue of the α-matrix. The β-matrix having a new mosaic arises under the inversion $0 \leftrightarrow 1$ in all binary numberings of columns and rows of the α-matrix with appropriate rearrangements of all cells with duplets.

Under the mentioned inversion $0 \leftrightarrow 1$, all α-matrix crystals (Fig. 2.1.4) are transformed into β-matrix crystals (or matrix β-crystals), that is, multi-block matrices having the β-matrix as their repeated block (Fig. 2.1.9).

The β-matrix is the sum of 4 sparse matrices (Fig. 2.1.10), the set of which is closed under multiplication and determines the same multiplication table of the algebra of four-dimensional split-quaternions of Cockle as the α-matrix under its similar dyadic-shift decomposition (Fig. 2.1.4).

$$\begin{bmatrix} 1 & -1 & 1 & -1 \\ -1 & 1 & -1 & 1 \\ -1 & -1 & 1 & 1 \\ -1 & -1 & 1 & 1 \end{bmatrix} = \begin{vmatrix} 1,0,0,0 \\ 0,1,0,0 \\ 0,0,1,0 \\ 0,0,0,1 \end{vmatrix} + \begin{vmatrix} 0,\ 0,1,0 \\ 0,\ 0,0,1 \\ -1,\ 0,0,0 \\ 0,-1,0,0 \end{vmatrix} + \begin{vmatrix} 0,0,0,-1 \\ 0,0,-1,0 \\ 0,-1,0,0 \\ -1,0,0,0 \end{vmatrix} + \begin{vmatrix} 0,-1,0,0 \\ -1,\ 0,0,0 \\ 0,\ 0,0,1 \\ 0,\ 0,1,0 \end{vmatrix} = e_0+e_1+e_2+e_3$$

*	e_0	e_1	e_2	e_3
e_0	e_0	e_1	e_2	e_3
e_1	e_1	$-e_0$	e_3	$-e_2$
e_2	e_2	$-e_3$	e_0	$-e_1$
e_3	e_3	e_2	e_1	e_0

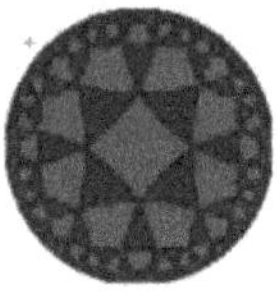

Fig. 2.1.10. The dyadic-shift decomposition of the β-matrix (from Fig. 2.1.8) into 4 sparse matrices e_0, e_1, e_2, e_3, whose set is closed concerning multiplication. The multiplication table for this set is shown, which is identical to the multiplication table in the case of the α-matrix (Fig. 2.1.4) and matches the multiplication table of the Cockle split-quaternion algebra.

Let's now compare actions of the α-matrix **A** and the β-matrix **B** on an arbitrary four-dimensional vector, for example, the vector $[2, 3, 4, 5]^T$:

$$\mathbf{A} \times [2, 3, 4, 5]^T = [-4, -4, -2, 2]^T, \mathbf{B} \times [2, 3, 4, 5]^T = [-2, 2, 4, 4]^T \quad (2.2)$$

The expression (2.2) illustrates a general situation: actions of the α-matrix and the β-matrix, which are binary-complementary to each other, on an arbitrary four-dimensional vector always generate vectors whose sets of coordinates are mirror-complementary and opposite in signs. For example, if the vector $[-2, 2, 4, 4]^T$ in (2.2) is read backwards and its signs are reversed, then it coincides with the set of coordinates of the first vector $[-4, -4, -2, 2]^T$. One should note that this interrelation of two vectors is structurally analogical to the interrelation of complementary 4-plets on two strands of double helical DNA: if one DNA strand contains a sequence of four nucleotides, for example, ACGT, then the second strand contains the complementary sequence TGCA; in addition, the nucleotide sequences on two DNA strands are read in the opposite direction, that is, the TGCA sequence is read as ACGT. It was shown above in § 1.2 that complementary nucleobases can be considered as opposite by its signs (A = –T and C = –G) from the point of view of the well-known mathematical method of complements.

The binary inversion $0 \leftrightarrow 1$ can be applied to α- and β-matrix crystals many times providing their mutual transformation into each other in the algebraic connection with the principle "like begets like". The α-matrix **A** and the β-matrix **B** have the following interrelated properties, which show, in particular, that matrices A/2 and B/2 are oblique projectors:

$$\mathbf{A} \times \mathbf{B} = \mathbf{A}^2,\ \mathbf{B} \times \mathbf{A} = \mathbf{B}^2,\ \mathbf{A}^2 = 2\mathbf{A},\ \mathbf{B}^2 = 2\mathbf{B},\ (\mathbf{A}/2)^2 = \mathbf{A}/2,\ (\mathbf{B}/2)^2 = \mathbf{B}/2 \quad (2.3)$$

2.1.3. *Cross-like decompositions of the matrix α- and β-crystals*

Let us continue to study algebraic properties of the α- and β-matrix crystals (Figs. 2.1.3 and 2.1.9, respectively). This section demonstrates that binary-complementary columns in each of such crystals have additional algebraic interrelations with the algebra of hyperbolic numbers. Fig. 2.1.11 shows a special kind of decomposition of the α-matrix into the sum of 4 sparse matrices c_0, c_1, c_2, c_3.

This decomposition is based on binary-complementary columns, which are enumerated by complementary binary numbers, that is, by numbers 00–11 and 01–10. In each of the four sparse matrices c_0, c_1, c_2, c_3, non-zero columns have complementary binary numbering (in the decimal system, these are numbers 0 and 3, as well as numbers 1 and 2). As Fig. 2.1.11

$$\begin{bmatrix} 1 & 1 & -1 & -1 \\ 1 & 1 & -1 & -1 \\ 1 & -1 & 1 & -1 \\ -1 & 1 & -1 & 1 \end{bmatrix} = \begin{vmatrix} 1,0,0,0 \\ 1,0,0,0 \\ 0,0,0,-1 \\ 0,0,0,1 \end{vmatrix} + \begin{vmatrix} 0,1,0,0 \\ 0,1,0,0 \\ 0,0,1,0 \\ 0,0,-1,0 \end{vmatrix} + \begin{vmatrix} 0,0,-1,0 \\ 0,0,-1,0 \\ 0,-1,0,0 \\ 0,1,0,0 \end{vmatrix} + \begin{vmatrix} 0,0,0,-1 \\ 0,0,0,-1 \\ 1,0,0,0 \\ -1,0,0,0 \end{vmatrix} = c_0+c_1+c_2+c_3$$

×	c_0	c_3
c_0	c_0	c_3
c_3	c_3	c_0

×	c_1	c_2
c_1	c_1	c_2
c_2	c_2	c_1

Fig. 2.1.11. The cross-like decomposition of the α-matrix. Multiplication tables for the sets of the matrices c_0 and c_3, and also of the matrices c_1 and c_2, are shown at the bottom.

shows, these non-zero columns are only half non-zero: one half of each column consists of half of the corresponding column of the α-matrix (for example, the column with binary number 00), and the other half consists of half of the complementary column of the α-matrix (in this example, the column with binary number 11). This decomposition is called cross-like, since the resulting sparse matrices use binary-complementary columns of the α-matrix in a cross-like manner and have a cross-like appearance [48].

In this decomposition, the set of the matrices c_0 and c_3 is unexpectedly closed concerning multiplication and defines the multiplication table of two-dimensional algebra of hyperbolic numbers (Fig. 2.1.11). The same holds for the set of the matrices c_1 and c_2. One can note that in the matrices c_0 and c_3 their non-zero columns are based on complementary columns of the α-matrix with numberings 00 and 11. In the matrices c_1 and c_2, their non-zero columns are based on complementary columns of the α-matrix with numberings 01 and 10.

Fig. 2.1.12 shows a similar cross-like decomposition of the β-matrix into the sum of 4 sparse matrices y_0, y_1, y_2, y_3.

By analogy with the cross-like decomposition of the α-matrix (Fig. 2.1.11), the set of the matrices y_0 and y_3 is closed concerning multiplication and defines the same multiplication table of two-dimensional algebra of hyperbolic numbers (Fig. 2.1.12). The same holds for the set of the matrices y_1 and y_2.

$$\begin{bmatrix} 1 & -1 & 1 & -1 \\ -1 & 1 & -1 & 1 \\ -1 & -1 & 1 & 1 \\ -1 & -1 & 1 & 1 \end{bmatrix} = \begin{vmatrix} 1,0,0,0 \\ -1,0,0,0 \\ 0,0,0,1 \\ 0,0,0,1 \end{vmatrix} + \begin{vmatrix} 0,-1,0,0 \\ 0,\ 1,0,0 \\ 0,\ 0,1,0 \\ 0,\ 0,1,0 \end{vmatrix} + \begin{vmatrix} 0,\ 0,\ 1,0 \\ 0,\ 0,-1,0 \\ 0,-1,\ 0,0 \\ 0,-1,\ 0,0 \end{vmatrix} + \begin{vmatrix} 0,0,0,-1 \\ 0,0,0,\ 1 \\ -1,0,0,\ 0 \\ -1,0,0,\ 0 \end{vmatrix} = y_0+y_1+y_2+y_3$$

*	y_0	y_3
y_0	y_0	y_3
y_3	y_3	y_0

*	y_1	y_2
y_1	y_1	y_2
y_2	y_2	y_1

Fig. 2.1.12. The cross-like decomposition of the β-matrix. Multiplication tables for the sets of the matrices y_0 and y_3, and also of the matrices y_1 and y_2, are shown at the bottom.

This algebraic similarity of the internal structure of the α- and β-matrices, generating each other by the indicated inversion 0 ↔ 1, additionally shows the importance of binary-complementarity relations and the "like begets like" principle for the genetic coding system in general.

The cross-like decompositions can be also applied to the α- and β-matrix crystals (Figs. 2.1.4 and 2.1.9, respectively) having larger sizes: $(2^n \times 2^n)$ under $n = 3, 4, \ldots$ In these cases, a similar connection of pairs of sparse complementary matrices with the multiplication table of hyperbolic numbers is always revealed. Figs. 2.1.13 and 2.1.14 show examples of this for the case of (8 × 8)-matrices from the families of the α- and β-crystals. In both cases, such decomposition leads to 8 sparse matrices, where each

1	1	−1	−1	1	1	−1	−1
1	1	−1	−1	1	1	−1	−1
1	−1	1	−1	1	−1	1	−1
1	−1	1	−1	1	−1	1	−1
1	1	−1	−1	1	1	−1	−1
1	1	−1	−1	1	1	−1	−1
1	−1	1	−1	1	−1	1	−1
1	−1	1	−1	1	−1	1	−1

e_0 =

1	0	0	0	0	0	0	0
1	0	0	0	0	0	0	0
1	0	0	0	0	0	0	0
−1	0	0	0	0	0	0	0
0	0	0	0	0	0	0	−1
0	0	0	0	0	0	0	−1
0	0	0	0	0	0	0	−1
0	0	0	0	0	0	0	1

e_1 =

0	1	0	0	0	0	0	0
0	1	0	0	0	0	0	0
0	−1	0	0	0	0	0	0
0	1	0	0	0	0	0	0
0	0	0	0	0	0	−1	0
0	0	0	0	0	0	−1	0
0	0	0	0	0	0	1	0
0	0	0	0	0	0	−1	0

e_2 =

0	0	−1	0	0	0	0	0
0	0	−1	0	0	0	0	0
0	0	1	0	0	0	0	0
0	0	−1	0	0	0	0	0
0	0	0	0	0	1	0	0
0	0	0	0	0	1	0	0
0	0	0	0	0	−1	0	0
0	0	0	0	0	1	0	0

e_3 =

0	0	0	−1	0	0	0	0
0	0	0	−1	0	0	0	0
0	0	0	−1	0	0	0	0
0	0	0	1	0	0	0	0
0	0	0	0	1	0	0	0
0	0	0	0	1	0	0	0
0	0	0	0	1	0	0	0
0	0	0	0	−1	0	0	0

e_4 =

0	0	0	0	1	0	0	0
0	0	0	0	1	0	0	0
0	0	0	0	1	0	0	0
0	0	0	0	−1	0	0	0
0	0	0	−1	0	0	0	0
0	0	0	−1	0	0	0	0
0	0	0	−1	0	0	0	0
0	0	0	1	0	0	0	0

e_5 =

0	0	0	0	0	1	0	0
0	0	0	0	0	1	0	0
0	0	0	0	0	−1	0	0
0	0	0	0	0	1	0	0
0	0	−1	0	0	0	0	0
0	0	−1	0	0	0	0	0
0	0	1	0	0	0	0	0
0	0	−1	0	0	0	0	0

e_6 =

0	0	0	0	0	0	−1	0
0	0	0	0	0	0	−1	0
0	0	0	0	0	0	1	0
0	0	0	0	0	0	−1	0
0	1	0	0	0	0	0	0
0	1	0	0	0	0	0	0
0	−1	0	0	0	0	0	0
0	1	0	0	0	0	0	0

e_7 =

0	0	0	0	0	0	0	−1
0	0	0	0	0	0	0	−1
0	0	0	0	0	0	0	−1
0	0	0	0	0	0	0	1
1	0	0	0	0	0	0	0
1	0	0	0	0	0	0	0
1	0	0	0	0	0	0	0
−1	0	0	0	0	0	0	0

×	e_0	e_7
e_0	e_0	e_7
e_7	e_7	e_0

×	e_1	e_6
e_1	e_1	e_6
e_6	e_6	e_1

×	e_2	e_5
e_2	e_2	e_5
e_5	e_5	e_2

×	e_3	e_4
e_4	e_4	e_3
e_3	e_3	e_4

Fig. 2.1.13. The cross-like decomposition of the (8 × 8)-matrix of the α-crystal type (Fig. 2.1.4) into 8 sparse matrices e_0, e_1, e_2, e_3, e_4, e_5, e_6, e_7. Each of the sets of two complementary sparse matrices corresponds to the same multiplication table of hyperbolic numbers shown at the bottom.

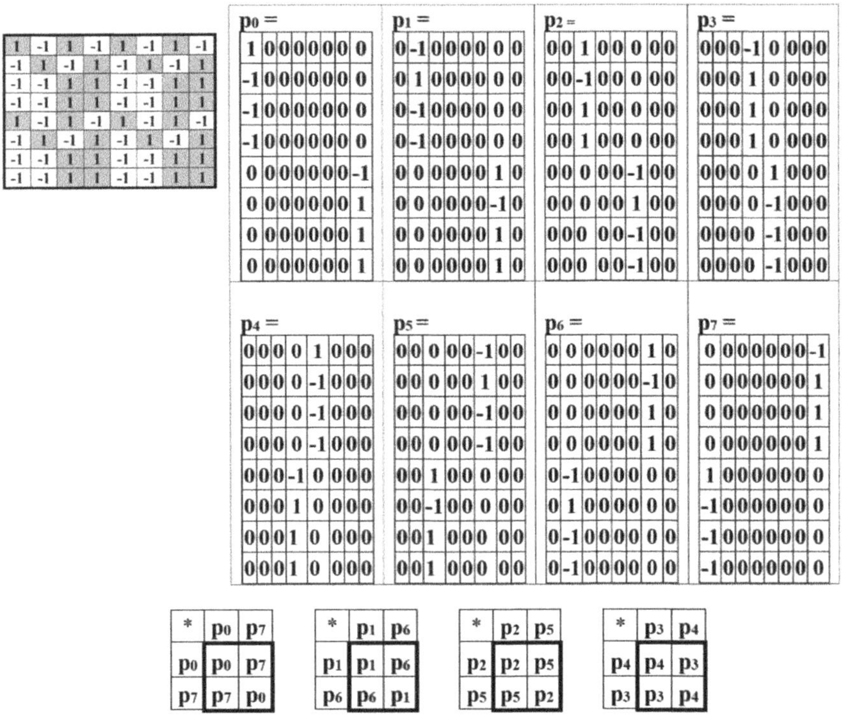

Fig. 2.1.14. The cross-like decomposition of the (8 × 8)-matrix of the β-crystals type (Fig. 2.1.9) into 8 sparse matrices p_0, p_1, p_2, p_3, p_4, p_5, p_6, p_7. Each of the sets of two complementary sparse matrices corresponds to the same multiplication table of hyperbolic numbers shown at the bottom.

of the sets of two complementary sparse matrices is closed concerning multiplication and defines the same multiplication table of two-dimensional algebra of hyperbolic numbers.

In their internal structure, the matrix α- and β-crystals have interrelations with hyperbolic numbers. The inversion operation 0 ↔ 1 transforms the tensor family of matrices $[C, A; T, G]^{(n)}$ to the tensor family of matrices $[G, T; A, C]^{(n)}$, which has other black-and-white mosaics of n-plets with strong and weak roots. Just as in molecular genetics you have nucleotide complementarity of two strands of DNA, so in matrix genetics you have algebraic binary-complementarity of two tensor families of genetic matrices, which are based on the multi-level system of binary oppositions containing the nucleotide complementarity of DNA strands as a particular case.

2.2. Hadamard matrices, Walsh functions, and the problem of noise immunity in genetics

In Chapter 1 (Fig. 1.4.2), a connection of the genetic code — via parameters of its degeneration — with Rademacher functions was described, which are defined by the following equations:

$$r_n(x) = \text{sign}(\sin 2^n \pi x),\ n = 1, 2, 3 \tag{2.4}$$

Rademacher functions take only two values «+1» and «−1» and form incomplete orthogonal systems of functions, related to complete orthogonal systems of Walsh (or Walsh-Hadamard) functions: Walsh functions can be represented as results of the multiplications of Rademacher functions with each other [26, 83–85]. The system is complete, for example, 8 Walsh functions means that any eight-dimensional vector can be represented as their superposition. Complete orthogonal systems of Walsh functions are sets of rows in appropriate Hadamard matrices. Walsh functions and Hadamard matrices play one of the most important roles in the theory of discrete signal processing, spectral analysis, quantum computing, and some other fields (a web search of different applications of Hadamard matrices gives 44,000 publications in the period 1978–2005) [86]. In particular, Chapter 14 of this book is devoted to the theme of logic holography based on Walsh functions.

By definition, a Hadamard matrix of dimension "n" is the $(n \times n)$-matrix $H(n)$ with elements "+1" and "−1". It satisfies the condition $H(n) \times H(n)^T = n \times I_n$, where $H(n)^T$ is the transposed matrix and I_n is the identity $(n \times n)$-matrix. Hadamard matrices are widely used in error-correcting codes such as the Reed-Muller code and Hadamard codes; in the theory of compression of signals and images; in spectral analysis and multi-channel spectrometers with Hadamard transformations; in quantum computers with Hadamard gates; in a realization of Boolean functions by means of spectral methods; in the theory of planning of multiple-factor experiments and in many other branches of science and technology.

If the family of genetic alphabets is associated — as it is shown — with the Rademacher functions due to mentioned biomolecular binary oppositions, is it possible that this family also contains an additional type of biomolecular binary opposition that connects the genetic alphabets with Walsh functions and Hadamard matrices? Yes, such an additional type of the binary opposition exists, that leads to a special T-algorithm (or an U-algorithm) transforming the mentioned genetic matrices (Fig. 1.4.2), into genetic Hadamard matrices, whose mosaic rows represent complete orthogonal systems of Walsh functions [12, 13, 49, 87]. Let us explain this.

The additional genetic binary opposition is that the letter T in DNA (and correspondingly the letter U in RNA) is a very special letter in the 4-letter alphabet of nucleobases in the following two aspects:

1) Each of the three nitrogenous bases A, C, G has one amino group NH_2, but the fourth basis T/U does not (Fig. 1.1.1). From the viewpoint of the existence of the amino group (which is very important for genetic functions) the letters A, C, G are identical to each other and the letter U is opposite to them;
2) The letter T is a single letter in DNA, which is replaced in RNA by the letter U.

This uniqueness of the letter T can be utilized in genetic computers of organisms. Considering the unique status of the letter T in this genetic alphabet, the formal "T-algorithm" was noted. By definition, the T-algorithm contains two steps: 1) on the first step, each of the triplets in the black-and-white genetic matrix (for example, in the matrix $[C\ A;\ T\ G]^{(3)}$ in Fig. 1.4.2) should change its own color into the opposite color each time when the letter T stands in an odd position (in the first or in the third position) inside the triplet; 2) on the second step, black triplets and white triples are interpreted as entries "+1" and "−1" correspondingly. For example, the white triplet TTA (see Fig. 1.4.2) should get the black color (and its matrix cell should be marked in black) because of the letter T in its first position; for this reason, the triplet TTA is interpreted finally as "+1". Or the white triplet TTT should not change its color because of the letter T in its first and third positions (the color of this triplet is changed twice according to the described algorithm); for this reason, the triplet TTT is interpreted finally as "−1". The triplet ACG does not change its color because the letter T is absent in this triplet.

Let's first look at the transformation of two genetic (4 × 4)-matrices $[C, A; T, G]^{(2)}_{12}$ and $[C, A; T, G]^{(2)}_{21}$ of 16 duplets under the T-algorithm. Fig. 2.2.1 shows these matrices, which differ from each other by the permutation of positions 1–2 in all duplets of the first matrix into positions 2–1 in duplets of the second matrix (both matrices were already considered above in Figs. 1.4.2 and 2.1.2). Mosaic rows of these matrices have a meander-like character and correspond to meander-like Rademacher functions. By the T-algorithm, the mosaic matrices $[C, A; T, G]^{(2)}_{12}$ and $[C, A; T, G]^{(2)}_{21}$ are transformed into mosaic matrices H_{12} and H_{21} (Fig. 2.2.1).

Black and white cells of each of the matrices in Fig. 2.2.1 reflect the binary opposition and therefore can be represented by entries +1 and −1 in them. In this case, numeric representations of these four symbolic matrices arise, which are shown in Fig. 2.2.2. For the transformed numeric matrices,

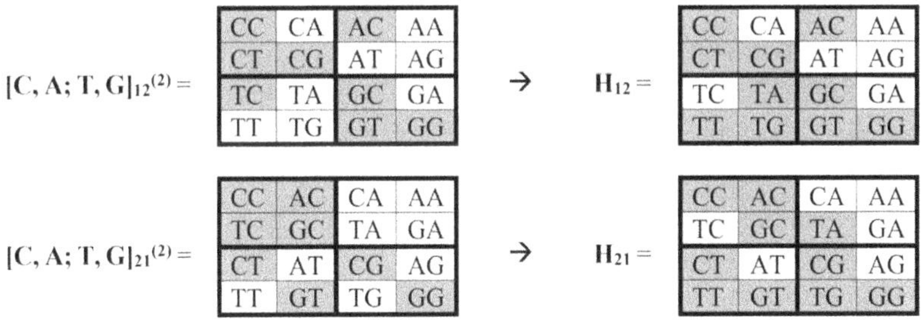

Fig. 2.2.1. The transformation of genetic matrices $[C, A; T, G]^{(2)}_{12}$ and $[C, A; T, G]^{(2)}_{21}$ into new mosaic matrices H_{12} and H_{21} by the T-algorithm.

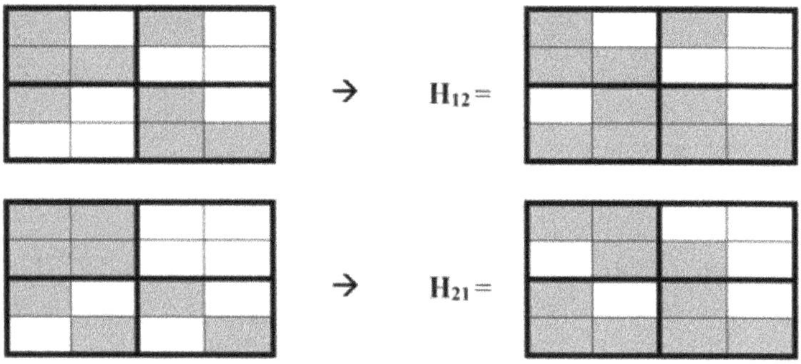

Fig. 2.2.2. Numeric representations of symbolic matrices from Fig. 2.2.1. Black (white) cells mean entries "+1" ("−1).

we preserve the same denotations H_{12} and H_{21}, which were used for their symbolic analogues in Fig. 2.2.1.

Each of the numeric matrices H_{12} and H_{21} in Fig. 2.2.2 is a Hadamard matrix because it satisfies the definition of a Hadamard matrix:

$$H_{12} \times H_{12}^{T} = 4 \times I_4, \quad H_{21} \times H_{21}^{T} = 4 \times I_4 \tag{2.5}$$

Determinants of these Hadamard matrices are equal to $2^4 = 16$.

Analogically, by means of the T-algorithm, all the genetic matrices of 64 triplets $[C\ A; T\ G]_{123}^{(3)}$, $[C\ A; T\ G]_{231}^{(3)}$, $[C\ A; T\ G]_{312}^{(3)}$, $[C\ A; T\ G]_{321}^{(3)}$, $[C\ A; T\ G]_{213}^{(3)}$, $[C\ A; T\ G]_{132}^{(3)}$ (Figs. 1.4.2, 1.6.1, 1.6.2) are transformed into relevant numeric matrices H_{123}, H_{231}, H_{312}, H_{321}, H_{213}, H_{132} in Fig. 2.2.3. Determinants of all these genetic matrices are equal to $2^{12} = 4096$.

Each of the six matrices in Fig. 2.2.3 is a Hadamard matrix because it satisfies the definition of a Hadamard matrix:

$$\begin{aligned} &H_{123} \times H_{123}^{T} = 8 \times I_8, H_{231} \times H_{231}^{T} = 8 \times I_8, H_{312} \times H_{312}^{T} = 8 \times I_8, \\ &H_{321} \times H_{321}^{T} = 8 \times I_8, H_{213} \times H_{213}^{T} = 8 \times I_8, H_{132} \times H_{132}^{T} = 8 \times I_8 \end{aligned} \tag{2.6}$$

In evolutionary biology, the Hadamard transform is used to estimate phylogenetic trees from molecular data for understanding the relationship among organisms [88–90]. A Hadamard transform applied to a vector

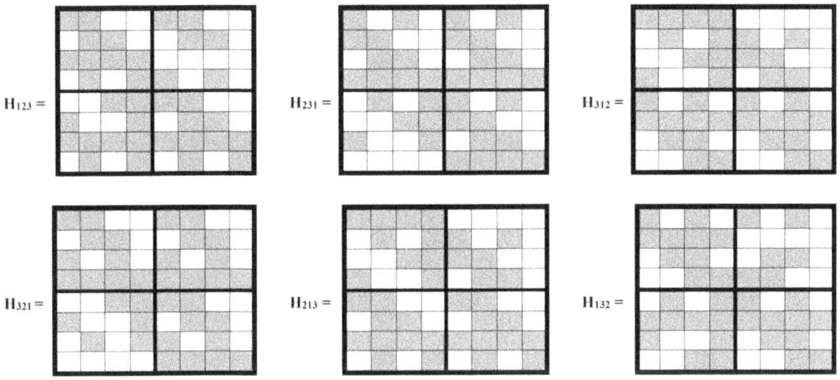

Fig. 2.2.3. The genetic Hadamard matrices H_{123}, H_{231}, H_{312}, H_{321}, H_{213}, H_{132}, which are obtained from the matrices $[C, A; T, G]^{(3)}_{123}$, $[C, A; T, G]^{(3)}_{231}$, $[C, A; T, G]^{(3)}_{312}$, $[C, A; T, G]^{(3)}_{321}$, $[C, A; T, G]^{(3)}_{213}$, $[C, A; T, G]^{(3)}_{132}$ (Figs. 1.4.2, 1.6.1, 1.6.2) by means of the T-algorithm. Black and white cells mean entries "+1" and "−1" correspondingly.

(or matrix) of site pattern frequencies obtained from a DNA multiple sequence alignment can be used to generate another vector that carries information about the tree topology. The invertible nature of the phylogenetic Hadamard transform also allows the calculation of site likelihoods from a tree topology vector, allowing one to use the Hadamard transform for maximum likelihood estimation of phylogenetic trees. The invertible nature of the phylogenetic Hadamard transform does provide an elegant tool for mathematical phylogenetics [91, 92].

Revealing the connection between the organization of binary-structured alphabets of a genetic coding system with Hadamard matrices and complete orthogonal systems of Walsh functions is important because of their role in many scientific methods and theories, as noted above. In particular, as noted in [86], many tens of thousands of works are devoted to diverse applications of Hadamard matrices for signal processing. For example, complete systems of Walsh functions are used to build noise-immune coding of information on the Mariner and Voyager spacecraft that transmit photos of Mars, Jupiter, Saturn, Uranus, and Neptune to Earth. Sequency analysis is based on Hadamard matrices and Walsh functions [27, 28, 93, 94], which is one of the important types of spectral analysis in communications on discrete signals and has found extensive applications in radio engineering, acoustics, optics, etc. [95]. Sequency theory gave rise to fundamentally new communication systems, instantaneous image processing, underwater acoustic high-speed cinema, and receivers and transmitters of non-sinusoidal electromagnetic waves. In particular, on the basis of sequency analysis, the absorption problem is bypassed radio waves and acoustic waves, which is also important for bioinformatics [96].

Future comparative studies of the role of Hadamard matrices and Walsh functions in these sciences and genetics may be useful for their mutual enrichment including disclosure of informational patents of living nature.

2.3. Decompositions of genetic Hadamard matrices, Hamilton quaternions, and genetic octetons

One can make the dyadic-shift decomposition of each of the genetic Hadamard matrices H_{12} and H_{21} of 16 duplets (Fig. 2.2.2) by analogy with the described decompositions of the numeric representations of matrices $[C, A; T, G]_{12}^{(2)}$ and $[C, A; T, G]_{21}^{(2)}$ (Figs. 1.5.2 and 2.1.4). Fig. 2.3.1 shows the dyadic-shift decomposition of the Hadamard matrix H_{12} into the sum of 4

$$\begin{vmatrix} 1 & -1 & 1 & -1 \\ 1 & 1 & -1 & -1 \\ -1 & 1 & 1 & -1 \\ 1 & 1 & 1 & 1 \end{vmatrix} = \begin{vmatrix} 1 & 0 & 0 & 0 \\ 0 & 1 & 0 & 0 \\ 0 & 0 & 1 & 0 \\ 0 & 0 & 0 & 1 \end{vmatrix} + \begin{vmatrix} 0 & -1 & 0 & 0 \\ 1 & 0 & 0 & 0 \\ 0 & 0 & 0 & -1 \\ 0 & 0 & 1 & 0 \end{vmatrix} + \begin{vmatrix} 0 & 0 & 0 & -1 \\ 0 & 0 & -1 & 0 \\ 0 & 1 & 0 & 0 \\ 1 & 0 & 0 & 0 \end{vmatrix} + \begin{vmatrix} 0 & 0 & 1 & 0 \\ 0 & 0 & 0 & -1 \\ -1 & 0 & 0 & 0 \\ 0 & 1 & 0 & 0 \end{vmatrix} = e_0+e_1+e_2+e_3$$

*	e_0	e_1	e_2	e_3
e_0	e_0	e_1	e_2	e_3
e_1	e_1	$-e_0$	e_3	$-e_2$
e_2	e_2	$-e_3$	$-e_0$	e_1
e_3	e_3	e_2	$-e_1$	$-e_0$

Fig. 2.3.1. The decomposition of the genetic Hadamard matrix H_{12} (from Fig. 2.2.2) into the set of sparse orthogonal matrices e_0, e_1, e_2, e_3, which is closed under multiplication. A corresponding multiplication table is the multiplication table of the algebra of Hamilton quaternions.

$$\begin{vmatrix} 1 & 1 & -1 & -1 \\ -1 & 1 & 1 & -1 \\ 1 & -1 & 1 & -1 \\ 1 & 1 & 1 & 1 \end{vmatrix} = \begin{vmatrix} 1 & 0 & 0 & 0 \\ 0 & 1 & 0 & 0 \\ 0 & 0 & 1 & 0 \\ 0 & 0 & 0 & 1 \end{vmatrix} + \begin{vmatrix} 0 & 1 & 0 & 0 \\ -1 & 0 & 0 & 0 \\ 0 & 0 & 0 & -1 \\ 0 & 0 & 1 & 0 \end{vmatrix} + \begin{vmatrix} 0 & 0 & -1 & 0 \\ 0 & 0 & 0 & -1 \\ 1 & 0 & 0 & 0 \\ 0 & 1 & 0 & 0 \end{vmatrix} + \begin{vmatrix} 0 & 0 & 0 & -1 \\ 0 & 0 & 1 & 0 \\ 0 & -1 & 0 & 0 \\ 0 & 0 & 0 & 0 \end{vmatrix} = k_0+k_1+k_2+k_3$$

*	k_0	k_1	k_2	k_3
k_0	k_0	k_1	k_2	k_3
k_1	k_1	$-k_0$	k_3	$-k_2$
k_2	k_2	$-k_3$	$-k_0$	k_1
k_3	k_3	k_2	$-k_1$	$-k_0$

Fig. 2.3.2. The decomposition of the genetic Hadamard matrix H_{21} (from Fig. 2.2.2) into the set of sparse orthogonal matrices k_0, k_1, k_2, k_3, which is closed under multiplication. Its multiplication table is the multiplication table of the algebra of Hamilton quaternions. The matrix k_0 is the identity matrix.

sparse matrices e_0, e_1, e_2, e_3, each of which is orthogonal. It is unexpected but this set of 4 sparse matrices is closed in relation to multiplication and defines a corresponding multiplication table, which is known as the multiplication table of basis elements of the algebra of Hamilton quaternions [https://en.wikipedia.org/wiki/Quaternion].

Fig. 2.3.2 shows the dyadic-shift decomposition of the Hadamard matrix H_{21} (Figs. 2.2.1, 2.2.2) into the sum of the other 4 sparse matrices k_0, k_1, k_2, k_3, each of which is orthogonal. It turns out that this set of 4 sparse matrices is also closed in relation to multiplication and defines the same multiplication table of the algebra of Hamilton quaternions.

One can make the dyadic-shift decomposition of each of the six Hadamard genomatrices H_{123}, H_{231}, H_{312}, H_{321}, H_{213}, H_{132}, which are shown in Fig. 2.2.3, by analogy with the described decompositions of the relevant genomatrices R_{123}, R_{231}, R_{312}, R_{321}, R_{213}, R_{132} in Figs. 1.5.3 and 1.6.3. In the result, six new different sets of basic sparse matrices h_0, h_1, h_2, h_3, h_4, h_5, h_6, h_7 arise, where each of the sparse matrices is orthogonal (h_0 is the identity matrix). Fig. 2.3.3 shows these sparse matrices for the particular case of the matrix H_{123}. It is unexpected but each of the six sets of sparse matrices is closed in relation to multiplication. Moreover, each of these six sets h_0, h_1, h_2, h_3, h_4, h_5, h_6, h_7 corresponds to the same multiplication table, shown in Fig. 2.3.3 for the particular case of the matrix H_{123}.

$$H_{123} = h_0 + h_1 + h_2 + h_3 + h_4 + h_5 + h_6 + h_7 =$$

$$\begin{bmatrix} 1&0&0&0&0&0&0&0\\ 0&1&0&0&0&0&0&0\\ 0&0&1&0&0&0&0&0\\ 0&0&0&1&0&0&0&0\\ 0&0&0&0&1&0&0&0\\ 0&0&0&0&0&1&0&0\\ 0&0&0&0&0&0&1&0\\ 0&0&0&0&0&0&0&1 \end{bmatrix} + \begin{bmatrix} 0&1&0&0&0&0&0&0\\ -1&0&0&0&0&0&0&0\\ 0&0&0&1&0&0&0&0\\ 0&0&-1&0&0&0&0&0\\ 0&0&0&0&0&1&0&0\\ 0&0&0&0&-1&0&0&0\\ 0&0&0&0&0&0&0&1\\ 0&0&0&0&0&0&-1&0 \end{bmatrix} + \begin{bmatrix} 0&0&-1&0&0&0&0&0\\ 0&0&0&-1&0&0&0&0\\ 1&0&0&0&0&0&0&0\\ 0&1&0&0&0&0&0&0\\ 0&0&0&0&0&0&-1&0\\ 0&0&0&0&0&0&0&-1\\ 0&0&0&0&1&0&0&0\\ 0&0&0&0&0&1&0&0 \end{bmatrix} + \begin{bmatrix} 0&0&0&-1&0&0&0&0\\ 0&0&1&0&0&0&0&0\\ 0&1&0&0&0&0&0&0\\ -1&0&0&0&0&0&0&0\\ 0&0&0&0&0&0&0&-1\\ 0&0&0&0&0&0&1&0\\ 0&0&0&0&0&1&0&0\\ 0&0&0&0&-1&0&0&0 \end{bmatrix} +$$

$$\begin{bmatrix} 0&0&0&0&1&0&0&0\\ 0&0&0&0&0&1&0&0\\ 0&0&0&0&0&0&-1&0\\ 0&0&0&0&0&0&0&-1\\ -1&0&0&0&0&0&0&0\\ 0&-1&0&0&0&0&0&0\\ 0&0&1&0&0&0&0&0\\ 0&0&0&1&0&0&0&0 \end{bmatrix} + \begin{bmatrix} 0&0&0&0&0&1&0&0\\ 0&0&0&0&-1&0&0&0\\ 0&0&0&0&0&0&0&-1\\ 0&0&0&0&0&0&1&0\\ 0&-1&0&0&0&0&0&0\\ 1&0&0&0&0&0&0&0\\ 0&0&0&1&0&0&0&0\\ 0&0&-1&0&0&0&0&0 \end{bmatrix} + \begin{bmatrix} 0&0&0&0&0&0&-1&0\\ 0&0&0&0&0&0&0&-1\\ 0&0&0&0&-1&0&0&0\\ 0&0&0&0&0&-1&0&0\\ 0&0&1&0&0&0&0&0\\ 0&0&0&1&0&0&0&0\\ 1&0&0&0&0&0&0&0\\ 0&1&0&0&0&0&0&0 \end{bmatrix} + \begin{bmatrix} 0&0&0&0&0&0&0&-1\\ 0&0&0&0&0&0&1&0\\ 0&0&0&0&0&-1&0&0\\ 0&0&0&0&1&0&0&0\\ 0&0&0&1&0&0&0&0\\ 0&0&-1&0&0&0&0&0\\ 0&1&0&0&0&0&0&0\\ -1&0&0&0&0&0&0&0 \end{bmatrix}$$

*	1	h_1	h_2	h_3	h_4	h_5	h_6	h_7
1	1	h_1	h_2	h_3	h_4	h_5	h_6	h_7
h_1	h_1	-1	h_3	$-h_2$	h_5	$-h_4$	h_7	$-h_6$
h_2	h_2	h_3	-1	$-h_1$	$-h_6$	$-h_7$	h_4	h_5
h_3	h_3	$-h_2$	$-h_1$	1	$-h_7$	h_6	h_5	$-h_4$
h_4	h_4	h_5	h_6	h_7	-1	$-h_1$	$-h_2$	$-h_3$
h_5	h_5	$-h_4$	h_7	$-h_6$	$-h_1$	1	$-h_3$	h_2
h_6	h_6	h_7	$-h_4$	$-h_5$	h_2	h_3	-1	$-h_1$
h_7	h_7	$-h_6$	$-h_5$	h_4	h_3	$-h_2$	$-h_1$	1

Fig. 2.3.3. The dyadic-shift decomposition of the Hadamard matrix H_{123} (Fig. 2.2.3) into the sum of 8 sparse matrices $h_0, h_1, \ldots, h_7$. The multiplication table for this closed set of sparse matrices is shown at the bottom. In the table, the identity matrix h_0 is shown by the symbol 1.

This multiplication table in Fig. 2.3.3 coincides with the multiplication table of the basis elements of eight-dimensional algebra, which is conditionally called the algebra of genetic octetons. So, the genetic (4×4)-Hadamard matrices H_{12} and H_{21} (Figs. 2.3.1, 2.3.2) turn out to be two different matrix representations of Hamiltonian quaternions with unit coordinates. Correspondingly, the genetic (8×8)-Hadamard matrices H_{123}, H_{231}, H_{312}, H_{321}, H_{213}, H_{132} (Figs. 2.2.3, 2.3.3) turn out to be six different matrix representations of genetic octetons with unit coordinates.

The question of the invariance of the multiplication tables for the basic elements of the named algebras of hypercomplex numbers for different permutations of elements in the genomatrices seems to be interesting in connection with the phenomenon of metamorphosis of organisms. For example, in the metamorphosis of a butterfly, which includes four stages (egg — caterpillar — pupa — butterfly), the pupa does not feed on anything and has a fixed (!) atomic composition, but due to a genetically determined rearrangement of elements, the pupa miraculously turns into a new organism — a butterfly.

Let us consider the possible consequences of the connection of the genetic coding system with Hamilton's quaternions and genetic octetons.

2.4. Some applications of Hamiltonian quaternions and genetic octetons for modeling biological phenomena

For some reason, our three-dimensional physical space is arranged in such a way that its geometric features correspond to Hamilton quaternions. And now it turns out that the system of genetic coding for some reason is arranged in accordance with Hamilton's quaternions and genetic octetons. Hamilton quaternions are closely related to Pauli matrices, electromagnetic field theory (Maxwell wrote his equations in the language of Hamilton quaternions), special relativity, spin theory, quantum mechanical theory of chemical valence, etc. Thousands of works have been devoted to quaternions in physics in the 20th century [97].

Matrix genetics has led to the discovery of an important bridge between physics, biology and computer science for their mutual enrichment. In our research, a new example of the incomprehensible effectiveness of mathematics has been obtained: abstract mathematical structures, deduced by mathematicians at the tip of a pen more than 150 years ago, turn out to be embodied in the information basis of living matter — a system of structured genetic alphabets. And what mathematicians discover through painful thought (like Hamilton, who spent 10 years of continuous thought to discover his quaternions) is already represented in the genetic coding system.

Hamilton quaternions and Cockle split-quaternions are united by the fact that the entire set of Clifford algebras for the two-dimensional case $Cl_{1,1}(R)$, $Cl_{2,0}(R)$, $Cl_{0,2}(R)$ is isomorphic to just these two four-dimensional algebras of hypercomplex numbers (http://www.mi.ras.ru/noc/11_12/cllifalg04.12.11.pdf). This is of interest to the international community of "Cliffordists" who seek to interpret Clifford algebras as a universal language of geometry and physics. Hamilton's quaternions are known as a tool in mathematics and physics. We draw attention to the fact that the Hamilton quaternion with unit coordinates through its matrix representations (there is more than one such representation) is deeply related to computer science for two reasons:

1) its (4×4)-matrix representations are Hadamard matrices, which play an important role in algorithms and codes of error-correcting information transmission;
2) in different (4×4)-matrix representations of Hamilton quaternions, the placement of their four coordinates always corresponds to the mentioned principle of dyadic shifts, based on one of the main operations of computer technology — the modulo-2 addition of binary numbers. Fig. 2.4.1 demonstrates these properties on specific examples of (4×4)-matrix representations of Hamiltonian quaternions.

Location of coordinates *a*, *b*, *c*, *d* in each matrix in Fig. 2.4.1 corresponds to the above principle of the structure of matrices of dyadic shifts (Figs. 1.5.1, 1.5.2) described, for example, in [26].

$$\begin{bmatrix} a & b & c & d \\ -b & a & -d & c \\ -c & d & a & -b \\ -d & -c & b & a \end{bmatrix} ; \begin{bmatrix} a & b & -c & d \\ -b & a & d & c \\ c & -d & a & b \\ -d & -c & -b & a \end{bmatrix} ; \begin{bmatrix} a & -b & c & d \\ b & a & -d & c \\ -c & d & a & b \\ -d & -c & -b & a \end{bmatrix}$$

Fig. 2.4.1. Examples of the matrix representations of Hamilton quaternions $a + bi + cj + dk$, where *a*, *b*, *c*, *d* are real numbers; i, j, k are imaginary units of these quaternions. For unit values of coordinates (a = b = c = d = 1), all these matrices are Hadamard matrices.

The named quaternions (H_{12}, H_{21}) and genetic octetons (H_{123}, H_{231}, H_{312}, H_{321}, $H_{21}3$, H_{132}) with unit coordinates (we denote them as H_4 and H_8, respectively), being normalized to the unit determinant, under their exponentiations form cyclic groups with periods 6 and 24 respectively:

$$(2^{-1} \times H_4)^n = (2^{-1} \times H_4)^{n+6}; \ (2^{-1.5} \times H_8)^n = (2^{-1.5} \times H_8)^{n+24} \qquad (2.7)$$

The properties of these cyclic groups are such that they can be used to model the genetically inherited properties of living organisms. For example, one can show the coincidence of the properties of genetically inherited color perception with the 6-membered cyclic group of the Hamiltonian quaternion $(2^{-1} \times H_4)^n$ (2.7). Let's depict all the members of this cyclic group on a circular dial (the black cells of the matrices contain the numbers "+0.5", the white ones "−0.5", and E is the identity matrix) (Fig. 2.4.2).

This 6-member group of Hamiltonian quaternions has the following properties:

1) The quaternions opposite on the circle are opposite in sign and add up to zero, and therefore are mutually complementary;

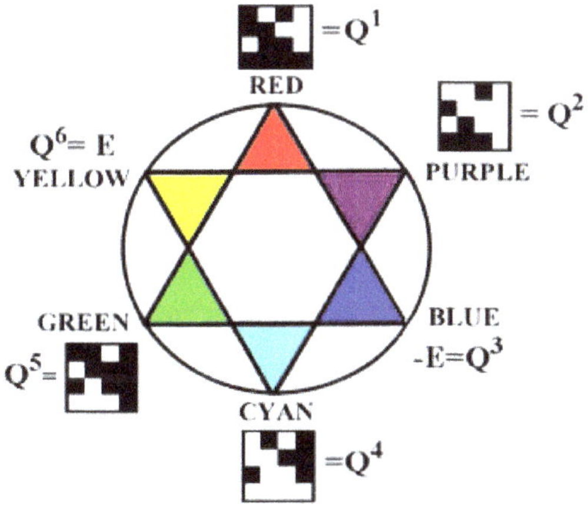

Fig. 2.4.2. Newton's color wheel from the psychophysics of color perception and the correspondence to it of members of the cyclic group of the normalized Hamilton quaternion with unit coordinates $Q^n = (0.5\ H_4)^n$, the period of which is 6.

2) Each quaternion on the circle is the sum of two neighboring quaternions;
3) The sum of the quaternions at the vertices of each of the two triangles of the "Star of David" is zero.

These properties of the cyclic group of the considered Hamiltonian quaternion are identical to the phenomenological properties of inherited color perception. The latter is based on three primary colors (red, blue, green) and three additional colors (cyan, yellow, magenta) lying at the vertices of two triangles of the "Star of David" on the well-known Newton's color wheel (https://web.mit.edu/22.51/www/Extras/color_theory/color.html):

1) Opposite colors on the circle are mutually complementary and, when superimposed, cancel each other out;
2) Each color on the circle is the sum of two neighboring colors;
3) Three basic colors, as well as three additional colors, extinguish each other when superimposed.

The color mixing task can now be formally solved in the language of the Hamiltonian quaternion cyclic group, if we assign a member of this group to each color in accordance with Fig. 2.4.2. For example, what color will you get if you mix 3 parts red, 2 parts yellow and 5 parts blue? The answer is obtained by adding the corresponding quaternions: $3Q^1 + 2Q^6 + 5Q^3 = 3Q^2$, i.e., 3 parts magenta (here we use the designation $Q = 0.5 \times H_4$).

In psychophysics, it is known that color is not a physical property of an object, but an inherited reaction of a person to a light effect coming from an object, taking into account the general light environment. Why is our color perception arranged precisely in accordance with the color wheel and the cyclic group of Hamilton's quaternions? Possible answer: because the organism processes light information on the basis of inherited unified algorithms for processing genetic and physical information with the participation of the genetically inherited cyclic group of Hamilton quaternions, with which the system of genetic alphabets itself is connected.

Fig. 2.4.3. 24-hour dial of the day with its division into day and night 12-hour halves.

As you know, our body is a huge choir of genetically inherited rhythmic processes. Even proteins are cyclically broken down into amino acids and reassembled. The phases of activity and passivity of physiological systems are coordinated with each other and, according to the ideas of ancient and modern medicine, determine the division of a day into 24 hours by a person [98]. The biorhythmic organization of inherited processes is distributed throughout the body of various species, including those without a nervous system. Bio-rhythmology is important for medicine, ergonomics, etc. Fig. 2.4.3 shows a 24-hour clock in which the numbers correspond to the exponents of this normalized genetic octeton ($B = 2^{-1.5} H_8$) with unit coordinates.

Genetic octetons in powers opposite on the dial differ only in the sign $B^n = -B^{n+12}$, by analogy with the opposite of day and night hours (for example, in ordinary life, instead of "14 o'clock" they often say "2 o'clock in the afternoon", naming the opposite value in this dial number). This cyclic group of genetic octetons also has other interesting properties, for example: $B^n = B^{n-4} + B^{n+4}$; $B^n = -(B^{n-8} + B^{n+8})$; $B^n = 2^{-0.5} \times (B^{n-3} + B^{n+3})$.

To this topic of genetic octetons, we can add that molecular genetics has examples of the special phenomenological role of the number 8. For example, in eukaryotic cells, DNA strands are wound around nucleosomes, each of which is an octameric rod consisting of four types of histone proteins: H2A, H2B, H3 and H4. More precisely, a single nucleosome is an ensemble of 8 histones in which a pair of histones of each of these four types is present (there are complementarity relationships between these histones, reminiscent of the complementarity relations between pairs of nitrogenous bases in DNA). The DNA molecule is wound around this octameric rod. The octameric structure of nucleosomes plays a central role in DNA packaging at all levels. In molecular genetics, there is a well-known concept of the histone code [99]. It is possible that the mathematical formalisms of the described genetic hypercomplex numbers will be useful for modeling the phenomenological facts of this fundamental concept and for its development. It can be added that in the field of molecular genetics, to ensure certain physiological processes, entire families of proteins must work in concert (a separate, isolated protein is not a living system). At the same time, these families should contain certain amounts of proteins of different types. The genetic system successfully solves this numerical problem of genetic determination for the amounts of proteins of each type within separate families for different biological processes. The hypercomplex numbers described in this chapter can be additionally used for mathematical modeling of such genetic phenomena. Generally speaking, one can think that in the field of molecular genetics and inherited physiological structures, these hypercomplex systems can serve as a natural basis for providing numerical patterns in inherited genetic sets of information elements.

Genetic Hadamard matrices determine special variants of sequency spectral analysis, proposed in works [27, 28, 93, 94], and spatial and temporal filtering of physiological processes and their correlation analysis. The data of the developed "matrix genetics" suggest that this genetic sequency analysis can serve as a key to important aspects of bioinformatics. As it is known, the wide use of Hadamard matrices and their systems of Walsh functions in digital informatics is determined by their special properties. Firstly, Walsh step functions are implemented in electronic devices more easily than trigonometric and many other functions. Secondly, the use of Walsh functions makes it possible to process digital signals using only addition and subtraction operations without multiplication and division (see, for example, [26, 100]). Since division operations are not required in this case, algebras without division can be used in digital computer science, among which are extensions of quaternion and split-quaternion algebras to cases of higher 2^n-dimensional algebras. This additionally distinguishes computer science from theoretical physics, where the operations of multiplication and division are essential, and where, in this regard, the attention of physicists is drawn to division algebras, primarily to Cayley octonion algebras. For the area of molecular genetics we are considering, it is essential that in it the operations of addition and subtraction can be organized by means of a simple connection or separation of molecular elements, which, obviously, is much simpler than organizing the operations of multiplication and division on molecules.

It should be noted that there are some works on the application of Walsh functions with their useful properties for the spectral analysis of genetic sequences and the creation of new genetic algorithms [101–107]. It is known that in the field of multi-channel digital communication, multi-dimensional numerical systems are used to provide functional advantages. For example, 2D complex numbers are used for digital processing of 2D signals [66, Chapter 8]; in this case, the real part of the complex numbers corresponds to the signal of the first channel, and the imaginary part of the complex numbers corresponds to the signal of the second channel. But biological organisms are multi-channel information systems. For example, the retina of the eye has an innate set of receptors that provides multi-channel transmission to the nervous system of information about the images projected onto the retina. For the efficient operation of information systems with many parallel channels for transmitting streams of independent initial signals, a cooperative principle of accounting for these streams is required. This principle can be implemented in the form of an algebraic system of hypercomplex numbers. We believe that the described genetic hypercomplex numbers are used in inherited multi-channel bioinformatic systems for similar purposes.

Why are Hamilton quaternions and genetic octetons associated with the genetic coding system? Probably, a deep answer to this question will be obtained by taking into account not only the geometric properties of quaternions and genetic octetons, but also their information essence and suitability for efficient information processing as special kinds of Hadamard matrices with special systems of Walsh functions in them, etc. Let us point out possible reasons why the molecular genetic system of alphabets turns out to be associated with these associative algebras of hypercomplex numbers, in contrast to the systems of theoretical physics of inanimate matter, where much attention is paid to non-associative algebras of Cayley octonions and split-octonions. The supposed reason is that associativity is an important condition in the computer science of error-correcting coding, where many kinds of codes (for example, cyclic codes) are widely used, which are based on the concept of algebraic groups and Galois fields. At the same time, the very definitions of the concepts of a Galois group and fields contain the conditions for the associativity of their elements. In other words, in the field of noise immunity coding, the associative matrix algebras of mentioned hypercomplex numbers have an advantage over non-associative algebras of Cayley octonions and split-octonions.

In connection with mentioned cyclic shifts in described genetic matrices, one can mention so-named cyclic codes, which are based on cyclic shifts [84]. Cyclic codes are considered usually as one of the most interesting codes in the field of digital technique due to their mathematical properties. Some modern publications in the field of molecular genetics analyze the question about a possible important participation of cyclic codes in systems of genetic coding [14, 108–111].

2.5. Genetic intelligence, inherited intellectual abilities, and genomorphic artificial intelligence

All living organisms have inherited abilities for some intellectual activity. For example, a spider possesses its inherited intellectual ability to weave its spider web — without any training — using up to 7 kinds of filaments arranged in a certain order inside its spider web, which is tied to random external supports. Weaving birds weave nests of an inherited complex shape from hundreds of random straws. Cooperative work of bees on the construction of bee honeycombs provides one of many other examples of the inherited ability of organisms to act intellectually. Organisms devoid of nerve cells also have intellectual abilities, for example, slime molds and mold fungi, which are able to find optimal paths in labyrinths [112–115]. When creating artificial intelligence systems, scientists should look into algorithms and principles of genetic inheritance of such intellectual abilities from wildlife.

Inheritance of physiological systems are connected with genetic intelligence. By genetic intelligence, we understand that part of the intellectual potentialities of living organisms, which allows one, on the basis of genetic information in DNA and RNA, to build, for example — from one fertilized cell — an organism, having trillions of cells where parental characteristics are reproduced by a multi-channel noise-resistant manner, despite strong noises and frequent changing conditions of nutrition and external influences.

Genetic intelligence deals with a systematic growth in the course of ontogenesis of the number of body parameters and a corresponding increase in the dimension of its configuration space of states. In mathematics, the tensor product of matrices is used to increase the dimension of a vector space. The genetic coding system, together with the family of DNA alphabets, is structured in accordance with the tensor product of genetic matrices, which links, for example, matrix representations of different DNA alphabets into a single tensor family of alphabetic matrices [12, 13, 116].

One can believe that structural basis of genetic intelligence is connected, in particular, with orthogonal matrices, which arise many times in this book under analysis of the genetic code system. Orthogonal matrices describe vector transformations of rotations and mirror reflection, which conserve metric characteristics. One of the arguments in favor of the important role of orthogonal transformations in the structural foundations of genetic

Fig. 2.5.1. The kinematic scheme of human body movements is built on orthogonal transformations of rotation in the joints and mirror reflection.

intelligence can be the fact that the entire inherited kinematic scheme of the human body is built precisely on rotations in the joints and mirror reflections (Fig. 2.5.1).

The study of the genetic coding system using matrix-tensor methods leads to the disclosure of living nature patents and hidden universal rules in the field of genetic intelligence. We believe that thanks to these studies, a new class of artificial intelligence machines will emerge based on the use of structural and emergent features of genetic intelligence systems. This type of artificial intelligence, the development of which is stimulated by the materials of this book, we call "genomorphic artificial intelligence" to distinguish it from many other known types.

The results described in different chapters of this book are connected to the problem of genetic intelligence, which is useful not only for the possible creation of genomorphic artificial intelligence, but also for the development of algebraic biology, new approaches in regeneration of biological tissues and organs, new approaches in digital antenna arrays, and in some other fields.

2.6. Some concluding remarks

This chapter additionally shows a close connection of the genetic code system with binary opposition principles, dyadic groups of binary numbers, and some algebras of hypercomplex numeric systems. The well-known theme of holographic features of non-local informatics in biological bodies gets new essential additions due to analysis of the matrix representations of the DNA alphabets. This analysis is based on the application of algorithms of the method of the bit-reversible holography known in noise immunity communication technologies. These results concern the problem of amazing noise immunity properties of the genetic code system.

The application of the named algorithms to the considered genetic matrices transforms them into multi-block matrices connected with the algebra of split-quaternion used in the Poincaré conformal disk model to describe hyperbolic motions in Lobachevsky's hyperbolic geometry. Simultaneously, it was unexpectedly discovered that these matrix representations of the genetic code system are related to the theme of the holographic quantum error-correcting codes with a tensor network structure. These codes are intensively developed in the USA in the field of quantum communications and the consideration of space-time as a quantum error-correcting code.

The theme of noise immunity properties of the genetic code system is added by the described connections of the genetic matrices with Hadamard matrices and Walsh functions, which are the basis of many noise immunity codes in modern communication technologies. The considered Hadamard matrices have a hidden relation to Hamilton quaternions and the presented genetic octetons, which can be used for modeling biological structures. In particular, the cyclic groups of the normalized Hamilton quaternion with unit coordinates and also the normalized genetic octeton model some known biological phenomena.

The described problem of genetic intelligence and inherited intellectual abilities of organisms is one of the great challenges of modern science. The presented tensor-matrix approaches to this problem seem to be useful towards solving it, which can give humanity many wonderful technologies in artificial intelligence and other fields. The study of genetic noise immunity and genetic intelligence should be continued, taking into account the results shown in this chapter.

Chapter 3

Theory of Resonances and the Concept of Multi-Resonance Genetics

The founder of the theory of resonance in structural chemistry, L. Pauling, argued for the importance of resonance patterns in the organization of living systems. Any living organism is a great chorus of coordinated oscillatory processes. From the formal point of view, a biological organism is an oscillatory system with a great number of degrees of freedom. Such systems are studied in the theory of oscillations using matrix mathematics of their resonance characteristics. This chapter is devoted to a new approach for modeling genetically inherited structures and processes in living organisms using mathematical tools of the theory of resonances. This approach reveals hidden relationships in several genetic phenomena and gives rise to a new class of bio-mathematical models, which contribute to a convergence of biology with physics and informatics. In addition, problems of biorhythms and cyclic processes are analyzed. Perspectives on applications of the phenomena of vibrational mechanics for modeling in biology are discussed.

3.1. The topic of resonances in biological structures

The general idea about vibration bases of the world, including living matter, was known in ancient China and ancient India. This chapter describes new data on this ancient topic, which consist of discovering the structural intersection of the world of genetic phenomena with the world of mathematics of resonances of vibrational systems with many degrees of freedom [36, 117–119].

In modern science, the concept of resonance plays a fundamental and interdisciplinary role. Quantum mechanics began in 1900 due to works by M. Planck, who analyzed a great set of resonant oscillators inside a cavity and as a result obtained his famous law of electromagnetic radiation emitted by a black body in thermal equilibrium. One can say that Planck has represented matter as a set of vibrating oscillators and set the task to study the equilibrium established as a result of the exchange of energy between the oscillators and radiation.

After more than 50 years of successful development of quantum mechanics, E. Schrodinger emphasized the basic meaning of resonances: "*The one thing which one has to accept and which is the inalienable consequence of the wave-equation as it is used in every problem, under the most various forms, is this: that the interaction between two microscopic physical systems is controlled by a peculiar law of resonance*" [120]. In considering an exact balance in nature between bundles of energy, lost by one system and gained by another, he noted: "*I maintain that it can in all cases be understood as a resonance phenomenon*" [ibid, p. 114]. He wrote in his resonance concept of quantum interactions that chemical reactions, including photochemical reactions, can be explained on the basis of resonances. His book [121] said that the chromosome is an aperiodic crystal since its atoms are connected to each other by forces of the same nature as atoms in crystals. But vibrations and resonances play a very important role in the physics of crystals. One can hope that a resonance approach can also usefully serve in genetics.

Pauling used ideas of resonances in quantum mechanical systems in his theory of resonance in structural chemistry. His book [122] about this theory is the most cited among scientific books of the twentieth century. The theory was developed to explain the formation of hybrid bonds in molecules. The actual molecule, as Pauling proposed, is a sort of hybrid, a structure that resonates between the two alternative extremes; and whenever there is a resonance between the two forms, the structure is stabilized. His theory uses the fundamental principle of minimal energy because — in resonant combining of parts into a single unit — each member of the ensemble requires less energy for performing its own work than when working individually.

N. Tesla believed that the law of resonance is the most general natural law and that all connections between phenomena are established exclusively through various kinds of simple and complex resonances. In particular, his ideas stimulated much research about resonances regarding biological bodies (see, for example, [123]).

In classical mechanics, the concept of resonances also has wide theoretical and engineering applications due to vibrational phenomena of a resonant synchronization of oscillatory processes, vibrational separation and structuring of multiphase systems, vibro-transportation of substances, vibro-transmission of energy within systems, *etc.* [124–127]. Results, presented in this chapter, give a basis for wider use of these phenomena in modeling biological phenomena.

Genetic molecules DNA and RNA exist on principles of quantum mechanics, but they encode structures of living macro-organisms, which are subjects of classical mechanics. By this reason, mathematics of genetic systems should be appropriate to quantum mechanics and classical mechanics simultaneously. Mathematics of resonances of oscillatory systems is appropriate for quantum and classical mechanics since such mathematics uses in both cases the same property of matrices to express resonances.

This chapter continues the theme of the important role of resonances in living organisms. A unified mathematical approach is developed for the analysis of a number of systems of genetic coding and genetic phenomena. This approach reveals hidden relationships and regularities in these systems and leads to a new class of bio-mathematical models using matrices of resonances. These results support the feasibility of understanding the genetic system and genetic phenomena based on mathematics of special systems of resonances. Energetic principles of biological organization are also under attention in this approach because frequency characteristics of vibro-systems are associated with energetics.

The proposed approach is correlated with the following situation in modern science. In the past century, science has discovered that the molecular basis of genetic coding (DNA structures, etc.) is identical in all species of organisms. A new understanding of life appeared: "*Life is a partnership between genes and mathematics*" [6]. All physiological systems of the body should be structurally coordinated with the genetic code for their reproductions in descendants. It is obvious that unifying mathematical approaches are needed for the simulation of the genetic unity of different structures of organisms. Every organism is endowed with the inherited ability to tune into resonances and to use resonances as carriers of information. Our speech and singing are examples of this because they use resonances of our voice apparatus, which is an oscillatory system with many degrees of freedom. According to the classics of structural linguistics (Roman Jakobson and others), our language did not come out of nowhere, but is a superstructure over the oldest language — the genetic language [13, 126, 127]. This is one of the reasons to investigate the genetic system, including genetic alphabets, from the standpoint of the mathematics of resonances, which is based on matrix analysis.

One can add that the informational genetic system possesses noise immunity properties, but the general theory of noise immunity encoding and transmitting of information is also based on matrix methods. One of many examples of using matrix mathematics is the technology of noise-immune transmission of information from Mars to Earth, which bears high-quality photographs of the surface of Mars, in conditions of strong distortions of carrier electromagnetic signals passing through millions of kilometers of noises and interference. The named mathematical applications are important reasons for attracting attention to the matrix analysis of structures of the genetic code, including the possibility of using genetic systems of resonances for the transmission of hereditary information.

The presented concept of resonance genetics brings together biology with physics and informatics since systems of resonances allow providing a reception and transmission of information (our inherited abilities of acoustic communication through resonance mechanisms of speech and singing are examples of this). The described study reveals that the molecular genetic ensembles are structurally related to known formalisms of the mathematical theory of noise immunity coding of information (functions of Rademacher and Walsh, Hadamard matrices, etc., as described in Chapter 2).

This chapter pays its main attention to matrices of vibro-systems since we analyze organisms as a set of heritable systems of resonances. Our results give a basis for wider use of vibro-phenomena, which are well known in classical mechanics for modeling biological phenomena. They can also pull together genetics and quantum mechanics, which is based on frequencies and resonant characteristics of objects of the quantum-mechanical world.

3.2. Oscillatory processes and matrix representations of resonances

The background of this study includes well-known genetic and physiological phenomena, such as Mendel's laws, molecular structures of the genetic code, and the psychophysical Weber-Fechner law. In this chapter, the main research method to study the mentioned phenomena is their mathematical modeling by means of classical mathematics of theory of oscillations, allowing analysis of the resonant characteristics of oscillatory systems with many degrees of freedom (see, for example, [128]). From a formal point of view, a living organism is an oscillatory system with a great number of degrees of freedom. We use matrices to study genetic phenomena. Our unified approach based on matrix representations of resonances allows us to reveal hidden relationships in different genetic and inherited physiological phenomena, and to discover new possibilities of convergence of biology with physics and informatics.

Any living organism is a great chorus of coordinated oscillatory processes (mechanical, electric, piezoelectric, biochemical, etc.), which relate to their genetic inheritance along chains of generations. In the ontogenetic development of an organism from embryo to adult, the number of oscillatory processes in the chorus is greatly increased while maintaining their mutual consistency not only at each stage of development, but also at different stages. Since ancient times, chrono-medicine believes that all diseases are the result of disturbances in the ordered set of these oscillatory processes. But how to study this inherited consistency of many oscillatory processes? To study oscillatory systems with many degrees of freedom, a matrix apparatus for modeling their resonant properties is traditionally used, which represents a system of differential equations in a concise and convenient form. Below, applications of this matrix apparatus are presented to study structural consistency of different parts of the genetic code system.

Matrices possess a wonderful property to express resonances, which sometimes is called their main quality [17; 129, p. 21 and 26]. Physical resonance phenomenon is familiar to everyone. The expression $y = A \times S$ models the transmission of a signal S via an acoustic system A, represented by a relevant matrix A. If an input signal is a resonant tone, then the output signal will repeat it with a precision up to a scale factor $y = \lambda \times S$ by analogy with a situation when a musical string sounds in unison with the neighboring vibrating string. In the case of a matrix A, its number of resonant tones S_i corresponds to its size. They are called its eigenvectors, and the scale factors λ_i with them are called its eigenvalues or, briefly, spectrum A. Frequencies $\omega_i = \lambda_i^{0.5}$ [128] are defined as natural frequencies of the system, and the corresponding eigenvectors are defined as its own forms of oscillations (or, simply, natural oscillations).

These free undamped oscillations occur in the system in the absence of the friction forces in it and in the absence of external excitation forces. Behavior of the system in conditions of free oscillations determines its behavior in many other conditions. In this context, one of the main tasks of the theory of oscillations is a determination of natural frequencies (mathematically, eigenvalues of corresponding matrices) and the natural forms of oscillations of bodies. To find all the eigenvalues λ_i (i.e., spectrum system A) and eigenvectors of the matrix A, which are defined by the matrix equation $A \times s = \lambda \times s$, the "characteristic equation" of the matrix A is analyzed: $\det(A - \lambda E) = 0$, where E is the identity matrix. The characteristic equation together with its eigenvalues and eigenvectors is fundamental in the theory of mechanical, electrical and other oscillations at macroscopic or microscopic levels. The theory of oscillations is described in many books with some terminological differences in them sometimes. In this chapter, its terminology is borrowed mainly from a detailed book [128], which additionally contains many examples of matrix analysis of oscillatory systems. Below we talk in detail about eigenvalues and eigenvectors of matrices, remembering their closed connection with the theory of resonances of oscillatory systems with many degrees of freedom.

Let us recall the essence of the eigenvalues and eigenvectors by means of the matrix A in Fig. 3.2.1, which acts on vectors [x, y]. In this case almost any vector is transformed into a new vector [x, y] × A by changing its direction. The exceptions are those vectors [x, y] which belong to two orthogonal dotted lines and are called "eigenvectors" of the matrix A; they conserve their direction under action of the matrix A, but their lengths are scaled with factors λ_i, which are called "eigenvalues" of the matrix A (each eigenvalue corresponds to its own direction of eigenvectors).

Let us illustrate this with an example of an oscillatory spring-mass system with two degrees of freedom and the parameters shown in Fig. 3.2.2 together with the corresponding (2 × 2)-matrix whose spectrum consists of two eigenvalues λ_1 and λ_2. Each of them is equal to the square of the

$$A = \begin{vmatrix} 4, & 0.6 \\ 0.6 & 2 \end{vmatrix}$$

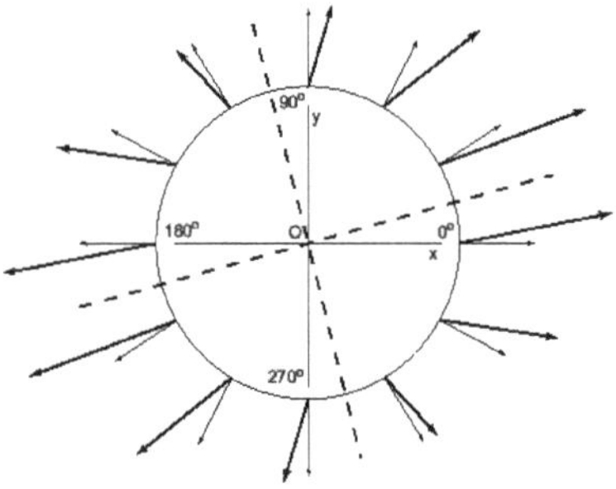

Fig. 3.2.1. Illustration of actions of the matrix A by vectors [x, y] (from [130]).

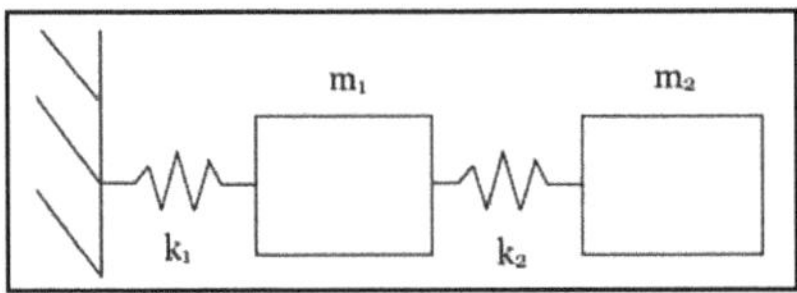

$$\begin{vmatrix} k_1 + k_2 - \lambda m_1 & -k_2 \\ -k_2 & k_2 - \lambda m_2 \end{vmatrix} \cdot \begin{vmatrix} x_1 \\ x_2 \end{vmatrix} = 0$$

Fig. 3.2.2. The example of an oscillatory spring-mass system with two degrees of freedom. The characteristic equation of the corresponding (2 × 2)-matrix is shown on the right. Symbols m_1 and m_2 denote masses, symbols k_1 and k_2 denote stiffness coefficients of springs.

oscillation frequency ω_1^2 and ω_2^2 at the resonant tone that corresponds to the eigenvector of the given eigenvalue.

Corresponding eigenvalues $\lambda_{1,2}$ in the case of this oscillatory system is: $\lambda_{1,2} = \omega_{1,2}^2 = \{(k_2m_1 + k_1m_2 + k_2m_2) \pm ((k_2m_1 + k_1m_2 + k_2m_2)^2 - 4k_1k_2 \times m_1 \times m_2)^{0.5}\}/2m_1m_2$. The real matrix of this system is symmetrical. Matrices, which are relevant to the various problems of the theory of oscillations, are usually symmetrical real matrices [128]. Such matrices have real eigenvalues and their eigenvectors are orthogonal.

The theory of resonances of oscillatory systems with many degrees of freedom expresses the resonant frequencies through the eigenvalues of the matrices representing these systems. With the tensor (Kronecker) product of two such matrices leading to a system with an increased number of degrees of freedom and resonances (Fig. 1.2.2), the eigenvalues of the new matrix are the pairwise product of the eigenvalues of the original (or parental) matrices: if the original matrices V and W have the eigenvalues λ_i and μ_j respectively, then in their tensor product $Q = V \otimes W$ all eigenvalues are equal to $\lambda_i \times \mu_j$ (figuratively speaking, λ_i and μ_j are inherited in this tensor way). This algorithmic process of tensor generation of oscillatory systems with increasing degrees of freedom and inherited matrix eigenvalues can be continued indefinitely, culminating in oscillatory systems with an arbitrarily large number of degrees of freedom.

This tensor "inheritance" of the eigenvalues of the matrices of oscillatory systems turns out to be an effective model for the inheritance of traits in Mendelian genetics, reproducing the structure of its Punnett squares. This indicates that the genetic system is built on special tensor systems of resonances. This algebraic model is described in the next subsection.

3.3. The analogy between Punnett squares and tables of the tensor inheritance of eigenvalues of matrices

Features of the tensor inheritance of eigenvalues of the original matrices (or "parental" matrices) in the result of their tensor product can be conveniently represented in the form of "tables of inheritance". The top row of Fig. 3.3.1

shows the example of two simplest cases, conventionally referred to as monohybrid and dihybrid cases of a tensor hybridization of vibro-systems. In the first case, the tensor product of two (2 × 2)-matrices V and W, which have the same set of eigenvalues A and a, gives the (4 × 4)-matrix Q = V ⊗ W with its 4 eigenvalues A × A, A × a, A × a, a × a. In the second case, the tensor product of (4 × 4)-matrices, having the same set of eigenvalues AB, Ab, aB, ab, gives (16 × 16)-matrix with 16 eigenvalues, represented in the tabular form.

One can see that the internal content of the table of inheritance for the dihybrid case (Fig. 3.3.1, top) is equal to [AA, Aa; Aa, aa] ⊗ [BB, Bb; Bb, bb]; in other words, the spectrum of the dihybrid vibro-system is equal to the tensor product of spectra of two monohybrid vibro-systems. Similar tables of inheritance for n-hybrid cases (n = 3, 4, …) of the tensor hybridization of vibro-systems can be constructed by analogy.

		spectrum of W	
		A	a
spectrum of V	A	AA	Aa
	a	aA	aa

		maternal spectrum			
		AB	Ab	aB	ab
pat. spectrum	AB	AABB	AABb	AaBB	AaBb
	Ab	AABb	AAbb	AaBb	Aabb
	aB	AaBB	AaBb	aaBB	aaBb
	ab	AaBb	Aabb	aaBb	aabb

		maternal gametes	
		A	a
paternal gametes	A	AA	Aa
	a	aA	aa

		maternal gametes			
		AB	Ab	aB	ab
pat. gam.	AB	AABB	AABb	AaBB	AaBb
	Ab	AABb	AAbb	AaBb	Aabb
	aB	AaBB	AaBb	aaBB	aaBb
	ab	AaBb	Aabb	aaBb	aabb

Fig. 3.3.1. Comparison of Punnett squares and tables of inheritance of eigenvalues of matrices under the tensor product. Top row: examples of tables of inheritance of eigenvalues under the tensor product in cases of (2 × 2)-matrices (left) and (4 × 4)-matrices (cases of monohybrid and dihybrid hybridizations). Bottom row: examples of Punnett squares for monohybrid (left) and dihybrid crosses of organisms under the laws of Mendel. Abbreviations «pat. spectrum» and «pat. gam.» mean «paternal spectrum» and «paternal gametes».

It is important that these tables of the tensor inheritance for spectra of vibro-systems are identical to Punnett squares for poly-hybrid crosses of organisms (Fig. 3.3.1). They represent Mendel's laws of inheritance of traits under poly-hybrid crosses. But in Punnett squares, instead of eigenvalues of matrices and their combinations, there exist similar combinations of dominant and recessive alleles of genes from parent reproductive cells — gametes.

This formal analogy — between Punnett squares of combinations of alleles and tables of tensor inheritance for eigenvalues of matrices of vibro-systems — generates the following idea: alleles of genes and their combinations can be interpreted as eigenvalues of $(2^n \times 2^n)$-matrices from tensor families of matrices of oscillatory systems. For genetic systems, this model approach focuses attention on the possible importance of a particular class of mutually related resonances from tensor families of matrices, which play the role of biological "matrix archetypes".

In this modeling approach, each allele of a gene, which has a polyatomic structure, is characterized by a single number: an eigenvalue of a matrix of an oscillatory system with a corresponding number of degrees of freedom. It resembles the phenomenon, known in vibration mechanics since the time of Christiaan Huygens, of self-synchronization of a plurality of pendulums mounted on a common movable platform: the self-synchronization provides that all the pendulums begin to oscillate with a single common frequency, although initially each of them could have its own natural frequency of oscillation.

This approach is also associated with dyadic groups of binary numbers and matrices of dyadic shifts, known in the theory of digital signal processing [26, 27, 93]. Taking into account binary-oppositional attributes of homozygosity and heterozygosity for alleles of genes, one can represent combinations of individual alleles by means of symbols "0" and "1". This generates numerical representations of Punnett squares as matrices of dyadic shifts [131]. But it is known that, if a system shows its connection with dyadic shifts, it indicates a connection of its structural organization (in our case, it is the genetic system) with modulo-2 addition, which is the basic logical operation in computers.

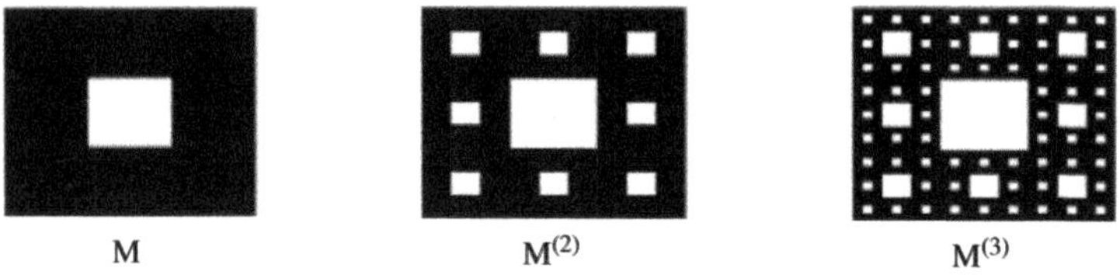

Fig. 3.3.2. The example of a fractal carpet of Sierpinski produced in the tensor family of matrices $M^{(n)} = [1, 1, 1; 1, 0, 1; 1, 1, 1]^{(n)}$, where (n) means tensor power. Black and white elements of mosaics correspond to elements 1 and 0, respectively.

One can add that, starting from 1906, Punnett lattices were traditionally considered simply as tables, and only after more than 100 years these lattices began to be considered as matrices [131]. It is reminiscent of the historical situation with the Heisenberg tables, from which quantum mechanics in the form of Heisenberg matrix mechanics began: they, too, were not initially perceived as matrices, and it took the work of M. Born and P. Jordan in order to see that the Heisenberg tables could and should be interpreted like matrices.

One can else note that the tensor product has the property of inheritance of mosaic structure of the original matrix under its tensor exponentiation. This property connects this operation with fractals [18]. Fig. 3.3.2 shows an example of the formation of fractal patterns, which depend on the mosaic of the original matrix.

3.4. Genetic alphabets and tensor systems of resonances

Considering that resonances can be carriers of information (in particular, human speech and singing, built on resonance mechanisms, confirm this), let us turn to the hypothesis that genetic alphabets are also built based on systems of resonances or, more precisely, on systems of eigenvalues and eigenvectors of tensor families of $[2^n \times 2^n]$-matrices [117–119]. Below we represent obtained results in favor of this hypothesis.

As it is known, any square matrix, representing an oscillatory system with many degrees of freedom, can be transformed into an appropriate diagonal matrix that shares the same fundamental properties of the underlying matrix. Diagonalizing such matrix is equivalent to finding the matrix's eigenvalues, which turn out to be precisely the entries of the diagonalized matrix. Similarly, the eigenvectors make up the new set of axes corresponding to the diagonal matrix (see details in [17, Chapter 3]; https://mathworld.wolfram.com/MatrixDiagonalization.html). By definition, a diagonal matrix is a matrix in which the entries outside the main diagonal are all zero. The tensor product of diagonal matrices always generates a diagonal matrix.

Let us assume that amino nucleobases C and A of DNA and also keto nucleobases T and G are eigenvalues of corresponding (2 × 2)-matrices and so they can be located on diagonals of the corresponding diagonal matrices [C, 0; 0, A] and [T, 0; 0, G] (Fig. 3.4.1). Briefly, diagonal matrices are conditionally written in the form of vectors, whose coordinates are diagonal members of the diagonal matrices and which are marked by the index d, for example, the matrix [C, 0; 0, A] is briefly written as $[C, A]_d$. Tensor products of the two

$$\begin{vmatrix} C, 0 \\ 0, A \end{vmatrix} = [C, A]_d \, ; \quad \begin{vmatrix} T, 0 \\ 0, G \end{vmatrix} = [T, G]_d$$

$[C,A]_d \otimes [C,A]_d \otimes [C,A]_d = [CCC, CCA, CAC, CAA, ACC, ACA, AAC, AAA]_d$

$[C,A]_d \otimes [C,A]_d \otimes [T,G]_d = [CCT, CCG, CAT, CAG, ACT, ACG, AAT, AAG]_d$

$[C,A]_d \otimes [T,G]_d \otimes [C,A]_d = [CTC, CTA, CGC, CGA, ATC, ATA, AGC, AGA]_d$

$[C,A]_d \otimes [T,G]_d \otimes [T,G]_d = [CTT, CTG, CGT, CGG, ATT, ATG, AGT, AGG]_d$

$[T,G]_d \otimes [C,A]_d \otimes [C,A]_d = [TCC, TCA, TAC, TAA, GCC, GCA, GAC, GAA]_d$

$[T,G]_d \otimes [C,A]_d \otimes [T,G]_d = [TCT, TCG, TAT, TAG, GCT, GCG, GAT, GAG]_d$

$[T,G]_d \otimes [T,G]_d \otimes [C,A]_d = [TTC, TTA, TGC, TGA, GTC, GTA, GGC, GGA]_d$

$[T,G]_d \otimes [T,G]_d \otimes [T,G]_d = [TTT, TTG, TGT, TGG, GTT, GTG, GGT, GGG]_d$

Fig. 3.4.1. Representations of 4 nucleobases and 64 triplets in the form of eigenvalues of corresponding diagonal matrices. Above: the original diagonal (2 × 2)-matrices with pairs of nucleobases C and A, and T and G. Bottom: the alphabet of 64 triplets in the form of 8 sets of 8 diagonal members belonging to 8 diagonal (8 × 8)-matrices generated by the shown tensor product between the diagonal (2 × 2)-matrices.

diagonal matrices $[C, A]_d$ and $[T, G]_d$ in all possible combinations in three represent the entire alphabet of 64 triplets in the ordered form of diagonal members of 8 diagonal (8×8)-matrices (the octet of diagonal members in Fig. 3.4.1 below). Here each triplet is an eigenvalue of one of eight (8×8)-matrices and corresponds to its own eigenvector.

The tensor products of these two diagonal matrices $[C, A]_d$ and $[T, G]_d$ in different combinations of three represent the entire set of 64 triplets in the form of 8 octets, each of which lies on the diagonal of its appropriate diagonal matrix (Fig. 3.4.1, bottom). In this model, triplets with different sequences of identical letters (for example, CCA and CAC) belong to different eigenvectors, which correspond to their different meanings in genetic coding despite their identical contents of nucleobases.

The 8 octets, containing 64 eigenvalues at the bottom of Fig. 3.4.1, precisely coincide with 8 rows in the genetic matrix of 64 triplets in Chapter 1 (Fig. 1.2.1), which was obtained through binary-oppositional traits of 4 nucleobases without any mention about eigenvalues, resonances, etc. It means that the set of 64 genetic triplets can be interpreted in an alternative manner as a set of 64 eigenvalues of matrices, which are obtained by the shown tensor product and which relate to the theory of resonances of oscillatory systems with 8 degrees of freedom. But Fig. 1.2.1 demonstrated very unexpected features of the degeneracy of the genetic code: pairs of octets of genetic triplets in its lines 1–2, 3–4, 5–6, 7–8 encode the same sets of amino acids and stop codons in the most ancient and symmetrical dialect of the genetic code. Correspondingly, the pairs of octets of eigenvalues in rows 1–2, 3–4, 5–6, 7–8 in Fig. 3.4.1 encode the same sets of amino acids and stop codons. It gives some pieces of evidence that the degeneracy of the genetic code is based on multi-resonance systems.

In Chapter 1, it was described that the alphabet of 64 triplets is divided by nature into two equal subsets based on strong and weak roots [22]. Members of these subsets are located in the genetic matrix $[C, A; T, G]^{(3)}$ very symmetrically that define symmetric black-and-white mosaics in Fig. 1.4.2, where the mosaic of each row has a meander-like character and correspond to one of the Rademacher functions. Correspondingly, each of the 8 octets of eigenvalues (Fig. 3.4.1), representing triplets with strong and weak roots, is divided into two equal subsets of black and white eigenvalues, representing these two kinds of genetic triplets (Fig. 3.4.2).

In Chapter 2, the additional binary opposition connecting with the unique status of the nucleobase T in DNA (and correspondingly the nucleobase U in RNA) in the 4-letter alphabet of nucleobases was described. Based on this binary opposition, the mentioned T-algorithm arises to transform the genetic mosaic matrices $[C, A; T, G]^{(n)}$, whose mosaics correspond to Rademacher functions, into new mosaic matrices, whose mosaics correspond to Walsh functions (Figs. 2.2.1–2.2.3). By analogy, one can consider the eigenvalue, which represents the nucleobase T, as having unique status in the set of eigenvalues; in this case, similar algorithmic transformation of black-and-white octets of eigenvalues (Fig. 3.4.2) into the same octets is possible but with new black-and-white mosaics corresponding to Walsh functions. Briefly speaking, the structural features of DNA alphabets testify that the complete orthogonal system of Walsh functions can be used in modeling of biological bodies on the basis of multi-resonance systems in accordance with the concept of multi-resonance genetics.

[CCC, CCA, CAC, CAA, ACC, ACA, AAC, AAA]d	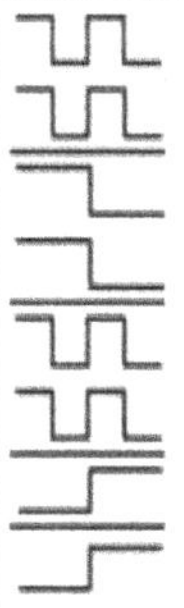	[+1,+1, -1,-1,+1,+1, -1,-1]d
[CCT, CCG, CAT, CAG, ACT, ACG, AAT, AAG]d		[+1,+1, -1,-1,+1,+1, -1,-1]d
[CTC, CTA, CGC, CGA, ATC, ATA, AGC, AGA]d		[+1,+1,+1,+1,-1, -1, -1,-1]d
[CTT, CTG, CGT, CGG, ATT, ATG, AGT, AGG]d		[+1,+1,+1,+1,-1, -1, -1,-1]d
[TCC, TCA, TAC, TAA, GCC, GCA, GAC, GAA]d		[+1,+1, -1,-1,+1,+1, -1,-1]d
[TCT, TCG, TAT, TAG, GCT, GCG, GAT, GAG]d		[+1,+1, -1,-1,+1,+1, -1,-1]d
[TTC, TTA, TGC, TGA, GTC, GTA, GGC, GGA]d		[-1,-1,-1,-1, +1,+1,+1,+1]d
[TTT, TTG, TGT, TGG, GTT, GTG, GGT, GGG]d		[-1, -1,-1,-1,+1,+1,+1,+1]d

Fig. 3.4.2. The 8 octets of black-and-white eigenvalues representing — in a form of entries of diagonal matrices — genetic triplets with strong and weak roots by analogy with Fig. 1.4.2. Middle: meander character of sequences of black and white eigenvalues in each octet is shown. Right: the representation of each octet of eigenvalues in the form of a corresponding Rademacher function consisting of elements of "+1" and "−1".

3.5. The Mendel's law of segregation, segregations inside genetic alphabets, and resonances

In Chapter 1, the binary-oppositional traits ("purine-pyrimidine", "amino-keto", "strong hydrogen bonds-weak hydrogen bonds") of nucleobases A, C, G, T were noted. Each of these traits can be interpreted as connected with its own resonance characteristics. For example, it is obvious that purines may have resonances that differ from the resonances of pyrimidines due to differences in the structure of the molecules. In this light, each of the mentioned pairs of binary-oppositional traits can be treated as a pair of oppositional resonance characteristics or eigenvalues of the corresponding matrix of an oscillatory system [36, 118].

Genetics arose from Mendel's discovery of mathematical regularities in the phenomena of inheritance of traits in poly-hybrid crosses of organisms. Mendelian genetics explained these phenomenological regularities, which are observed at the level of the whole organism, on the basis of the idea of dominant and recessive alleles of genes from parent reproductive cells — gametes. By analogy with the Mendelian theory, below a notion of dominant and recessive resonances of genetic molecules is described to explain some phenomenological regularities of genetic alphabets, which are based in DNA on nucleobases A, C, G, and T [113, 132]. In particular, the well-known phenomenon of segregation of the set of 64 triplets into two equal subsets containing 32 triplets with strong roots and 32 triplets with weak roots is analyzed (see details in Chapter 1, Figs. 1.4.1–1.4.3).

For a simpler explanation, it is comfortable to return to the three sub-alphabets of the alphabet of 4 nucleobases (Table 1.2.1 in Chapter 1) but with the following clarifying change: each of the considered traits of nucleobases is represented as an eigenvalue of the corresponding resonance. In Table 3.5.1, these eigenvalues of traits are denoted by the following symbols (traditionally used in genetics for nucleobases): purine — R, pyrimidine — Y, amino — M, keto — K, strong hydrogen bonds — S, weak hydrogen bonds — W.

We postulate that the resonances and the corresponding oscillatory energies are key factors for encoding. One can see from Table 3.5.1 that for

Table 3.5.1. The division of the four-letter alphabet of DNA (RNA) nucleobases into three binary sub-alphabets in accordance with three binary-oppositional traits, which are interpreted as eigenvalues of molecular resonances with their symbols R, Y, M, K, S and W.

TRAITS (or INDICATORS)	G	A	C	T(U)
1) purine (R), pyrimidine (Y)	R	R	Y	Y
2) amino (M), keto (K)	K	M	M	K
3) strong hydrogen bonds (S), weak hydrogen bonds (W)	S	W	S	W

an unambiguous definition of each of the nucleobases by these eigenvalues it is enough to take into account two of these three pairs of eigenvalues. We will characterize each nucleobase in terms of the pairs of eigenvalues "M or K" and "Y or R". In quantum mechanics, which is appropriate for the genetic molecular level, eigenvalues of two resonances of an oscillator with two degrees of freedom lie on the main diagonal of the diagonal (2×2)-matrix, which characterizes the oscillator. Let us use such matrix representation. For example, as one can see in Table 3.5.1, adenine A is the single nucleobase with the pair of eigenvalues M and R and in such bi-resonance representation one can write adenine A in a matrix form A = [M, 0; 0, R] (for brevity, we will denote this matrix by the symbol MR). Accordingly, for other nucleobases their bi-resonance representations are the following: cytosine C = [M, 0; 0, Y] (or MY), guanine G = [K, 0; 0, R] (or KR), thymine T = [K, 0; 0, Y] (or KY).

Corresponding state spaces of duplets and triplets are constructed by means of the tensor product of state spaces of the nucleobases, which are components of duplets and triplets. For example, the state space of the duplet CA is characterized by the diagonal matrix (4×4): [M, 0; 0 Y] ⊗ [M, 0; 0, R] = [MM, 0, 0, 0; 0, MR, 0, 0; 0, 0, YM, 0; 0, 0, 0, YR] and state space of the duplet AC, which has the inverse order of the same letters, is characterized by another diagonal matrix (4×4): [M, 0; 0, R] ⊗ [M, 0; 0, Y] = [MM, 0, 0, 0; 0, MY, 0, 0; 0, 0, RM, 0; 0, 0, 0, RY]. When recording sequences of nucleobases in their bi-resonance representations, we will separate two-letter symbols of adjacent nucleobases by means of a vertical dash: for example, the duplet CA will be denoted as MY|MR.

<table>
<tr><td>CC
MY|MY</td><td>CA
MY|MR</td><td>AC
MR|MY</td><td>AA
MR|MR</td></tr>
<tr><td>CT
MY|KY</td><td>CG
MY|KR</td><td>AT
MR|KY</td><td>AG
MR|KR</td></tr>
<tr><td>TC
KY|MY</td><td>TA
KY|MR</td><td>GC
KR|MY</td><td>GA
KR|MR</td></tr>
<tr><td>TT
KY|KY</td><td>TG
KY|KR</td><td>GT
KR|KY</td><td>GG
KR|KR</td></tr>
</table>

<table>
<tr><td></td><td></td><td colspan="2">Maternal gametes</td></tr>
<tr><td></td><td></td><td>H</td><td>h</td></tr>
<tr><td rowspan="2">Paternal gametes</td><td>h</td><td>hH</td><td>hh</td></tr>
<tr><td>H</td><td>HH</td><td>Hh</td></tr>
</table>

Fig. 3.5.1. The segregation of the set of 16 duplets into subsets of 8 strong duplets (black) and 8 weak duplets (white); under each duplet, a bi-resonance representation of its two nucleobases are shown. Right: the Punnet square for monohybrid crosses of organisms; combinations of alleles H and h, which contain at least one dominant allele H, are marked in black.

Fig. 3.5.1 shows the well-known segregation of the alphabet of 16 duplets into two sub-alphabets of 8 strong duplets and 8 weak duplets, which correspond to the degeneracy of the genetic code and the segregation of the alphabet of 64 triplets into two sub-alphabets of 32 triplets with strong roots and 32 triplets with weak roots (as described in Chapter 1, Figs. 1.4.1 and 1.4.2). Each of the two families of 4 duplets with their first letter from the complementary pair C and G contains 3 strong duplets and 1 weak duplet (Fig. 3.5.1, left). In contrast, each of the two other families of 4 duplets with their first letter from another complementary pair A and T contains 1 strong duplet and 3 weak duplets. But the same ratio 3:1 is realized in Mendel's law of segregation of traits in the case of monohybrid crosses of organisms, whose gametes contain one dominant allele of a corresponding gene and one recessive allele of the same gene (here we mean the situation of complete dominance, where the effect of the dominant allele completely masks the effect of the recessive allele). This case of crosses of organisms is traditionally shown by means of the Punnet square with the same mosaic ratio 3:1 (Fig. 3.5.1).

Based on this analogy — between segregations of traits in the case of molecular duplets and in the case of whole organisms under the law of Mendel — the following statement is possible [132, 133]:

The principle of dominant and recessive factors plays its key genetic role in the segregation phenomena not only at the level of whole organisms but also at the level of ensembles of genetic molecules; the mentioned resonances of genetic molecules play roles of dominant and recessive factors in laws of segregation of traits in molecular-genetic ensembles.

This statement allows the explanation of some features of segregations inside genetic alphabets. In addition, it allows the thinking that Mendel's laws, which are true at supra-chromosome levels, did not come out of nowhere, but have precursors in the form of phenomena of segregation at sub-chromosome levels. In this case, some analogies of phenomena of segregation in populations of whole organisms (for example, the division into male and female organisms with their sexual interrelations) can exist in the world of biological molecules, including proteins [132–133]. It could lead to new points of view about the existence of appropriate oppositional groups of proteins and about similar interrelations among them, etc. It seems to be logical to think on the whole that regularities of the world of macro-organisms did not come out of nowhere but reflect and reproduce regularities of the bio-molecular world. We show now an example of positive applications of this statement.

In Mendelian genetics, those dominant and recessive factors, which are responsible for segregation at the level of organisms, are called "alleles". In the presented approach — for terminological unity — those dominant and recessive factors, which are responsible for segregation at the deeper level of molecular-genetic alphabets, are called "pro-alleles". We suppose that the mentioned molecular resonances of nucleobases play the role of such «pro-alleles». Let us show how the idea of dominant and recessive resonances of nucleobases (Table 3.5.1) explains the segregation of strong and weak duplets in the mentioned four families of duplets, presented in Fig. 3.5.1. One can see in Fig. 3.5.1 that — in each family of 4 duplets in tabular quadrants — the second letter in each duplet defines a strong or weak character of the duplet.

Initially one can look at two families of duplets with the first letter C or G in them (they are called the C-family and the G-family of duplets). Each of these families contains 3 strong duplets (CC, CT, CG and GC, GT, GG correspondingly) and 1 weak duplet (CA and GA) (Fig. 3.5.1, left). In both mentioned duplets CA and GA, their second letter — adenine A — is the carrier of two resonance pro-alleles M and R. By analogy with combinations of dominant and recessive alleles in the case of monohybrid crosses of organisms in Mendelian genetics (Fig. 3.5.1, right), one should conclude that — in the C-family and G-family of duplets — the resonances M and R are recessive pro-alleles that belong to the second nucleobases of all the duplets.

Correspondingly, two oppositional resonances, presented by eigenvalues K and Y, play the role of dominant pro-alleles in the second nucleobases of all the duplets. By analogy with the Mendelian scheme of complete dominance, one can assume for these families that any pair of resonance pro-alleles, which contains at least one of the dominant pro-alleles K and Y, defines the dominant character of the second nucleobase of a duplet and, by this reason, the dominant character of the duplet. In each of the C-family and G-family, 3 mentioned duplets contain these dominant resonances, presented by eigenvalues K and Y, in their second nucleobases (CC, CT, CG and GC, GT, GG) and only 1 duplet does not (CA and GA, which are weak duplets).

Now let us turn to the other two families of duplets with the first letter A or T in them (the A-family and the T-family of duplets). Each of these families has the opposite ratio of quantities of strong and weak duplets (Fig. 3.5.1): 3 weak duplets (AA, AT, AG and TA, TT, TG correspondingly) and 1 strong duplet (AC and TC). In both named duplets AC and TC, their second nucleobase — cytosine C — is the carrier of two pro-alleles M and Y, which should be interpreted as recessive pro-alleles. Then two opposite resonances, presented by eigenvalues K and R, play the role of dominant pro-alleles in the second nucleobases of all duplets of these families. By analogy with the Mendelian scheme of complete dominance, one can assume again that any pair of resonance pro-alleles, which contains at least one of the dominant pro-alleles K and R, defines the dominant character of the appropriate second nucleobase of a duplet. In each of the A-family and T-family, 3 named duplets contain these dominant resonances K and R in their second nucleobases (AA, AT, AG and TA, TT, TG correspondingly) and only 1 duplet does not (AC and TC).

It should be especially noted that resonances presented by eigenvalues Y and R correspondingly play the oppositional roles of dominant and recessive pro-alleles in the C-family and G-family of duplets, but in the A-family and T-family, they play the inverse roles of recessive and dominant pro-alleles. In Mendelian genetics it is well known that dominant and recessive characteristics are not absolute features, but rather are relative: the same trait is dominant or recessive depending on the biological environment of its manifestation. For example, a trait of baldness is dominant in men and recessive in women. Therefore, to explain the phenomenon of the reverse segregation concerning strong and weak duplets in these families, it is sufficient to take by analogy the following. The C-family and G-family of duplets are conditionally "male" families, in which the resonance Y of the second letter of each duplet is one of the dominant pro-alleles and the resonance R is one of the recessive pro-alleles for the considered phenomenon. In contrast, the A-family and T-family of duplets are conditionally "female" families, in which the same resonances Y and R are correspondingly the recessive and dominant pro-alleles of the second letter of each duplet. In other words, the first letter of each duplet defines what kind of resonances Y or R in the second nucleobase of the duplet will be. This modeling approach has shown that members of the molecular alphabet of 16 duplets are resonance hybrid entities, which are similar to hybrids in Mendelian genetics of organisms.

Modern science explains Mendel's laws on the basis of division of pairs of homological chromosomes. Mendel's laws do not spread to levels below the level of chromosomes. Correspondingly, the rules of segregation of the molecular alphabet of 16 duplets, which are described in this chapter, have an independent character in relation to Mendel's laws since they cannot be deduced from them [132–133].

The binary-oppositional structure of resonance characteristics of members of the genetic alphabet A, C, G, T(U) (Table 3.5.1) can be connected with binary numbers for using logic of their dyadic groups in operating genetic binary computers on the basis of resonances in organisms. Correspondingly, each of the pairs of binary-oppositional resonances "R or

Y", "M or K", "S or W" of this alphabet (Table 3.5.1) can correspond to its own binary units: Y = 0 and R = 1; M = 0 and K = 1; S = 0 and W = 1.

The theory of discrete signal processing uses so-called ($2^n \times 2^n$) matrices of dyadic shifts, in which each of the three dyadic groups of binary n-bit numbers are employed for binary numerations of its rows, columns and cells [26]. Fig. 3.5.2 shows the simplest examples of such ($2^n \times 2^n$) matrices, where $n = 1, 2$, for two dyadic groups: 0, 1 and 00, 01, 10, 11. Numeration of any matrix cell is given by modulo-2 addition of numerations of its row and column. Let us use these classical numeric matrices of dyadic shifts to represent the specificity of genetic alphabets, members of which are carriers of the binary-oppositional resonance characteristics.

For matrices of genetic alphabets in Fig. 3.5.2, one can denote binary numerations of columns by means of the first pair Y = 0 and R = 1 and binary numerations of rows by means of the second pair M = 0 and K = 1. In this case, we get a coordinate grid, in which each cell contains those nucleobases A, C, G or T, which are carriers of resonances depicted in symbolic numerations of the row and column of the cell. For example, in these resonant coordinates, the intersection of resonances M and R from numerations of rows and columns always defines in appropriate cells adenine A, which is the carrier of this pair of resonances M and R (Table 3.5.1). In addition, one can see from the (2×2)-matrix in Fig. 3.5.2 that in cells, each of the nucleobases has its own binary representation: C = G = 0, A = T = 1. From this standpoint, each of the duplets in its cell is represented by a corresponding binary combination: for example, the duplet AG is read as binary number 10, which coincides with the binary numeration of its cell in (4×4)-matrix in Fig. 3.5.2. One can check that such binary readings of all duplets in their cells in Fig. 3.5.2 coincide with binary numerations of their cells in classical matrices of dyadic shifts.

In computer technology, binary elements 0 and 1 are physically realized through using two types of signal amplitudes (e.g., oppositional in polarity) or two kinds of laser beams, etc., but in the considered genetic case, the binary opposition of the resonance characteristics gives an opportunity to consider genetic systems as binary computers on resonances.

	Y(0)	R(1)
M(0)	MY C (0)	MR A (1)
K(1)	KY T (1)	KR G (0)

	YY (00)	YR (01)	RY (10)	RR (11)
MM (00)	MY|MY CC (00)	MY|MR CA (01)	MR|MY AC (10)	MR|MR AA (11)
MK (01)	MY|KY CT (01)	MY|KR CG (00)	MR|KY AT (11)	MR|KR AG (10)
KM (10)	KY|MY TC (10)	KY|MR TA (11)	KR|MY GC (00)	KR|MR GA (01)
KK (11)	KY|KY TT (11)	KY|KR TG (10)	KR|KY GT (01)	KR|KR GG (00)

Fig. 3.5.2. Matrices of dyadic-group representations of resonance traits for genetic alphabets of 4 nucleobases and 16 duplets. Numerations of rows and columns use binary-oppositional resonances (Table 3.5.1) with the conformity M = 0, K = 1, Y = 0, R = 1. Strong duplets are marked in black.

3.6. Octets of eigenvalues and the T-algorithm

Let us return to the 8 black-and-white octets of eigenvalues, representing 64 triplets with strong and weak roots in Fig. 3.4.2, for additional study of structural features of the genetic coding system. In Chapter 2, the T-algorithm was described, which transforms black-and-white mosaics of dispositions of entries with strong and weak roots in the genetic matrices (Figs. 2.2.1–2.2.3). Applying the T-algorithm to the mosaic octets in Fig. 3.4.2, you get that each of the octets receives a mosaic, which corresponds to one of the Walsh functions shown in Fig. 3.6.1, right. The whole set of these 8 functions represents the complete set of orthogonal Walsh functions, which has many applications in spectral analysis, error-correcting codes, etc.

The described set of results supports the following statement: Alphabets of the genetic code are alphabets of eigenvalues and eigenvectors of matrices of oscillatory systems (figuratively speaking, the genetic code is the code of systems of resonances); accordingly, one can think that genetic texts are written in the language of the resonances.

It is not for nothing that a genetically inherited organism is a chorus of coordinated oscillatory processes.

[CCC, CCA, CAC, CAA, ACC, ACA, AAC, AAA]d	[+1, +1, −1, −1, +1, +1, −1, −1]d
[CCT, CCG, CAT, CAG, ACT, ACG, AAT, AAG]d	[−1, +1, +1, −1, −1, +1, +1, −1]d
[CTC, CTA, CGC, CGA, ATC, ATA, AGC, AGA]d	[+1, +1, +1, +1, −1, −1, −1, −1]d
[CTT, CTG, CGT, CGG, ATT, ATG, AGT, AGG]d	[−1, +1, −1, +1, +1, −1, +1, −1]d
[TCC, TCA, TAC, TAA, GCC, GCA, GAC, GAA]d	[−1, −1, +1, +1, +1, +1, −1, −1]d
[TCT, TCG, TAT, TAG, GCT, GCG, GAT, GAG]d	[+1, −1, −1, +1, −1, +1, +1, −1]d
[TTC, TTA, TGC, TGA, GTC, GTA, GGC, GGA]d	[+1, +1, +1, +1, +1, +1, +1, +1]d
[TTT, TTG, TGT, TGG, GTT, GTG, GGT, GGG]d	[−1, +1, −1, +1, −1, +1, −1, +1]d

Fig. 3.6.1. The numeric representation of the 8 octets from Fig. 3.4.2 coincides with a complete orthogonal system of Walsh functions for the eight-dimensional case, if the unique status of thymine T is taken into account within digitizing of triplets by the T-algorithm. Black (white) color corresponds to triplets represented by "+1" ("−1").

3.7. Resonances and the basic psychophysical law of Weber-Fechner

The living body is a single whole. Our resonant approach models inherited phenomena from different areas of physiology in the unified academic language of the theory of resonant oscillations of vibro-systems with many degrees of freedom. This approach uses the fact that the structural organization of many inherited phenomena is associated with (2 × 2)-matrices of vibro-systems with two degrees of freedom. We will show this using the example of the basic psychophysical law of Weber-Fechner (and then we will move on to morphogenesis).

It is profitable for an organism to have the same typical algorithms at different levels of its functioning for a mutual optimal coordination of its parts. By this reason we study possibilities to simulate different innate phenomena on the general basis of formalisms for the theory of resonances of vibro-systems with many degrees of freedom. Let us turn to the basic psychophysical Weber-Fechner law, to which different types of inherited sensory perception are subordinated: sight, hearing, smell, touch, taste, etc.

The Weber-Fechner law declares the following: the intensity of the perception is proportional to the logarithm of stimulus intensity; it is expressed by the equation-

$$p = k \times \ln(x/x_0) = k \times \{\ln(x) - \ln(x_0)\}, \qquad (3.1)$$

where p — the intensity of perception, x — stimulus intensity, x_0 — threshold stimulus, ln — natural logarithm, and k — a weight factor. Because of this law, the power of sound in technology is measured on a logarithmic scale in decibels.

The Weber-Fechner law is based on the natural logarithm (3.1). Historically the natural logarithm was formerly termed the hyperbolic logarithm, as it corresponds to the area under a hyperbola [134]. The logarithmic law (3.1) is simply modeled on the basis of natural resonance frequencies of a particular class of oscillatory systems with 2 degrees of freedom. As is known, the natural logarithm can be defined for any positive real number "a" as the area under the hyperbola $y = 1/x$ from 1 to a (Fig. 3.7.1, left). It means that two points of the hyperbola with their coordinates $(x, 1/x)$ and $(x_0, 1/x_0)$, where $x > 1$ and $x_0 > 1$, define values of natural logarithms $\ln(x)$ and $\ln(x_0)$. Subtraction $\ln(x) - \ln(x_0) = \ln(x/x_0)$ expresses the intensity of perception p in the expression (3.1) of the Weber–Fechner law (Fig. 3.7.1, right). A change of a stimulus intensity x_1 into a new stimulus intensity x_2 corresponds to a hyperbolic rotation, which transforms points of this hyperbola into each other and defines an appropriate change of

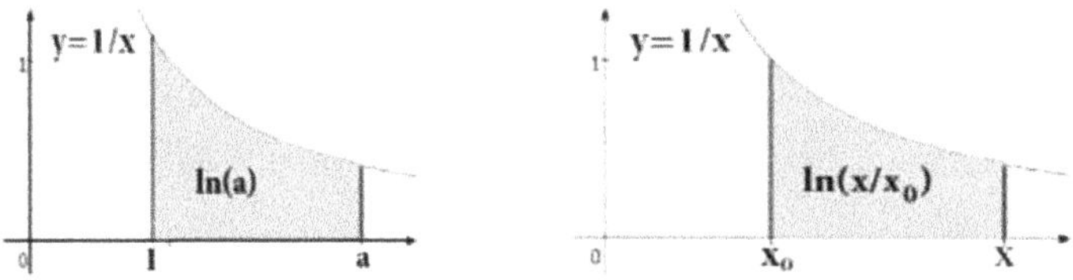

Fig. 3.7.1. Natural logarithm as a square under hyperbola $y = 1/x$. Left: $\ln(a)$ is equal to the square under the hyperbola from 1 to a. Right: $\ln(x/x_0)$ is equal to the square under hyperbola from x_0 to x.

intensity of perception: $\Delta p = k \times \ln(x_2/x_1)$. But the same points $(x, 1/x)$ and $(x_0, 1/x_0)$ are defined by diagonal matrices $[x, 0; 0, 1/x]$ and $[x_0, 0; 0, 1/x_0]$ of two vibro-systems with eigenvalues, which are related reciprocally (correspondingly, natural resonance frequencies in every vibro-system are also reciprocal to each other).

One can suppose that the innate Weber-Fechner law is the law especially for the nervous system. But it is not so since its meaning is much wider because it holds true in many kinds of lower organisms without a nervous system: "*This law is applicable to chemo-tropical, helio-tropical and geo-tropical movements of bacteria, fungi and antherozoids of ferns, mosses and phanerogams... The Weber-Fechner law, therefore, is not the law of the nervous system and its centers, but the law of protoplasm in general and its ability to respond to stimuli*" [135, p. 126].

Here one can also recall the known phenomenon of conformational fluctuations of enzyme macromolecules on sound waves and other frequencies. In connection with this phenomenon, the possible importance for life of "*fantastic pictures of musical interactions of biochemical systems, cells, organs*" related to "*biochemical aesthetics*" has been noted in [136, p. 75].

Our research is associated with "biochemical aesthetics". An organism can be seen as a musical synthesizer with multiple settings of inherited resonant modes [12, 36, 117–119, 137]. Music is a game with acoustic resonances, to which people are remarkably predisposed. Throughout tens of thousands of years, people create musical instruments, adjusting them to specific systems of resonances. Over the centuries, people have learned to combine individual instruments and singers into orchestras and choirs as coordinated oscillating systems with an increased number of degrees of freedom. Leibniz declared that music is arithmetic of soul, which computes without being aware of it. Taking into account that music is represented by systems of resonances, one can reformulate this declaration: systems of resonances are the arithmetic of soul, which computes without being aware of it.

3.8. Hyperbolic rotations, hyperbolic numbers, and resonances

The description in the previous paragraph of the Weber-Fechner law drew attention to the fact that this biological law is associated with the transformation of one point of the hyperbola $y = 1/x$ to another point, that is, with the sliding of the hyperbola along itself. Such transformations are called hyperbolic rotations, and they are known in physics, for example, as Lorentz transformations in special relativity connected with Minkowski geometry. Hyperbolic rotations are described by the following symmetric (2×2)-matrices whose components are real numbers:

$$\begin{vmatrix} a, & b \\ b, & a \end{vmatrix}, \quad \text{where } a^2 - b^2 = 1 \tag{3.2}$$

Hyperbolic rotations are usually expressed by a symmetric matrix (3.3) through hyperbolic cosine «cosh» and hyperbolic sine «sinh» [138–140]:

$$\begin{vmatrix} \cosh x, & \sinh x \\ \sinh x, & \cosh x \end{vmatrix}, \quad \text{where } (\cosh x)^2 - (\sinh x)^2 = 1 \tag{3.3}$$

The bisymmetric matrix $[a, b; b, a]$ in (3.2) is a matrix representation of two-dimensional hyperbolic numbers $z = a + b\mathrm{j}$, where "a" and "b" are real numbers and imaginary unit "j" satisfies the condition $\mathrm{j}^2 = +1$ (in contrast to complex numbers $c = a + b\mathrm{i}$, where imaginary unit "i" satisfies the condition $\mathrm{i}^2 = -1$) [58, 141]. Hyperbolic numbers have synonymous names in literature: split complex numbers, double numbers, perplex numbers. Like one-dimensional real numbers, these two-dimensional numbers can be added, subtracted, multiplied, divided (with some reservations) with each other, each time receiving a number of the same type. In doing so, they are represented by points and vectors in their two-dimensional metric space, where the vectors have length.

To show that the bisymmetric matrix $[a, b; b, a]$ represents hyperbolic numbers $z = a + b\mathrm{j}$, it is sufficient to analyze its following decomposition:

$$\begin{vmatrix} \mathrm{a}, & \mathrm{b} \\ \mathrm{b}, & \mathrm{a} \end{vmatrix} = \mathrm{a}\begin{vmatrix} 1, & 0 \\ 0, & 1 \end{vmatrix} + \mathrm{b}\begin{vmatrix} 0, & 1 \\ 1, & 0 \end{vmatrix} \tag{3.4}$$

In the right part of the decomposition (3.4), the first sparse matrix is the identity matrix and the second sparse matrix is the matrix representation of the imaginary unit j, because its square gives the identity matrix ($[0, 1; 1, 0]^2 = [1, 0; 0, 1]$) by analogy with the condition $j^2 = +1$. Symmetric matrices that represent hyperbolic numbers have real eigenvalues and orthogonal eigenvectors (which distinguish them from non-symmetric matrix representations of complex numbers). Correspondingly, such symmetric matrices are very appropriate for the concept of multi-resonance genetics since they define a special class of matrices in the theory of resonances of oscillatory systems with many degrees of freedom.

Certain connections of hyperbolic rotations to inherited biological phenomena exist not only with regard to the Weber-Fechner law. For example, the article [33] on the results of 20 years of research by the author of animal and human locomotion states that the spatio-temporal organization of the locomotion is associated with hyperbolic turns and the geometry of Minkowski. In this regard, Smolyaninov put forward the "locomotor theory of relativity" and wrote about the relativistic brain and relativistic biomechanics.

Another example regarding hyperbolic rotations in biology concerns a well-known phyllotaxis phenomena, which have been studied for over 150 years. These phenomena are demonstrated by typical spiral lattices in the forms of plant and animal organisms ("the laws of phyllotaxis") (see [142] and http://www.goldenmuseum.com/0604Phillotaxis_engl.html). The quantitative characteristics of these lattices are expressed by the Fibonacci numbers F_n and their ratios F_{n+1}/F_n. For example, the number of left and right spirals in the head of a sunflower are neighboring members of the Fibonacci series.

During the growth of some organisms, their phyllotaxis lattices are transformed with the transition to other Fibonacci ratios F_{k+1}/F_k. Studies of the growth rearrangements of these bio-lattices have shown that they correspond to transformations of hyperbolic rotations such

Fig. 3.8.1. Biological examples of phyllotaxis patterns: (a) Common sunflower (taken from https://en.wikipedia.org/wiki/Common_sunflower, permission granted under the Creative Commons Attribution-Share Alike 2.5 Generic license); (b) Pine cone (taken from https://en.wikipedia.org/wiki/Pine, permission granted under the Creative Commons Attribution 2.0 Generic license); (c) *Aloe polyphylla* (taken from https://en.wikipedia.org/wiki/Phyllotaxis, permission granted under the GNU Free Documentation License); (d) Romanesco broccoli (*Brassica oleracea*) (taken from https://en.wikipedia.org/wiki/Romanesco_broccoli, permission granted under the Creative Commons Attribution-Share Alike 4.0 International license).

as Lorentz transformations in the special theory of relativity [34, 35]. In these publications, it is stated that living matter is structurally associated with Minkowski's geometry.

Continuing the theme of algebraic features of phyllotaxis phenomena, the next section will present some data on the connection of phyllotaxis phenomena with two-dimensional hyperbolic numbers and the mathematical theory of resonances of oscillatory systems.

3.9. Phyllotaxis laws, matrix representations of hyperbolic numbers, and resonances

The inherited morphogenetic laws of phyllotaxis (leaf arrangement) deal with a widespread spiral stacking of body parts of plants and animals associated with numbers of the Fibonacci series $F_{n+2} = F_n + F_{n+1}$ where $F_0 = 0$, $F_1 = 1$ (Fig. 3.9.1). If such additive series begins not with 0 and 1 but with two other numbers, for example, with 2 and 1 (in this case the series of numbers 2, 1, 3, 4, 7, 11, ... appears), they are termed the series of Lucas numbers.

n	0	1	2	3	4	5	6	7	8	9	10	11	12	...
F_n	0	1	1	2	3	5	8	13	21	34	55	89	144	...

Fig. 3.9.1. The initial fragment of the Fibonacci series $F_{n+2} = F_n + F_{n+1}$

Fibonacci numbers are strongly related to the golden ratio (or the golden section) $\varphi = (1+5^{0.5})/2$. Binet's formula (3.5) expresses the *n*th Fibonacci number in terms of *n* and the golden ratio, and implies that the ratio of two consecutive Fibonacci numbers tends to the golden ratio as *n* increases:

$$F_n = (\varphi^n - (-\varphi^{-1}))/5^{0.5} \tag{3.5}$$

In biology, it has long been known that, for example, in many plant objects the spiral arrangement of their parts form ordered patterns (shoots of plants and trees, seeds in the heads of sunflowers, scales of coniferous cones and pineapples, etc.). Such spiral arrangement of body parts is also realized in animal bodies, for example, in mollusk shells and tentacles, canals and zooids of some jellyfishes, the arrangement of new buds along the trunk of the hydra, the sequence of chambers in bodies of some foraminifera, etc. (see reviews in books [13, 142, 143]). In other words, the laws of phyllotaxis appear in inherited morphological patterns on a variety of levels and branches of biological evolution. These patterns are determined by overlapping left- and right-oriented spiral lines, whose quantities are usually equal to adjacent Fibonacci numbers. Phyllotaxis on plant stems is expressed in the fact that the sequence of their leaves forms a spiral pattern. The angle that exists between two alternate leaves on a stem is called the angle of divergence. The divergence angle $360°/\varphi^2 \approx 137.5°$ is called the ideal phyllotaxis angle or the Fibonacci angle [142, § 1.3.3]. The laws of phyllotaxis are among the best known in the field of mathematical biology; for example, the bibliography in [142] includes about 1,000 publications devoted to them. Phyllotaxis phenomena and their patterns interest not only biologists, but also mathematicians and physicists.

$$Q = \begin{vmatrix} 0, 1 \\ 1, 1 \end{vmatrix}; \quad Q^2 = \begin{vmatrix} 1, 1 \\ 1, 2 \end{vmatrix}; \quad Q^n = \begin{vmatrix} F_{n-1}, F_n \\ F_n, F_{n+1} \end{vmatrix}$$

Fig. 3.9.2. The Fibonacci matrix Q and its powers.

Algorithmic models of phyllotaxis based on the so-called Fibonacci matrix Q, sometimes termed the "growth matrix", are known (Fig. 3.9.2) [142]. Its raising to integer powers always gives a symmetric matrix, all components of which are Fibonacci numbers F_n. The eigenvalues of the Fibonacci matrix are equal to the golden ratio φ and $-\varphi^{-1}$. The ratio of the coordinates for the eigenvectors of the Fibonacci matrix is also equal to the golden ratio.

For the stated concept of multi-resonance genetics, the matrix $Q^2 = [1, 1; 1, 2]$ has a special interest: its eigenvalues are equal to positive numbers φ^2 and φ^{-2}. Correspondingly, an oscillatory system with two degrees of freedom, described by this matrix, has resonant frequencies φ and φ^{-1}, which can participate in a coordination of oscillatory genetic systems growing step-by-step in ontogenesis. The eigenvalue φ^2 is equal to the value that appears in the expression for the ideal phyllotaxis angle $360°/\varphi^2$. Together with the second eigenvalue φ^{-2}, it determines the coordinates $(\varphi^2, \varphi^{-2})$ of a point on the hyperbola $y = 1/x$ of the model space (Fig. 3.7.1) and the corresponding value of the hyperbolic logarithm $\ln(\varphi^2)$.

Phyllotaxis angles are part of the total angle of the circle $360° = 2\pi$ radians. Therefore, we can single out the corresponding second point on this hyperbola with coordinates $(2\pi, 1/2\pi)$, which are treated in the described approach as eigenvalues of the second oscillating system with two degrees of freedom. This second point defines the value of the hyperbolic logarithm $\ln(2\pi)$ according to Fig. 3.7.1. The difference α of the two named logarithms given by the two considered points on the hyperbola $y = 1/x$ is equal to $\alpha = \ln(2\pi) - \ln(\varphi^2) = \ln(2\pi/\varphi^2)$, where the value $2\pi/\varphi^2$ is the value of the ideal phyllotaxis angle mentioned above. This expression is similar to expression (3.1) of the Weber-Fechner psychophysical law $k \times \ln(x/x_0)$. This indicates a formal relationship between the genetically inherited basic psychophysical law and the morphogenetic laws of phyllotaxis. We emphasize that the highlighted points on the hyperbola $y = 1/x$ are not just mathematical points, but in the described model they represent physical factors — the resonant frequencies of two vibro-systems — associated with energy and energy principle organizations.

In the genetic spiral of phyllotaxis, an approximation to its ideal angle $2\pi \times \varphi^{-2}$ is usually realized, which has the form $2\pi \times F_n/F_{n+2}$, where F_n are the Fibonacci numbers (this expression becomes the value of the ideal angle as $n \to \infty$). Phyllotaxis divergence angles $2\pi \times F_n/F_{n+2}$ are modeled respectively using matrices of oscillating systems with mutually inverse eigenvalues $2\pi \times F_n/F_{n+2}$ and $(2\pi \times F_n/F_{n+2})^{-1}$; the diagonal form of these matrices is: $[2\pi \times F_n/F_{n+2}, 0; 0, (2\pi \times F_n/F_{n+2})^{-1}]$.

But is anything known about the existence in nature of vibration systems, whose resonant frequencies are associated with the golden section φ? Yes, it is. The website ScienceDaily published an article with the title "The Golden Section is Open in the Quantum World" [144] about the following. Research of cobalt niobate, which has magnetic properties, has revealed that "*the chain of atoms acts like a nanoscale guitar string. ... The tension comes from the interaction between spins causing them to magnetically resonate. For these interactions we found a series of resonant notes: the first two notes show a perfect relationship with each other. Their frequencies (pitch) are in the ratio of 1.618..., which is the golden ratio famous from art and architecture.*"

We note in passing that the diagonal matrices $[x, 0; 0, 1/x]$ with mutually reciprocal eigenvalues, which play an important role in the above models of the Weber-Fechner laws and phyllotaxis, are also known in physics, where they express the Lorentz transformations of the pseudo-Euclidean plane given in the basis of two isotropic vectors.

Now let us continue the theme of algebraic features of phyllotaxis phenomena by using two-dimensional hyperbolic numbers and their matrix representations, described in § 3.8, for modeling phyllotaxis structures in connection with the theory of resonances. One should recall that in the history of development of the notion "number", after the establishment of real numbers, multi-dimensional hypercomplex numbers appeared. First of all, in modern science, two-dimensional numbers have played the role of the magic tool for development of theories and calculations in the field of problems of heat, light, sounds, fluctuations, elasticity, gravitation, magnetism, electricity, current of liquids, and the quantum-mechanical phenomena. In modern theoretical physics, systems of four-dimensional Hamilton quaternions and eight-dimensional hypercomplex numbers (mainly, Cayley's octonions and split-octonions) are important mathematical objects. But in algebraic biology, multi-dimensional number systems have not been practically used till recent times.

The phyllotaxis laws provide an interesting opportunity to introduce into algebraic biology two-dimensional hyperbolic numbers and their representations in the form of symmetric matrices always having positive eigenvalues, which are appropriate for the theory of resonances with many degrees of freedom.

To characterize phyllotaxis of botanical objects, usually two parameters are indicated: number of left spirals and number of right spirals. Such phyllotaxis structures are described by ratios of neighboring Fibonacci numbers:

$$F_{n+1}/F_n\text{: } 2/1,\ 3/2,\ 5/3,\ 8/5,\ 13/8,\ 21/13,\ 34/21,\ \ldots \qquad (3.6)$$

The sequence (3.6), which is termed the "parastichy sequence", in the limit at $n \to \infty$ tends to the golden ratio φ, which is a mathematical symbol of self-reproduction for many centuries (Leonardo da Vinci, Kepler, etc.). It is known that many authors note the golden section in different physiological systems: cardiovascular system, respiratory system, electric activities of brain, etc.

The sequence of transitions in the parastichy sequence (3.6) from one member to the next, which can be sometimes used for description of ontogenetic transformations of phyllotaxis patterns, is described by expression (3.7) where φ is the golden ratio:

$$(F_{n+1}/F_n) \to (F_{n+2}/F_{n+1})\text{: } 2/1 \to 3/2 \to 5/3 \to 8/5 \to 13/8 \to 21/13 \ldots \to \varphi \qquad (3.7)$$

It should be emphasized that two kinds of spirals — left and right spirals — are simultaneously realized in phyllotaxis objects. Traditionally number "*a*" of left spirals and number "*b*" of right spirals are expressed separately by one-dimensional real numbers "*a*" and "*b*". But if in the phyllotaxis phenomenon these two real numbers always exist jointly and form the mutual parastichy sequence (3.6), they can be interpreted as two parts of a single two-dimensional hyperbolic number "*a*+*bj*", whose matrix representation has two positive eigenvalues and is connected with the theory of resonances of an oscillatory system with two degrees of freedom. In this case, the Fibonacci series of real numbers (Fig. 3.9.1) turns into the Fibonacci series of hyperbolic numbers (3.8):

$$F_{n+1} + jF_n: 2 + j,\ 3 + 2j,\ 5 + 3j,\ 8 + 5j,\ 13 + 8j,\ 21 + 13j,\ 34 + 21j,\ \ldots \quad (3.8)$$

The series of Fibonacci hyperbolic numbers (3.8) is additive since their members are interrelated by the additive condition: $(F_{n+1}+jF_n)+(F_{n+2}+jF_{n+1})= F_{n+3} + jF_{n+2}$. Ratios of adjacent Fibonacci hyperbolic numbers from (3.8) form a new sequence of hyperbolic numbers (3.9) tending — as n increases — to the hyperbolic number $\varphi + 0j$, which coincides with the real golden section φ because its imaginary part is equal to zero. The sequence (3.9) is called the parastichy sequence of hyperbolic numbers. In this approach [80], the sequence (3.6) of phyllotaxis ratios of real Fibonacci numbers is transformed into the sequence of hyperbolic Fibonacci numbers (3.9).

$$(F_{n+2} + jF_{n+1})/(F_{n+1} + jF_n): (3 + 2j)/(2 + j),\ (5 + 3j)/(3 + 2j),\ (8 + 5j)/(5 + 3j),$$
$$(13 + 8j)/(8 + 5j),\ \ldots \rightarrow \varphi + 0j \quad (3.9)$$

The Fibonacci hyperbolic numbers (3.8) have their (2 × 2)-matrix representations, which define the additive series of matrices (3.10):

$$\begin{vmatrix} F_{n+1}, F_n \\ F_n, F_{n+1} \end{vmatrix} : \begin{vmatrix} 2, 1 \\ 1, 2 \end{vmatrix},\ \begin{vmatrix} 3, 2 \\ 2, 3 \end{vmatrix},\ \begin{vmatrix} 5, 3 \\ 3, 5 \end{vmatrix},\ \begin{vmatrix} 8, 5 \\ 5, 8 \end{vmatrix},\ \begin{vmatrix} 13, 8 \\ 8, 13 \end{vmatrix},\ \ldots \quad (3.10)$$

Now let us describe the results of the study of eigenvalues of the bisymmetric matrices in (3.10). Each of these matrices $[F_{n+1}, F_n; F_n, F_{n+1}]$ has two eigenvalues, which are equal to two Fibonacci numbers again: F_{n+2} and F_{n-1}. One can note that these eigenvalues are the sum and the difference of the

Fibonacci components of the original hyperbolic number $F_{n+1} + jF_n$ since $F_{n+2} = F_{n+1} + F_n$ and $F_{n-1} = F_{n+1} - F_n$. The ratio F_{n+2}/F_{n-1} of such eigenvalues defines a new sequence (3.11) of Fibonacci ratios, which tend to φ^3 as n increases:

$$F_{n+2}/F_{n-1}: 3/1, 5/1, 8/2, 13/3, 21/5, 34/8, 55/13, \ldots \rightarrow \varphi^3 \qquad (3.11)$$

Such pair of eigenvalues F_{n+2} and F_{n-1} can be considered as components of a new hyperbolic number $F_{n+2} + jF_{n-1}$. In this case, the sequence of ratios (3.11) is transformed into additive sequences (3.12, 3.13) reflecting linear notations of appropriate hyperbolic numbers and their matrix presentations:

$$F_{n+2} + jF_{n-1}: 3 + j, 5 + j, 8 + j2, 13 + j3, 21 + j5, 34 + j8, \ldots \qquad (3.12)$$

$$\begin{vmatrix} F_{n+2}, F_{n-1} \\ F_{n-1}, F_{n+2} \end{vmatrix} : \begin{vmatrix} 3, 1 \\ 1, 3 \end{vmatrix}, \begin{vmatrix} 5, 1 \\ 1, 5 \end{vmatrix}, \begin{vmatrix} 8, 2 \\ 2, 8 \end{vmatrix}, \begin{vmatrix} 13, 3 \\ 3, 13 \end{vmatrix}, \begin{vmatrix} 21, 5 \\ 5, 21 \end{vmatrix}, \ldots \qquad (3.13)$$

Each of the symmetric matrices $[F_{n+2}, F_{n-1}; F_{n-1}, F_{n+2}]$ of the sequence (3.13) has two eigenvalues, which are again equal to two Fibonacci numbers multiplied by a factor 2 (twice the Fibonacci numbers): $2F_{n+1}$ and $2F_n$. Ratios $2F_{n+1}/2F_n$ of such eigenvalues form a sequence, which is identical to the initial parastichy sequence (3.6).

This procedure of analysis of the eigenvalues of old and new sequences of symmetric matrices, representing hyperbolic numbers by analogy with sequences (3.8, 3.10–3.13), can be repeated as long as desired, obtaining a hierarchy of eigenvalues of the matrices based on Fibonacci numbers multiplied by a factor 2 at corresponding steps of the iterative procedure.

The matrix form of the hyperbolic Fibonacci numbers is a matrix operator that symmetrically transforms on the hyperbolic plane any figure (as a system of points) into a new figure. One can add that symmetric matrix representations of hyperbolic numbers are related not only to the theory of resonances but also to the theory of solitons of the sine-Gordon equation, which are the only relativistic type of solitons; they were put forward for the role of the fundamental type of solitons of living matter in the book [145].

The proposed approach connects real with hyperbolic Fibonacci numbers. It expands modern possibilities of constructing new applications and theories based on Fibonacci numbers and matrix representations of multi-dimensional numbers. One of the interesting directions is the application of hyperbolic numbers with Fibonacci entries in generalized crystallography, genetic biomechanics, and biochemical aesthetics [12, 146].

3.10. Morphogenesis, metric tensors, and the morpho-resonance field

Morphogenetic processes on different lines and branches of biological evolution sometimes demonstrate an amazing structural commonality. Examples of this generality are the law of homological series of N. I. Vavilov and the phenomena of phyllotaxis.

Many researchers devoted their works to the problems of morphogenetic parallelisms (or independent similarities not related to the dictates of environmental conditions or functions) and non-Euclidean bio-symmetries. A characteristic feature of inherited biological surfaces is their curvilinear character. The life of organisms is in many ways connected precisely with surfaces, an example of which are the germ layers that give rise to various organs and tissues (https://en.wikipedia.org/wiki/Germ_layer).

In mathematics, curvilinear surfaces are studied by means of differential Riemannian geometry and tensor analysis using the key concept of the metric tensor (see, for example, [147–152]). The mentioned metric tensor defines the metric in an infinitely small part of the surface by specifying the distance between its two infinitely close elements. Specifying a system (or a field) of metric tensors on a surface determines its "internal" geometry, allowing one to calculate arc lengths, angles between curves, and areas of regions on the surface, regardless of its location in space. Therefore, it is natural to try to build a general theory of biological morphogenesis with the involvement of metric tensors.

By definition, the metric tensor in an n-dimensional affine space with the introduced scalar multiplication operation is given by a nondegenerate symmetric matrix $||g_{ij}||$, $g_{ij} = g_{ji}$ [151], which is satisfied by the structure of the symmetric real (2×2)-matrices of vibro-systems considered in this chapter. The coordinates g_{ij} of the metric tensor are pairwise scalar products of the vectors of the frame on which it is built. If a square root is extracted from a symmetric matrix that is a metric tensor, then a new symmetric matrix is obtained, the columns of which are the vectors of this frame (and which, in turn, can be treated as a new metric tensor). In order for a symmetric real matrix to be interpreted as a tensor, it is necessary to specify a group of transformations, with respect to which it acts as a tensor. For example, for the symmetric (2×2)-matrices under consideration, such a group is the group of plane rotations (or a wider group of motions), whose transformations leave the value of the scalar products of the reference vectors unchanged, although the coordinates of the vectors themselves change during these transformations.

Let us consider in more detail the modeling of two-dimensional curvilinear surfaces embedded in a three-dimensional Euclidean space. The metric tensor at the points of this surface (it can vary from point to point) is given by a (2×2)-matrix, which has a diagonal form in orthogonal local coordinates on this surface. This view is similar to those diagonal (2×2)-matrices of vibro-systems with two degrees of freedom, which were used above in the analysis of the molecular genetic system, Weber-Fechner laws, and phyllotaxis. The diagonal terms of the metric tensor are also positive, as are the eigenvalues of these matrices of vibro-systems. In other words, there is an isomorphism between the set of these (2×2)-matrices of vibro-systems and the set of metric tensors of two-dimensional curvilinear surfaces: for each matrix of one set that has two orthogonal eigenvectors and two eigenvalues, there is the analogical matrix of the second set with the same eigenvectors and eigenvalues. This isomorphism makes it possible to "encode" or set metric tensors of surfaces through similar matrices of vibro-systems with two degrees of freedom, and therefore through resonant frequencies of vibro-systems (that is, the coding of morphogenesis through systems of resonant frequencies). This allows modeling of biomorphological surfaces from the standpoint of the concept of multi-resonance genetics.

An effective theory of morphological surfaces should be able not only to model static surfaces, but also changing of these surfaces during growth

Fig. 3.10.1. Examples of parallelisms of forms and movements in different organisms associated with channel surfaces. (a) ammonite shell (taken from https://commons.wikimedia.org/wiki/File:Clymenia_laevigata.JPG, permission granted under the Creative Commons Attribution 3.0 Unported license); (b) seahorse (see https://en.wikipedia.org/wiki/Seahorse, permission granted under the Creative Commons Attribution-Share Alike 3.0 Unported license); (c) the foraminifera *Talamophora* (reproduced with permission from [153]); (d) a tendril of a climber plant (taken from https://commons.wikimedia.org/wiki/File:DirkvdM_natural_spiral.jpg, permission granted under the GNU Free Documentation License); (e, f) an elephant trunk and twisted tail of a reptile, respectively (taken from https://www.researchgate.net/publication/337834607_Nonlinear_Concave_Spiral_Waves_in_Active_Media_Transferring_Energy/figures?lo=1, permission granted under the Creative Commons Attribution license).

and other natural transformations, when at each step of movement, a new form is realized, smoothly arising from the previous ones in a regular way. Let us outline the beginnings of such theory [36, 118, 119], which allows modeling of curvilinear biological surfaces and their growth transformations in a unified way within the framework of the concept of resonance genetics. It proceeds from the key role of the resonant frequencies of vibro-systems with two (or more) degrees of freedom for the genetic inheritance of morphological surfaces and their natural biological transformations. In this regard, this theory is conditionally called morpho-resonance. It uses the named isomorphism between the set of (2×2)-matrices of vibro-systems and the set of metric tensors of two-dimensional curvilinear surfaces embedded in three-dimensional Euclidean space.

It is known that the metric tensor or metric at a point on the surface is represented as the so-called first quadratic form of the differentials of the coordinates du, dv on the surface:

$$I = Edu^2 + 2Fdudv + Gdv^2, \tag{3.14}$$

where E, F, G are shape coefficients. Geometrically, this form coincides with the square of the element of arc length of the curve on the surface. In the orthogonal coordinate system (u, v) on the surface, the coefficient $F = 0$, and the shape is simplified:

$$I = Edu^2 + Gdv^2 \tag{3.15}$$

These coefficients E and G are the eigenvalues of the metric tensor at a given point on the surface. They coincide with the eigenvalues of some (2×2)-matrix of a vibro-system with two degrees of freedom due to the isomorphism noted above. Accordingly, these coefficients can be encoded (set) by the resonant frequencies of this vibration system.

The first quadratic form of the surface defines its internal geometry, and the second quadratic form is used to characterize its curvature in space:

$$II = Ldu^2 + 2Mdudv + Ndv^2, \tag{3.16}$$

where L, M, N are the coefficients of this form. Its geometric meaning lies in the fact that it characterizes the deviation of the surface from the tangent plane at the point under consideration.

On curved surfaces, there are many geometric types of lines: geodesic, asymptotic, etc. All of them can be analyzed to identify their possible connection with the concept of resonance genetics. But at the first stage of development of the morpho-resonance theory, our main attention is given to the principal lines of curvature of surfaces. The network of these principal lines of curvature forms an orthogonal system, which is convenient to use as a coordinate system. In building mechanics, the theory of thin shells is built precisely on local coordinate systems based on the principal lines of curvature, which greatly simplifies the equations [154, p. 162].

Recall the information about curvatures and principal lines of curvature. The ratio II/I of the second quadratic form to the first is called the normal curvature of the surface at a given point. It reaches its maximum and minimum values k_1 and k_2 along two orthogonal directions, called principal ones (we do not consider singular points at which the value does not depend on the direction as, for example, at points of a plane or sphere). A line of curvature on a surface is one that touches the principal direction at every point. If two orthogonal families of curvature lines are used as a system of surface coordinate lines, then the average coefficients of the first and second quadratic forms are equal to zero: $F = M = 0$. In this case, the values of the principal curvatures k_1 and k_2 at a point are calculated from the ratio of the coefficients of the second and first quadratic forms from (3.15) and (3.16):

$$k_1 = L/E, \; k_2 = N/G \tag{3.17}$$

Specifying two quadratic forms defines a curved smooth surface. Of these, only the first form is positive definite and is comparable with positive-definite (2×2)-matrices of vibro-systems with two degrees of freedom, which allows you to set the internal geometry (i.e., metric) of the surface. The coefficients of the second quadratic form, which in the general case is not positive definite, cannot be directly specified (coded) through the resonant frequencies of such vibro-systems.

How are curvatures k_1 and k_2 and resonant frequencies related in our model approach? The answer to this question is given by expressions (3.17), in which these curvatures change in inverse proportion to the change in the coefficients E and G of the first quadratic form. In this case, it is possible to selectively change any of the curvatures by changing the resonant frequency that determines the value of the corresponding coefficient of the first form. Indeed, with a change in the value of the resonant frequency corresponding to the coefficient E (or G) in the indicated matrix isomorphism, by virtue of (3.17) the curvature k_1 (or k_2) selectively changes at the considered point of the surface (it is assumed that the behavior of L and N in the numerators of these expressions is not essential). This resonant frequency of a vibro-system with two degrees of freedom acts as a physical regulator of curvature at a point on the surface (even though the surface metric in a small neighborhood of the point is determined by both resonant frequencies of the given vibro-system). Let us link this to the examples of channel surfaces in Fig. 3.10.1.

Channel surfaces in geometry are called surfaces formed by the movement of a circle of variable radius, in which the center of the circle moves along a given curve, and the plane of the circle always remains perpendicular to this curve [149, p. 315]. Surfaces of this type can be used as models of many biological objects: blood vessels, plant shoots, etc. For channel surfaces, the generating circles act as the principal lines of curvature, and the lines orthogonal to them are the principal lines of curvature of the second family. For example, in the described model approach, the growth curvature of the spiral tendril of a plant (Fig. 3.10.1d) is interpreted as being given by a corresponding change in at least one of the two resonant frequencies that regulate the corresponding curvature k_1 or k_2 at points on the surface. Of course, when this regulating resonant frequency changes, not only the curvature changes, but also the metric tensor of the channel surface, which is consistent with the phenomenology of the considered growth transformations. So, from the point of view of morpho-resonance theory, the transformation of some channel surfaces into others (Fig. 3.10.1d) is, first of all, a smooth restructuring of the network of main curvature lines through setting one or two resonant frequencies of some regulating vibration system with two degrees of freedom. Fig. 3.10.1 shows that during the muscular movements of the elephant's trunk and the seahorse's tail, configurations are realized that are also presented in growth movements and static morphological structures. This suggests that muscle movements have prototypes in growth movements and are organized in accordance with certain archetypes of biological shaping. Studies of biological forms and movements are related to a number of applied aspects, including those in architectural bionics and robotics.

Let us add that the networks of curvature lines of surfaces are associated with Möbius transformations (in the literature they are also called conformal-geometric, circular, inverse radius, Liouville, or Kelvin transformations). The latter transform the surfaces in such a way that the curvature lines of the original surfaces pass into the lines of curvature of the new ones (for comparison, projective-geometric surface transformations transform the curvature lines of the original surfaces into lines of a different type). Möbius transformations preserve the magnitudes of the angles, transform spheres into spheres, and are locally similar. Many phenomena of biological symmetries are based on these Möbius transformations and their cyclic groups in multi-block bodies, which was the subject of a doctoral dissertation [32, 155]. Speaking of conformal-geometric topics, we recall that in the previous chapters the reader has already met with the conformal Poincaré model of hyperbolic geometry and its connections with the genetic system.

Möbius transformations are locally similar: transformations of a small neighborhood of any point on the surface are scale transformations, preserving the lines of curvature with a simple scale of the curvatures k_1 and k_2 at each point (but the shape of the surface as a whole can change significantly due to different scaling at its different points); from the point of view of morpho-resonance theory, this local-similarity feature of Möbius transformations simplifies the resonance regulation of growth. Another reason for the biological significance of Möbius transformations can be related to the important role of the network of principal lines of curvature in the theory of thin shells of structural mechanics, where their use as local coordinate systems greatly simplifies the equations of mechanics that relate the shape of shells to stresses and strains under loads [154, p. 162]. We also recall that Maxwell's equations of electrodynamics are invariant under the group of Möbius transformations; comprehension of this fact is one of the problems of physics, for the solution of which different authors proposed various versions of the conformal theory of relativity. The mathematical apparatus of quantum mechanics is also closely related to Möbius symmetries and there are several variants of conformal quantum field theory (https://en.wikipedia.org/wiki/Conformal_field_theory). Fig. 3.10.2 shows examples of modeling ontogenetic transformations of biological bodies based on Möbius transformations. These graphics resemble some illustrations in the classical book of D'Arcy Thompson [156], who developed the notion of the space curvature of the biological body; he used curvilinear coordinates to demonstrate the idea that a regular mathematical curvature of the shape of an organism of one species reveals its relationship with the shape of an organism of another species.

The product of two metric tensors generates a new metric tensor related to a Riemannian space of correspondingly increased dimension. The same applies to curvature tensors. By specifying algorithms or rules for changing

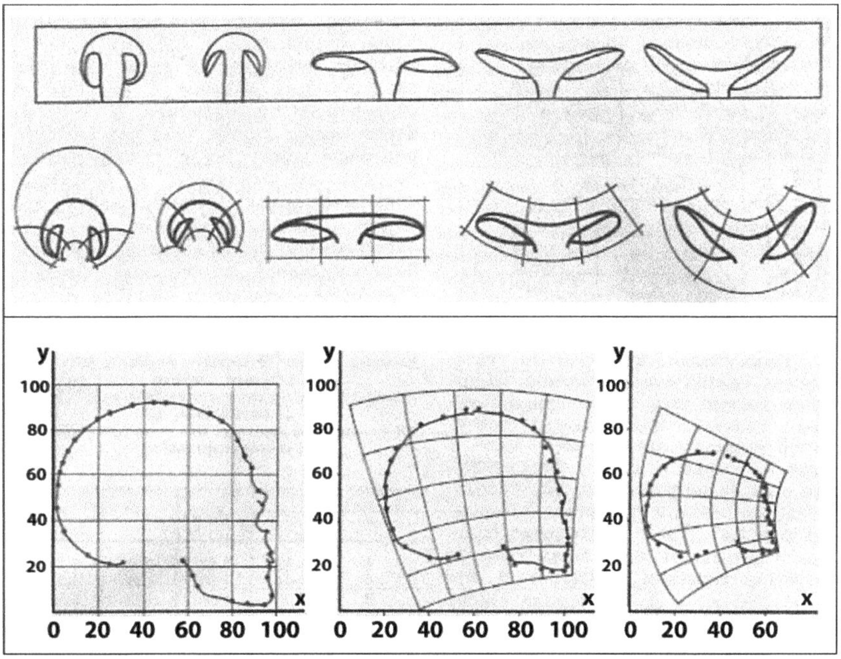

Fig. 3.10.2. Examples of Möbius transformations. Top: growth transformations of the fly agaric cap *Amanita* and their modeling using Möbius transformations. Below: Möbius transformations in modeling ontogenetic transformations of the human skull (reproduced with permission from [157]).

metric and curvature tensors, one can build models of morphogenesis based on the formalisms of Riemannian geometry and think that it is the genetic transmission of systems of these tensors from generation to generation that provides morphological similarity of generations. We believe that in bioinformatics and biological evolution of organisms, to a certain extent, the following principle is realized: "systems of resonant frequencies encode more general systems of resonant frequencies", which includes the principle "metric tensors encode metric tensors of higher orders".

Physics and mathematics have gained valuable experience in the use of metric tensors, in particular in connection with the concept of a tensor field. "*We say that we are given a tensor field if at each point M of space a certain tensor of constant valence is given, but otherwise, generally speaking, changing from point to point. We will call this tensor the field tensor*" [151, p. 46]. There are precedents in physics where physical fields are identified with certain tensor fields. Thus, A. Einstein and M. Grossman identified the gravitational field with the field of metric tensors of Riemann geometry. Since Riemannian geometry is defined by specifying a doubly covariant symmetric tensor, any physical problem that reduces to the study of such a tensor field can be formulated as a problem of Riemannian geometry. In particular, tensor fields of this type are used to describe elastic, optical, thermodynamic, dielectric, piezomagnetic, and other properties of anisotropic bodies. In connection with the concept of resonance genetics, some of these physical applications of Riemannian geometry can be used in the study of the phenomena of biological morphogenesis and the development of the theory of the morphogenetic field as a theory of the morpho-resonance field.

In biology, to explain the phenomena of morphogenesis, there have long been ideas of a certain morphogenetic field located inside the organism. Its versions are also known under other names: embryonic field, biological field, cellular fields, etc. At the same time, some authors believed that it could be a fundamentally new type of physical field, not yet known to science. A historical overview of attempts to develop the theory of the morphogenetic field can be found, for example, in the book [61].

The concept of multi-resonant genetics has led to the emergence of a new version of the morphogenetic field under the name of "morpho-resonant field". A morpho-resonant field is a tensor field of oscillatory or wave processes that exists inside the body and develops over time under consistent resonant frequencies of many vibro-systems with many degrees of freedom [36, 119]. In this author's version of the morphogenetic field, the existence of types of physical fields unknown to science is not assumed. It should be noted that this version differs from the versions of other authors also in that for the first time it connects the morphogenetic field with the structural features of the molecular genetic system. The formalisms of the theory of resonances of vibro-systems with many degrees of freedom, tensor analysis and differential geometry of curvilinear surfaces serve as a mathematical apparatus for the development of the theory of a morpho-resonance field.

But how can this class of models cover the situation, for example, of plant phyllotaxis, when periodically in the growth zone, which ensures the formation of a shoot, the formation of a new leaf organ suddenly appears? To model such situations, one can develop the model of cyclic changes in metric tensors in the morpho-resonance field occurring at the corresponding stages of a repeating shaping cycle.

It is known in embryology that serious disorders of the embryo, for example, microsurgical removal of its parts, in many cases do not prevent it from growing into a completely normal organism [61]. What is the criterion for the correctness of the finite configuration of an organism growing from an embryo? Such a criterion can be the coordination of the system of resonant characteristics of an adult organism with the resonant patterns of its molecular-genetic system, including DNA. The theme of the morpho-resonance field is associated with the doctrine of energy-informational evolution based on bio-antenna arrays, which is presented below in this book and is related to the emergent properties of systems of antenna operating on resonance mechanisms.

Genetic and other biological molecules must obey the principles of quantum mechanics. Its formalisms are based on the eigenvalues of matrices, and it operates with the frequency and resonance characteristics of objects in the quantum mechanical world. As is known, quantum mechanics arose and developed largely as a science of resonances in the microcosm (starting with the work of Planck in 1900). Thus, the concept of multi-resonance genetics and the morpho-resonance field considers genetic phenomena in the same language of oscillation frequencies and resonances on which quantum mechanics is based. In general, from the standpoint of the described approach, morphogenesis is defined as a restructuring and development of a system of morpho-resonance field tensors or as a restructuring and development of a system of consistent resonant frequencies of oscillatory processes in the body. The curvature of the shape of a biological body is explained in this case largely by the action of the so-called vibrational and wave forces known in vibrational and wave mechanics, a brief review of which is given in the next paragraph.

3.11. Vibrational mechanics and biological phenomena

The concept of multi-resonant genetics draws attention to the possible biological significance of the phenomena of vibrational and wave mechanics in connection with the behavior of parts of biological bodies at the subcellular, cellular and supracellular levels. Vibrational and wave mechanics are widely used in engineering and are saturated with amazing phenomena of vibrational separation and structurization of multiphase media, vibration transportation, vibration transfer of energy from one part of the system to another, etc. [124–127]. Acquaintance with some of these phenomena gives the impression of getting into a world with other physical laws. It is not for nothing that the popular science article [126] is called: "Vibration changes the laws of mechanics". These phenomena are associated with factors which, following the Nobel Laureate in Physics P. Kapitsa, are conditionally called "vibrational forces".

In a liquid, vibrating bodies can attract and repel (https://en.wikipedia.org/wiki/Bjerknes_force), and pulsating gas bubbles can merge and break up (many videos on the Internet are devoted to this). Barely noticeable or almost imperceptible vibrations, which can have high frequencies and small amplitudes, provide, for example, the following phenomena:

- the upper position of the inverted pendulum begins stably;
- a heavy metal ball "floats" on a layer of sand;
- the pile is easily immersed in the ground under its own weight;
- a massive body or a layer of bulk material moves up an inclined plane;
- a rope takes the form of a vertical stem if the appropriate vibration is applied to its base, etc.

In living organisms with their many components, vibrating in concert at different frequencies and amplitudes and obeying the wave principles of quantum physics at the micro level, many phenomena are known that also occur as if under the influence of mysterious forces. Such, for example, are the amazingly organized multi-stage processes of cell division. Or the searching in the molecular soup by histone proteins of four types of those histones, which should be complementary to them, for the gradual formation of dimers, tetramers, and then octameric histone rods, on which DNA is wound. All these searches and unions of single-species histones, first into pairs, and then into pairs of pairs, occur in a huge crowd of biomolecules of many other species and their fragments, and occur with high velocities despite charge-shielding effects and other interfering circumstances.

The phenomenon of vibro-transfer of energy among parts of an oscillatory system is known: a rotary electromotor operates stably, when it is disconnected from the power supply, if it is standing on a mutual vibro-platform with another rotary electromotor of similar resonant characteristics, which is connected to a power supply (self-synchronization by resonant interactions). Taking into account possibilities of such energy transferring, living organisms can be seen as resonance consumers of energy of surrounding electromagnetic waves coming from space and the depths of the Earth. Photosynthesis, which is based on absorbing solar energy of light waves, is probably only one example of the biological consumption of energy from external wave sources based on resonant mechanisms (a resonant "vampirism" of energy and information in organisms). A lot of data from homeopathy and physiological phenomena of ultraweak influences testify in favor of resonant organization of living matter.

Vibrational mechanics also allows simulating the evolutionary coordination and selection in the creation of a single organism from many pieces. Let us return to the example of rotary motors on a total mobile platform. Around a single electric motor, connected to the power supply, one can set — on the total mobile platform — many other rotary motors without the electrical power supply, which have different resonance characteristics. Through the self-synchronization by resonant interactions, only those motors, which are capable of resonance coordination with the base motor, will work as parts of a single functional colony, and the remaining motors will not work and will be correspondingly eliminated from the total working ensemble. Resonant combining parts into a single whole is based on the fundamental physical principle of minimum energy: each member of the ensemble requires less energy for performing its own work than when working individually.

For many morphogenetic processes, as well as for other physiological processes, their mutual synchronization is important. Vibrational mechanics provides a vivid example of the possibilities of synchronizing a multitude of oscillatory processes in the form of a well-known phenomenon of synchronizing oscillating pendulums mounted on a common moving platform [158]. This phenomenon was discovered by the founder of wave optics H. Huygens in the 17th century.

Inside a living organism, its structural water apparently plays the role of such a common mobile platform, which is required for synchronization. An illustrative example of the morphogenetic and general physiological role of structural water is given by jellyfish, which consists of 99% water, but despite this, it is a full-fledged living organism with a genetic system of inheritance of physiological functions, coordination of all parts into a single organism, etc. This structural water is also a candidate for the role of a unifying vibro-platform for vibro-transfer of energy among different parts of a living body. The physical features of structural water, associated with resonance interactions in it, are currently being studied in laboratories around the world (see, for example, [159, 160]).

An important role in vibro-connections among parts of an organism belongs also to the cytoskeleton that works in coordination with boundary water and membranes [161].

The ability of resonant processes to influence shaping is known in physics by the examples of Chladni figures [162–165]. They use the imposition of external vibrations on a resonating plate coated with powder particles to demonstrate resonant shaping. These examples are far from exhausting the formative potential of resonance systems, especially in cases where the vibration is not imposed from the outside, but each particle of the ensemble itself actively vibrates with its own vibration parameters and is in resonant interactions with other particles.

Both living and non-living objects are in general oscillatory systems with many degrees of freedom. But — from the point of view of multi-resonance genetics — living organisms are distinguished by special inherited tensor-matrix families of resonances in them. These biological families of resonances constitute a relatively narrow class of conceivable resonance systems that must be taken into account when understanding many biological problems: onto- and phylogenetic development, biological time, aging, etc.

3.12. Stochastic resonances and biological phenomena

Many processes in living bodies are stochastic due to the stochastic nature of interactions between individual molecules in cells. Analysis of stochastic properties of biological processes and structures is one of the main themes of this book. Results of study of stochastic rules in biological structures are presented and discussed in some subsequent chapters. Due to the importance of stochastic processes in living bodies, this last section of the chapter on resonances is specially devoted to stochastic resonances.

The concept of stochastic resonance is a real physical mechanism, which allows us to rethink vibrational and sound processes in living bodies, as well as the results of external vibrations and sound harmonic effects on organisms. These influences have been used for medical purposes since ancient times, defining well-known fields such as vibrational medicine and music therapy (see, for example, [166, 167]). This type of resonance was first described 40 years ago [168–170]. It is about the phenomenon of amplification in nonlinear systems of weak periodic signals flowing against the background of noise, the energy of which is partially pumped into these signals. At the time of this writing, over 13,000 publications have been devoted to the topic of stochastic resonance, including a series of publications in the journal *Nature*. In particular, stochastic resonance methods are used in medical clinics and centers in Germany, the USA and other countries. Accounting for stochastic resonances seems useful for understanding the phenomenon of the dictatorial influence of genetic DNA molecules on the formation of inherited macrostructures [25, 171].

The phenomenon of stochastic resonance speaks to the usefulness of noise — in some cases — for the transmission of information, contrary to the popular opinion that noise is always a hindrance. For research and application purposes, the occurrence of stochastic resonance can be controlled by adjusting the appropriate parameters, such as the intensity of the excitation noise, the structure, and parameters of nonlinear systems. A number of authors note that, from a fundamental point of view, applications of the theory of stochastic resonances to the study of processing information by biological systems are the most interesting, and that there is reason to believe the following: in the process of evolution, living organisms have adapted to use the unremovable internal noise and the noise of the environment to optimally identify useful information [172]. This judgment is based on the discovery by different authors of the use of stochastic resonance mechanisms in the sensory systems of animal organisms, for example, in crayfish and crickets [173–175]. Stochastic resonance is also used to explain the physiological phenomenon of synesthesia and the mutual influence of different sensory channels on each other; for example, the phenomenon of a significant improvement in threshold speech perception during electro-tactile stimulation of the index finger [176]. The authors of the same work believe that stochastic resonance plays a key role in the processing of auditory information. The existence of stochastic resonance has been demonstrated not only at the macroscopic level, but also at the molecular level [177, 178]. Some authors discuss the possible role of stochastic resonance in the hypersensitivity of some organisms to the detection of weak signals in a noisy environment, as well as in the influence of electromagnetic fields on human and animal health [179].

Since the genetic system in a living being has a significant specificity, the concept of "genetic stochastic resonance" was introduced as a special case of stochastic resonance [25, 171]. Among the specific features of the genetic system, what should be taken into account in matters of genetic stochastic resonance in inherited biological systems include, for example, the biological dissymmetry of left and right isomers discovered by Pasteur, which distinguishes living from non-living structures, as well as the helical nature of both DNA and the set of inherited biostructures at different levels [180]. It is not for nothing that since the time of W. Goethe, spirals have been called "the curves of life" lines [181]. The DNA double helix is associated with helical antennas used in engineering. These antennas generate and receive electromagnetic waves of circular polarization, which make it possible to exchange information regardless of the spatial orientation of the transmitter and receiver, for example, regardless of the rotation of the spacecraft during space communications. The idea of the DNA double helix as a helical antenna emitting and receiving waves of circular polarization is associated with the above-mentioned inherited dissymmetry of left and right forms in living bodies and selective exchange of wave information between biomolecules. Stochastic resonance in such chiral systems with their periodic processes and noise of circular polarization has its own specifics, which are studied within the concept of "genetic stochastic resonance", including in connection with the problems of epigenetics in the noise environment of the body [171]. We suppose that the mechanisms of stochastic resonance are important participants in the language of cells considered in the books [182, 183].

Several genetically inherited physiological mechanisms can also be considered as manifestations of the instinctive use of stochastic resonance mechanisms by a living organism. For example, humans and all warm-blooded organisms have an innate response to body cooling. It consists of involuntary muscle tremors that create muscle and neural noise and lead to the release of the hormone irisin. The release of this hormone signals the cells to generate increased heat (https://www.gazeta.ru/health/2014/02/06_a_5884077.shtml). Another physiological example of interest in connection with genetic stochastic resonance is Parkinson's disease, in which involuntary body tremor arises. This tremor can be considered as some kind of protective reaction of the body to a change in the state of its physiological systems, which it tries to overcome by additional activation of the mechanisms of stochastic resonance due to the generation of periodic bodily movements and noises. The enchanting effect of music on the body should also be studied taking into account the mechanisms of stochastic resonance. In our opinion, many well-known physiological phenomena and traditions of medicine, culture and religions associated with cyclic processes and noises need to be considered and comprehended in connection with genetic stochastic resonance, including, for example, the following:

- instinctive motion sickness by the mother of the baby to calm it down;
- a dance tradition that exists among all peoples and is based on rhythmic movements and rhythmic music;
- the tradition of shamans to move rhythmically to the beat of a tambourine to enter special states of "communication with spirits".

3.13. Some concluding remarks

The concept of multi-resonance genetics appeared as a result of the discovery of structural analogies between genetic phenomena and the mathematical theory of resonances of vibro-systems with many degrees of freedom [36, 118]. Due to this, promising intersections of biology with physics and informatics were revealed. All natural objects possess resonance properties. From the standpoint of the described model approach, genetic physiology is associated with a relatively narrow class of systems of resonances, which are related with eigenvalues and eigenvectors of $(2^n \times 2^n)$-matrices from tensor families based on tensor products of (2×2)-matrices. The presented results argue in favor of specificity of the biological system for inherited varieties of resonances expanding during ontogeny. Hydrogen bonds, which are one of the most important components of life and which are considered in the next chapter, play the role of a promotional factor for intermolecular vibrational energy relaxation, and as a driving force for the occurrences of specific reaction channels in binary molecular complexes [184].

Genetic molecules belong to the microworld and therefore are subordinate to the principles of quantum mechanics. Quantum mechanics operates with frequency and resonance characteristics of quantum-mechanical objects; its mathematics uses eigenvalues of matrices. In general, quantum mechanics emerged and developed largely as a science about resonances in the microworld. Thus, the concept of multi-resonance genetics creates models of genetic phenomena in the same language of frequencies and resonances, on which models in quantum mechanics are based. In addition to this, it uses the same matrix language on which "matrix mechanics" of W. Heisenberg was created; it is historically the first form of quantum mechanics, which retains its value to this day. The concept of multi-resonance genetics can facilitate a convergence of biology and quantum mechanics. The creator of the theory of resonance in structural chemistry, Pauling, was right when he presupposed an important meaning of resonances in the organization of living matter [122].

Chapter 4 Universal Rules of the Stochastic Organization of Genomic DNAs

The revelation of compositional principles in the organization of long DNA sequences is one of the crucial tasks in the study of biosystems. In the last few years, huge amounts of genetic data from different organisms have become available. In view of this, there is an increasing importance to reveal hidden construction principles and symmetries in long DNA sequences (for comparison, the DNA text of the human genome consists of approximately 3 billion nucleotide bases) [42]. The following quotation, although quite old, describes the current situation to the point: "*What will we have when these genomic sequences are determined? ... We are in the position of Johann Kepler when he first began looking for patterns in the volumes of data that Tycho Brahe had spent his life accumulating. We have the program that runs the cellular machinery, but we know very little about how to read it*" [7]. Kepler is mentioned here not without reason. The history of science shows the importance of cognitive forms of presentation of phenomenological data to find regularities or laws in this phenomenology. The work by Kepler is the classic example of an important meaning of a cognitive form of presentation of phenomenological data. He did not make his own astronomical observations, but he found the cognitive form of presentation in the huge astronomical data from the collection of Tycho Brahe. This discovered form, which was connected to the general idea of movements along ellipses, allowed him to formulate the famous Kepler's laws of planetary movements relative to the Sun. Owing to this cognitive form, Kepler and Newton led us to the law of Newtonian attraction. Our book provides evidence that, in order to unlock the secrets of the genetic coding system, an effective approach is to represent its interrelated ensembles in the cognitive form of algebraic matrices.

Since the middle of the last century, the so-called Chargaff's rules have been known, which concern "*a grammar of biology*" [185, 186]. The first rule holds that a double-stranded DNA molecule globally contains equally pyrimidine and purine bases and, more specifically, that the amount of guanine (purine) should be equal to cytosine (pyrimidine) and the amount of adenine (purine) should be equal to thymine (pyrimidine). This rule was theoretically confirmed by the double-helix model of DNA by Crick and Watson [187].

Chargaff's second rule states that the same parity is approximately valid for each of the two long DNA strands alone. According to Albrecht-Buehler [188], this rule applies to eukaryotic chromosomes, bacterial chromosomes, double-stranded DNA viral genomes, and archaeal chromosomes provided they are long enough. Perez extended in [189] Chargaff's second rule to codons: more precisely, for the sequences divided into triplets (codons) the parity of complementary bases takes place in each of these three positions. The work [190] investigated, in following Prabhu's ideas [191], the parity of k-nucleotides and their inverted complementary k-nucleotides in long DNA sequences: "Inversion Symmetry (IS): the counts of a k-mer of nucleotides on a chromosomal strand are almost equal to those of its inverse (reverse-complement) string". For $k = 1$ it is equivalent to Chargaff's second rule.

Many works of different authors analyzed, reviewed, and discussed Chargaff's second rule and questions about the grammar of biology (see, for example, [42, 190, 192–196]). Different chromosomes of a biological species can greatly differ in their length, characteristics, and quantities of genes within them, the cytogenetic bands (which show the biochemical specificity of the different parts of chromosomes), etc. However, Chargaff's second rule and its generalizations are almost identical for different chromosomes (see, for example, [197]). It was noted that Chargaff's second parity rule can reveal general properties common to all species and have remarkable implications of some unknown mechanism that seems to be present [188].

This chapter presents a generalization of Chargaff's second rule jointly with universal rules of stochastic organization of genomic DNA in eukaryotes and prokaryotes described in works [69, 70, 116, 153, 154, 171, 198, 233, 234]. The data of this chapter contribute to understanding the "*program that runs the cellular machinery*" [7].

4.1. Representation of genomic DNAs as *n*-plet sequences

One of the founders of quantum mechanics, who also introduced the term "quantum biology," P. Jordan noted the main difference between living and inanimate objects: inanimate objects are controlled by the average random movement of their millions of particles, whose individual influence is negligible, while in a living organism selected — genetic — molecules have a dictatorial influence on the whole living organism. Besides this, he claimed that life's missing laws were the rules of chance and probability of the quantum world [199]. From the standpoint of Jordan's statement, the study of probabilities or percentages of n-plets (monoplets, duplets, triplets, etc., that is, oligomers with lengths n) in long DNA sequences is important for discovering hidden biological laws and for developing quantum biology.

This chapter presents results of the study of probabilistic (stochastic) organization of single-stranded DNAs in genomes of higher and lower organisms relative to DNA alphabets of 4 nucleotides, 16 duplets, 64 triplets, etc. Any DNA sequence can be considered from the point of view of different DNA alphabets: the alphabet of 4 nucleotides — cytosine C, adenine A, thymine T, and guanine G; the alphabet of 16 duplets; the alphabet of 64 triplets; the alphabet of 256 tetraplets, etc. It allows representing any long DNA sequence in the form of multilayered texts, in which each nth layer is a sequence of n-plets (or oligomers of fixed length n); in other words, each nth layer is a separate DNA text written in its alphabet of 4^n n-plets. For example, in the sequence ACCTGTAACG…, the first layer is a text which is based on the alphabet of 4 nucleotides (A-C-C-T-G-…), and the second layer is a

```
TCTGACCTGAGGAGAACTGTGCTCCGCCTTCAGAGTACCACCGAAATCTGTGCAGAGGACAACGCAGCTC
CGCCCTCGCGGTGCTCTCCGGGTCTGTGCTGAGGAGAACGCAACTCCGCCGTTGCAAAGGCGCGCCGCGC
CGGCGCAGGCGCAGAGAGGCGCGCCGCGCCGGCGCAGGCGCAGAGAGGCGCGCCGCGCCGGCGCAGGCGC
AGAGAGGCGCGCCGCGCCGGCGCAGGCGCAGAGAGGCGCGCCGCGCCGGCGCAGGCGCAGAGAGGCGCGC
CGCGCCGGCGCAGGCGCAGACACATGCTAGCGCGTCGGGGTGGAGGCGTGGCGCAGGCGCAGAGAGGCGC
GCCGCGCCGGCGCAGGCGCAGAGACACATGCTACCGCGTCCAGGGGTGGAGGCGTGGCGCAGGCGCAGAG
AGGCGCACCGCGCCGGCGCAGGCGCAGAGACACATGCTAGCGCGTCCAGGGGTGGAGGCGTGGCGCAGGC
GCAGAGACGCAAGCCTACGGGCGGGGTTGGGGGGCGTGTGTTGCAGGAGCAAAGTCGCACGGCGCCGG
GCTGGGGCGGGGGAGGGTGGCGCCGTGCACGCGCAGAAACTCACGTCACGGTGGCGCGGCGCAGAGACG
GGTAGAACCTCAGTAATCCGAAAAGCCGGGATCGACCGCCCCTTGCTTGCAGCCGGGCACTACAGGACCC
GCTTGCTCACGGTGCTGTGCCAGGGCGCCCCCTGCTGGCGACTAGGGCAACTGCAGGGCTCTCTTGCTTA
GAGTGGTGGCCAGCGCCCCCTGCTGGCGCCGGGGCACTGCAGGGCCCTCTTGCTTACTGTATAGTGGTGG
CACGCCGCCTGCTGGCAGCTAGGGACATTGCAGGGTCCTCTTGCTCAAGGTGTAGTGGCAGCACGCCCAC
CTGCTGGCAGCTGGGGACACTGCCGGGCCCTCTTGCTCCAACAGTACTGGCGGATTATAGGGAAACACCC
GGAGCATATGCTGTTTGGTCTCAGTAGACTCCTAAATATGGGATTCCTGGGTTTAAAAGTAAAAAATAAA
TATGTTTAATTTGTGAACTGATTACCATCAGAATTGTACTGTTCTGTATCCCACCAGCAATGTCTAGGAA
TGCCTGTTTCTCCACAAAGTGTTTACTTTTGGATTTTTGCCAGTCTAACAGGTGAAGCCCTGGAGATTCT
TATTAGTGATTTGGGCTGGGGCCTGGCCATGTGTATTTTTTTAAATTTCCACTGATGATTTTGCTGCATG
GCCGGTGTTGAGAATGACTGCGCAAATTTGCCGGATTTCCTTTGCTGTTCCTGCATGTAGTTTAAACGAG
ATTGCCAGCACCGGGTATCATTCACCATTTTTCTTTTCGTTAACTTGCCGTCAGCCTTTTCTTTGACCTC
TTCTTTCTGTTCATGTGTATTTGCTGTCTCTTAGCCCAGACTTCCCGTGTCCTTTCCACCGGGCCTTTGA
```

	0	1
0	%C	%A
1	%T	%G

→ $P_1 =$

	0	1
0	0.208499	0.291001
1	0.291756	0.208744

	00	01	10	11
00	%CC	%CA	%AC	%AA
01	%CT	%CG	%AT	%AG
10	%TC	%TA	%GC	%GA
11	%TT	%TG	%GT	%GG

→ $P_2 =$

	00	01	10	11
00	0.054090	0.072739	0.050327	0.095037
01	0.071341	0.010305	0.074290	0.071368
10	0.060080	0.063123	0.044024	0.060083
11	0.095680	0.072859	0.050459	0.054195

Fig. 4.1.1. Top: typical view of a short fragment of genomic nucleotide sequence in a single strand of DNA from GenBank. Below: matrices on the left contain percentage symbols of 4 nucleotides and 16 duplets; the matrices $\mathbf{P}_1$ and $\mathbf{P}_2$ show percentage values of each of the nucleotides and duplets correspondingly in the 1-textual and 2-textual representations of the DNA nucleotide sequence of human chromosome №1.

text which is based on the alphabet of 16 duplets (AC-CT-GT-AA-CG-...), the third is a text which is based on the alphabet of 64 triplets (ACC-TGT-AAC-...), etc. Calculating in the mentioned DNA the percentages of each member of the DNA alphabet of 4 nucleotides (%C, %A, %T, %G), we obtain a table of their probabilities (Fig. 4.1.1, the first row). Then we represent the same DNA as a text of two-letter words based on the alphabet of 16 duplets and, calculating the percentage of each type of duplets, we get another table of percentages (Fig. 4.1.1, the second row). Similarly, presenting the same DNA as a text of three-letter words based on the alphabet of 64 triplets and as a text of four-letter words based on the alphabet of 256 tetraplets, we obtain the table of percentages of 64 triplets and 256 tetraplets (Figs. 4.1.2 and 4.1.3, respectively). This procedure can be prolonged for higher values

	000	001	010	011	100	101	110	111
000	%CCC	%CCA	%CAC	%CAA	%ACC	%ACA	%AAC	%AAA
001	%CCT	%CCG	%CAT	%CAG	%ACT	%ACG	%AAT	%AAG
010	%CTC	%CTA	%CGC	%CGA	%ATC	%ATA	%AGC	%AGA
011	%CTT	%CTG	%CGT	%CGG	%ATT	%ATG	%AGT	%AGG
100	%TCC	%TCA	%TAC	%TAA	%GCC	%GCA	%GAC	%GAA
101	%TCT	%TCG	%TAT	%TAG	%GCT	%GCG	%GAT	%GAG
110	%TTC	%TTA	%TGC	%TGA	%GTC	%GTA	%GGC	%GGA
111	%TTT	%TTG	%TGT	%TGG	%GTT	%GTG	%GGT	%GGG

→

$P_3 =$

	000	001	010	011	100	101	110	111
000	0.013845	0.018782	0.015236	0.018615	0.011825	0.019768	0.014466	0.036928
001	0.018532	0.002906	0.017889	0.021037	0.016224	0.002537	0.023753	0.019875
010	0.017584	0.012749	0.002508	0.002272	0.013175	0.019420	0.014406	0.022369
011	0.020088	0.020880	0.002591	0.002911	0.023881	0.017807	0.016136	0.018480
100	0.015877	0.019642	0.011031	0.019859	0.012545	0.014563	0.009615	0.019596
101	0.022258	0.002330	0.019392	0.012844	0.014374	0.002531	0.013271	0.017560
110	0.019724	0.019810	0.014573	0.019469	0.009561	0.011152	0.012560	0.015998
111	0.037246	0.018843	0.019881	0.018954	0.014452	0.015337	0.011855	0.013822

Fig. 4.1.2. The upper matrix contains percentage symbols of 64 triplets. The bottom matrix P_3 shows percentage values of all triplets in the 3-textual representation of the DNA nucleotide sequence of human chromosome №1.

.0033	.0055	.0042	.0044	.0040	.0056	.0032	.0070	.0030	.0042	.0040	.0053	.0032	.0059	.0055	.0149
.0041	.0010	.0044	.0058	.0047	.0010	.0040	.0044	.0041	.0005	.0051	.0054	.0047	.0006	.0095	.0071
.0050	.0029	.0008	.0006	.0036	.0039	.0049	.0059	.0037	.0032	.0006	.0006	.0040	.0070	.0037	.0066
.0048	.0058	.0006	.0009	.0057	.0047	.0045	.0058	.0049	.0044	.0007	.0007	.0071	.0057	.0049	.0047
.0052	.0057	.0027	.0038	.0010	.0006	.0003	.0006	.0033	.0045	.0032	.0063	.0042	.0049	.0039	.0078
.0058	.0009	.0035	.0028	.0007	.0003	.0005	.0008	.0049	.0005	.0064	.0035	.0046	.0007	.0049	.0059
.0048	.0036	.0046	.0051	.0005	.0004	.0008	.0006	.0047	.0056	.0033	.0050	.0031	.0037	.0047	.0057
.0073	.0044	.0053	.0058	.0006	.0010	.0005	.0010	.0096	.0040	.0051	.0044	.0046	.0047	.0040	.0041
.0046	.0049	.0042	.0051	.0024	.0045	.0028	.0078	.0030	.0041	.0029	.0038	.0023	.0037	.0029	.0073
.0057	.0006	.0051	.0052	.0037	.0004	.0056	.0036	.0047	.0008	.0033	.0046	.0030	.0005	.0046	.0048
.0057	.0040	.0005	.0006	.0030	.0055	.0026	.0041	.0032	.0027	.0006	.0005	.0025	.0031	.0032	.0057
.0068	.0058	.0006	.0006	.0070	.0039	.0032	.0029	.0037	.0048	.0006	.0008	.0041	.0036	.0036	.0051
.0051	.0063	.0034	.0063	.0041	.0049	.0031	.0061	.0023	.0031	.0017	.0035	.0031	.0042	.0023	.0052
.0077	.0006	.0063	.0038	.0049	.0006	.0045	.0057	.0038	.0004	.0032	.0027	.0042	.0010	.0034	.0052
.0073	.0078	.0038	.0051	.0037	.0046	.0042	.0051	.0030	.0028	.0029	.0042	.0023	.0024	.0030	.0046
.0150	.0072	.0055	.0045	.0059	.0057	.0043	.0054	.0055	.0032	.0040	.0042	.0032	.0039	.0030	.0033

Fig. 4.1.3. The matrix of percentage values of each of the 256 tetraplets in the DNA of human chromosome №1. Each tetraplet is located in the same cell in the matrix $[C, A; T, G]^{(4)}$ in Fig. 1.4.2 of Chapter 1.

of n, but for this book, it is enough to consider $n = 1, 2, 3, 4$. Figs. 4.1.1–4.1.3 show appropriate probabilities for the DNA of the first human chromosome, which contains a sequence of about 250 million nucleotides (initial data on this chromosome were taken from GenBank: https://www.ncbi.nlm.nih.gov/nuccore/NC_000001.11).

One should be reminded here that genomic sequences in the GenBank sites usually contain some letters N, indicating that there can be any nucleotide in this place (https://www.ncbi.nlm.nih.gov/books/NBK21136/). For this reason, the total amount of all nucleotides A, T, C, G, which are calculated for the sequence from GenBank, is slightly less than the complete length of the DNA sequence, which is indicated in GenBank. But practically this is not essential for the resulting values of percentages of separate nucleotides in the analyzed genomic sequences, each of which contains many millions of nucleotides (Fig. 4.11 shows a short fragment of genomic DNA).

One should note that structures of matrices, representing percentages of n-plets in Figs. 4.1.1–4.1.3, are identical to structures of matrices $[C, A; T, G]^{(n)}$, representing DNA alphabets of 4 nucleobases, 16 duplets, 64 triplets, and 256 tetraplets in a strictly ordered form in Chapter 1 (Fig. 1.2.1). But in cells of these matrices, the percentages of the corresponding n-plets are indicated instead of the symbols of the n-plets. One can recall that the columns and rows of the matrices of DNA alphabets are enumerated by binary numbers based on the following binary-oppositional molecular features of nucleotides: the columns are numbered due to binary indicators "pyrimidine-or-purine" ($C = T = 0$, $A = G = 1$), and the rows are numbered due to binary indicators "amino-or-keto" ($C = A = 0$, $T = G = 1$). Comparative analysis of hidden algebraic interconnections of these phenomenological percentage matrices has led to revealing universal rules of stochastic organization of genomic DNAs [70, 116]. These phenomenological universal rules, described below, are in line with the mentioned supposition of Jordan about the existence of such rules.

4.2. Positional tetra-groupings rule of percentage composition of n-plets

At first glance, the sets of values in the percentage matrices (Figs. 1.4.1–1.4.3) are quite chaotic and the percentage matrices have no numeric interrelations. They have the following features regarding the percentage of separate n-plets:

- The percentages of individual n-plets are not equal to the product of the percentages of their constituent nucleotides. For example, %CG = 0.0103 is not equal to %C × %G = 0.2085 × 0.2087 = 0.0435;
- Percentage of n-plets significantly depends on the order of letters in them. For example, the percentage of duplets CG and GC, having the same letter composition, differ several times: %CG = 0.0103, and %GC = 0.0440.

But in this seeming chaos, many universal rules for n-plet groupings were discovered that are valid for all genomes. For example, the values %C = 0.2085, %G = 0.2087, %A = 0.2910, %T = 0.2917 showed themselves in the block organization of percentages of different n-plets in various layers (or n-textual representations) of the genomic DNA sequence as one can calculate from data of percentage matrices in Figs. 4.1.1–4.1.3:

- The total sum Σ%CN of percentages of all 4 duplets CN (hereinafter, the symbol N denotes any of the nucleotides A, T, C, and G), which start with the nucleotide C, is equal to %C, that is, Σ%CN ≈ %CC + %CA + %CT + %CG ≈ 0.0541 + 0.0727 + 0.0713 + 0.0103 ≈ 0.2085 ≈ %C;
- The total sum Σ%NC of percentages of all 4 duplets NC, which have the nucleotide C at their second position, is practically equal to %C, that is, Σ%NC ≈ %CC + %AC + %TC + %GC = 0.0541 + 0.0503 + 0.0601 + 0.0440 ≈ 0.2085 ≈ %C as well;

- The total sum Σ%CNN of percentages of all 16 triplets CNN, which have the nucleotide C at their first position, is practically equal to %C, that is, Σ%CNN ≈ 0.0284 ≈ %C as well;
- The total sum Σ%NCN of percentages of all 16 triplets NCN, which have the nucleotide C at their second position, is practically equal to %C, that is, Σ%NCN ≈ 0.0285 ≈ %C as well;
- The total sum Σ%NNC of percentages of all 16 triplets NNC, which have the nucleotide C at their third position, is practically equal to %C, that is, Σ%NNC ≈ 0.0285 ≈ %C as well;
- The total sum Σ%CNNN of percentages of all 64 tetraplets CNNN, which have the nucleotide C at their first position, is practically equal to %C, that is, Σ%CNNN ≈ 0.0285 ≈ %C as well;
- The total sum Σ%NCNN of percentages of all 64 tetraplets NCNN, which have the nucleotide C at their second position, is practically equal to %C, that is, Σ%NCNN ≈ 0.0285 ≈ %C as well;
- The total sum Σ%NNCN of percentages of all 64 tetraplets NNCN, which have the nucleotide C at their third position, is practically equal to %C, that is, Σ%NNCN ≈ 0.0285 ≈ %C as well;
- The total sum Σ%NNNC of percentages of all 64 tetraplets NNNC, which have the nucleotide C at their fourth position, is practically equal to %C, that is, Σ%NNCN ≈ 0.0285 ≈ %C as well.

Similar equalities turn out to be valid also for the total sums Σ of the considered *n*-plets with nucleotides A, T, G at the analogical positions, as shown in Fig. 4.2.1.

Briefly speaking, the following relations (4.1) hold true — with a high level of accuracy — regarding percentages of the nucleotides C, G, A, and T, and the considered *n*-plets in the human chromosome №1:

$$\%C \approx \Sigma\%CN \approx \Sigma\%NC \approx \Sigma\%CNN \approx \Sigma\%NCN \approx \Sigma\%NNC \approx$$
$$\approx \Sigma\%CNNN \approx \Sigma\%NCNN \approx \Sigma\%NNCN \approx \Sigma\%NNNC;$$
$$\%G \approx \Sigma\%GN \approx \Sigma\%NG \approx \Sigma\%GNN \approx \Sigma\%NGN \approx \Sigma\%NNG \approx$$
$$\approx \Sigma\%GNNN \approx \Sigma\%NGNN \approx \Sigma\%NNGN \approx \Sigma\%NNNG;$$

%C ≈ 0.2085	%G ≈ 0.2087	%A ≈ 0.2910	%T ≈ 0.2918
Σ%CN ≈ 0.2085	Σ%GN ≈ 0.2088	Σ%AN ≈ 0.2910	Σ%TN ≈ 0.2917
Σ%NC ≈ 0.2085	Σ%NG ≈ 0.2087	Σ%NA ≈ 0.2910	Σ%NT ≈ 0.2918
Σ%CNN ≈ 0.2084	Σ%GNN ≈ 0.2088	Σ%ANN ≈ 0.2910	Σ%TNN ≈ 0.2917
Σ%NCN ≈ 0.2085	Σ%NGN ≈ 0.2088	Σ%NAN ≈ 0.2910	Σ%NTN ≈ 0.2917
Σ%NNC ≈ 0.2085	Σ%NNG ≈ 0.2087	Σ%NNA ≈ 0.2910	Σ%NNT ≈ 0.2918
Σ%CNNN ≈ 0.2085	Σ%GNNN ≈ 0.2088	Σ%ANNN ≈ 0.2910	Σ%TNNN ≈ 0.2917
Σ%NCNN ≈ 0.2085	Σ%NGNN ≈ 0.2087	Σ%NANN ≈ 0.2910	Σ%NTNN ≈ 0.2918
Σ%NNCN ≈ 0.2085	Σ%NNGN ≈ 0.2088	Σ%NNAN ≈ 0.2910	Σ%NNTN ≈ 0.2918
Σ%NNNC ≈ 0.2085	Σ%NNNG ≈ 0.2087	Σ%NNNA ≈ 0.2910	Σ%NNNT ≈ 0.2918

Fig. 4.2.1. Percentages of nucleotides C, G, A, T, and the sum Σ of percentage of *n*-plets with these nucleotides at certain positions for the case of DNA of the human chromosome №1. The symbol N denotes any of the nucleotides.

$$\begin{aligned}
&\%A \approx \Sigma\%AN \approx \Sigma\%NA \approx \Sigma\%ANN \approx \Sigma\%NAN \approx \Sigma\%NNA \approx \\
&\quad \approx \Sigma\%ANNN \approx \Sigma\%NANN \approx \Sigma\%NNAN \approx \Sigma\%NNNA; \\
&\%T \approx \Sigma\%TN \approx \Sigma\%NT \approx \Sigma\%TNN \approx \Sigma\%NTN \approx \Sigma\%NNT \approx \\
&\quad \approx \Sigma\%TNNN \approx \Sigma\%NTNN \approx \Sigma\%NNTN \approx \Sigma\%NNNT
\end{aligned} \tag{4.1}$$

These equalities (4.1) can also be written in the form (4.2) of the equality of four-dimensional vectors of genomic percentages, the coordinates of which are the percentages of individual nucleotides or the percentage sums of *n*-plets with nucleotides at certain positions:

$$\begin{aligned}
&[\%A, \%T, \%C, \%G] \approx [\Sigma\%AN, \Sigma\%TN, \Sigma\ \%CN, \Sigma\%GN] \approx \\
&[\Sigma\%NA, \Sigma\%NT, \Sigma\%NC, \Sigma\%NG] \approx \\
&[\Sigma\%ANN, \Sigma\%TNN, \Sigma\%CNN, \Sigma\%GNN] \approx \\
&[\Sigma\%NAN, \Sigma\%NTN, \Sigma\%NCN, \Sigma\%NGN] \approx \\
&[\Sigma\%NNA, \Sigma\%NNT, \Sigma\%NNC, \Sigma\%NNG] \approx \\
&[\Sigma\%ANNN, \Sigma\%TNNN, \Sigma\%CNNN, \Sigma\%GNNN] \approx \\
&[\Sigma\%NANN, \Sigma\%NTNN, \Sigma\%NCNN, \Sigma\%NGNN] \approx \\
&[\Sigma\%NNAN, \Sigma\%NNTN, \Sigma\%NNCN, \Sigma\%NNGN] \approx \\
&[\Sigma\%NNNA, \Sigma\%NNNT, \Sigma\%NNNC, \Sigma\%NNNG].
\end{aligned} \tag{4.2}$$

Knowing the percentages of nucleotides %A, %T, %C, and %G, it is possible to predict with high accuracy the sums of percentages of n-plets of the noted groupings. The ability of such predictions based on equalities (4.1) or (4.2) holds not only for the considered human chromosome №1 but also for many eukaryotic and prokaryotic genomes, including the following (see numerical data in [154]):

- all 24 human chromosomes;
- all chromosomes of *Drosophila*, mouse, worm, many plants;
- 19 genomes of bacteria and archaea;
- many extremophiles, living in extreme conditions, for example, radiation with a level 1,000 times higher than fatal for humans.

One should note that percentages of nucleotides %A, %T, %C, and %G can be essentially different in various genomes. For example, the genome of bacteria *Bradyrhizobium japonicum* contains %A ≈ 0.1819, %T ≈ 0.1815, %C ≈ 0.3184, and %G ≈ 0.3182 in contrast to the considered case of the human chromosome №1. This indicates a universal cooperative organization of percentage compositions of n-plets in genomic DNA texts, which can be comfortably studied using the genetic matrices of alphabets of n-plets (Fig. 4.1.1–4.1.3).

The four columns in Fig. 4.2.1 show that in each of the presented layers of the genomic DNA text, there exist corresponding tetra-groupings of n-plets with the same percentage sums. The following **rule of probabilities** can be formulated based on such results:

- All n layers (or n-textual presentations) of a long DNA text, each of which consists of a sequence of 4^n n-plets, has approximately the same sum of percentages of all those n-plets, which contain the considered nucleotide (C, G, A, or T) at a fixed m^{th} position ($m \leq n$); here $n = 1, 2, 3, \ldots$ but is not too large compared to the length of DNA.

By this rule, the percentage sums of n-plets in such groupings in different layers of a long DNA text are equal to the percentage of a corresponding nucleotide in the first layer of the DNA text, although the percentage values of individual n-plets, that are summands in these sums, can differ significantly. For example, in the second layer (or the 2-textual presentation) of DNA sequence of human chromosome № 1, the sum of the percentages of all 4 duplets with nucleotide C in the first position ($m = 1$) is equal to $\Sigma\%CN \approx \%CC + \%CA + \%CT + \%CG \approx 0.0541 + 0.0727 + 0.0713 + 0.0103 \approx 0.2085$. In the same second layer, the sum of the percentages of all 4 duplets with a nucleotide C in the second position ($m = 2$) is equal to the same number, although the summands in this sum are significantly different: $\Sigma\%NC \approx \%CC + \%AC + \%TC + \%GC = 0.0541 + 0.0503 + 0.0601 + 0.0440 \approx 0.2085$. These two equal total values are equal to the percentage of nucleotide C in the first layer of the given genomic DNA text: $\%C \approx 0.2085$.

This rule of probabilities can be also formulated in the form of an exclusion principle as a supposed universal principle of exclusion in genetics:

- Sums of percentages of all those n-plets, which contain the considered nucleotide (C, G, A, or T) at a fixed m^{th} position ($m \leq n$), cannot be essentially different from each other in various n layers of any genomic DNA text (here $n = 1, 2, 3, \ldots$ but is not too large compared to the length of the DNA text).

This rule of the constancy in the sums of percentages in different layers of long DNA texts (in the indicated tetra-groupings of n-plets), which is relatively independent of the percentage values of separate n-plets, is called the **Gestalt genetic rule** of percentages of n-plets for their C-, G-, A-, and T-groupings in multilayer DNA texts of genomes. The names "Gestalt genetic phenomena" and "Gestalt genetic rules" are due to a similarity of these genetic phenomena with the phenomena of Gestalt psychology (this similarity is described below in more detail). In Gestalt psychology, the term "Gestalt" is traditionally capitalized; due to this tradition, we also capitalize this term in the name "Gestalt genetics".

Each of the tetra-groupings of n-plets, which is defined by a disposition of attributive nucleotides C, G, A, and T at a certain position m ($m \leq n$)

in these n-plets, we call as an m-positional tetra-grouping. For example, a tetra-grouping corresponding to sets ANN, TNN, CNN, and GNN is called as a 1-positional tetra-grouping; a tetra-grouping corresponding to NAN, NTN, NCN, NGN is called as a 2-positional tetra-grouping, and so on. The described phenomena, reflected in the Gestalt genetic rule, have analogies with the phenomena of holography, in which it is possible to reconstruct the image of a whole object from the image of its piece (see Chapter 2 in this book about algebraic holography in genetics). Indeed, the knowledge of the sums of the percentages of n-plets in m-positional tetra-groupings of one layer of a long DNA text gives knowledge about the sums of the percentages of n-plets in m-positional tetra-groupings of other layers.

Returning for a moment to the tensor family of matrices [C, A; T, G]$^{(n)}$ (Fig. 1.2.1), let us consider conditional (or model) percentages of n-plets in a tensor family of percentage matrices [%C, %A; %T, %G]$^{(n)}$ ≈ [0.2085, 0.2910; 0.2917, 0.2089]$^{(n)}$. Figures 4.2.2 and 4.2.3 show the obtained matrices for (n) = 2, 3, whose percentage entries significantly differ from real percentages of n-plets shown above in Figs. 2.3–2.4.

But unexpectedly the Gestalt rule holds for these model percentages of n-plets by the expression (4.3) as well:

$$\begin{aligned}&[\%A, \%T, \%C, \%G] \approx [\Sigma\%AN, \Sigma\%TN, \Sigma\%CN, \Sigma\%GN] \approx\\ &[\Sigma\%NA, \Sigma\%NT, \Sigma\ \%NC, \Sigma\%NG] \approx\\ &[\Sigma\%ANN, \Sigma\%TNN, \Sigma\%CNN, \Sigma\%GNN] \approx\\ &[\Sigma\%NAN, \Sigma\%NTN, \Sigma\%NCN, \Sigma\%NGN] \approx\\ &[\Sigma\%NNA, \Sigma\%NNT, \Sigma\%NNC, \Sigma\%NNG] \approx [0.2910, 0.2918,\\ &\qquad 0.2085, 0.2087] \qquad (4.3)\end{aligned}$$

%CC	%CA	%AC	%AA		0.0435	0.0607	0.0607	0.0847
%CT	%CG	%AT	%AG		0.0608	0.0435	0.0849	0.0607
%TC	%TA	%GC	%GA	=	0.0608	0.0849	0.0435	0.0607
%TT	%TG	%GT	%GG		0.0851	0.0609	0.0609	0.0436

Fig. 4.2.2. The matrix of model percentages for 16 duplets, which is the second tensor power of the matrix [0.2085, 0.2910; 0.2917, 0.2089] (all values are rounded to the fourth decimal place).

%CCC	%CCA	%CAC	%CAA	%ACC	%ACA	%AAC	%AAA
%CCT	%CCG	%CAT	%CAG	%ACT	%ACG	%AAT	%AAG
%CTC	%CTA	%CGC	%CGA	%ATC	%ATA	%AGC	%AGA
%CTT	%CTG	%CGT	%CGG	%ATT	%ATG	%AGT	%AGG
%TCC	%TCA	%TAC	%TAA	%GCC	%GCA	%GAC	%GAA
%TCT	%TCG	%TAT	%TAG	%GCT	%GCG	%GAT	%GAG
%TTC	%TTA	%TGC	%TGA	%GTC	%GTA	%GGC	%GGA
%TTT	%TTG	%TGT	%TGG	%GTT	%GTG	%GGT	%GGG

.00906382	.01265035	.01265035	.01765605	.01265035	.01765605	.01765605	.02464250
.01268315	.00907448	.01770183	.01266522	.01770183	.01266522	.02470639	.01767681
.01268315	.01770183	.00907448	.01266522	.01770183	.02470639	.01266522	.01767681
.01774773	.01269806	.01269806	.00908514	.02477045	.01772264	.01772264	.01268011
.01268315	.01770183	.01770183	.02470639	.00907448	.01266522	.01266522	.01767681
.01774773	.01269806	.02477045	.01772264	.01269806	.00908514	.01772264	.01268011
.01774773	.02477045	.01269806	.01772264	.01269806	.01772264	.00908514	.01268011
.02483467	.01776859	.01776859	.01271298	.01776859	.01271298	.01271298	.00909582

Fig. 4.2.3. The matrix of model percentages for 64 triplets, which is the third tensor power of the matrix [0.2085, 0.2910; 0.2917, 0.2089].

It should be especially noted that in the case of a tensor family of model percentage matrices $[\%C, \%A; \%T, \%G]^{(n)}$, which are quite different from phenomenological percentage matrices, we have the absolute accuracy of fulfillment of the Gestalt rule.

It gives additional pieces of evidence that fundamental genetic phenomena, reflected in the formulated Gestalt rule, relate to the algebraic operation of the tensor product, which is so important in quantum mechanics and quantum informatics.

4.3. DNA epi-chains and the Gestalt rule of percentages of *n*-plets

This section presents some results of the analogical study of percentages of *n*-plets in special sub-sequences of long nucleotide sequences in single-stranded DNA. These sub-sequences are termed «DNA epi-chains» [153, 198].

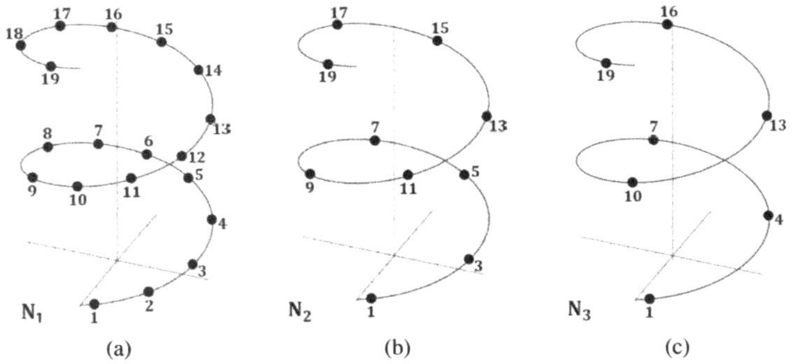

Fig. 4.3.1. Schematic representations of a single-stranded DNA and its initial epi-chains of numerated nucleotides, denoted by black circles. (a) a sequence N_1 of numerated nucleotides of the DNA strand; (b) an epi-chain of the second-order N_2 having nucleotides with numbers 1-3-5-7-...; (c) an epi-chain of the third-order N_3 nucleotide numbers 1-4-7-10-...

The author's results testify that the above-described equalities (4.1) and (4.2) for total sums of percentages of n-plets hold for these epi-chains as well.

By definition, in a nucleotide sequence N_1 of any DNA strand with sequentially numbered nucleotides 1, 2, 3, 4, ... (Fig. 4.3.1a), epi-chains of different orders k are such sub-sequences that contain only those nucleotides whose numeration differ from each other by natural number $k = 1, 2, 3, 4, \ldots$ For example, in any single-stranded DNA, one can consider its epi-chain of the second-order N_2, in which its nucleotide sequence numbers differ by k = 2: an epi-chain N_2 contains nucleotides with numerations 1, 3, 5, ... (Fig. 4.3.1b). By analogy, an epi-chain of the third-order N_3 is connected with k = 3 and contains a sub-sequence of nucleotides with numerations 1, 4, 7, 10, ... (Fig. 4.3.1c).

Each genomic DNA epi-chain of kth order (if $k = 2, 3, 4, \ldots$) contains k times fewer nucleotides than the DNA strand and has its own arrangements of nucleobases A, T, C, and G. Each of these epi-chains contains different percentages of corresponding n-plets. But unexpectedly the total sums of percentages of n-plets in C-, G-, A-, and T-groupings practically coincide for each of the epi-chains and for the complete genomic DNA text (at this stage of the research, the percentages of n-plets in epi-chains are studied only in cases of epi-chains with relatively small orders of k).

%CC	%CA	%AC	%AA		0.05409	0.07274	0.05033	0.09504
%CT	%CG	%AT	%AG		0.07134	0.01031	0.07429	0.07137
%TC	%TA	%GC	%GA	≈	0.06008	0.06312	0.04402	0.06008
%TT	%TG	%GT	%GG		0.09568	0.07286	0.05046	0.05419

Epi-chain of the 2nd order				Epi-chain of the 3rd order				Epi-chain of the 4th order			
0.0492	0.0524	0.0538	0.0985	0.0487	0.0562	0.0579	0.0912	0.0487	0.0558	0.0574	0.0940
0.0591	0.0478	0.0800	0.0588	0.0571	0.0464	0.0849	0.0570	0.0566	0.0475	0.0831	0.0565
0.0612	0.0788	0.0443	0.0613	0.0608	0.0826	0.0411	0.061	0.0579	0.0835	0.0445	0.0579
0.0988	0.0530	0.0539	0.0492	0.0918	0.0565	0.0578	0.0489	0.0943	0.0561	0.0577	0.0487

Fig. 4.3.2. Matrices of percentages of 16 duplets in the single-stranded DNA of the human chromosome №1. Upper row: percentage matrices of the genomic DNA ($k = 1$). Bottom row: percentage matrices for the DNA epi-chains of the second, third, and fourth orders ($k = 2, 3, 4$).

To illustrate this result, Fig. 4.3.2 shows the percentage matrices for epi-chains of the second, third and fourth orders ($k = 2, 3, 4$) in the DNA of the human chromosome №1.

One can see from data in Fig. 4.3.2 that for considered epi-chains, different percentage matrices contain essentially different percentages in corresponding separate cells. For example, %CG = 0.0103 in the complete single-stranded DNA; %CG = 0.0478 in the epi-chain of the 2nd order; %CG = 0.0464 in the epi-chain of the 3rd order; %CG = 0.0475 in the epi-chain of the 4th order. But in each of these four percentage matrices the equalities (4.2) are realized with a high level of accuracy:

$$[\%A, \%T, \%C, \%G] \approx [\Sigma\%AN, \Sigma\ \%TN, \Sigma\%CN, \Sigma\%GN] \approx$$
$$[\Sigma\%NA, \Sigma\ \%NT, \Sigma\%NC, \Sigma\%NG] \approx [0.2910, 0.2918, 0.2085, 0.2087] \quad (4.4)$$

It illustrates that in the considered DNA epi-chains, the sums of the percentages in the tetra-groupings of n-plets in each of the epi-chains practically do not depend on the percentage of its separate n-plets and coincide with corresponding sums in the genomic DNA text. This can be considered as the separate Gestalt rule for genomic epi-chains (in addition to the above-presented gestalt rule of percentages of n-plets in long multilayer DNA texts).

4.4. Graphical representation of rules of *n*-plet tetra-groupings in genomic DNA stochastics

Let us return to the table (Fig. 4.2.1) of tetra-groupings for the sums of genomic percentages of *n*-plets in the layers of the DNA text and see how the *n*-plets of these groupings are located in the cells of the initial tensor family of genetic matrices [C, A; T, G]$^{(n)}$. Taking into account the belonging of each *n*-plet to one of the 4 columns of the table, that is, to one of the C-, G-, A-, and T-groupings, we introduce the following symbolic notation for each of the *n*-plets:

- *n*-plets of the first column (C-column) are denoted by the symbol ♠;
- *n*-plets of the second column (G-column) are denoted by the symbol ♣;
- *n*-plets of the third column (A-column) are denoted by the symbol ♥;
- *n*-plets of the fourth column (T-column) are denoted by the symbol ♦.

Figures 4.4.1–4.4.4 show the arrangement of *n*-plets belonging to C-, G-, A-, T-groupings of *n*-plets in the matrices of the tensor family [C, A; T, G]$^{(n)}$ under $n = 1, 2, 3, 4$.

One can see that all mosaic matrices in Figs. 4.4.1–4.4.4 are bisymmetric about arrangements of black and red symbols. They also possess meander-like patterns: each column and each row consist of sequences of fragments

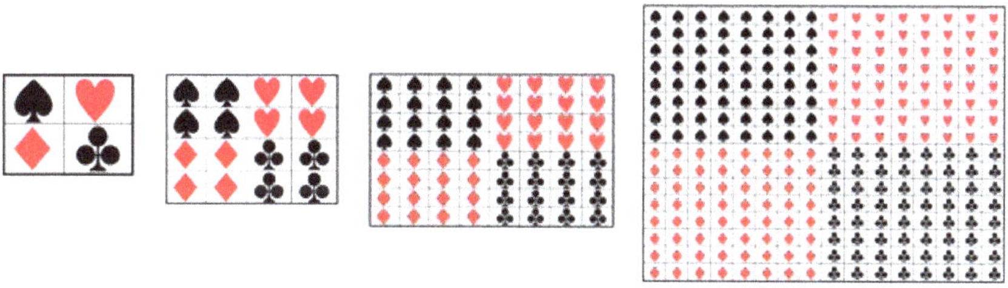

Fig. 4.4.1. The arrangements of *n*-plets, which start with the nucleotides C, G, A, or T, in the matrices of the tensor family [C, A; T, G]$^{(n)}$ under $n = 1, 2, 3, 4$ (see explanations in the text).

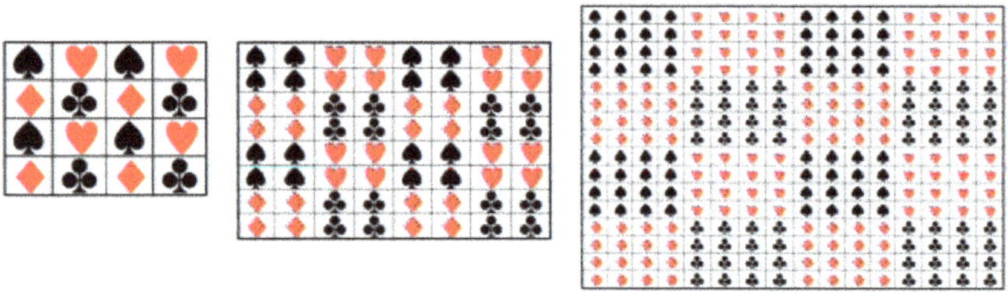

Fig. 4.4.2. The arrangements of n-plets, which have the nucleotides C, G, A, or T at their second position, in the matrices of the tensor family $[C, A; T, G]^{(n)}$ under $n = 2, 3, 4$ (see explanations in the text).

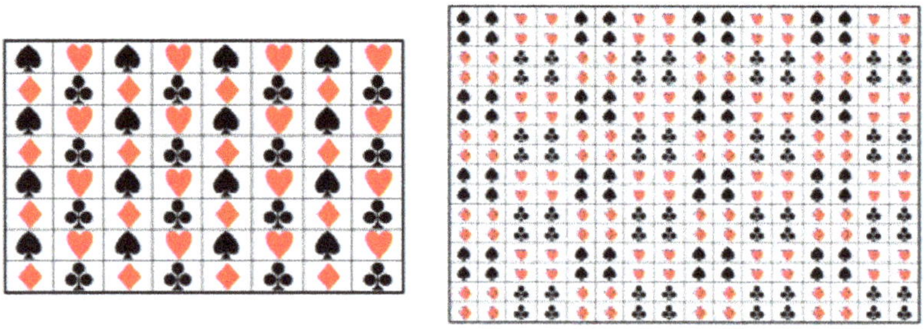

Fig. 4.4.3. The arrangements of 64 triplets and 256 tetraplets, which have the nucleotides C, G, A, or T at their third position, in the matrices of the tensor family $[C, A; T, G]^{(n)}$ under $n = 3, 4$ (see explanations in the text).

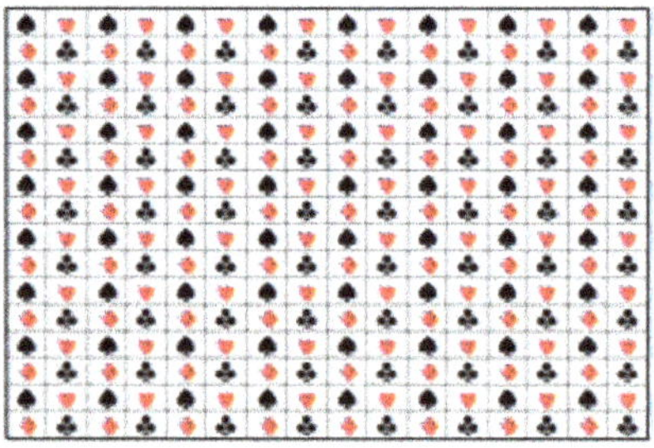

Fig. 4.4.4. The arrangements of 256 tetraplets, which have the nucleotides C, G, A, T at their fourth position, in the matrix $[C, A; T, G]^{(4)}$ (see explanations in the text).

of equal lengths with black and red symbols. These meander-like patterns resemble meander functions of Rademacher, which is well known in the theory of noise-immune coding of information and which were noted in Chapter 1.

4.5. Projectors and tetra-groupings of the genomic percentages of *n*-plets in genetic matrices

Let us continue studying properties of the percentage content of *n*-plets in the multilayer presentation of long DNA texts. This section is devoted to another type of tetra-groupings called projector tetra-groupings of *n*-plets since they are related to oblique projectors. The projector tetra-groupings are connected with the corresponding projector Gestalt rule described below.

Our visual perception and a "projection method" in the drawing are based on projections of external objects on the retina and a drawing plane (Fig. 4.5.1). In mathematics, such operations of projections are expressed using square matrices, which are called "projection operators" (or, briefly, "projectors").

The notion of projection operators is one of the important notions in many non-biological sciences: physics, including quantum mechanics; mathematics; computer science and informatics, including the theory of digital codes; chemistry; mathematical logic, etc. The matrix P (Fig. 4.5.2)

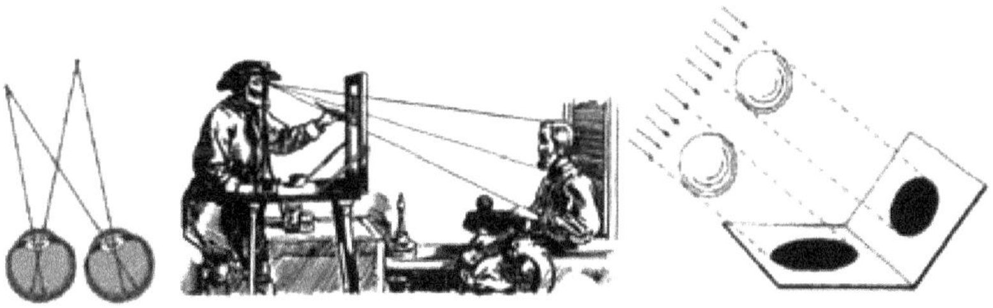

Fig. 4.5.1. Examples of projection. Left: the path of light rays entering the eyes. Middle: ray projection in the old manner of painting (famous painting by A. Dürer, 1525). Right: oblique and orthogonal projection.

$$P = \begin{bmatrix} 1 & 0 & 0 \\ 0 & 1 & 0 \\ 0 & 0 & 0 \end{bmatrix}. \quad P\begin{pmatrix} x \\ y \\ z \end{pmatrix} = \begin{pmatrix} x \\ y \\ 0 \end{pmatrix}.$$

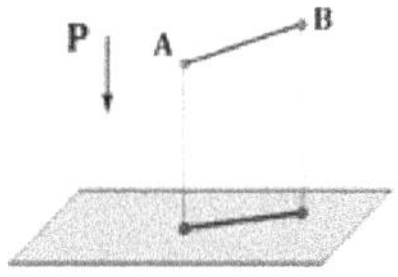

Fig. 4.5.2. Example of a matrix operator P, which provides a projection operation.

is an example of a projector, making a projection of vector [x, y, z] on the plane [x, y, 0]:

Projectors are expressed using square matrices (http://mathworld.wolfram.com/ProjectionMatrix.html). A necessary condition for a matrix P to be a projection operator is $P^2 = P$. A set of projectors is separated into two subsets:

- Orthogonal projectors, which are expressed by symmetric matrices (orthogonal projections should not be confused with orthogonal transformations, which preserve vector lengths). Their theory is well developed and has a lot of applications;
- Oblique projectors, which are expressed by non-symmetric matrices. Their theory and applications are developed much more weakly. They are connected unexpectedly with genetic structures, and will be the main objects of attention in this section.

Now let us describe another type of tetra-groupings of n-plets in different layers of genomic DNA texts, whose percentage sums conserve constant values both in different layers and also in different projector tetra-groupings of the nth layer ($n = 2, 3, 4,\ldots$, but much less than the length of the genomic DNA text considered). We conventionally call tetra-groupings of this new type the projector tetra-groupings to distinguish them from the above-considered m-positional tetra-groupings presented in expressions (4.1) and (4.2) and additionally called hyperbolic tetra-groupings [155]. The reason for the name "projector" for these tetra-groupings will be explained below.

Let us return to the percentage matrices of 16 duplets and 64 triplets in the human chromosome №1 (Figs. 4.1.1 and 4.1.2). To facilitate a visual presentation of locations of these 4 projector tetra-groupings of n-plets in the genetic matrices, the percentage matrices from Figs. 4.1.1 and 4.1.2 are transformed in corresponding symbolic matrices in Fig. 4.5.3 with 4 types of playing card symbols in their cells: spades (♠), hearts (♥), diamonds (♦), and clubs (♣). Each of these 4 symbols is located in those cells, which contain percentages of n-plets, belonging to an appropriate grouping.

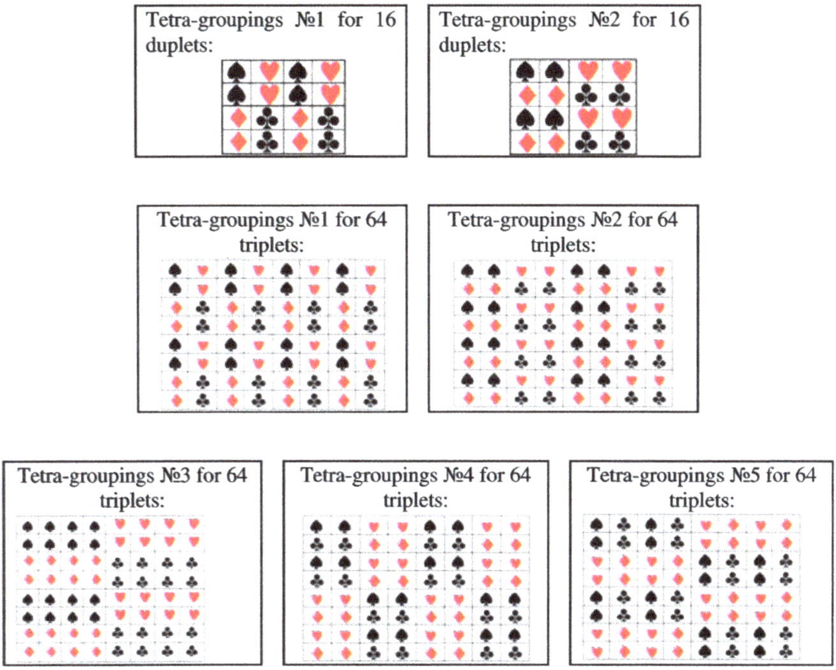

Fig. 4.5.3. Arrangements of n-plets from 4 different projector tetra-groupings in matrices for 16 duplets and 64 triplets from Fig. 4.1.1 and 4.1.2. In each matrix, corresponding cells, containing percentages of one of the 4 groupings of n-plets, are denoted by one of 4 symbols of playing cards.

The amazing result is the constancy of percentage sums Σ%♥, Σ%♦, Σ%♠, and Σ%♣ of n-plets, which are denoted by the same symbol, in all matrices with different mosaics of tetra-groupings shown in Fig. 4.5.3:

$$\Sigma\%\heartsuit \approx \text{constant},\ \Sigma\%\diamondsuit \approx \text{constant},\ \Sigma\%\spadesuit \approx \text{constant},\ \Sigma\%\clubsuit \approx \text{constant}. \quad (4.5)$$

For example, for the human chromosome №1, whose duplets percentage (4×4)-matrix is shown in Fig. 4.3.2, the following sums exist for the projector tetra-grouping №1 of 16 duplets shown in Fig. 4.5.3:

$$\Sigma\%\heartsuit = \%CA + \%CG + \%AA + \%AG \approx 0.07274 + 0.01031 + 0.09504 + 0.07137 \approx \mathbf{0.250},$$

$$\Sigma\%\diamondsuit = \%TC + \%TT + \%GC + \%GT \approx 0.06008 + 0.09568 + 0.04402 + 0.05046 \approx \mathbf{0.250},$$

$$\Sigma\%♠ = \%CC + \%CT + \%AC + \%AT \approx 0.05409 + 0.07134 + 0.05033 + 0.07429 \approx \mathbf{0.250},$$
$$\Sigma\%♣ = \%TA + \%TG + \%GA + \%GG \approx 0.06312 + 0.07286 + 0.06008 + 0.05419 \approx \mathbf{0.250}. \quad (4.6)$$

The tetra-grouping №2 for 16 duplets, shown in Fig. 4.5.3, has another combination of 4 duplets in each of its 4 groupings but the same values of percentage sums for these duplets exist by analogy with (4.6):

$$\Sigma\%♥ = \%AC + \%AA + \%GC + \%GA \approx 0.05033 + 0.09504 + 0.04402 + 0.06008 \approx \mathbf{0.250},$$
$$\Sigma\%♦ = \%CT + \%CG + \%TT + \%TG \approx 0.07134 + 0.01031 + 0.09568 + 0.07286 \approx \mathbf{0.250},$$
$$\Sigma\%♠ = \%CC + \%CA + \%TC + \%TA \approx 0.05409 + 0.07274 + 0.06008 + 0.06312 \approx \mathbf{0.250},$$
$$\Sigma\%♣ = \%AT + \%AG + \%GT + \%GG \approx 0.07429 + 0.07137 + 0.05046 + 0.05419 \approx \mathbf{0.250}. \quad (4.7)$$

A similar rule of total percentage constancy holds for all five (8 × 8)-matrices in Fig. 4.5.3, whose mosaics present different projector tetra-groupings for the alphabet of 64 triplets. The reader can check this by direct summation of corresponding percentages. We illustrate this constancy here only for the total percentage Σ%♠ of 16 triplets marked by the symbol ♠, using mosaic matrices in Fig. 4.5.3 and percentage matrices in Fig. 4.1.2.

The tetra-grouping №1 for 64 triplets in Fig. 4.5.3 gives the following:

$$\Sigma\%♠ = \%CCC + \%CCT + \%CAC + \%CAT + \%ACC + \%ACT + \%AAC + \%AAT + \%TTC + \%TCT + \%TAC + \%TAT + \%GCC + \%GCT + \%GAC + \%GAT$$

$$\approx 0.0138 + 0.0185 + 0.0152 + 0.0179 + 0.0118 + 0.0162 + 0.0145 + 0.0238 + 0.0159 + 0.0223 + 0.0110 + 0.0194 + 0.0125 + 0.0144 + 0.0096 + 0.0133 \approx \mathbf{0.250}. \quad (4.8)$$

The projector tetra-grouping №2 for 64 triplets in Fig. 4.5.3 gives the following:

$$\Sigma\%\spadesuit = \%CCC + \%CCA + \%ACC + \%ACA + \%CTC + \%CTA + \%ATC + \%ATA + \%TCC + \%TCA + \%GCC + \%GCA + \%TTC + \%TTA + \%GTC + \%GTA$$
$$\approx 0.0138 + 0.0188 + 0.0118 + 0.0198 + 0.0176 + 0.0127 + 0.0132 + 0.0194 + 0.0159 + 0.0196 + 0.0125 + 0.0146 + 0.0197 + 0.0198 + 0.0096 + 0.0112 \approx \mathbf{0.250}. \quad (4.9)$$

The tetra-grouping №3 for 64 triplets in Fig. 4.5.3 gives the following:

$$\Sigma\%\spadesuit = \%CCC + \%CCA + \%CAC + \%CAA + \%CCT + \%CCG + \%CAT + \%CAG + \%TCC + \%TCA + \%TAC + \%TAA + \%TCT + \%TCG + \%TAT + \%TAG$$
$$\approx 0.0138 + 0.0188 + 0.0152 + 0.0186 + 0.0185 + 0.0029 + 0.0179 + 0.0210 + 0.0159 + 0.0196 + 0.0110 + 0.0199 + 0.0223 + 0.0023 + 0.0194 + 0.0128 \approx \mathbf{0.250}. \quad (4.10)$$

The tetra-grouping №4 for 64 triplets in Fig. 4.5.3 gives the following:

$$\Sigma\%\spadesuit = \%CCC + \%CCA + \%ACC + \%ACA + \%CTC + \%CTA + \%ATC + \%ATA + \%TAC + \%TAA + \%GAC + \%GAA + \%TTA + \%TGA + \%GGC + \%GGA$$
$$\approx 0.0138 + 0.0188 + 0.0118 + 0.0198 + 0.0176 + 0.0127 + 0.0132 + 0.0194 + 0.0110 + 0.0096 + 0.0196 + 0.0146 + 0.0195 + 0.0126 + 0.0160 \approx \mathbf{0.250}. \quad (4.11)$$

The projector tetra-grouping №5 for 64 triplets in Fig. 4.5.3 gives the following:

$$\begin{aligned}\Sigma\%\spadesuit &= \%CCC + \%CCT + \%CAC + \%CAT + \%ATC + \\ &\quad \%ATT + \%AGC + \%AGT + \%TCC + \%TCT + \%TAC + \%TAT \\ &\quad + \%GTC + \%GTT + \%GGC + \%GGT \\ &\approx 0.0138 + 0.0185 + 0.0152 + 0.0179 + 0.0132 + 0.0239 \\ &\quad + 0.0144 + 0.0161 + 0.0159 + 0.0223 + 0.0110 + 0.0194 \\ &\quad + 0.0096 + 0.0145 + 0.0126 + 0.0119 \\ &\approx \mathbf{0.250}. \end{aligned} \tag{4.12}$$

In the particular numeric example of the human chromosome №1, the expressions (4.6–4.12) show that $\Sigma\%\heartsuit \approx \Sigma\%\diamondsuit \approx \Sigma\%\spadesuit \approx \Sigma\%\clubsuit \approx 0.250$. But in the general case, for genomic DNA texts, the following two approximate equalities are fulfilled:

$$\Sigma\%\heartsuit \approx \Sigma\%\diamondsuit, \qquad \Sigma\%\spadesuit \approx \Sigma\%\clubsuit \tag{4.13}$$

For example, in the case of the bacterial genome of *Bradyrhizobium japonicum* (see Appendix 4.I) the following equalities are fulfilled for projector tetra-groupings of n-plets in all 7 mosaic matrices of duplets and triplets in Fig. 4.5.3:

$$\Sigma\%\heartsuit \approx \Sigma\%\diamondsuit \approx 0.277, \qquad \Sigma\%\spadesuit \approx \Sigma\%\clubsuit \approx 0.223 \tag{4.14}$$

It should be noted that there is a significant difference of the Gestalt phenomena in the case of m-positional tetra-groupings, which are related to the expressions (4.1) or (4.2), and in the case of projector tetra-groupings, which are related to the expression (4.13). In m-positional tetra-groupings, values of percentage sums of n-plets in each of the 4 groupings are equal to the percentage of appropriate nucleotides A, T, C, or G. In projector tetra-groupings, values of percentage sums of n-plets in each of the 4 groupings have no obvious connection with the percentage of separate nucleotides.

We should mention that only some examples of projector tetra-groupings of n-plets are presented in this book. More projector tetra-groupings of n-plets exist in the set of 256 tetraplets. It seems natural to suppose that projector tetra-groupings exist also in sets of 4^n kinds of n-plets under $n = 5, 6, 7, \ldots$ but the authors did not check this supposition.

This genomic phenomenon of the constancy of summary percent (4.14) in different projector tetra-groupings, having different percentages of separate n-plets in the described layers of long DNA texts, is the basis of a corresponding **Gestalt genetic rule for projector tetra-groupings** [154, 155].

Figure 4.5.4 shows that arrangements of 4 playing cards, presenting Gestalt projector groupings in (4×4)-matrices for 16 duplets and in (8×8)-matrices of 64 triplets (Fig. 4.5.3), have a connection based on an algorithm of the tensor product by the (2×2)-matrix [1, 1; 1, 1], which is the matrix presentation of the hyperbolic number with unit coordinates.

A list of connections of oblique projection operators with the percentage content of n-plets in the multilayer presentation of genomic DNA texts is far from being limited to the tetra-groupings noted above in this book.

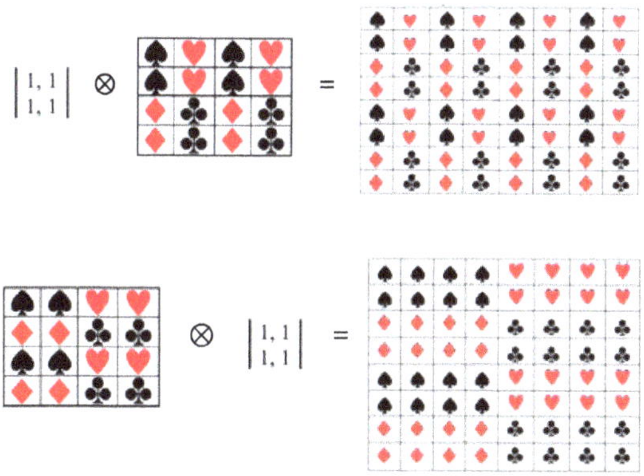

Fig. 4.5.4. Tensor interrelations between some projector (4×4)- and (8×8)-matrices from Fig. 4.5.3. The symbol $\otimes$ refers to the tensor product.

4.6. Projector decompositions of the matrices bearing projector tetra-groupings

This section shows a special decomposition of each of the symbolic matrices in Fig. 4.5.3 and explains why the tetra-groupings of *n*-plets in these matrices are conditionally called the projector groupings. The presented data additionally indicate that the probability distribution of *n*-plets in genomic DNAs is far from being random, but obeys algebraic laws.

Figure 4.6.1 shows a decomposition of the symbolic matrix of tetra-grouping №1 for 16 duplets from Fig. 4.5.3, which presents this matrix as the sum of four symbolic sparse matrices. Arrangements of symbols in these sparse matrices are not accidental but correspond to arrangements of non-zero entries in a set of four numeric matrices e_0, e_1, e_2, and e_3, each of which is an oblique projector since $e_0^2 = e_0$, $e_1^2 = e_1$, $e_2^2 = e_2$, and $e_3^2 = e_3$. Besides the set of these projectors e_0, e_1, e_2, and e_3 is closed relative to multiplication: multiplication of any two of these projectors gives a projector from this set.

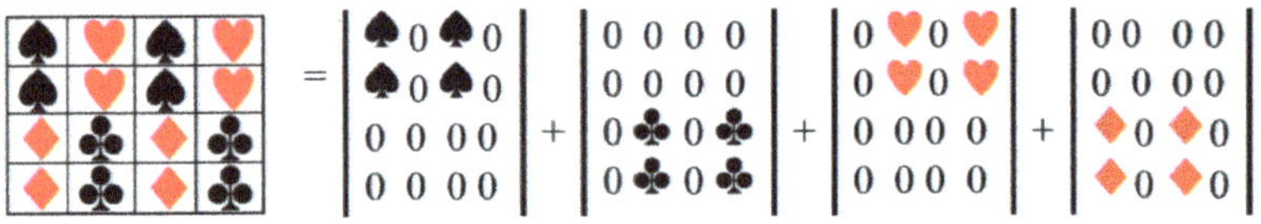

$$e_0+e_1+e_2+e_3 = \begin{vmatrix}1&0&1&0\\1&0&1&0\\0&0&0&0\\0&0&0&0\end{vmatrix} + \begin{vmatrix}0&0&0&0\\0&0&0&0\\0&1&0&1\\0&1&0&1\end{vmatrix} + \begin{vmatrix}0&1&0&1\\0&1&0&1\\0&0&0&0\\0&0&0&0\end{vmatrix} + \begin{vmatrix}0&0&0&0\\0&0&0&0\\1&0&1&0\\1&0&1&0\end{vmatrix};$$

*	e_0	e_1	e_2	e_3
e_0	e_0	e_2	e_2	e_0
e_1	e_3	e_1	e_1	e_3
e_2	e_0	e_2	e_2	e_0
e_3	e_3	e_1	e_1	e_3

Fig. 4.6.1. Upper: the decomposition of the symbolic matrix of the projector tetra-grouping №1 for 16 duplets (from Fig. 4.5.3) into the sum of four symbolic sparse matrices. Bottom: corresponding four numeric sparse matrices, each of which is an oblique projector, define the multiplication table of a non-commutative algebra of four-dimensional numbers (shown on the right).

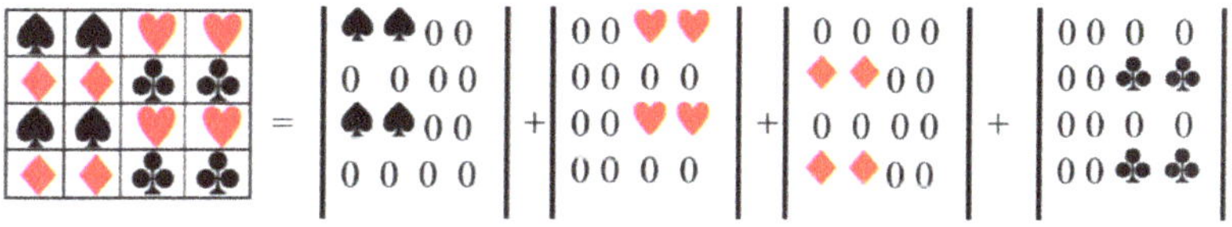

$$e_0+e_1+e_2+e_3=\begin{vmatrix}1&1&0&0\\0&0&0&0\\1&1&0&0\\0&0&0&0\end{vmatrix}+\begin{vmatrix}0&0&1&1\\0&0&0&0\\0&0&1&1\\0&0&0&0\end{vmatrix}+\begin{vmatrix}0&0&0&0\\1&1&0&0\\0&0&0&0\\1&1&0&0\end{vmatrix}+\begin{vmatrix}0&0&0&0\\0&0&1&1\\0&0&0&0\\0&0&1&1\end{vmatrix};$$

*	e_0	e_1	e_2	e_3
e_0	e_0	e_2	e_2	e_0
e_1	e_3	e_1	e_1	e_3
e_2	e_0	e_2	e_2	e_0
e_3	e_3	e_1	e_1	e_3

Fig. 4.6.2. Upper: the decomposition of the symbolic matrix of the projector tetra-grouping №2 for 16 duplets (from Fig. 4.5.3) into the sum of four symbolic sparse matrices. Bottom: corresponding four numeric sparse matrices, each of which is an oblique projector; they define the same multiplication table of the non-commutative algebra of four-dimensional numbers as in Fig. 4.6.1.

The multiplication table, which is defined by these 4 oblique projectors, coincides with the multiplication table of a non-commutative algebra of four-dimensional numbers (Fig. 4.6.1, bottom row). Correspondingly each of these projectors e_0, e_1, e_2, and e_3 is a base unit of this algebra.

Figure 4.6.2 shows a decomposition of the symbolic matrix of the projector tetra-grouping №2 for 16 duplets from Fig. 4.5.3, which presents this matrix as the sum of four symbolic sparse matrices. Arrangements of symbols in these sparse matrices correspond to arrangements of non-zero entries in a set of four numeric matrices e_0, e_1, e_2, and e_3, each of which is an oblique projector since $e_0^2 = e_0$, $e_1^2 = e_1$, $e_2^2 = e_2$, and $e_3^2 = e_3$. Besides, the set of these projectors e_0, e_1, e_2, and e_3 is closed relative to multiplication as well. The multiplication table, which is defined by these 4 oblique projectors, is the same multiplication table of the non-commutative algebra of four-dimensional numbers, which was shown in Fig. 4.6.1.

Figure 4.6.3 shows a decomposition of the symbolic matrix of the projector tetra-grouping №1 for 64 triplets from Fig. 4.5.3, which presents this matrix as the sum of 8 symbolic sparse matrices. Arrangements of symbols in these sparse matrices correspond to arrangements of non-zero

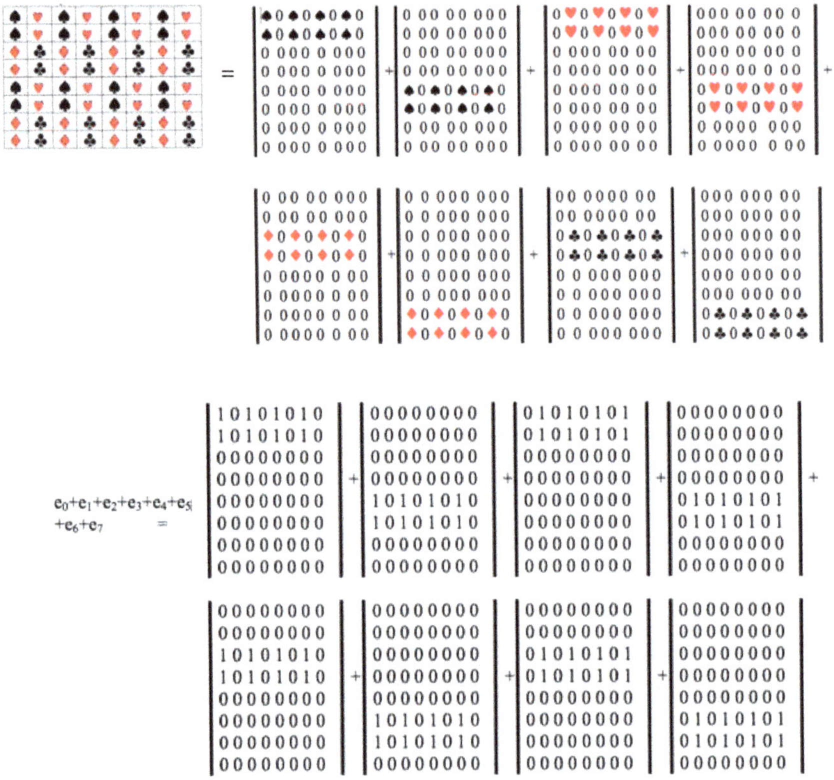

•	e_0	e_1	e_2	e_3	e_4	e_5	e_6	e_7
e_0	e_0	e_0	e_2	e_2	e_0	e_0	e_2	e_2
e_1	e_1	e_1	e_3	e_3	e_1	e_1	e_3	e_3
e_2	e_0	e_0	e_2	e_2	e_0	e_0	e_2	e_2
e_3	e_1	e_1	e_3	e_3	e_1	e_1	e_3	e_3
e_4	e_4	e_4	e_6	e_6	e_4	e_4	e_6	e_6
e_5	e_5	e_5	e_7	e_7	e_5	e_5	e_7	e_7
e_6	e_4	e_4	e_6	e_6	e_4	e_4	e_6	e_6
e_7	e_5	e_5	e_7	e_7	e_5	e_5	e_7	e_7

Fig. 4.6.3. Upper: the decomposition of the symbolic matrix of the projector tetra-grouping №1 for 64 triplets (from Fig. 4.5.3) into the sum of 8 symbolic sparse matrices. Middle: corresponding 8 numeric sparse matrices, each of which is an oblique projector. Bottom: the multiplication table of a non-commutative algebra of eight-dimensional numbers, which is defined by these 8 sparse matrices and which has 4 sub-algebras marked by bold borders.

entries in a shown set of 8 numeric matrices e_0, e_1, e_2, e_3, e_4, e_5, e_6, and e_7, each of which is an oblique projector satisfying the criterion $e_i^2 = e_i$ ($i = 0, 1, 2, \ldots, 7$). Besides, the set of these projectors e_0, e_1, ..., e_7 is closed relative to multiplication as well. A multiplication table, which is defined by these 8 oblique projectors, is the multiplication table of a non-commutative algebra of eight-dimensional numbers.

Figure 4.6.4 shows a decomposition of the symbolic matrix of the projector tetra-grouping №2 for 64 triplets from Fig. 4.5.3, which presents this matrix as the sum of 8 symbolic sparse matrices. Arrangements of symbols in these sparse matrices correspond to arrangements of non-zero entries in a shown set of 8 numeric matrices e_0, e_1, e_2, e_3, e_4, e_5, e_6, and e_7, each of which is an oblique projector satisfying the criterion $e_i^2 = e_i$ ($i = 0, 1, 2, \ldots, 7$). Besides, the set of these projectors e_0, e_1, ..., e_7 is closed relative to multiplication as well. A multiplication table, which is defined by these 8 oblique projectors, is the multiplication table of a non-commutative algebra of eight-dimensional numbers.

Figure 4.6.5 shows a decomposition of the symbolic matrix of the projector tetra-grouping №3 for 64 triplets from Fig. 4.5.3, which presents this matrix as the sum of 8 symbolic sparse matrices. Arrangements of symbols in these sparse matrices correspond to arrangements of non-zero entries in a shown set of 8 numeric matrices e_0, e_1, e_2, e_3, e_4, e_5, e_6, e_7, each of which is an oblique projector satisfying the criterion $e_i^2 = e_i$ ($i = 0, 1, 2, \ldots, 7$). Besides. the set of these projectors e_0, e_1, ..., e_7 is closed relative to multiplication as well. A multiplication table, which is defined by these 8 oblique projectors, is the multiplication table of a non-commutative algebra of eight-dimensional numbers.

Figure 4.6.6 shows a decomposition of the symbolic matrix of the projector tetra-grouping №4 for 64 triplets from Fig. 4.5.3, which presents this matrix as the sum of 8 symbolic sparse matrices. Arrangements of symbols in these sparse matrices correspond to arrangements of non-zero entries in a shown set of 8 numeric matrices e_0, e_1, e_2, e_3, e_4, e_5, e_6, e_7, each of which is an oblique projector satisfying the criterion $e_i^2 = e_i$ ($i = 0, 1, 2, \ldots, 7$). Besides, the set of these projectors e_0, e_1, ..., e_7 is closed relative to multiplication as well. A multiplication

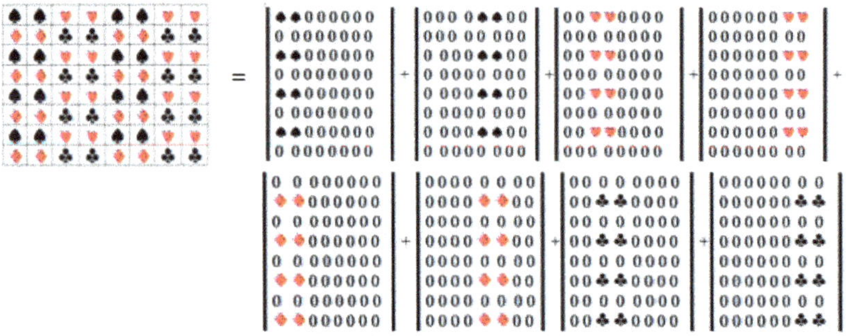

$$e_0+e_1+e_2+e_3+e_4+e_5+e_6+e_7 = \begin{vmatrix}11000000\\00000000\\11000000\\00000000\\11000000\\00000000\\11000000\\00000000\end{vmatrix} + \begin{vmatrix}00001100\\00000000\\00001100\\00000000\\00001100\\00000000\\00001100\\00000000\end{vmatrix} + \begin{vmatrix}00110000\\00000000\\00110000\\00000000\\00110000\\00000000\\00110000\\00000000\end{vmatrix} + \begin{vmatrix}00000011\\00000000\\00000011\\00000000\\00000011\\00000000\\00000011\\00000000\end{vmatrix} +$$

$$\begin{vmatrix}00000000\\11000000\\00000000\\11000000\\00000000\\11000000\\00000000\\11000000\end{vmatrix} + \begin{vmatrix}00000000\\00001100\\00000000\\00001100\\00000000\\00001100\\00000000\\00001100\end{vmatrix} + \begin{vmatrix}00000000\\00110000\\00000000\\00110000\\00000000\\00110000\\00000000\\00110000\end{vmatrix} + \begin{vmatrix}00000000\\00000011\\00000000\\00000011\\00000000\\00000011\\00000000\\00000011\end{vmatrix}$$

*	e_0	e_1	e_2	e_3	e_4	e_5	e_6	e_7
e_0	e_0	e_1	e_2	e_3	e_0	e_1	e_2	e_3
e_1	e_0	e_1	e_2	e_3	e_0	e_1	e_2	e_3
e_2	e_0	e_1	e_2	e_3	e_0	e_1	e_2	e_3
e_3	e_0	e_1	e_2	e_3	e_0	e_1	e_2	e_3
e_4	e_4	e_5	e_6	e_7	e_4	e_5	e_6	e_7
e_5	e_4	e_5	e_6	e_7	e_4	e_5	e_6	e_7
e_6	e_4	e_5	e_6	e_7	e_4	e_5	e_6	e_7
e_7	e_4	e_5	e_6	e_7	e_4	e_5	e_6	e_7

Fig. 4.6.4. Upper: the decomposition of the symbolic matrix of the projector tetra-grouping №2 for 64 triplets (from Fig. 4.5.3) into the sum of 8 symbolic sparse matrices. Middle: corresponding 8 numeric sparse matrices, each of which is an oblique projector. Bottom: the multiplication table of a non-commutative algebra of eight-dimensional numbers, which is defined by these 8 sparse matrices and which has 4 sub-algebras marked by bold borders.

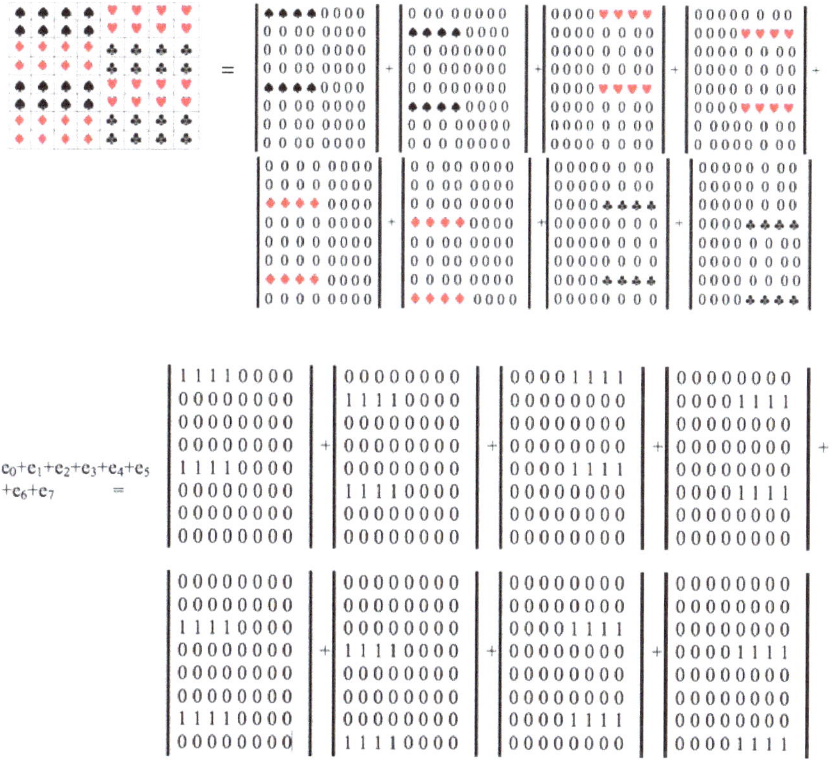

*	e_0	e_1	e_2	e_3	e_4	e_5	e_6	e_7
e_0	e_0	e_0	e_2	e_2	e_0	e_0	e_2	e_2
e_1	e_1	e_1	e_3	e_3	e_1	e_1	e_3	e_3
e_2	e_0	e_0	e_2	e_2	e_0	e_0	e_2	e_2
e_3	e_1	e_1	e_3	e_3	e_1	e_1	e_3	e_3
e_4	e_4	e_4	e_6	e_6	e_4	e_4	e_6	e_6
e_5	e_5	e_5	e_7	e_7	e_5	e_5	e_7	e_7
e_6	e_4	e_4	e_6	e_6	e_4	e_4	e_6	e_6
e_7	e_5	e_5	e_7	e_7	e_5	e_5	e_7	e_7

Fig. 4.6.5. Upper: the decomposition of the symbolic matrix of the projector tetra-grouping №3 for 64 triplets (from Fig. 4.5.3) into the sum of 8 symbolic sparse matrices. Middle: corresponding 8 numeric sparse matrices, each of which is an oblique projector. Bottom: the multiplication table of a non-commutative algebra of eight-dimensional numbers, which is defined by these 8 sparse matrices and which has 4 sub-algebras marked by bold borders.

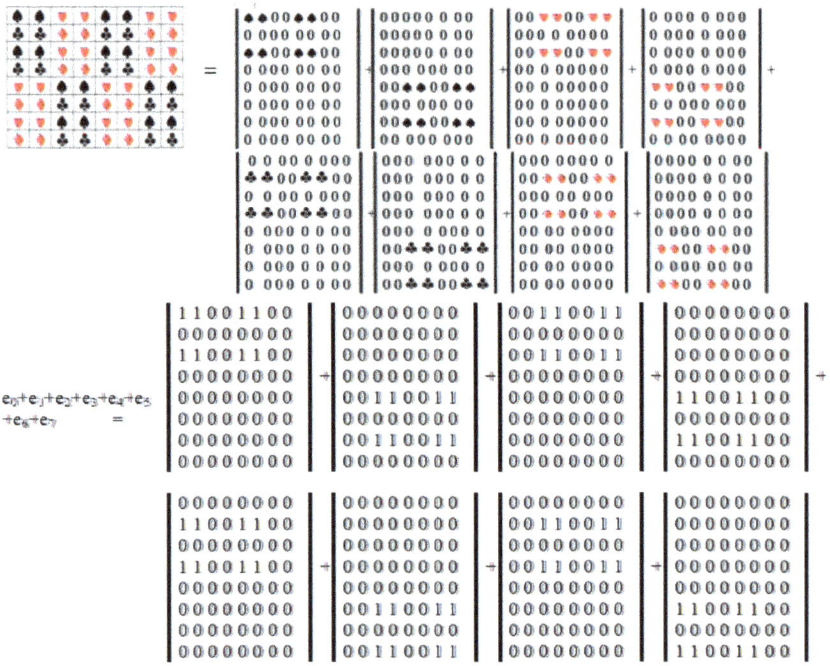

*	e_0	e_1	e_2	e_3	e_4	e_5	e_6	e_7
e_0	e_0	e_2	e_2	e_0	e_0	e_2	e_2	e_0
e_1	e_3	e_1	e_1	e_3	e_3	e_1	e_1	e_3
e_2	e_0	e_2	e_2	e_0	e_0	e_2	e_2	e_0
e_3	e_3	e_1	e_1	e_3	e_3	e_1	e_1	e_3
e_4	e_4	e_6	e_6	e_4	e_4	e_6	e_6	e_4
e_5	e_7	e_5	e_5	e_7	e_7	e_5	e_5	e_7
e_6	e_4	e_6	e_6	e_4	e_4	e_6	e_6	e_4
e_7	e_7	e_5	e_5	e_7	e_7	e_5	e_5	e_7

Fig. 4.6.6. Upper: the decomposition of the symbolic matrix of the projector tetra-grouping №4 for 64 triplets (from Fig. 4.5.3) into the sum of 8 symbolic sparse matrices. Middle: corresponding 8 numeric sparse matrices, each of which is an oblique projector. Bottom: the multiplication table of a non-commutative algebra of eight-dimensional numbers, which is defined by these 8 sparse matrices.

table, which is defined by these 8 oblique projectors, is the multiplication table of a non-commutative algebra of eight-dimensional numbers.

Figure 4.6.7 shows a decomposition of the symbolic matrix of the projector tetra-grouping №5 for 64 triplets from Fig. 4.5.3, which presents

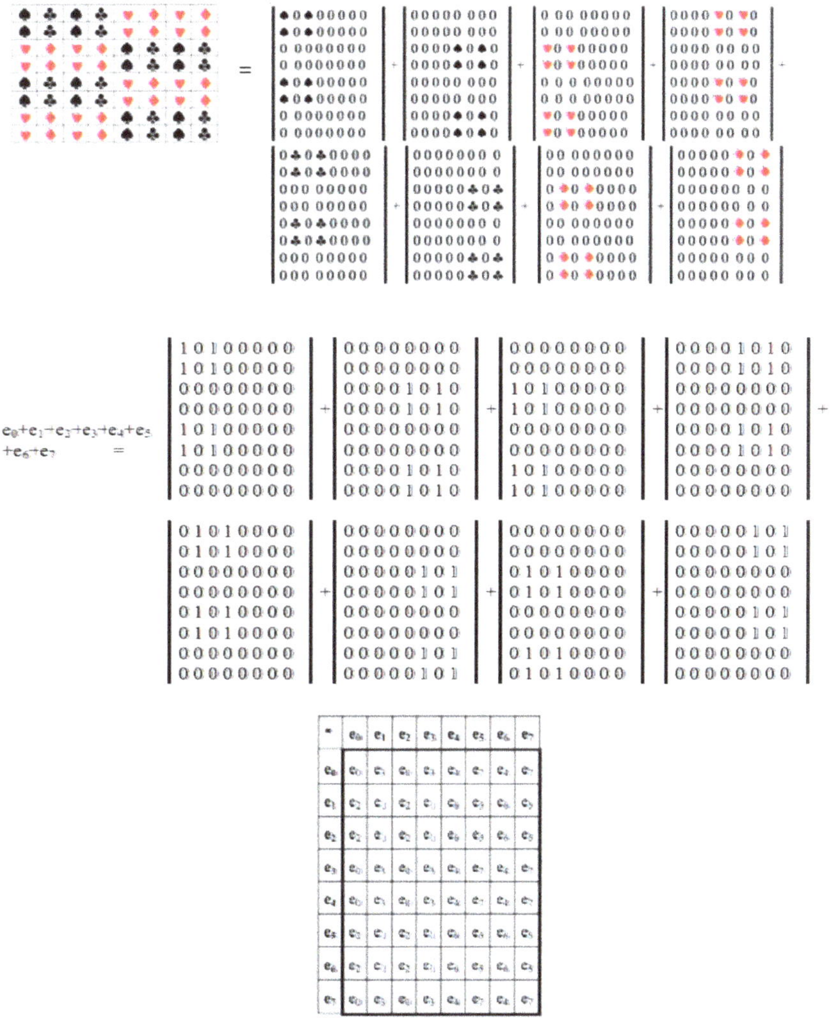

Fig. 4.6.7. Upper: the decomposition of the symbolic matrix of the projector tetra-grouping №5 for 64 triplets (from Fig. 4.5.3) into the sum of 8 symbolic sparse matrices. Middle: corresponding 8 numeric sparse matrices, each of which is an oblique projector. Bottom: the multiplication table of a non-commutative algebra of eight-dimensional numbers, which is defined by these 8 sparse matrices.

this matrix as the sum of 8 symbolic sparse matrices. Arrangements of symbols in these sparse matrices correspond to arrangements of non-zero entries in a shown set of 8 numeric matrices e_0, e_1, e_2, e_3, e_4, e_5, e_6, e_7, each of which is an oblique projector satisfying the criterion $e_i^2 = e_i$ ($i = 0, 1, 2, \ldots, 7$). Besides, the set of these projectors e_0, e_1, …, e_7 is closed relative to multiplication as well. A multiplication table, which is defined by these 8 oblique projectors, is the multiplication table of a non-commutative algebra of eight-dimensional numbers.

As far as the authors can judge, the oblique projectors, which are shown in Figs. 4.6.1–4.6.7, have not been previously used in science (the book describes only a few examples of this wide class of oblique projectors). They are conditionally called as genetic Gestalt projectors. They have interesting algebraic properties that allow them to be used as operators in modeling the generation of stably structured states of multi-parameter systems from their unstructured states. In particular, such modeling is very interesting for algebraic biology in studying structural states of biological systems. Their algebraic properties are briefly described in the next section.

4.7. Some algebraic properties of genetic Gestalt projectors

To begin with, consider a 4-parameter system, whose state vector $\overline{X} = [x_0(t), x_1(t), x_2(t), x_3(t)]$ has coordinates in the form of 4 functions $x_0(t)$, $x_1(t)$, $x_2(t)$, and $x_3(t)$, which are independent from each other (that is, we consider a "chaotic" set of initial 4 coordinates). Let us study, for example, the action on this state vector $\overline{X}$ of the genetic Gestalt projectors e_0, e_1, e_2, and e_3, presented in Fig. 4.6.1:

$$[x_0(t), x_1(t), x_2(t), x_3(t)] \times e_0 = [x_0(t) + x_1(t), x_0(t) + x_1(t), 0, 0], \qquad (4.15)$$

$$[x_0(t), x_1(t), x_2(t), x_3(t)] \times e_1 = [0, x_2(t) + x_3(t), 0, x_2(t) + x_3(t)], \qquad (4.16)$$

$$[x_0(t), x_1(t), x_2(t), x_3(t)] \times e_2 = [0, x_0(t) + x_1(t), 0, x_0(t) + x_1(t)], \qquad (4.17)$$

$$[x_0(t), x_1(t), x_2(t), x_3(t)] \times e_3 = [x_2(t) + x_3(t), 0, x_2(t) + x_3(t), 0]. \qquad (4.18)$$

Expressions (4.15)–(4.18) show that under the action of these oblique projectors, the state vector $\overline{X} = [x_0(t), x_1(t), x_2(t), x_3(t)]$ with unbound coordinates is transformed into a state vector with strongly bound and ordered coordinates. For example, the expression (4.15) shows that the new state vector has two coordinates with an identical function $x_0(t) + x_1(t)$, and the other two coordinates are zero. If $x_0(t)$, $x_1(t)$, $x_2(t)$, $x_3(t)$ are cyclic or wave functions, then two non-zero coordinates of the state vector are synchronously and cyclically changed in time, and the new state vector of the 4-parameter system is also cyclically changed in the two-dimensional subspace defined by these two coordinates. One should separately note the arising symmetry in the set of coordinates of the new state vector.

Below an example of similar results of the ordering and synchronizing action of genetic Gestalt projectors on eight-dimensional state vectors, having random sets of coordinates, will be shown. But firstly, we note the importance of modeling the consistency of genetically inherited cyclic processes in living bodies.

The coordinated cyclic behavior of genetically inherited parts is one of the characteristic features of living bodies. Many mysterious biological phenomena are associated with this. For example, as it is known, the first thing that comes into motion in a three-week-old human embryo, which has a size of 4 millimeters only, is the heart; it begins to pulsate due to inner reasons since at this stage it has no nerve connections and there is no blood to pump. The living organism as a whole is a huge chorus of coordinated cyclic processes. Tibetan pulse diagnosis is based on the relationship of cyclic processes in the body, and so on.

Let us consider one more example of an 8-parameter system, whose state vector $\overline{X} = [x_0(t), x_1(t), x_2(t), x_3(t), x_4(t), x_5(t), x_6(t), x_7(t)]$ has coordinates, for example, in the form of 8 functions $x_0(t)$, $x_1(t)$, $x_2(t)$, $x_3(t)$, $x_4(t)$, $x_5(t)$, $x_6(t)$, and $x_7(t)$, which are independent from each other. Expressions (4.19)–(4.26) show results of the action of the genetic Gestalt projectors e_0, e_1, e_2, e_3, e_4, e_5, e_6, and e_7, presented in Fig. 4.6.3, on this state vector $\overline{X}$:

$$\bar{X} \times e_0 = [x_0(t) + x_1(t), 0, x_0(t) + x_1(t), 0, x_0(t) + x_1(t), 0, x_0(t) + x_1(t), 0], \quad (4.19)$$
$$\bar{X} \times e_1 = [x_4(t) + x_5(t), 0, x_4(t) + x_5(t), 0, x_4(t) + x_5(t), 0, x_4(t) + x_5(t), 0], \quad (4.20)$$
$$\bar{X} \times e_2 = [0, x_0(t) + x_1(t), 0, x_0(t) + x_1(t), 0, x_0(t) + x_1(t), 0, x_0(t) + x_1(t)], \quad (4.21)$$
$$\bar{X} \times e_3 = [0, x_4(t) + x_5(t), 0, x_4(t) + x_5(t), 0, x_4(t) + x_5(t), 0, x_4(t) + x_5(t)], \quad (4.22)$$
$$\bar{X} \times e_4 = [x_2(t) + x_3(t), 0, x_2(t) + x_3(t), 0, x_2(t) + x_3(t), 0, x_2(t) + x_3(t), 0], \quad (4.23)$$
$$\bar{X} \times e_5 = [x_6(t) + x_7(t), 0, x_6(t) + x_7(t), 0, x_6(t) + x_7(t), 0, x_6(t) + x_7(t), 0], \quad (4.24)$$
$$\bar{X} \times e_6 = [x_2(t) + x_3(t), 0, x_2(t) + x_3(t), 0, x_2(t) + x_3(t), 0, x_2(t) + x_3(t), 0], \quad (4.25)$$
$$\bar{X} \times e_7 = [0, x_6(t) + x_7(t), 0, x_6(t) + x_7(t), 0, x_6(t) + x_7(t), 0, x_6(t) + x_7(t)]. \quad (4.26)$$

The expressions (4.19)–(4.26) confirm the ordering, synchronizing, and symmetrizing action of genetic Gestalt projectors on eight-dimensional state vectors, having random sets of coordinates.

The tensor product of these genetic Gestalt projectors, when acting on a state vector with a random set of coordinates $x_i(t)$, also generates a vector with an ordered, synchronized and symmetrized set of coordinates. This makes it possible to extend this algorithm for generating ordered, synchronized and symmetrized state vectors to the general case of 2^n-dimensional vectors. For example, the action of the tensor product of two projectors e_0 and e_3 from Fig. 4.6.1 on 16-dimensional vectors, having random sets of coordinates, gives the following result:

$$[x_0, x_1, x_2, x_3, x_4, x_5, x_6, x_7, x_8, x_9, x_{10}, x_{11}, x_{12}, x_{13}, x_{14}, x_{15}] \times (e_0 \otimes e_3) =$$
$$= [x_2 + x_3 + x_6 + x_7, 0, x_2 + x_3 + x_6 + x_7, 0, 0, 0, 0, 0, x_2 + x_3 +$$
$$x_6 + x_7, 0, x_2 + x_3 + x_6 + x_7, 0, 0, 0, 0, 0] \quad (4.27)$$

Such genetic Gestalt projectors are algebraic tools for modeling the fundamental ability of biological bodies to create a genetically inherited biological order from disorder. Regardless of the chaotic set of coordinates of the initial state vector, the result of the action of these Gestalt projectors is always the same: ordering, synchronization, and symmetrization of coordinates in the new state vector. This interconnected ordering, cyclicity, and symmetry of the coordinates of the emerging state vector of a multi-parameter system can be considered as a kind of genetic Gestalt for further developing Gestalt genetics and algebraic biology. It should be additionally noted that here an interesting theme of information waves of collective probability arises in the layers of genomic DNA texts and genomic DNA epi-chains; this topic should be investigated in future.

4.8. Stochastics of genomic DNAs and rules of percentage equalities in complementary columns

The stochastic organization of genomic DNAs gives valuable data to the theme of complementary replications as the important principle in molecular and matrix genetics. This theme was presented in §§ 1.5 ad 1.7 of Chapter 1, where the notions of complementary binary numbers and complementary columns in genetic matrices of DNA alphabets of n-plets were introduced and used. Two binary numbers that are converted into each other under interconverting $0 \leftrightarrow 1$ are called complementary. For example, the pairs of complementary binary numbers are the following: 000–111, 001–110, 010–101, 011–100. Each pair contains one even number and one odd number. Accordingly, those pairs of columns (rows) that are enumerated by complementary binary numbers are called complementary. In the genetic matrices in Figs. 1.4.2, 4.1.1–4.1.3, complementary columns are located mirror-symmetric in the left and right halves of the matrices, and complementary rows are located mirror-symmetric in the upper and lower halves.

But what can one say in the case of genomic DNAs about n-plet percentages presented in complementary columns of the genetic matrices (like in Figs. 4.1.1–4.1.3)? In the case of stochastics of genomic DNAs, are there any numerical rules for the interrelation of percentages in two columns belonging to a complementary pair and containing percentages of their corresponding n-plets?

Yes, such impressive rules exist in all tested genomic DNAs. They consist of approximate equality — with high accuracy — of the sums of genomic probabilities of n-plets, which belong to different two columns enumerated by bit-converted binary numbers. Table 4.8.1 shows examples of such rules for all pairs of complementary columns in the percentage matrices of 16 duplets, 64 triplets, and 256 tetraplets (Figs. 4.1.1–4.1.3).

Table 4.8.1. The sums of percentages in the complementary columns of the matrices of 16 duplets, 64 triplets, and 256 tetraplets for human chromosome № 1 (see percentage matrices in Figs. 4.1.1–4.1.3). Pairs of the complementary columns with bit-converted numbers are marked by bold frames. In each frame, both amounts are practically the same, despite the strong difference in the values of the summands. Binary numbers of columns are shown on the left.

Columns	Percentage sums in matrix columns for 16 duplets
11	0.05409 + 0.07134 + 0.06008 + 0.09568 = **0.28119**
00	0.09504 + 0.07137 + 0.06008 + 0.05419 = **0.28068**
10	0.07274 + 0.01031 + 0.06312 + 0.07286 = **0.21903**
01	0.05033 + 0.07429 + 0.04402 + 0.05046 = **0.21910**

Columns	Percentage sums in matrix columns for 64 triplets
111	0.01385 + 0.01853 + 0.01758 + 0.02009 + 0.01588 + 0.02226 + 0.01972 + 0.03725 = **0.16516**
000	0.03693 + 0.01988 + 0.02237 + 0.01848 + 0.01960 + 0.01756 + 0.01600 + 0.01382 = **0.16464**
110	0.01878 + 0.00291 + 0.01275 + 0.02088 + 0.01964 + 0.00233 + 0.01981 + 0.01884 = **0.11594**
001	0.01447 + 0.02375 + 0.01441 + 0.01614 + 0.00962 + 0.01327 + 0.01256 + 0.01185 = **0.11607**
101	0.01524 + 0.01789 + 0.00251 + 0.00259 + 0.01103 + 0.01939 + 0.01457 + 0.01988 = **0.10310**
010	0.01977 + 0.00254 + 0.01942 + 0.01781 + 0.01456 + 0.00253 + 0.01115 + 0.01534 = **0.10312**
100	0.01861 + 0.02104 + 0.00227 + 0.00291 + 0.01986 + 0.01284 + 0.01947 + 0.01895 = **0.11595**
011	0.01183 + 0.01622 + 0.01317 + 0.02388 + 0.01255 + 0.01437 + 0.00956 + 0.01445 = **0.11603**

(Continued)

Table 4.8.1. (*Continued*)

Columns	Percentage sums in matrix columns for 256 tetraplets
1111	0.0033 + 0.0041 + 0.0050 + 0.0048 + 0.0052 + 0.0058 + 0.0048 + 0.0073 + 0.0046 + 0.0057 + 0.0057 + 0.0068 + 0.0051 + 0.0077 + 0.0073 + 0.0150 = 0.0982
0000	0.0149 + 0.0071 + 0.0066 + 0.0047 + 0.0078 + 0.0059 + 0.0057 + 0.0041 + 0.0073 + 0.0048 + 0.0057 + 0.0051 + 0.0052 + 0.0052 + 0.0046 + 0.0033 = 0.0980
1110	0.0055 + 0.0010 + 0.0029 + 0.0058 + 0.0057 + 0.0009 + 0.0036 + 0.0044 + 0.0049 + 0.0006 + 0.0040 + 0.0058 + 0.0063 + 0.0006 + 0.0078 + 0.0072 = 0.0670
0001	0.0055 + 0.0095 + 0.0037 + 0.0049 + 0.0039 + 0.0049 + 0.0047 + 0.0040 + 0.0029 + 0.0046 + 0.0032 + 0.0036 + 0.0023 + 0.0034 + 0.0030 + 0.0030 = 0.0671
1101	0.0042 + 0.0044 + 0.0008 + 0.0006 + 0.0027 + 0.0035 + 0.0046 + 0.0053 + 0.0042 + 0.0051 + 0.0005 + 0.0006 + 0.0034 + 0.0063 + 0.0038 + 0.0055 = 0.0555
0010	0.0059 + 0.0006 + 0.0070 + 0.0057 + 0.0049 + 0.0007 + 0.0037 + 0.0047 + 0.0037 + 0.0005 + 0.0031 + 0.0036 + 0.0042 + 0.0010 + 0.0024 + 0.0039 = 0.0556
1100	0.0044 + 0.0058 + 0.0006 + 0.0009 + 0.0038 + 0.0028 + 0.0051 + 0.0058 + 0.0051 + 0.0052 + 0.0006 + 0.0006 + 0.0063 + 0.0038 + 0.0051 + 0.0045 = 0.0604
0011	0.0032 + 0.0047 + 0.0040 + 0.0071 + 0.0042 + 0.0046 + 0.0031 + 0.0046 + 0.0023 + 0.0030 + 0.0025 + 0.0041 + 0.0031 + 0.0042 + 0.0023 + 0.0032 = 0.0602
1011	0.0040 + 0.0047 + 0.0036 + 0.0057 + 0.0010 + 0.0007 + 0.0005 + 0.0006 + 0.0024 + 0.0037 + 0.0030 + 0.0070 + 0.0041 + 0.0049 + 0.0037 + 0.0059 = 0.0555
0100	0.0053 + 0.0054 + 0.0006 + 0.0007 + 0.0063 + 0.0035 + 0.0050 + 0.0044 + 0.0038 + 0.0046 + 0.0005 + 0.0008 + 0.0035 + 0.0027 + 0.0042 + 0.0042 = 0.0555
1010	0.0056 + 0.0010 + 0.0039 + 0.0047 + 0.0006 + 0.0003 + 0.0004 + 0.0010 + 0.0045 + 0.0004 + 0.0055 + 0.0039 + 0.0049 + 0.0006 + 0.0046 + 0.0057 = 0.0476
0101	0.0040 + 0.0051 + 0.0006 + 0.0007 + 0.0032 + 0.0064 + 0.0033 + 0.0051 + 0.0029 + 0.0033 + 0.0006 + 0.0006 + 0.0017 + 0.0032 + 0.0029 + 0.0040 = 0.0476
1001	0.0032 + 0.0040 + 0.0049 + 0.0045 + 0.0003 + 0.0005 + 0.0008 + 0.0005 + 0.0028 + 0.0056 + 0.0026 + 0.0032 + 0.0031 + 0.0045 + 0.0042 + 0.0043 = 0.0490
0110	0.0042 + 0.0005 + 0.0032 + 0.0044 + 0.0045 + 0.0005 + 0.0056 + 0.0040 + 0.0041 + 0.0008 + 0.0027 + 0.0048 + 0.0031 + 0.0004 + 0.0028 + 0.0032 = 0.0488
1000	0.0070 + 0.0044 + 0.0059 + 0.0058 + 0.0006 + 0.0008 + 0.0006 + 0.0010 + 0.0078 + 0.0036 + 0.0041 + 0.0029 + 0.0061 + 0.0057 + 0.0051 + 0.0054 = 0.0668
0111	0.0030 + 0.0041 + 0.0037 + 0.0049 + 0.0033 + 0.0049 + 0.0047 + 0.0096 + 0.0030 + 0.0047 + 0.0032 + 0.0037 + 0.0023 + 0.0038 + 0.0030 + 0.0055 = 0.0674

One can note that in all matrices of percentage tetra-groupings with black-and-red mosaics in Fig. 4.5.3, any two complementary columns have a mirror-antisymmetric mosaic: any two appropriate cells of these columns possess oppositional colors. The same holds true for any two complementary rows.

The described numeric rules of percentages in complementary even and odd columns additionally confirm the importance of the principle of complementarity and of the opposition of even-odd (Yin-Yang) numbers for structural organization not only for DNA but also for different parts of the genetic system. Correspondingly, complementary replication is a systemic phenomenon in genetic organization. The complementary replication of DNA is a manifestation of only one of the parts of a holistic bio-algebraic genetic complex of complementary replications manifesting at different levels of organization in the living in different forms. One can think that this bio-algebraic complex is responsible for the implementation of the ancient principle "like begets like" at different levels of biological organization in the course of biological evolution.

4.9. Stochastics of genomic DNAs and rules of percentage equalities in complementary rows

But what can one say in the case of genomic DNAs about n-plet percentages presented in complementary rows of the genetic matrices (like in Figs. 4.1.1–4.1.3)? It turns out that similar rules exist for complementary rows: the sums of genomic probabilities of n-plets, which belong to two different rows enumerated by bit-converted binary numbers, are practically identical. Figure 4.9.1 shows examples of such rules for all pairs of the complementary rows in the percentage matrices of 16 duplets, 64 triplets, and 256 tetraplets (Figs. 4.1.1–4.1.3).

Row	Percentage sums in matrix rows for 16 duplets
11	0.05409 + 0.07274 + 0.05033 + 0.09504 = 0.27219
00	0.09568 + 0.07286 + 0.05046 + 0.05419 = 0.27319
10	0.07134 + 0.01031 + 0.07429 + 0.07137 = 0.22730
01	0.06008 + 0.06312 + 0.04402 + 0.06008 = 0.22731

Rows	Percentage sums in matrix rows of 64 triplets
111	0.01385 + 0.01878 + 0.01524 + 0.01861 + 0.01183 + 0.01977 + 0.01447 + 0.03693 = 0.14946
000	0.03725 + 0.01884 + 0.01988 + 0.01895 + 0.01445 + 0.01534 + 0.01185 + 0.01382 = 0.15039
110	0.01853 + 0.00291 + 0.01789 + 0.02104 + 0,01622 + 0.00254 + 0.02375 + 0.01988 = 0.12275
001	0.01972 + 0.01981 + 0.01457 + 0.01947 + 0.00956 + 0.01115 + 0.01256 + 0.01600 = 0.12285
101	0.01758 + 0.01275 + 0.00251 + 0.00227 + 0,01317 + 0.01942 + 0.01441 + 0.02237 = 0.10448
010	0.02226 + 0.00233 + 0.01939 + 0.01284 + 0.01437 + 0.00253 + 0.01327 + 0.01756 = 0.10456
100	0.02009 + 0.02088 + 0.00259 + 0.00291 + 0.02388 + 0.01781 + 0.01614 + 0.01848 = 0.12277
011	0.01588 + 0.01964 + 0.01103 + 0.01986 + 0.01255 + 0.01456 + 0.00962 + 0.01960 = 0.12273

Rows	Percentage sums in matrix rows of 256 tetraplets
1111	0.0033 + 0.0055 + 0.0042 + 0.0044 + 0.0040 + 0.0056 + 0.0032 + 0.0070 + 0.0030 + 0.0042 + 0.0040 + 0.0053 + 0.0032 + 0.0059 + 0.0055 + 0.0149 = 0.0830
0000	0.0150 + 0.0072 + 0.0055 + 0.0045 + 0.0059 + 0.0057 + 0.0043 + 0.0054 + 0.0055 + 0.0032 + 0.0040 + 0.0042 + 0.0032 + 0.0039 + 0.0030 + 0.0033 = 0.0838
1110	0.0041 + 0.0010 + 0.0044 + 0.0058 + 0.0047 + 0.0010 + 0.0040 + 0.0044 + 0.0041 + 0.0005 + 0.0051 + 0.0054 + 0.0047 + 0.0006 + 0.0095 + 0.0071 = 0.0664
0001	0.0073 + 0.0078 + 0.0038 + 0.0051 + 0.0037 + 0.0046 + 0.0042 + 0.0051 + 0.0030 + 0.0028 + 0.0029 + 0.0042 + 0.0023 + 0.0024 + 0.0030 + 0.0046 = 0.0666
1101	0.0050 + 0.0029 + 0.0008 + 0.0006 + 0.0036 + 0.0039 + 0.0049 + 0.0059 + 0.0037 + 0.0032 + 0.0006 + 0.0006 + 0.0040 + 0.0070 + 0.0037 + 0.0066 = 0.0569
0010	0.0077 + 0.0006 + 0.0063 + 0.0038 + 0.0049 + 0.0006 + 0.0045 + 0.0057 + 0.0038 + 0.0004 + 0.0032 + 0.0027 + 0.0042 + 0.0010 + 0.0034 + 0.0052 = 0.0580
1100	0.0048 + 0.0058 + 0.0006 + 0.0009 + 0.0057 + 0.0047 + 0.0045 + 0.0058 + 0.0049 + 0.0044 + 0.0007 + 0.0007 + 0.0071 + 0.0057 + 0.0049 + 0.0047 = 0.0659
0011	0.0051 + 0.0063 + 0.0034 + 0.0063 + 0.0041 + 0.0049 + 0.0031 + 0.0061 + 0.0023 + 0.0031 + 0.0017 + 0.0035 + 0.0031 + 0.0042 + 0.0023 + 0.0052 = 0.0648
1011	0.0052 + 0.0057 + 0.0027 + 0.0038 + 0.0010 + 0.0006 + 0.0003 + 0.0006 + 0.0033 + 0.0045 + 0.0032 + 0.0063 + 0.0042 + 0.0049 + 0.0039 + 0.0078 = 0.0579
0100	0.0068 + 0.0058 + 0.0006 + 0.0006 + 0.0070 + 0.0039 + 0.0032 + 0.0029 + 0.0037 + 0.0048 + 0.0006 + 0.0008 + 0.0041 + 0.0036 + 0.0036 + 0.0051 = 0.0570

Fig. 4.9.1. The sums of percentages in the complementary rows of the matrices of 16 duplets, 64 triplets, and 256 tetraplets for human chromosome № 1 (see percentage matrices in Figs. 4.1.1–4.1.3). Pairs of the complementary rows with bit-converted numbers are marked by bold frames. In each frame, both amounts are practically the same, despite the strong difference in the values of the summands. Binary numbers of rows are shown on the left.

1010	0.0058 + 0.0009 + 0.0035 + 0.0028 + 0.0007 + 0.0003 + 0.0005 + 0.0008 + 0.0049 + 0.0005 + 0.0064 + 0.0035 + 0.0046 + 0.0007 + 0.0049 + 0.0059 = 0.0466
0101	0.0057 + 0.0040 + 0.0005 + 0.0006 + 0.0030 + 0.0055 + 0.0026 + 0.0041 + 0.0032 + 0.0027 + 0.0006 + 0.0005 + 0.0025 + 0.0031 + 0.0032 + 0.0057 = 0.0476
1001	0.0048 + 0.0036 + 0.0046 + 0.0051 + 0.0005 + 0.0004 + 0.0008 + 0.0006 + 0.0047 + 0.0056 + 0.0033 + 0.0050 + 0.0031 + 0.0037 + 0.0047 + 0.0057 = 0.0563
0110	0.0057 + 0.0006 + 0.0051 + 0.0052 + 0.0037 + 0.0004 + 0.0056 + 0.0036 + 0.0047 + 0.0008 + 0.0033 + 0.0046 + 0.0030 + 0.0005 + 0.0046 + 0.0048 = 0 0563
1000	0.0073 + 0.0044 + 0.0053 + 0.0058 + 0.0006 + 0.0010 + 0.0005 + 0.0010 + 0.0096 + 0.0040 + 0.0051 + 0.0044 + 0.0046 + 0.0047 + 0.0040 + 0.0041 = 0.0666
0111	0.0046 + 0.0049 + 0.0042 + 0.0051 + 0.0024 + 0.0045 + 0.0028 + 0.0078 + 0.0030 + 0.0041 + 0.0029 + 0.0038 + 0.0023 + 0.0037 + 0.0029 + 0.0073 = 0.0664

Fig. 4.9.1. (*Continued*)

4.10. Gestalt percentage rules for sets of *n*-plets starting with certain *k*-plets ($k < n$)

Above, in § 4.2, we described *m*-positional tetra-groupings of *n*-plets, which are defined by a disposition of attributive nucleotides C, G, A, and T at a certain position *m* ($m \leq n$) inside these *n*-plets. Now in this section, we describe other positional tetra-groupings of *n*-plets where a role of attributive elements is played not by separate nucleotides but by separate duplets, or separate triplets, and so on. As will be revealed, these new tetra-groupings of *n*-plets obey Gestalt rules but have their own values for percentage sums in each of the groupings. Let us explain this by some examples using percentages of *n*-plets in different layers of the DNA text of the human chromosome №1 (Figs. 4.1.1–4.1.3).

For the beginning, consider the percentage contents of positional tetra-groupings, each of which is defined by one of 16 duplets, which is its attributive positional element. In a general case, we call such tetra-groupings as *mm*-positional tetra-groupings or, in individual cases, as a CA-tetra-grouping, or a TA-tetra-grouping, etc. Below special denotations are used for percentage sums of *n*-plets, which are explained by the following examples (here N refers to any of the nucleotides A, T, C, and G):

- Σ%TAN means a percentage sum of all 4 triplets which start with the duplet TA in an analyzed DNA text presented as a sequence of triplets;

- Σ%NTA means a percentage sum of all 4 triplets which end with the duplet TA in an analyzed DNA text presented as a sequence of triplets;
- Σ%TANN means a percentage sum of all 16 tetraplets which start with the duplet TA in an analyzed DNA text presented as a sequence of tetraplets;
- Σ%NTAN means a percentage sum of all 16 tetraplets which have the duplet TA in the middle in a DNA text presented as a sequence of tetraplets;
- Σ%NNTA means a percentage sum of all 16 tetraplets which end with the duplet TA in an analyzed DNA text presented as a sequence of tetraplets.

Let us calculate these percentage sums of *n*-plets for corresponding layers of the DNA text of the human chromosome №1 using the data presented in Figs. 4.1.1–4.1.3 about percentages of separate duplets, triplets, and tetraplets. This calculation gives equalities shown in Fig. 4.10.1.

Data in Fig. 4.10.1 show that percentage sums in each of these *mm*-tetra-groupings are equal to a percentage of that duplet, which plays the role of the attributive positional element for these tetra-groupings. The presented equality of these percentage sums in all tetra-groupings of each row of Fig. 4.10.1 occurs even though summands in them in each case are very different. Other genomes, analyzed in [70], have similar Gestalt properties for *mm*-positional tetra-groupings in the layers of their DNA texts. This new genomic Gestalt phenomenon and the corresponding Gestalt rule for *mm*-positional tetra-groupings (Fig. 4.10.1) are analogs of the Gestalt phenomenon and the Gestalt rule described for *m*-positional tetra-groupings in § 4.2 and its expressions (4.1) and (4.2).

Now consider percentage contents of 64 *mmm*-positional tetra-groupings, each of which is defined by one of 64 triplets, which is its attributive positional element. The analogical denotations are used for these tetra-groupings:

- Σ%TAAN means a percentage sum of all 4 tetraplets, which start with the triplet TAA in an analyzed DNA text presented as a sequence of triplets;

%AA ≈ Σ%AAN ≈ Σ%NAA ≈ Σ%AANN ≈ Σ%NAAN ≈ Σ%NNAA ≈ 0.095
%AT ≈ Σ%ATN ≈ Σ%NAT ≈ Σ%ATNN ≈ Σ%NATN ≈ Σ%NNAT ≈ 0.074
%AC ≈ Σ%ACN ≈ Σ%NAC ≈ Σ%ACNN ≈ Σ%NACN ≈ Σ%NNAC ≈ 0.050
%AG ≈ Σ%AGN ≈ Σ%NAG ≈ Σ%AGNN ≈ Σ%NAGN ≈ Σ%NNAG ≈ 0.071
%TA ≈ Σ%TAN ≈ Σ%NTA ≈ Σ%TANN ≈ Σ%NTAN ≈ Σ%NNTA ≈ 0.063
%TT ≈ Σ%TTN ≈ Σ%NTT ≈ Σ%TTNN ≈ Σ%NTTN ≈ Σ%NNTT ≈ 0.096
%TC ≈ Σ%TCN ≈ Σ%NTC ≈ Σ%TCNN ≈ Σ%NTCN ≈ Σ%NNTC ≈ 0.060
%TG ≈ Σ%TGN ≈ Σ%NTG ≈ Σ%TGNN ≈ Σ%NTGN ≈ Σ%NNTG ≈ 0.073
%CA ≈ Σ%CAN ≈ Σ%NCA ≈ Σ%CANN ≈ Σ%NCAN ≈ Σ%NNCA ≈ 0.073
%CT ≈ Σ%CTN ≈ Σ%NCT ≈ Σ%CTNN ≈ Σ%NCTN ≈ Σ%NNCT ≈ 0.071
%CC ≈ Σ%CCN ≈ Σ%NCC ≈ Σ%CCNN ≈ Σ%NCCN ≈ Σ%NNCC ≈ 0.054
%CG ≈ Σ%CGN ≈ Σ%NCG ≈ Σ%CGNN ≈ Σ%NCGN ≈ Σ%NNCG ≈ 0.010
%GA ≈ Σ%GAN ≈ Σ%NGA ≈ Σ%GANN ≈ Σ%NGAN ≈ Σ%NNGA ≈ 0.060
%GT ≈ Σ%GTN ≈ Σ%NGT ≈ Σ%GTNN ≈ Σ%NGTN ≈ Σ%NNGT ≈ 0.050
%GC ≈ Σ%GCN ≈ Σ%NGC ≈ Σ%GCNN ≈ Σ%NGCN ≈ Σ%NNGC ≈ 0.044
%GG ≈ Σ%GGN ≈ Σ%NGG ≈ Σ%GGNN ≈ Σ%NGGN ≈ Σ%NNGG ≈ 0.054

Fig. 4.10.1. Percentage sums are presented for *mm*-positional tetra-groupings related to 64 triplets and 256 tetraplets in the appropriate triplet- and tetraplet-layers of the DNA text of the human chromosome №1. Each of these tetra-groupings is defined by one of 16 duplets as its attributive positional element disposed of in the beginning, middle, or at the end of the *n*-plets. Numerical values of percentage sums are calculated based on data about the percentage of separate *n*-plets in Figs. 4.1.1–4.1.3.

- Σ%NTAA means a percentage sum of all 4 tetraplets, which end with the triplet TAA.

Let us calculate these percentage sums of *n*-plets for corresponding layers of the DNA text of the human chromosome №1 using the data presented in Figs 4.1.2 and 4.1.3 about the percentage of separate triplets and tetraplets. This calculation gives equalities shown in Fig. 4.10.2.

Data in Fig. 4.10.2 show that percentage sums in each of these *mmm*-tetra-groupings are equal to a percentage of the triplet, which plays the role of the attributive positional element for these tetra-groupings. The presented equality of these percentage sums in all tetra-groupings of each row of Figure 4.10.2 occurs even though summands in them in each case are very different. Other genomes, analyzed in [70], have similar Gestalt properties for such *mmm*-positional tetra-groupings in the appropriate layers of their DNA texts. This new genomic Gestalt phenomenon and the corresponding

%CCC ≈ Σ%CCCN ≈ Σ%NCCC ≈ 0.014	%ACC ≈ Σ%ACCN ≈ Σ%NACC ≈ 0.012
%CCG ≈ Σ%CCGN ≈ Σ%NCCG ≈ 0.003	%ACG ≈ Σ%ACGN ≈ Σ%NACG ≈ 0.003
%CGC ≈ Σ%CGCN ≈ Σ%NCGC ≈ 0.003	%AGC ≈ Σ%AGCN ≈ Σ%NAGC ≈ 0.014
%CGG ≈ Σ%CGGN ≈ Σ%NCGG ≈ 0.003	%AGG ≈ Σ%AGGN ≈ Σ%NAGG ≈ 0.018
%GCC ≈ Σ%GCCN ≈ Σ%NGCC ≈ 0.013	%TCC ≈ Σ%TCCN ≈ Σ%NTCC ≈ 0.016
%GCG ≈ Σ%GCGN ≈ Σ%NGCG ≈ 0.003	%TCG ≈ Σ%TCGN ≈ Σ%NTCG ≈ 0.002
%GGC ≈ Σ%GGCN ≈ Σ%NGGC ≈ 0.013	%TGC ≈ Σ%TGCN ≈ Σ%NTGC ≈ 0.015
%GGG ≈ Σ%GGGN ≈ Σ%NGGG ≈ 0.014	%TGG ≈ Σ%TGGN ≈ Σ%NTGG ≈ 0.019
%CCA ≈ Σ%CCAN ≈ Σ%NCCA ≈ 0.019	%ACA ≈ Σ%ACAN ≈ Σ%NACA ≈ 0.020
%CCT ≈ Σ%CCTN ≈ Σ%NCCT ≈ 0.019	%ACT ≈ Σ%ACTN ≈ Σ%NACT ≈ 0.016
%CGA ≈ Σ%CGAN ≈ Σ%NCGA ≈ 0.002	%AGA ≈ Σ%AGAN ≈ Σ%NAGA ≈ 0.022
%CGT ≈ Σ%CGTN ≈ Σ%NCGT ≈ 0.003	%AGT ≈ Σ%AGTN ≈ Σ%NAGT ≈ 0.016
%GCA ≈ Σ%GCAN ≈ Σ%NGCA ≈ 0.015	%TCA ≈ Σ%TCAN ≈ Σ%NTCA ≈ 0.020
%GCT ≈ Σ%GCTN ≈ Σ%NGCT ≈ 0.014	%TCT ≈ Σ%TCTN ≈ Σ%NTCT ≈ 0.022
%GGA ≈ Σ%GGAN ≈ Σ%NGGA ≈ 0.016	%TGA ≈ Σ%TGAN ≈ Σ%NTGA ≈ 0.019
%GGT ≈ Σ%GGTN ≈ Σ%NGGT ≈ 0.012	%TGT ≈ Σ%TGTN ≈ Σ%NTGT ≈ 0.020
%CAC ≈ Σ%CACN ≈ Σ%NCAC ≈ 0.015	%AAC ≈ Σ%AACN ≈ Σ%NAAC ≈ 0.014
%CAG ≈ Σ%CAGN ≈ Σ%NCAG ≈ 0.021	%AAG ≈ Σ%AAGN ≈ Σ%NAAG ≈ 0.020
%CTC ≈ Σ%CTCN ≈ Σ%NCTC ≈ 0.018	%ATC ≈ Σ%ATCN ≈ Σ%NATC ≈ 0.013
%CTG ≈ Σ%CTGN ≈ Σ%NCTG ≈ 0.021	%ATG ≈ Σ%ATGN ≈ Σ%NATG ≈ 0.018
%GAC ≈ Σ%GACN ≈ Σ%NGAC ≈ 0.010	%TAC ≈ Σ%TACN ≈ Σ%NTAC ≈ 0.011
%GAG ≈ Σ%GAGN ≈ Σ%NGAG ≈ 0.018	%TAG ≈ Σ%TAGN ≈ Σ%NTAG ≈ 0.013
%GTC ≈ Σ%GTCN ≈ Σ%NGTC ≈ 0.010	%TTC ≈ Σ%TTCN ≈ Σ%NTTC ≈ 0.020
%GTG ≈ Σ%GTGN ≈ Σ%NGTG ≈ 0.015	%TTG ≈ Σ%TTGN ≈ Σ%NTTG ≈ 0.019
%CAA ≈ Σ%CAAN ≈ Σ%NCAA ≈ 0.019	%AAA ≈ Σ%AAAN ≈ Σ%NAAA ≈ 0.037
%CAT ≈ Σ%CATN ≈ Σ%NCAT ≈ 0.018	%AAT ≈ Σ%AATN ≈ Σ%NAAT ≈ 0.024
%CTA ≈ Σ%CTAN ≈ Σ%NCTA ≈ 0.013	%ATA ≈ Σ%ATAN ≈ Σ%NATA ≈ 0.019
%CTT ≈ Σ%CTTN ≈ Σ%NCTT ≈ 0.020	%ATT ≈ Σ%ATTN ≈ Σ%NATT ≈ 0.024
%GAA ≈ Σ%GAAN ≈ Σ%NGAA ≈ 0.020	%TAA ≈ Σ%TAAN ≈ Σ%NTAA ≈ 0.020
%GAT ≈ Σ%GATN ≈ Σ%NGAT ≈ 0.013	%TAT ≈ Σ%TATN ≈ Σ%NTAT ≈ 0.019
%GTA ≈ Σ%GTAN ≈ Σ%NGTA ≈ 0.011	%TTA ≈ Σ%TTAN ≈ Σ%NTTA ≈ 0.020
%GTT ≈ Σ%GTTN ≈ Σ%NGTT ≈ 0.014	%TTT ≈ Σ%TTTN ≈ Σ%NTTT ≈ 0.037

Fig. 4.10.2. Percentage sums are presented for *mmm*-positional tetra-groupings related to 256 tetraplets in the tetraplet-layer of the DNA text of the human chromosome №1. Each of these tetra-groupings is defined by one of 64 triplets as its attributive positional element disposed of in the beginning or at the end of the tetraplets. Numerical values of percentage sums are calculated based on data about the percentage of separate triplets and tetraplets in Figs. 4.1.2 and 4.1.3.

Gestalt rule for *mmm*-positional tetra-groupings (Fig. 4.10.2) are also analogs of the Gestalt phenomenon and the Gestalt rule described for *m*-positional tetra-groupings in § 4.2 and its expressions (4.1) and (4.2).

Taking into account results of the work [42] regarding additional tests of different genomic DNAs, one can state that similar Gestalt phenomena exist also for layers of *n*-plets with $n = 5, 6, 7, \ldots, 50$ of genomic DNA texts of various types of organisms.

4.11. Stochastics of genomic DNAs for *n*-plets having strong or weak roots

This section is devoted to one more kind of the rules of percentages in different *n*-plet layers of genomic DNAs sequences. Let us turn back to the genetic matrices having black-and-white mosaics in Fig. 1.4.2 of Chapter 1. Each black (white) cell of these matrices contains *n*-plet with strong (weak) roots.

It was revealed that cooperative percentage organization of a set of different layers of genomic DNA nucleotide sequences is also connected with this structural division of the *n*-plet alphabets into equal sub-alphabets of *n*-plets with strong and weak roots [70]. To demonstrate this, let us initially calculate — separately for left and right halves of each of the mosaic matrices in Fig. 1.4.2 — a total percentage of *n*-plets in their black and white cells for the case of the human chromosome №1, whose *n*-plet percentages in different DNA texts are shown in Figs. 4.1.1–4.1.3. Results of such calculation are shown in Tables 4.11.1 and 4.11.2.

In line with phenomenological numeric data in Table 4.11.1, percentage sums of 4 duplets, 16 triplets, and 64 tetraplets with strong roots, which are located in black cells of the left halves of the mosaic matrices $[C, A; T, G]^{(n)}$, are approximately identical to each other with high accuracy (≈ **0.1958**). Simultaneously, percentage sums of 4 duplets, 16 triplets, and 64 tetraplets

Table 4.11.1. Comparison of sums of the chromosomal percentage of duplets, triplets, and tetraplets with strong and weak roots located in the left halves of the mosaic matrices in Figs. 4.1.1–4.1.3 for the case of the human chromosome № 1.

Left halves of genetic matrices [C, A; T, G]$^{(n)}$ from Figs. 4.1.1–4.13		
	Sums of percentage for *n*-plets with strong roots	**Sums of percentage for *n*-plets with weak roots**
Mosaic matrix of 16 duplets	%CC + %CT + %TC + %CG ≈ 0.05409 + 0.07134 + 0.06008 + 0.01031 ≈ 0.1958	%CA + %TT + %TA + %TG ≈ 0.07274 + 0.09568 + 0.06312 + 0.0728 ≈ 0.3044
Mosaic matrix of 64 triplets	%CCC + %CCT + %CTC + %CTT + %TCC + %TCT + %CCA + %CCG + %CTA + %CTG + %TCA + %TCG + %CGC + %CGT + %CGA + %CGG ≈ 0.0138 + 0.0185 + 0.0176 + 0.0201 + 0.0159 + 0.0223 + 0.0188 + 0.0029 + 0.0127 + 0.0209 + 0.0196 + 0.0023 + 0.0025 + 0.0026 + 0.0023 + 0.0029 ≈ 0.1957	%CAC + %CAT + %CAA + %CAG + %TAC + %TAT + %TGC + %TGT + %TAA + %TAG + %TGA + %TGG + %TTC + %TTT + %TTA + %TTG ≈ 0.0152 + 0.0179 + 0.0186 + 0.0210 + 0.0110 + 0.0194 + 0.0146 + 0.0199 + 0.0199 + 0.0128 + 0.0195 + 0.0190 + 0.0197 + 0.0372 + 0.0198 + 0.0188 ≈ 0.3043
Mosaic matrix of 256 tetraplets	The total sum of percentages of 64 tetraplets with strong roots ≈ 0.1958	The total sum of percentages of 64 tetraplets with weak roots ≈ 0.3044

Table 4.11.2. Comparison of sums of the chromosomal percentage of duplets, triplets, and tetraplets with strong and weak roots located in the right halves of the matrices in Figs. 4.1.1–4.1.3 for the case of the human chromosome № 1.

Right halves of genetic matrices [C, A; T, G]$^{(n)}$ from Figs. 4.1.1–4.1.3		
	Sums of percentage for *n*-plets with strong roots	**Sums of percentage for *n*-plets with weak roots**
Mosaic matrix of 16 duplets	%AC + %GC + %GT + %GG ≈ 0.05033 + 0.04402 + 0.05046 + 0.05419 ≈ 0.1990	%AA + %AT + %AG + %GA = 0.09504 + 0.07429 + 0.07137 + 0.06008 ≈ 0.3008
Mosaic matrix of 64 tripets	%ACC + %ACT + %ACA + %ACG + %GCC + %GCT + %GTC + %GTT + %GCA + %GCG + %GTA + %GTG + %GGC + %GGT + %GGA + %GGG ≈ 0.0118 + 0.0162 + 0.0198 + 0.0025 + 0.0125 + 0.0144 + 0.0096 + 0.0145 + 0.0146 + 0.0025 + 0.0112 +	%ATC + %ATT + %ATA + %ATG + %AAC + %AAT + %AGC + %AGT + %AAA + %AAG + %AGA + %AGG + %GAC + %GAT + %GAA + %GAG ≈ 0.0132 + 0.0239 + 0.0194 + 0.0178 + 0.0145 + 0.0238 + 0.0144 + 0.0161 + 0.0369 + 0.0199 + 0.0224 + 0.0185

Table 4.11.2. (*Continued*)

Right halves of genetic matrices [C, A; T, G]$^{(n)}$ from Figs. 4.1.1–4.1.3		
	Sums of percentage for n-plets with strong roots	**Sums of percentage for n-plets with weak roots**
	0.0153 + 0.0126 + 0.0119 + 0.0160 + 0.0138 ≈ 0.1992	+ 0.0096 + 0.0133 + 0.0196 + 0.0176 ≈ 0.3009
Mosaic matrix of 256 tetraplets	The total sum of percentages of 64 tetraplets in black cells ≈ 0.1990	The total sum of percentages of 64 tetraplets in white cells ≈ 0.3008

with weak roots, which are located in the right halves of these matrices, are also approximately identical to each other with high accuracy (≈ 0.3044).

In line with phenomenological numeric data in Table 4.11.2, percentage sums of 4 duplets, 16 triplets, and 64 tetraplets with strong roots, which are located in black cells of the right halves of the mosaic matrices [C, A; T, G]$^{(n)}$ (Fig. 1.4.2), are approximately identical to each other with high accuracy (≈ 0.1990). Simultaneously, percentage sums of 4 duplets, 16 triplets, and 64 tetraplets with weak roots, which are located in the right halves of these matrices, are also approximately identical to each other with high accuracy (≈ 0.3008).

Thus, the mysterious division of the DNA alphabets of n-plets into two equal sub-alphabets — with strong and weak roots — is associated with Gestalt genetics: four types of sub-alphabets of n-plets presented in Tables 4.11.1 and 4.11.2 (two in the left half of the mosaic matrices and two in their right half) form a new kind of tetra-groupings of n-plets, where the percentage sums in each of the four groupings are practically the same in different n-plet layers representing genomic DNA sequences.

Additional kinds of cooperative tetra-groupings of n-plets in Gestalt genetics are revealed under analysis of genomic percentage sums of n-plets in quadrants along the main diagonal of the matrices (Figs. 4.1.1–4.1.3) and also in quadrants along the second diagonal of these matrices (in other words, the black-and-white mosaic along each of the two diagonals has its own tetra-grouping with constant percentage sums in its 4 parts). For example, in considering human chromosome № 1, both quadrants along the main diagonal give the following practically invariant percentage sums of their n-plets in all the considered matrices:

- In the upper left quadrant, a percentage sum of n-plets in black cells is approximately equal to 0.1357, and that in white cells is approximately equal to 0.0727;
- In the lower right quadrant, a percentage sum of n-plets in black cells is approximately equal to 0.1486, and that in white cells is approximately equal to 0.0600.

Analogically, both quadrants along the second diagonal give the following practically invariant percentage sums of their n-plets in all the matrices for human chromosome № 1:

- In the upper right quadrant, a percentage sum of n-plets in black cells is approximately equal to 0.0503, and that in white cells is approximately equal to 0.2407;
- In the lower left quadrant, a percentage sum of n-plets in black cells is approximately equal to 0.0601, and that in white cells is approximately equal to 0.2317.

Now let us turn to one more phenomenological interrelation of different mosaic matrices $[C, A; T, G]^{(n)}$ in Figs. 4.1.1–4.1.3, which is based on percentage sums of n-plets with strong and weak roots in their different rows, whose mosaics correspond to Rademacher functions. The analysis revealed genomic percentage interrelations of different rows from these matrices. For example, in the case of human chromosome №1, using percentage data for rows of matrices in Figs. 4.1.1–4.1.3, you have the following interrelations between percentage sums of n-plets with strong and weak roots:

- The percentage sum of each of the two oppositional subsets of duplets in row 1 (upper row in the matrix of 16 duplets in Fig. 1.4.2) is approximately equal to the percentage sum of the corresponding subset of triplets in rows 1 and 2 (two upper rows in the matrix of 64 triplets in Fig. 1.4.2), and also is approximately equal to the percentage sum of the corresponding subset of tetraplets in rows 1–4 (four upper rows in the matrix of 256 tetraplets in Fig. 1.4.2). This numeric result is shown in Table 4.11.3;

- The percentage sum of each of the two oppositional subsets of duplets in row 2 of the matrix of 16 duplets in Fig. 1.4.2 is approximately equal to the percentage sum of the corresponding subset of triplets in rows 3 and 4 in the matrix of 64 triplets in Fig. 1.4.2, and also is approximately equal to the percentage sum of the corresponding subset of tetraplets in rows 5–8 of the matrix of 256 tetraplets in Fig. 1.4.2. This numeric result is shown in Table 4.11.4;
- The percentage sum of each of the two oppositional subsets of duplets in row 3 of the matrix of 16 duplets in Figs. 4.1.1–4.1.3 is approximately equal to the percentage sum of the corresponding subset of triplets in rows 5 and 6 in the matrix of 64 triplets in Figs. 4.1.1–4.1.3, and also is approximately equal to the percentage sum of the corresponding subset of tetraplets in rows 9–12 of the matrix of 256 tetraplets in Figs. 4.1.1–4.1.3. This numeric result is shown in Table 4.11.5;

Table 4.11.3. The percentage sums of those *n*-plets with strong and weak roots, which are located correspondingly in black and white matrix cells (Figs. 4.1.1–4.1.3) in named rows for the DNA of the human chromosome № 1. Corresponding percentage values for separate *n*-plets are shown in Figs. 4.1.1–4.1.3.

	Σ% of *n*-plets with strong roots	**Σ% of *n*-plets with weak roots**
Row 1 in matrix of 16 duplets	%CC + %AC ≈ 0.05409 + 0.05033 ≈ **0.1044**.	%CA + %AA ≈ 0.07274 + 0.09504 ≈ **0.1678**.
Rows 1–2 in matrix of 64 triplets	%CCC + %CCA + %ACC + %AC + %CCT + %CCG + %ACT + %ACG ≈ 0.0138 + 0.0188 + 0.0118 + 0.0198 + 0.0185 + 0.0029 + 0.0162 + 0.0025 ≈ **0.1043**.	%CAC + %CAA + %AAC + %AAA + %CAT + %CAG + %AAT + %AAG ≈ 0.0152 + 0.0186 + 0.0145 + 0.0369 + 0.0179 + 0.0210 + 0.0238 + 0.0199 ≈ **0.1678**.
Rows 1–4 in matrix of 256 tetraplets	Percentage sum of 32 tetraplets with strong roots ≈ **0.1044**.	Percentage sum of 32 tetraplets with weak roots ≈ **0.1678**.

Table 4.11.4. The percentage sums of those *n*-plets with strong and weak roots, which are located correspondingly in black and white matrix cells (Figs. 4.1.1–4.1.3) in named rows for the DNA of the human chromosome № 1. Corresponding percentage values for separate *n*-plets are shown in Figs. 4.1.1–4.1.3.

	Σ% of *n*-plets with strong roots	**Σ% of *n*-plets with weak roots**
Row 2 in matrix of 16 duplets	%CT + %CG ≈ 0.07134 + 0.01031 ≈ 0.0817.	%AT + %AG ≈ 0.07429 + 0.07137 ≈ 0.1457.
Rows 3–4 in matrix of 64 triplets	%CTC + %CTA + %CGC + %CGA + %CTT + %CTG + %CGT + %CGG ≈ 0.0176 + 0.0127 + 0.0025 + 0.0023 + 0.0201 + 0.0209 + 0.0026 + 0.0029 ≈ 0.0816.	%ATC + %ATA + %AGC + %AGA + %ATT + %ATG + %AGT + %AGG ≈ 0.0132 + 0.0194 + 0.0144 + 0.0224 + 0.0239 + 0.0178 + 0.0161 + 0.0185 ≈ 0.1457.
Rows 5–8 in matrix of 256 tetraplets	Percentage sum of 32 tetraplets with strong roots ≈ 0.0816.	Percentage sum of 32 tetraplets with weak roots ≈ 0.1456.

Table 4.11.5. The percentage sums of those *n*-plets with strong and weak roots, which are located correspondingly in black and white matrix cells (Figs. 4.1.1–4.1.3) in named rows for the DNA of the human chromosome № 1. Corresponding percentage values for separate *n*-plets are shown in Figs. 4.1.1–4.1.3.

	Σ% of *n*-plets with strong roots	**Σ% of *n*-plets with weak roots**
Row 3 in matrix of 16 duplets	%TC + %GC ≈ 0.06008 + 0.04402 ≈ 0.1041.	%TA + %GA ≈ 0.06312 + 0.06008 ≈ 0.1232.
Rows 5–6 in matrix of 64 triplets	%TCC + %TCA + %GCC + %GCA + %TCT + %TCG + %GCT + %GCG ≈ 0.0159 + 0.0196 + 0.0125 + 0.0146 + 0.0223 + 0.0023 + 0.0144 + 0.0025 ≈ 0.1041.	%TAC + %TAA + %GAC + %GAA + %TAT + %TAG + %GAT + %GAG ≈ 0.0110 + 0.0199 + 0.0096 + 0.0196 + 0.0194 + 0.0128 + 0.0133 + 0.0176 ≈ 0.1232.
Rows 9–12 in matrix of 256 tetraplets	Percentage sum of 32 tetraplets with strong roots ≈ 0.1041.	Percentage sum of 32 tetraplets with weak roots ≈ 0.1232.

- The percentage sum of each of the two oppositional subsets of duplets in row 4 of the matrix of 16 duplets in Figs. 4.1.1–4.1.3 is approximately equal to the percentage sum of the corresponding subset of triplets in rows 7 and 8 in the matrix of 64 triplets in Figs. 4.1.1–4.1.3, and also is approximately equal to the percentage sum of the corresponding subset of

Table 4.11.6. The percentage sums of those n-plets with strong and weak roots, which are located correspondingly in black and white matrix cells (Figs. 4.1.1–4.1.3) in named rows for the DNA of the human chromosome № 1. Corresponding percentage values for separate n-plets are shown in Figs. 4.1.1–4.1.3.

	Σ% of n-plets with strong roots	**Σ% of n-plets with weak roots**
Row 4 in matrix of 16 duplets	%GT + %GG ≈ 0.05046 + 0.05419 ≈ 0.1047.	%TT + %TG ≈ 0.09568 + 0.07286 ≈ 0.1685.
Rows 7–8 in matrix of 64 triplets	%GTC + %GTA + %GGC + %GGA + %GTT + %GTG + %GGT + %GGG ≈ 0.0096 + 0.0112 + 0.0126 + 0.0160 + 0.0145 + 0.0153 + 0.0119 + 0.0138 ≈ 0.1049.	%TTC + %TTA + %TGC + %TGA + %TTT + %TTG + %TGT + %TGG ≈ 0.0197 + 0.0198 + 0.0146 + 0.0195 + 0.0372 + 0.0188 + 0.0199 + 0.0190 ≈ 0.1685.
Rows 13–16 in matrix of 256 tetraplets	Percentage sum of 32 tetraplets with strong roots ≈ 0.1047.	Percentage sum of 32 tetraplets with weak roots ≈ 0.1685.

tetraplets in rows 13–16 of the matrix of 256 tetraplets in Figs. 4.1.1–4.1.3. This numeric result is shown in Table 4.11.6.

4.12. The interrelations of n-plets of hydrogen bonds in genomic DNAs

As it is known, one of the most important components of life is hydrogen bonds. It occurs in many biological structures, such as DNA [200]. In particular, hydrogen bonding plays the role of a promotional factor for intermolecular vibrational energy relaxation, and as a driving force for the occurrences of specific reaction channels in binary molecular complexes [184]. Hydrogen bonding is an important factor in the functioning of enzymes, which manage molecular processes in biological bodies [201, 202]. The ultrafast intermolecular hydrogen bond dynamics of water were revealed [203]. Hydrogen bonds determine many of the properties of water and ice. The Nobel Laureate Pauling emphasized: "*Although the hydrogen bond is not a strong bond, it has a great significance in determining the properties of substances... It has been recognized that hydrogen bonds restrain protein molecules to their native configurations, and I believe that as the methods of structural chemistry are further applied to physiological problems it will be found that the significance of the hydrogen bond for physiology is greater than that of any other single structural feature*" [122].

Complementary pairs of nucleobases A-T and C-G in double-stranded DNAs are linked by 2 and 3 hydrogen bonds, respectively. Along any genomic DNA, these hydrogen bonds form a long sequence consisting of these two numbers such as 322323332332... This section considers hidden rules of stochastic organization of such sequences of hydrogen bonds in genomic DNAs. Analysis of hydrogen bond sequences in DNAs is important because many secrets of biological bodies and their genetic inheritance are connected with hydrogen bonds. We consider hydrogen bonds in complementary DNA strands as an important factor of genetic coding.

4.12.1. *n-plets of hydrogen bonds in the genetic matrices of DNA alphabets*

Each of the nucleotide n-plets corresponds to an n-plet of its hydrogen bonds. For example, the triplet CAG corresponds to the triplet 323 of hydrogen bonds in accordance with conditional denotations A = T = 2 and C = G = 3. These denotations reflect the fact that complementary nucleotides A and T are linked by 2 hydrogen bonds and complementary nucleotides C and G by 3 hydrogen bonds. The term "hydrogen-n-plets" or, briefly, H-n-plets is used for n-plets of hydrogen bonds. Different nucleotide n-plets can correspond to the same H-n-plet. For example, 8 nucleotide triplets CAC, CAG, CTC, CTG, GAC, GAG, GTC, GTG correspond to the same H-triplet 323, which can be presented in a form of this concatenation of numbers 3 and 2 or in a form of their arithmetic product $3 \times 2 \times 3 = 18$. The presentation of H-n-plets in the concatenation form is useful for the described study below of percentages of different H-n-plets in sequences of genomic DNAs since, for example, H-triplets 323 and 332 are met with different percentages in such sequences.

Let us return to the genetic matrices, which were constructed on the basis of binary-oppositional traits of nucleobases and which represent DNA alphabets of nucleobase n-plets in Chapter 1 (Fig. 1.2.1). Fig. 4.12.1 shows

	0	1
0	C 3	A 2
1	T 2	G 3

	00	01	10	11
00	CC 33	CA 32	AC 23	AA 22
01	CT 32	CG 33	AT 22	AG 23
10	TC 23	TA 22	GC 33	GA 32
11	TT 22	TG 23	GT 32	GG 33

	000	001	010	011	100	101	110	111
000	CCC 333	CCA 332	CAC 323	CAA 322	ACC 233	ACA 232	AAC 223	AAA 333
001	CCT 332	CCG 333	CAT 322	CAG 323	ACT 232	ACG 233	AAT 222	AAG 223
010	CTC 323	CTA 322	CGC 333	CGA 332	ATC 223	ATA 222	AGC 233	AGA 232
011	CTT 322	CTG 323	CGT 332	CGG 333	ATT 222	ATG 223	AGT 232	AGG 233
100	TCC 233	TCA 232	TAC 223	TAA 222	GCC 333	GCA 332	GAC 323	GAA 322
101	TCT 232	TCG 233	TAT 222	TAG 223	GCT 332	GCG 333	GAT 322	GAG 323
110	TTC 223	TTA 222	TGC 233	TGA 232	GTC 323	GTA 322	GGC 333	GGA 332
111	TTT 222	TTG 223	TGT 232	TGG 233	GTT 322	GTG 323	GGT 332	GGG 333

Fig. 4.12.1. The square tables of DNA alphabets of 4 nucleotides, 16 duplets, and 64 triplets (compare with analogous matrices in Fig. 1.2.1). Digits inside matrix cells show H-*n*-plets in their concatenation form.

initial members of the tensor family of genetic matrices $[C, A; T, G]^{(n)}$, in which each of the nucleotide *n*-plets is accompanied by a corresponding H-*n*-plet. Similar matrices exist for DNA alphabets of tetraplets, pentaplets, and so on.

We emphasize that for the convenience of further presentation, we conditionally show H-*n*-plets in these matrices and below in this section, mainly as a concatenation of numbers of hydrogen bonds 3 and 2, but remember the following: Each such concatenation can be simultaneously considered as the arithmetic product of the numbers 3 and 2 that make it up and, in this case, the H-*n*-plets are represented by the tensor family of numerical matrices $[3, 2]^{(n)}$, in which the arithmetic multiplication of numbers serves as the linking operation. For example, in this case, $[3, 2]^{(2)} = [3 \times 3, 3 \times 2, 2 \times 3, 2 \times 2] = [9, 6, 6, 4]$.

One can consider these matrices as numeric matrices of H-n-plets only (deleting nucleobase symbols). In this case, the tensor family of H-n-plet matrices $[3, 2; 2, 3]^{(n)}$ arises whose components we initially consider in their concatenation form. Applying the dyadic-shift decomposition to these H-n-plet matrices, which was explained and explored in Chapter 1, each of the H-n-plet matrices is presented as a sum of a set of sparse matrices with appropriate coefficients. Each of these sets of sparse matrices occurs closed concerning multiplication and defines a multiplication table of a certain algebra of hypercomplex numbers known in mathematics. Fig. 4.12.2 shows that the matrix of 4 H-monoplets (from Fig. 4.12.1) is a sum of 2 sparse orthogonal matrices e_0 and e_1, the set of which defines the multiplication table of two-dimensional algebra of hyperbolic numbers. Correspondingly, this matrix presents hyperbolic number $3 + 2e_1$ where e_1 is the imaginary unit of hyperbolic numbers.

Figure 4.12.3 shows that the matrix of 16 H-duplets (from Fig. 4.12.1) is a sum of 4 sparse orthogonal matrices s_0, s_1, s_2, s_3 with coefficients. The set of these sparse matrices defines the multiplication table of algebra of

$$\begin{vmatrix} 3, 2 \\ 2, 3 \end{vmatrix} = 3 \begin{vmatrix} 1, 0 \\ 0, 1 \end{vmatrix} + 2 \begin{vmatrix} 0, 1 \\ 1, 0 \end{vmatrix} = 3e_0 + 2e_1;$$

×	e_0	e_1
e_0	e_0	e_1
e_1	e_1	e_0

Fig. 4.12.2. The dyadic-shift decomposition of the matrix of 4 H-monoplets (from Fig. 4.12.1). The multiplication table of two sparse orthogonal matrices e_0 and e_1 is shown on the right.

$$\begin{vmatrix} 33, 32, 23, 22 \\ 32, 33, 22, 23 \\ 23, 22, 33, 32 \\ 22, 23, 32, 33 \end{vmatrix} = 33 \begin{vmatrix} 1,0,0,0 \\ 0,1,0,0 \\ 0,0,1,0 \\ 0,0,0,1 \end{vmatrix} + 32 \begin{vmatrix} 0,1,0,0 \\ 1,0,0,0 \\ 0,0,0,1 \\ 0,0,1,0 \end{vmatrix} + 23 \begin{vmatrix} 0,0,1,0 \\ 0,0,0,1 \\ 1,0,0,0 \\ 0,1,0,0 \end{vmatrix} + 22 \begin{vmatrix} 0,0,0,1 \\ 0,0,1,0 \\ 0,1,0,0 \\ 1,0,0,0 \end{vmatrix} =$$

$$= 33s_0 + 32s_1 + 23s_2 + 22s_3;$$

*	s_0	s_1	s_2	s_3
s_0	s_0	s_1	s_2	s_3
s_1	s_1	s_0	s_3	s_2
s_2	s_2	s_3	s_0	s_1
s_3	s_3	s_2	s_1	s_0

Fig. 4.12.3. The dyadic-shift decomposition of the matrix of 16 H-duplets (from Fig. 4.12.1). The multiplication table of 4 sparse orthogonal matrices s_0, s_1, s_2, s_3 is shown on the right. In this multiplication table, bold lines mark the multiplication table of sub-algebra of two-dimensional hyperbolic numbers.

four-dimensional hyperbolic numbers. By this reason, this (4 × 4)-matrix of hydrogen bonds represents a four-dimensional hyperbolic number $33s_0 + 32s_1 + 23s_2 + 22s_3$ where s_1, s_2, s_3 represent corresponding imaginary units. This four-dimensional algebra has the sub-algebra of two-dimensional hyperbolic numbers on the basis of the closed set of two matrices s_0 and s_1.

Figure 4.12.4 shows that the matrix of 64 H-triplets (from Fig. 4.12.1) is a sum of 8 sparse matrices r_0, r_1, r_2, r_3, r_4, r_5, r_6, r_7 with coefficients. The set

$$\begin{vmatrix} 333,332,323,322,233,232,223,222 \\ 332,333,322,323,232,233,222,223 \\ 323,322,333,332,223,222,233,232 \\ 322,323,332,333,222,223,232,233 \\ 233,232,223,222,333,332,323,322 \\ 232,233,222,223,332,333,322,323 \\ 223,222,233,232,323,322,333,332 \\ 222,223,232,233,322,323,332,333 \end{vmatrix} = 333 \begin{vmatrix} 1,0,0,0,0,0,0,0 \\ 0,1,0,0,0,0,0,0 \\ 0,0,1,0,0,0,0,0 \\ 0,0,0,1,0,0,0,0 \\ 0,0,0,0,1,0,0,0 \\ 0,0,0,0,0,1,0,0 \\ 0,0,0,0,0,0,1,0 \\ 0,0,0,0,0,0,0,1 \end{vmatrix} + 332 \begin{vmatrix} 0,1,0,0,0,0,0,0 \\ 1,0,0,0,0,0,0,0 \\ 0,0,0,1,0,0,0,0 \\ 0,0,1,0,0,0,0,0 \\ 0,0,0,0,0,1,0,0 \\ 0,0,0,0,1,0,0,0 \\ 0,0,0,0,0,0,0,1 \\ 0,0,0,0,0,0,1,0 \end{vmatrix} +$$

$$+ 323 \begin{vmatrix} 0,0,1,0,0,0,0,0 \\ 0,0,0,1,0,0,0,0 \\ 1,0,0,0,0,0,0,0 \\ 0,1,0,0,0,0,0,0 \\ 0,0,0,0,0,1,0,0 \\ 0,0,0,0,1,0,0,0 \\ 0,0,0,0,0,0,0,1 \\ 0,0,0,0,0,0,1,0 \end{vmatrix} + 322 \begin{vmatrix} 0,0,0,1,0,0,0,0 \\ 0,0,1,0,0,0,0,0 \\ 0,1,0,0,0,0,0,0 \\ 1,0,0,0,0,0,0,0 \\ 0,0,0,0,0,0,0,1 \\ 0,0,0,0,0,0,1,0 \\ 0,0,0,0,0,1,0,0 \\ 0,0,0,0,1,0,0,0 \end{vmatrix} + 233 \begin{vmatrix} 0,0,0,0,1,0,0,0 \\ 0,0,0,0,0,1,0,0 \\ 0,0,0,0,0,0,1,0 \\ 0,0,0,0,0,0,0,1 \\ 1,0,0,0,0,0,0,0 \\ 0,1,0,0,0,0,0,0 \\ 0,0,1,0,0,0,0,0 \\ 0,0,0,1,0,0,0,0 \end{vmatrix} + 232 \begin{vmatrix} 0,0,0,0,0,1,0,0 \\ 0,0,0,0,1,0,0,0 \\ 0,0,0,0,0,0,0,1 \\ 0,0,0,0,0,0,1,0 \\ 0,1,0,0,0,0,0,0 \\ 1,0,0,0,0,0,0,0 \\ 0,0,0,1,0,0,0,0 \\ 0,0,1,0,0,0,0,0 \end{vmatrix} +$$

$$+ 223 \begin{vmatrix} 0,0,0,0,0,0,1,0 \\ 0,0,0,0,0,0,0,1 \\ 0,0,0,0,1,0,0,0 \\ 0,0,0,0,0,1,0,0 \\ 0,0,1,0,0,0,0,0 \\ 0,0,0,1,0,0,0,0 \\ 1,0,0,0,0,0,0,0 \\ 0,1,0,0,0,0,0,0 \end{vmatrix} + 222 \begin{vmatrix} 0,0,0,0,0,0,0,1 \\ 0,0,0,0,0,0,1,0 \\ 0,0,0,0,0,1,0,0 \\ 0,0,0,0,1,0,0,0 \\ 0,0,0,1,0,0,0,0 \\ 0,0,1,0,0,0,0,0 \\ 0,1,0,0,0,0,0,0 \\ 1,0,0,0,0,0,0,0 \end{vmatrix} = 333r_0+332r_1+323r_2+322r_3+ + 233r_4+232r_5+223r_6+222r_7$$

×	r_0	r_1	r_2	r_3	r_4	r_5	r_6	r_7
r_0	r_0	r_1	r_2	r_3	r_4	r_5	r_6	r_7
r_1	r_1	r_0	r_3	r_2	r_5	r_4	r_7	r_6
r_2	r_2	r_3	r_0	r_1	r_6	r_7	r_4	r_5
r_3	r_3	r_2	r_1	r_0	r_7	r_6	r_5	r_4
r_4	r_4	r_5	r_6	r_7	r_0	r_1	r_2	r_3
r_5	r_5	r_4	r_7	r_6	r_1	r_0	r_3	r_2
r_6	r_6	r_7	r_4	r_5	r_2	r_3	r_0	r_1
r_7	r_7	r_6	r_5	r_4	r_3	r_2	r_1	r_0

Fig. 4.12.4. The dyadic-shift decomposition of the matrix of 64 H-triplets (from Fig. 4.12.1). The multiplication table of 8 sparse orthogonal matrices r_0, r_1, r_2, r_3, r_4, r_5, r_6, r_7 is shown at the bottom. In this multiplication table, bold lines mark the multiplication tables of the sub-algebra of two-dimensional hyperbolic numbers (on the basis of the closed set of matrices r_0 and r_1) and the sub-algebra of four-dimensional hyperbolic numbers (on the basis of the closed set of matrices r_0, r_1, r_2, r_3).

of these sparse matrices defines the multiplication table of the algebra of eight-dimensional hyperbolic numbers. By this reason, the (8 × 8)-matrix of hydrogen bonds represent eight-dimensional hyperbolic number $333r_0 + 332r_1 + 323r_2 + 322r_3 + 233r_4 + 232r_5 + 223r_6 + 222r_7$ where $r_1, r_2, r_3, r_4, r_5, r_6, r_7$ represent corresponding imaginary units. This eight-dimensional algebra has the sub-algebra of four-dimensional hyperbolic numbers (on the basis of 4 matrices r_0, r_1, r_2, r_3) and also the sub-algebra of two-dimensional hyperbolic numbers (on the basis of 2 matrices r_0 and r_1).

Similar dyadic-shift decompositions of matrices of 256 H-tetraplets, 1024 H-pentaplets, etc., give sets of sparse matrices closed concerning multiplication, which define corresponding multiplication tables of algebras of 2^n-dimensional hyperbolic numbers. Each matrix algebra of 2^n-dimensional hyperbolic numbers has an hierarchical structure because it contains a set of sub-algebras of 2^{n-1}, 2^{n-2}, ..., two-dimensional hyperbolic numbers by analogy with shown examples in Figs. 4.12.2 and 4.12.3. (These 2^n-dimensional hyperbolic numbers are also called "2^n-dimensional hyperbolic matrions" [12, 13]).

One can briefly note that in the case of 64 triplets, these multiplication tables of eight-dimensional algebra and its two sub-algebras attract attention to a corresponding division of sets of 64 triplets and 20 amino acids into three characteristic subsets. As one can see in Figs. 1.2.1, 4.12.1 and 4.12.4, the sparse matrices p_0 and p_1, which define the multiplication table of the two-dimensional sub-algebra, contain those non-zero entries which correspond to a subset of 16 triplets coding 4 amino acids with strong roots — Pro, Arg, Ala, and Gly; they are arranged along the main diagonal of the matrix of 64 triplets (it is this subset of triplets that is distinguished by the maximum stability of its code connection with these amino acids in many known dialects of the genetic code, which differ slightly from each other by changing the code value of individual triplets [204, 205]). The sparse matrices p_0, p_1, p_2, and p_3, which define the multiplication table of the four-dimensional sub-algebra, contain those non-zero entries which correspond to the subset of 32 triplets and 10 amino acids located in both quadrants along the main diagonal of this matrix. Four other sparse matrices p_4, p_5, p_6, and p_7 contain those non-zero entries, which correspond to the 32 triplets and appropriate amino acids and stop codons located in both quadrants along the second diagonal of the matrix (see Figs. 1.2.1, 4.12.1 and 4.12.4).

4.12.2. *Rules of stochastic organization of hydrogen n-plets in genomic DNAs*

The described connection between alphabetic matrices of H-n-plets and algebras of 2^n-dimensional hyperbolic numbers provokes a question about a possible role of H-n-plets in stochastic organization of genomic DNAs of different species. The study of this question culminated in results presented below. Each long DNA nucleotide sequence, for example, CAAGTGCAT… is simultaneously a sequence of its hydrogen bonds 322323322… This hydrogen bond sequence can be represented as a sequence of H-monoplets 3-2-2-3-2-3-3-…, where the H-monoplet 3 exists with a percentage value "q" and the H-monoplet 2 exists with a percentage value "p". The same hydrogen bonds sequence can be represented as a sequence of H-duplets 32-23-23-…, where 4 kinds of H-duplets form an appropriate alphabet: 33, 32, 23, 22. Their percentages in the considered sequence are denoted as f(33), f(32), f(23), f(22). Each of these percentage values of H-duplets is a sum of percentages of 4 appropriate nucleotides in the duplet representation of the DNA nucleotide sequence (that is, for example, CA-AG-TG-…). For example, f(33) is a sum of corresponding percentages of 4 duplets CC, CG, GC, GG, each of which is characterized by the same H-duplet 33. The same hydrogen bonds sequence can be represented as a sequence of H-triplets 322-323-322-…, where 8 kinds of H-triplets form an appropriate alphabet: 333, 332, 323, 322, 233, 232, 223, 222. Their percentages in the considered sequence are denoted as f(333), f(332), …, f(222). Similar representations are possible for sequences of H-tetraplets, H-pentaplets, and so on.

Percentages (or probabilities) of each member of H-n-alphabets (for $n = 1, 2, 3, 4, 5$) in genomic DNA sequences have been studied for many eukaryotes and prokaryotes. This study has led to evidences in favor of the existence of general rules (or laws) concerning percentages of members of H-n-alphabets in genomic DNA sequences [206].

As one of the results, Fig. 4.12.5 graphically shows percentages of members of H-n-alphabets (n = 1, 2, 3, 4, 5) in DNA sequences of all 5 chromosomes of a plant *Arabidopsis thaliana*; initial data about DNA sequences of these chromosomes were taken from the GenBank: https://www.ncbi.nlm.nih.gov/genome/4. The lengths of these DNA sequences are the following: in the chromosome № 1, the length is equal to 30427671 bp; chromosome № 2 — 19698289 bp; chromosome № 3 — 23459830 bp; chromosome № 4 — 18585056 bp; chromosome № 5 — 26975502 bp.

Members of all these H-n-alphabets in Fig. 4.12.5 are given in a certain arrangement corresponding to arrangements of coordinates in the tensor family of vectors $[3, 2]^{(n)}$, where (n) means the tensor power and concatenation is used as the linking operation of vector coordinates under the tensor product. For example, the vector $[3, 2]^{(2)} = [33, 32, 23, 22]$ has the arrangement of its coordinates 33, 32, 23, 22, which is used in Fig. 4.12.5 for the appropriate graph in the case of the alphabet of H-duplets. Similar correspondences hold true for vectors $[3, 2]^{(3)}$, $[3, 2]^{(4)}$, $[3, 2]^{(5)}$ and appropriate graphics for percentages of members of other H-n-alphabets in Fig. 4.12.5.

In addition to graphs in Fig. 4.12.5, Table 4.12.1 shows numeric values for percentages of all members of these H-n-alphabets for all 5 chromosomes and for n = 1, 2, 3, 4 (the case n = 5 is not shown because its numeric table is very large).

One can see from Table 4.12.1 that the percentage of each member of these H-n-alphabets has approximately the same value in all 5 chromosomes or, in other words, it serves as a special invariant inside the complete set of chromosomes of *Arabidopsis thaliana*. It is obvious that the revealed practical identity of the percentages of H-n-plets in DNAs of all considered chromosomes defines a limitation for possible arrangements of DNA nucleotide n-plets which are connected with hydrogen bonds: only special classes of arrangements of the nucleotide n-plets in long DNAs can provide such regularities.

	Percentage of H-monoplets, 2 members (3, 2)	Percentage of H-duplets, 4 members (33, 32, 23, 22)	Percentage of H-triplets, 8 members (333, 332, …, 222)	Percentage of H-tetraplets, 16 members (3333, 3332, …, 2222)
1	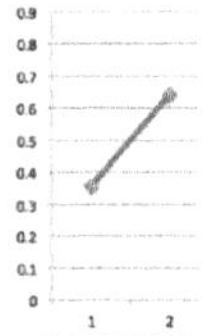 q=0.35873552 p=0.64126448	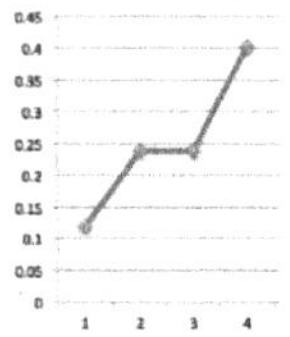	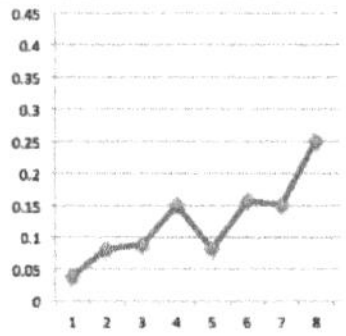	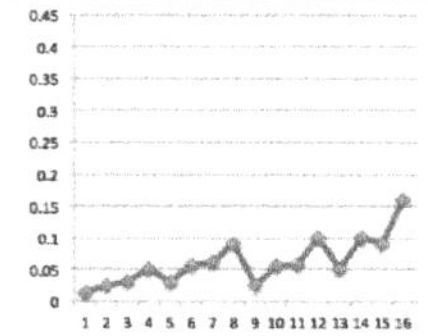
2	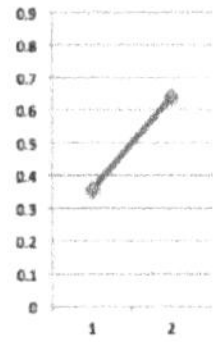 q=0.35864320 p=0.64135680	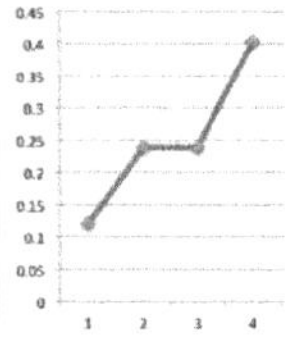	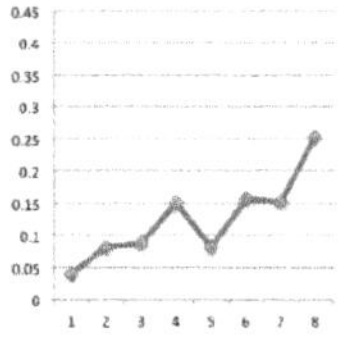	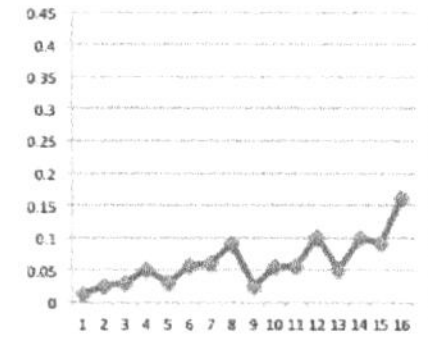
3	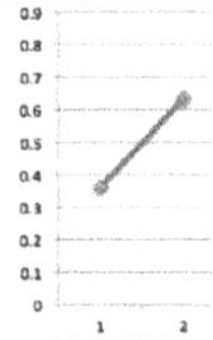 q=0.36331075 p=0.63668925	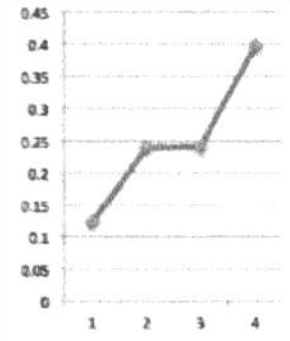	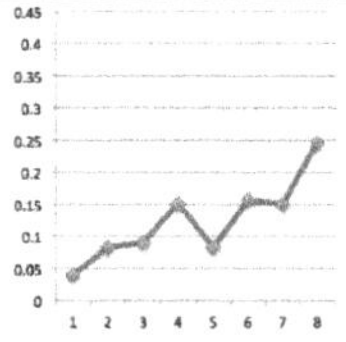	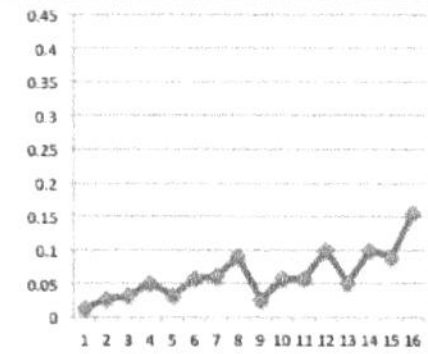
4	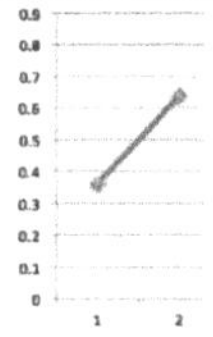 q=0.36204022 p=0.63795978	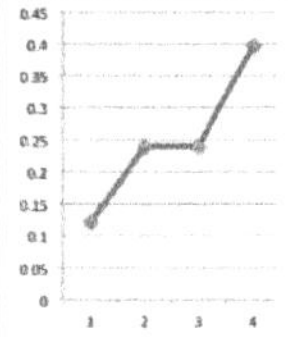	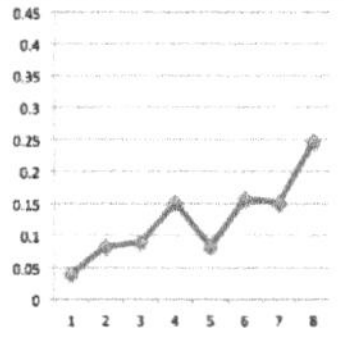	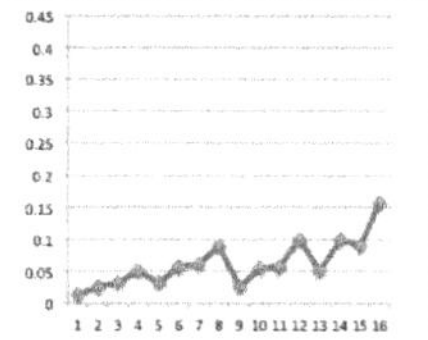
5	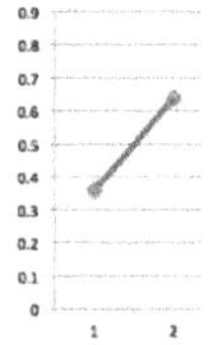 q=0.35938926 p=0.64061074	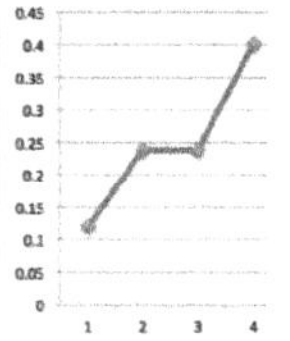	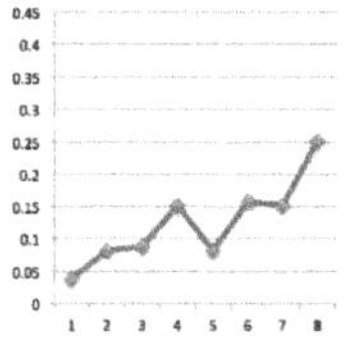	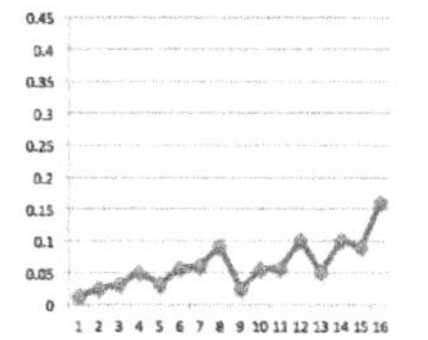

Fig. 4.12.5. Percentages of members of H-n-alphabets (n = 1, 2, 3, 4, 5) in DNA sequences of all 5 chromosomes of the plant *Arabidopsis thaliana*. The left column contains the sequence numbers of the chromosomes. Each point in the graphs shows a percentage of an appropriate member of H-n-alphabets. For each chromosome, its percentages q and p of H-monoplets 3 and 2, respectively, are shown in the second column. See explanations in the text.

Table 4.12.1. Percentages of all members of the H-*n*-alphabets ($n = 1, 2, 3, 4$) in H-*n*-plet sequences of DNAs in all 5 chromosomes of *Arabidopsis thaliana*. Symbols q and p denote percentages of H-monoplets 3 and 2, respectively; they are shown in the left column separately for each chromosome. The percentage values are rounded to the sixth decimal place. Traditional statistical symbols are also shown: $\bar{X}$ denotes the mean of the sample S = $\{x_1, x_2, \ldots, x_n\}$; s — sample standard deviation; V — coefficient of variation (see their formulas in http://www.real-statistics.com/descriptive-statistics/measures-variability/).

	Percentages of H-monoplets, 2 members: 3 and 2		**Percentages of H-duplets, 4 members: 33, 32, 23, 22**			
	3 (q)	**2 (p)**	**f(33)**	**f(32)**	**f(23)**	**f(22)**
Chr. 1	0.358736	0.641264	0.119788	0.239022	0.238874	0.402316
Chr. 2	0.358643	0.641357	0.120564	0.238138	0.238021	0.403277
Chr. 3	0.363311	0.636689	0.122972	0.240226	0.240451	0.396350
Chr. 4	0.362040	0.637960	0.121923	0.240142	0.240093	0.397843
Chr. 5	0.359389	0.640611	0.120390	0.239095	0.238902	0.401612
$\bar{X}$	0.360424	0.639576	0.121127	0.239325	0.239268	0.400280
s	0.001899	0.001899	0.001157	0.000779	0.000886	0.002694
V	0.005270	0.002970	0.009549	0.003254	0.003703	0.006730

	Percentages of H-triplets, 8 members: 333,332,323,322,233,232,223,222							
	f(333)	**f(332)**	**f(323)**	**f(322)**	**f(233)**	**f(232)**	**f(223)**	**f(222)**
Chr. 1	0.038514	0.081174	0.088018	0.150704	0.081231	0.157748	0.151366	0.251244
Chr. 2	0.039243	0.081345	0.087551	0.150357	0.081290	0.156850	0.150624	0.252741
Chr. 3	0.040181	0.082637	0.089748	0.150821	0.082845	0.157226	0.150879	0.245662
Chr. 4	0.039645	0.082675	0.089030	0.150815	0.082169	0.157718	0.150903	0.247044
Chr. 5	0.038976	0.081379	0.088099	0.150902	0.081370	0.157821	0.150821	0.250633
$\bar{X}$	0.039312	0.081842	0.088489	0.150720	0.081781	0.157473	0.150919	0.249465
s	0.000569	0.000668	0.000792	0.000192	0.000632	0.000376	0.000244	0.002668
V	0.014480	0.008166	0.008945	0.001273	0.007725	0.002387	0.001619	0.010694

	Percentages of H-tetraplets, 16 members: 3333, 3332, 3323, 3322, 3233, 3232, 3223, 3222, 2333, 2332, 2323, 2322, 2233, 2232, 2223, 2222							
	f(3333)	**f(3332)**	**f(3323)**	**f(3322)**	**f(3233)**	**f(3232)**	**f(3223)**	**f(3222)**
Chr. 1	0.013226	0.025347	0.030993	0.050243	0.031122	0.056975	0.060129	0.090644
Chr. 2	0.013814	0.025540	0.031157	0.050009	0.031066	0.056623	0.059948	0.090394

Table 4.12.1. *(Continued)*

Percentages of H-tetraplets, 16 members:
3333, 3332, 3323, 3322, 3233, 3232, 3223, 3222, 2333, 2332, 2323, 2322, 2233, 2232, 2223, 2222

	f(3333)	f(3332)	f(3323)	f(3322)	f(3233)	f(3232)	f(3223)	f(3222)
Chr. 3	0.013971	0.026168	0.032107	0.050762	0.032179	0.057320	0.061061	0.089600
Chr. 4	0.013625	0.026072	0.031796	0.050559	0.031881	0.057356	0.060887	0.090007
Chr. 5	0.013550	0.025449	0.031281	0.050195	0.031254	0.056887	0.060426	0.090554
$\bar{X}$	0.013637	0.025715	0.031467	0.050354	0.031500	0.057032	0.060490	0.090240
s	0.000253	0.000337	0.000418	0.000270	0.000447	0.000275	0.000427	0.000387
V	0.018529	0.013117	0.013274	0.005366	0.014180	0.004827	0.007059	0.004292
	2333	**2332**	**2323**	**2322**	**2233**	**2232**	**2223**	**2222**
Chr. 1	0.025264	0.056027	0.056957	0.100737	0.050154	0.100824	0.090685	0.160672
Chr. 2	0.025598	0.055669	0.056442	0.100473	0.050130	0.100413	0.090313	0.162411
Chr. 3	0.026132	0.056708	0.057532	0.100124	0.050654	0.100099	0.089708	0.155877
Chr. 4	0.025858	0.056428	0.057387	0.100305	0.050429	0.100297	0.090139	0.156975
Chr. 5	0.025397	0.056027	0.056796	0.100840	0.050101	0.100706	0.090241	0.160293
$\bar{X}$	0.025650	0.056172	0.057023	0.100496	0.050294	0.100468	0.090217	0.159246
s	0.000313	0.000360	0.000396	0.000265	0.000215	0.000265	0.000314	0.002435
V	0.012221	0.006409	0.006945	0.002639	0.004279	0.002642	0.003483	0.015294

Similar results (with some other levels of accuracy) have been obtained in the analysis of the following complete sets of nuclear chromosomes on the basis of initial data about their DNA sequences from the GenBank: a nematode *Caenorhabditis elegans*, fruit fly *Drosophila melanogaster*, house mouse *Mus musculus*, and *Homo sapiens*.

These results allow assuming the existence of the following general rule for eukaryotes about percentages of hydrogen bonds in complete sets of nuclear chromosomes:

- In complete sets of nuclear chromosomes of different eukaryotes, a percentage of any member of the H-n-alphabets has approximately the same value in all chromosomes.

Of course, further research is needed to define the degree of universality and precision of this rule. This rule has mathematical analogies with the main law of population genetics — the Hardy-Weinberg law as described in the next section.

4.12.3. *The algebraic model of probabilities of hydrogen n-plets in genomic DNAs*

In the proposed model approach, phenomenological probabilities (or percentages) of all members of H-n-alphabets $[3, 2]^{(n)}$ in a genomic DNA sequence under fixed n are modeled by numeric values of appropriate coordinates of 2^n-dimensional vector from the tensor family of percent vectors $[q, p]^{(n)}$ where q and p mean percentages of H-monoplets 3 and 2 correspondingly (these percentages q and p are shown for each chromosome in the second columns in Fig. 4.12.4). In this tensor family, the tensor product is based on concatenation of elements q and p in each of the vector coordinates (as noted above, each such concatenation can also be represented as the arithmetic product of the numbers 3 and 2 that make up this concatenation, but such a representation will be less visual for the presenting statistical characteristics of sets of H-plets). In other words, in this model approach, to get model values of percentages of all 2^n members of the H-n-alphabet in a genomic DNA, one should theoretically calculate the following 2^n coordinates of percent vectors from their tensor family using only knowledge about phenomenological values q and p:

— in the case of H-duplets alphabet, 4 coordinates of the percent vector $[q, p]^{(2)} = [qq, qp, pq, pp]$, which correspond to 4 coordinates of the H-duplets vector $[3, 2]^{(2)} = [33, 32, 23, 22]$;
— in the case of H-triplets alphabet, 8 coordinates of the percent vector $[q, p]^{(3)} = [qqq, qqp, qpq, qpp, pqq, pqp, ppq, ppp]$, which correspond to 8 coordinates of the H-triplets vector $[3, 2]^{(3)} = [333, 332, 323, 322, 233, 232, 223, 222]$;

— in the case of H-tetraplets alphabet, 16 coordinates of the percent vector $[q, p]^{(4)} = [qqqq, qqqp, qqpq, qqpp, qpqq, qpqp, qppq, qppp, pqqq, pqqp, pqpq, pqpp, ppqq, ppqp, pppq, pppp]$, which correspond to 16 coordinates of H-tetraplets vector $[q, p]^{(4)}$;

— and so on.

Briefly speaking, in line with the proposed model approach, for a prediction of approximate values for percentages of different members of H-n-alphabets in a long DNA H-n-plet sequence on the basis of knowledge about only two percentages q and p, it is enough to do the following:

- calculate phenomenological percentages q and p of H-monoplets 3 and 2 in the considered DNA H-n-plet sequence;
- theoretically calculate coordinates of vectors $[q, p]^{(n)}$ for appropriate values n and then match coordinates of these percent vectors with coordinates of corresponding H-n-plet vectors $[3, 2]^{(n)}$.

The expression (4.28) shows an example of such calculation of the percentage f(332) for the member 332 of the H-triplets alphabet under $q = 0.36$ and $p = 0.64$ shown in Table 4.12.1 ($q + p = 1$):

$$\mathrm{f}(332) = qqp = 0.36 \times 0.36 \times 0.64 = 0.0829 \tag{4.28}$$

Figure 4.12.6 shows nice correspondences between these model values (red points) of components of percent vectors $[q, p]^{(n)}$ and phenomenological values (blue points) of percentages for all members of H-n-alphabets ($n = 2$, 3, 4). One can see in these graphs that the model points in red turned out to be almost exactly superimposed on the phenomenological points in blue. It gives evidences in favor of the following: knowing only two probabilities q and p of H-monoplets 3 and 2 in a long DNA sequence, one can predict percentage values of all other members of H-n-plet alphabets for the analyzed DNA with an impressive level of accuracy. We presume that a similar model

	Alphabet of H-monoplets, 2 members (3, 2)	Alphabet of H-duplets, 4 members (33, 32, 23, 22)	Alphabet of H-triplets, 8 members (333, 332, …, 222)	Alphabet of H-tetraplets, 16 members (3333, 3332, …, 2222)
1	$q = 0.35873552$ $p = 0.64126448$	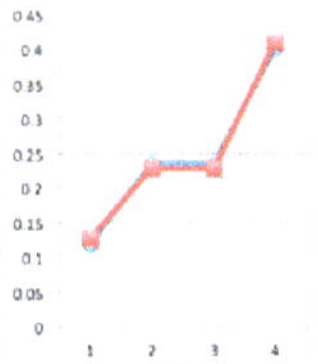 $[q, p]^{(2)}$	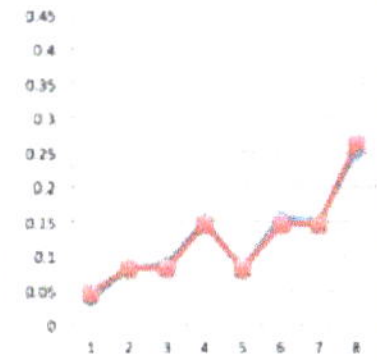 $[q, p]^{(3)}$	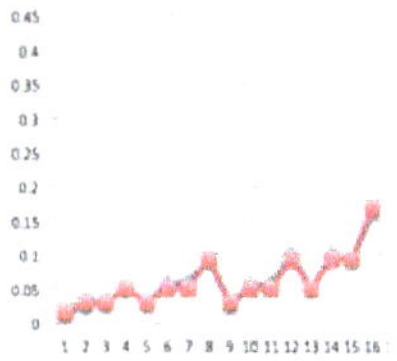 $[q, p]^{(4)}$
2	$q = 0.35864320$ $p = 0.64135680$	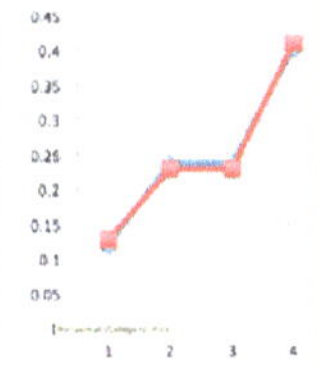 $[q, p]^{(2)}$	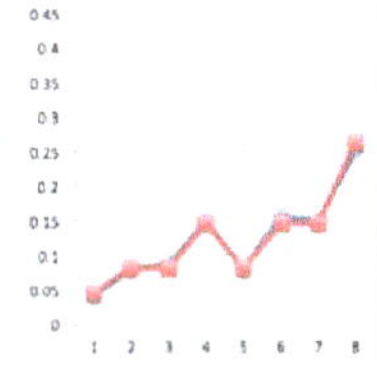 $[q, p]^{(3)}$	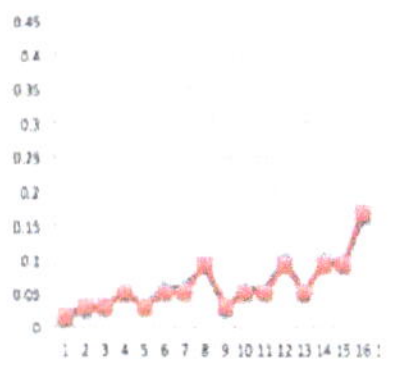 $[q, p]^{(4)}$
3	$q = 0.36331075$ $p = 0.63668925$	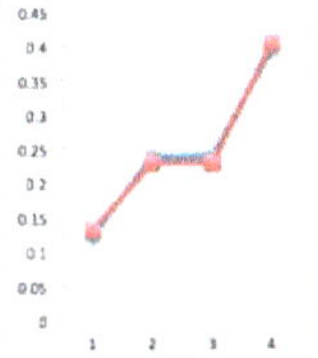 $[q, p]^{(2)}$	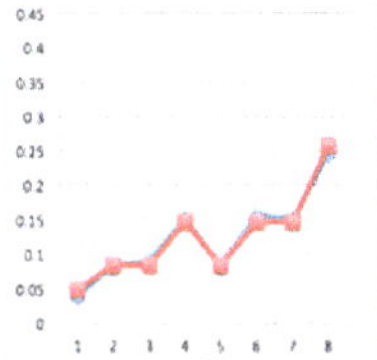 $[q, p]^{(3)}$	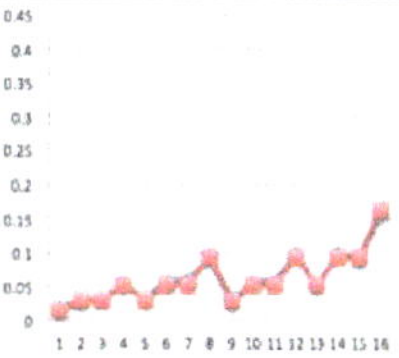 $[q, p]^{(4)}$
4	$q = 0.36204022$ $p = 0.63795978$	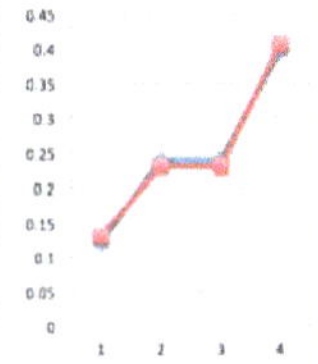 $[q, p]^{(2)}$	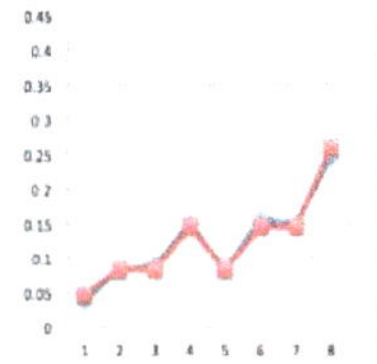 $[q, p]^{(3)}$	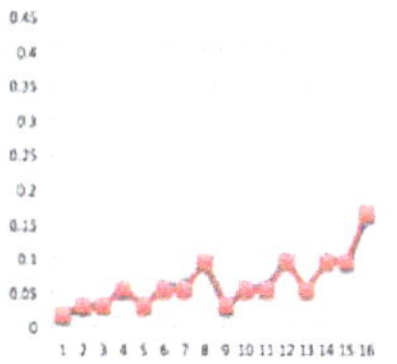 $[q, p]^{(4)}$
5	$q = 0.35938926$ $p = 0.64061074$	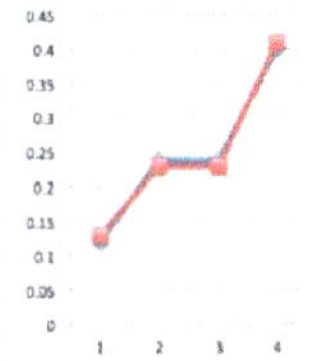 $[q, p]^{(2)}$	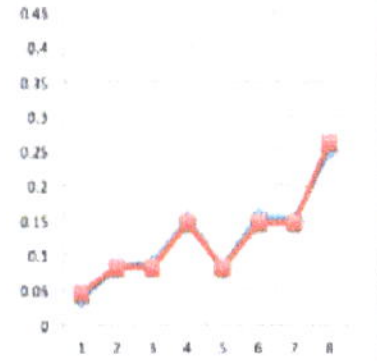 $[q, p]^{(3)}$	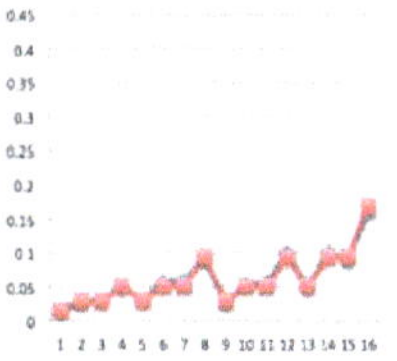 $[q, p]^{(4)}$

Fig. 4.12.6. The nice correspondence between coordinate values (red points) of model vectors $[q, p]^{(n)}$ and phenomenological values (blue points taken from Fig. 4.12.5) of percentages for all members of H-n-alphabets (n = 2, 3, 4) in DNA H-n-plet sequences of all 5 chromosomes of the plant *Arabidopsis thaliana*. The left column contains the sequence numbers of the chromosomes. See explanations in the text.

correspondence holds true also for $n = 6, 7, 8, \ldots$ (if n is not too large) but this should be studied in future research.

In addition to graphs in Fig. 4.12.6, Table 4.12.2 shows numeric values of coordinates of vectors $[q, p]^{(n)}$ for all 5 chromosomes under $n = 1, 2, 3, 4$. By a comparison of Tables 4.12.1 and 4.12.2, one can see that the proposed model approach gives nice results in modeling the phenomenological percentages.

Table 4.12.2. Coordinate values of percentage vectors $[q, p]^{(n)}$ are shown, which are used as models of phenomenological percentage values for all members of the H-n-alphabets ($n = 1, 2, 3, 4$) in H-n-plet DNA sequences of all 5 chromosomes of *Arabidopsis thaliana*. Symbols q and p denote percentages of H-monoplets 3 and 2, respectively, and they are shown in the left column separately for each chromosome. The model values are rounded to the sixth decimal place. Traditional statistical symbols are also shown: $\bar{X}$ denotes the mean of the sample $S = \{x_1, x_2, \ldots, x_n\}$; s — sample standard deviation; V — coefficient of variation (see their formulas in http://www.real-statistics.com/descriptive-statistics/measures-variability/).

	Alphabet of H-monoplets, 2 members (3, 2)		**Alphabet of H-duplets, 4 members (33, 32, 23, 22)**			
	Model percent values:					
	3 (*q*)	**2 (*p*)**	**33 (*qq*)**	**32 (*qp*)**	**23 (*pq*)**	**22 (*pp*)**
Chr. 1	0.358736	0.641264	0.128691	0.230044	0.230044	0.411220
Chr. 2	0.358643	0.641357	0.128625	0.230018	0.230018	0.411339
Chr. 3	0.363311	0.636689	0.131995	0.231316	0.231316	0.405373
Chr. 4	0.362040	0.637960	0.131073	0.230967	0.230967	0.406993
Chr. 5	0.359389	0.640611	0.129161	0.230229	0.230229	0.410382
$\bar{X}$	0.360424	0.639576	0.129909	0.230515	0.230515	0.409061
s	0.001899	0.001899	0.001371	0.000528	0.000528	0.002428
V	0.005270	0.002970	0.010553	0.002293	0.002293	0.005935

	Alphabet of H-triplets with 8 members (333, 332, 323, 322, 233, 232, 223, 222)							
	333	**332**	**323**	**322**	**233**	**232**	**223**	**222**
	Model percent values:							
	qqq	*qqp*	*qpq*	*qpp*	*pqq*	*pqp*	*ppq*	*ppp*
Chr. 1	0.046166	0.082525	0.082525	0.147519	0.082525	0.147519	0.147519	0.263701
Chr. 2	0.046130	0.082494	0.082494	0.147524	0.082494	0.147524	0.147524	0.263815

(Continued)

Table 4.12.2. *(Continued)*

Alphabet of H-triplets with 8 members (333, 332, 323, 322, 233, 232, 223, 222)								
	333	**332**	**323**	**322**	**233**	**232**	**223**	**222**
	Model percent values:							
	qqq	*qqp*	*qpq*	*qpp*	*pqq*	*pqp*	*ppq*	*ppp*
Chr. 3	0.047955	0.084040	0.084040	0.147276	0.084040	0.147276	0.147276	0.258097
Chr. 4	0.047454	0.083619	0.083619	0.147348	0.083619	0.147348	0.147348	0.259645
Chr. 5	0.046419	0.082742	0.082742	0.147487	0.082742	0.147487	0.147487	0.262895
$\bar{X}$	0.046825	0.083084	0.083084	0.147431	0.083084	0.147431	0.147431	0.261631
s	0.000742	0.000629	0.000629	0.000100	0.000629	0.000100	0.000100	0.002328
V	0.015849	0.007568	0.007568	0.000681	0.007568	0.000681	0.000681	0.008896

Alphabet of H-tetraplets with 16 members (3333, 3332, 3323, 3322, 3233, 3232, 3223, 3222, 2333, 2332, 2323, 2322, 2233, 2232, 2223, 2222)								
	Model percent values:							
	3333	**3332**	**3323**	**3322**	**3233**	**3232**	**3223**	**3222**
	qqqq	*qqqp*	*qqpq*	*qqpp*	*qpqq*	*qpqp*	*qppq*	*qppp*
Chr. 1	0.016561	0.029605	0.029605	0.052920	0.029605	0.052920	0.052920	0.094599
Chr. 2	0.016544	0.029586	0.029586	0.052908	0.029586	0.052908	0.052908	0.094615
Chr. 3	0.017423	0.030532	0.030532	0.053507	0.030532	0.053507	0.053507	0.093769
Chr. 4	0.017180	0.030274	0.030274	0.053346	0.030274	0.053346	0.053346	0.094002
Chr. 5	0.016682	0.029736	0.029736	0.053005	0.029736	0.053005	0.053005	0.094482
$\bar{X}$	0.016878	0.029947	0.029947	0.053137	0.029947	0.053137	0.053137	0.094293
s	0.000357	0.000385	0.000385	0.000244	0.000385	0.000244	0.000244	0.000344
V	0.021158	0.012857	0.012857	0.004588	0.012857	0.004588	0.004588	0.003649
	2333	**2332**	**2323**	**2322**	**2233**	**2232**	**2223**	**2222**
	pqqq	*pqqp*	*pqpq*	*pqpp*	*ppqq*	*ppqp*	*pppq*	*pppp*
Chr. 1	0.029605	0.052920	0.052920	0.094599	0.052920	0.094599	0.094599	0.169102
Chr. 2	0.029586	0.052908	0.052908	0.094615	0.052908	0.094615	0.094615	0.169199
Chr. 3	0.030532	0.053507	0.053507	0.093769	0.053507	0.093769	0.093769	0.164327
Chr. 4	0.030274	0.053346	0.053346	0.094002	0.053346	0.094002	0.094002	0.165643
Chr. 5	0.029736	0.053005	0.053005	0.094482	0.053005	0.094482	0.094482	0.168413

Table 4.12.2. (*Continued*)

Alphabet of H-tetraplets with 16 members (3333, 3332, 3323, 3322, 3233, 3232, 3223, 3222, 2333, 2332, 2323, 2322, 2233, 2232, 2223, 2222)								
Model percent values:								
	2333	**2332**	**2323**	**2322**	**2233**	**2232**	**2223**	**2222**
	pqqq	*pqqp*	*pqpq*	*pqpp*	*ppqq*	*ppqp*	*pppq*	*pppp*
$\bar{X}$	0.029947	0.053137	0.053137	0.094293	0.053137	0.094293	0.094293	0.167337
s	0.000385	0.000244	0.000244	0.000344	0.000244	0.000344	0.000344	0.001983
V	0.012857	0.004588	0.004588	0.003649	0.004588	0.003649	0.003649	0.011853

Percentages of members of H-n-alphabets (n = 1, 2, 3, 4, 5) in the following 19 genomes of bacteria and archaea from the full list in the article [194, p. 2] have also been calculated: *Aquifex aeolicus, Acidobacteria bacterium, Bradyrhizobium japonicum, Bacillus subtilis, Chlamydia trachomatis, Chromobacterium violaceum, Dehalococcoides ethenogenes, Escherichia coli, Flavobacterium psychrophilum, Gloeobacter violaceus, Helicobacter pilory, Methanosarcina acetivorans, Nanoarchaeum equitans, Syntrophus aciditrophicus, Streptomyces coelicolor, Sulfolobus solfataricus, Treponema denticola, Thermotoga maritima* and *Thermus thermophiles.* The calculated sets of percentages were also modeled on sets of coordinates of appropriate vectors $[q, p]^{(n)}$ with similar approximation results [206]. These results confirm that the proposed model approach on the basis of percent vectors $[q, p]^{(n)}$ can be used to get idealized models of percentages for the members of H-n-alphabets in genomic DNA H-n-plet sequences.

The obtained results allow thinking about the following universal rule for stochastic organization of H-n-plet sequences in genomic DNAs of eukaryotes and prokaryotes for limited values of n:

— In H-n-plet sequences of genomic DNAs of eukaryotes and prokaryotes, percentages of different H-n-plets are interconnected to one another and can be approximately calculated by the tensor matrix expression $[q, p]^{(n)}$ where q and p are phenomenological percentages of the H-monoplets 3 and 2 in the considered DNA.

This rule was tested for a wide set of genomic DNAs under $n = 1, 2, 3, 4, 5$ [206]. It remains unexplored regarding what values of n this rule is satisfied.

According to this rule, the genomic DNA of different organisms differ only in the percentages of q and p of their H-monoplets 3 and 2 (with accuracy up to the fulfillment of this rule). But, as shown above in Figs. 4.12.2–4.12.4, the H-n-plet alphabets, whose members are regularly located in the genetic matrices [C, A; T, G]$^{(n)}$, can be represented by 2^n-dimensional hyperbolic numbers, for example:

$$\begin{gathered} 3 + 2e_1, \\ 33 + 32s_1 + 23s_2 + 22s_3, \\ 333 + 332r_1 + 323r_2 + 322r_3 + 233r_4 + 232r_5 + 223r_6 + 222r_7 \end{gathered} \quad (4.29)$$

where e_1, s_1, s_2, s_3, r_1, r_2, r_3, r_4, r_5, r_6 are the imaginary units of corresponding 2^n-dimensional hyperbolic numbers.

Using this rule and replacing each H-n-plet in (4.29) by its percentage in a corresponding H-n-plet sequence of the analyzed genomic DNA, you get a holistic presentation of percentages for members of each H-n-plet alphabet in the form of a corresponding 2^n-dimensional hyperbolic number like (4.30):

$$\begin{gathered} q + p \times e_1,\ qq + qp \times s_1 + pq \times s_2 + pp \times s_3, \\ qqq + qqp \times r_1 + qpq \times r_2 + qpp \times r_3 + pqq \times r_4 + pqp \times r_5 + \\ ppq \times r_6 + ppp \times r_7 \end{gathered} \quad (4.30)$$

where e_1, s_1, s_2, s_3, r_1, r_2, r_3, r_4, r_5, r_6 are the imaginary units of corresponding 2^n-dimensional hyperbolic numbers. In (4.30), the H-n-plet percentages are presented in their concatenation forms for clear visualization but they can be also presented on the basis of their arithmetic product. For example, percentage duplets qq, qp, pq, pp can be presented as $q \times q$, $q \times p$, $p \times q$ and $p \times p$.

Genomic DNAs of different species have different H-n-plet sequences, different percentages q and p of H-monoplets, and different percentages of other H-n-plets. But percentage of H-n-plet compositions of different genomic DNAs can be represented in the universal form (4.30) of 2^n-dimensional hyperbolic numbers. In the general case, each genomic DNA is characterized by its individual set of such multi-dimensional numbers.

Algebras of 2^n-dimensional hyperbolic numbers have the operation of dividing one number by another, as a result of which a number of the same algebra appears. This allows transformation of hyperbolic number $q_1 + p_1$j, which characterizes one genomic DNA concerning its percentages of H-monoplets 3 and 2, into hyperbolic number $q_2 + p_2$j, which characterizes another genomic DNA concerning percentages of these H-monoplets (here j is the imaginary unit of hyperbolic numbers). When dividing these two hyperbolic numbers by each other, the third hyperbolic number appears:

$$(q_1 + p_1\mathrm{j})/(q_2 + p_2\mathrm{j}) = q_3 + p_3\mathrm{j}, \text{ where } q_1 + p_1 = q_2 + p_2 = q_3 + p_3 = 1 \quad (4.31)$$

This division of hyperbolic numbers is conveniently done in the matrix form of their representation shown in the expression (4.32):

$$\begin{vmatrix} q_1, p_1 \\ p_1, q_1 \end{vmatrix} \times \begin{vmatrix} q_2, p_2 \\ p_2, q_2 \end{vmatrix}^{-1} = \begin{vmatrix} q_3, p_3 \\ p_3, q_3 \end{vmatrix} \quad (4.32)$$

The tensor family of the shown matrix $[q_3, p_3; p_3, q_3]^{(n)}$ contains matrix representations of all 2^n-dimensional hyperbolic numbers, which arise under division of matrices of the tensor family $[q_1, p_1; p_1, q_1]^{(n)}$ by corresponding matrices of the tensor family $[q_2, p_2; p_2, q_2]^{(n)}$ (here the value of n is not too large to provide the implementation of the formulated rule). The described results give pieces of evidence that different genomic DNAs — in relation to stochastic characteristics of their H-n-plets — can be transformed into each other by matrix operators presenting 2^n-dimensional hyperbolic numbers. Biological evolution proceeds in such a way that this algebraic relationship of the genomic DNAs of different species is preserved. The evolution of genomes is associated with hyperbolic transformations of the probability vectors of n-plets of hydrogen bonds in genomic DNAs. These transformations can be described with the help of the orthogonal matrices based on binary-oppositional properties of DNA alphabets as in Figs. 4.12.2–4.12.4.

To compare many different genomic DNAs to one another regarding stochastic organization of their H-n-plet sequences, one can use a certain hyperbolic number, for example $q + pj = 0.5 + 0.5j$ as a reference. Such comparison can be useful to define a degree of evolutionary relatedness of different genomes.

4.12.4. *Hydrogen n-plets, Pascal's triangle, and the Hardy-Weinberg law*

Let us analyze in more detail the mentioned tensor family of percent vectors $[q, p]^{(n)}$ presenting stochastic organization of H-n-plet sequences of genomic DNA. One can see that in these vectors some values of coordinates, which indicate appropriate percentages, are repeated. For example, the set of coordinates of the vector $[q, p]^{(2)} = [qq, qp, pq, pp]$ has one numeric value qq, two identical values $qp = pq$, and one value pp; the set of coordinates of the vector $[q, p]^{(3)} = [qqq, qqp, qpq, qpp, pqq, pqp, ppq, ppp]$ has one value qqq, three identical values $qqp = qpq = pqq$, three identical values $qpp = pqp = ppq$, and one value ppp, and so on. These numbers of separate percentage values in the set of vectors $[q, p]^{(n)}$ form the Pascal's triangle (https://en.wikipedia.org/wiki/Pascal%27s_triangle), which is connected with Newton's binomial formula (https://www.lucaswillems.com/en/articles/5/newtons-binomial-formula) and also with Fibonacci numbers. Pascal's triangle is shown in Fig. 4.12.7. Similar connections with Pascal's triangle exist, of course, in the tensor family of vectors $\left[\sqrt{q}, \sqrt{p}\right]^{(n)}$ whose coordinates are amplitudes of probabilities q and p, used in the next section.

But effective applications of Newton's binomial formula are known for a long time in Mendelian genetics and in population genetics, first of all, in a connection with the Hardy-Weinberg law (https://en.wikipedia.org/wiki/Hardy%E2%80%93Weinberg_principle; https://en.wikipedia.org/wiki/Allele_frequency; [207, 208]). "*The Hardy-Weinberg law is the cornerstone of diploid population genetics. Yet it seems trivially obvious, a routine application*

	Numbers of identical values of coordinates in vectors $[q, p]^{(n)}$										
$n = 1$					1		1				
$n = 2$				1		2		1			
$n = 3$			1		3		3		1		
$n = 4$		1		4		6		4		1	
$n = 5$	1		5		10		10		5		1
…	.	.	.	.	.	.	.	.	.	.	.

Fig. 4.12.7. Pascal's triangle for numbers of identical values of coordinates in percent vectors $[q, p]^{(n)}$ presenting rules of the stochastic organization of H-n-plet sequences in genomic DNAs.

of the binomial theorem" [207, p. 821]. This law is also called the basis of mathematical constructions in population genetics and contemporary evolutionary theory. The Hardy-Weinberg law, known from 1908, states that allele and genotype percentages (or frequencies) in a population will remain constant from generation to generation in the absence of other evolutionary influences. "*These influences include genetic drift, mate choice, assortative mating, natural selection, sexual selection, mutation, gene flow, meiotic drive, genetic hitchhiking, population bottleneck, founder effect and inbreeding*" (https://en.wikipedia.org/wiki/Hardy%E2%80%93Weinberg_principle). By these reasons, the Hardy-Weinberg law is not a precise but an idealized model law in a general case.

This law uses the Hardy-Weinberg equation (4.33) to examine a simple genetic locus at which there are two alleles, "A" and "a":

$$q^2 + 2qp + p^2 = 1 \tag{4.33}$$

where q is the percentage (or frequency) of the "a" allele and p is the percentage of the "A" allele in the population. In the equation (4.33), q^2 represents the percentage of the homozygous genotype aa, p^2 represents that of AA, and $2qp$ that of the heterozygous genotype Aa. The sum of the allele percentages for all the alleles at the locus must be 1, so $p + q = 1$. If the p and q allele percentages are known, then the percentages of the three genotypes may be calculated using the Hardy-Weinberg equation (https://www.nature.com/scitable/definition/hardy-weinberg-equation-299). In population genetics studies, the Hardy-Weinberg equation (4.33) is used particularly to measure whether the observed genotype percentages in a population differ from the percentages predicted by the equation.

Now one should describe that in the analysis of the tensor families of H-n-alphabets $[3, 2]^{(n)}$ and the percent vectors $[q, p]^{(n)}$ in genomic DNAs, the similar equation (4.34) and the similar rule for percentages of members of H-n-plet alphabets are obtained though the obvious great difference exists between genotype percentages in population genetics and percentage compositions of H-n-plet sequences in genomic DNAs. As mentioned above, triplets with strong and weak roots have different properties concerning the degeneracy of the genetic code (Figs. 1.4.1 and 1.4.2). We will also distinguish in each of the H-n-alphabets $[3, 2]^{(n)}$ its members by their roots, that is, by their first two H-monoplets:

— members with a root composed of identical H-monoplets will be called homo-rooted members of the H-n-alphabet by analogy with the term "homozygous genotype" (for example, the H-tetraplets $\mathbf{\underline{33}23}$, $\mathbf{\underline{33}22}$, $\mathbf{\underline{33}32}$, $\underline{33}33$ belong to homo-rooted members relative to H-monoplet 3; the H-tetraplets $\underline{22}33$, $\underline{22}32$, $\underline{22}23$, $\underline{22}22$ belong to homo-rooted members relative to H-monoplet 2);

— members with a root composed of different H-monoplets will be called hetero-rooted members of the H-n-alphabet by analogy with the term "heterozygous genotype" (for example, the H-tetraplets $\mathbf{\underline{32}33}$, $\mathbf{\underline{32}22}$, $\mathbf{\underline{23}33}$, $\underline{23}22$ belong to hetero-rooted members of the H-tetraplets alphabet represented by components of the vector $[3, 2]^{(4)}$).

For each of the H-n-plets alphabets $[3, 2]^{(n)}$ ($n = 2, 3, 4, \ldots$), let us calculate the sums of percentages of its homo-rooted members and separately the sum of its hetero-rooted members in a genomic DNA. The percentages of all these H-n-plets are represented by appropriate components of the percent vectors $[q, p]^{(n)}$ where $q + p = 1$. Here q means a percentage of the H-monoplet 3, and p that of the H-monoplet 2 in the considered genomic DNA. The calculations give the following results.

In the case of the H-duplets alphabet, its 4 members are components of the vector $[3, 2]^{(n)} = [33, 32, 23, 22]$ and their percentages are represented by appropriate components of the percent vector $[q, p]^{(2)} = [qq, qp, pq, pp]$. Correspondingly the percentage of the homo-rooted member 33 is equal to q^2, that of the homo-rooted member 22 is equal to p^2 and the sum of percentages of the hetero-rooted members 32 and 23 is equal to $2qp$. The total sum of all these percentages should be equal to 1. This is expressed by the equation (4.34), which is completely analogical to the Hardy-Weinberg equation (4.33):

$$q^2 + 2qp + p^2 = 1 \tag{4.34}$$

In the case of the H-triplets alphabet, its 8 members are components of the vector $[3, 2]^{(3)} = [333, 332, 323, 322, 233, 232, 223, 222]$ and their percentages are represented by appropriate components of the percent vector $[q, p]^{(3)} = [qqq, qqp, qpq, qpp, pqq, pqp, ppq, ppp]$. In this case, the sum of percentages of homo-rooted members 333 and 332 is equal to q^2 again since $(qqq + qqp) = qq \times (q + p) = q^2$; the sum of percentages of homo-rooted members 223 and 222 is equal to p^2 again since $(ppq + ppp) = pp(q + p) = p^2$; the sum of percentages of all hetero-rooted members 323, 322, 233, 232 is equal to $2qp$ again since $(qpq + qpp + pqq + pqp) = qp(q + p) + pq(q + p) = 2qp$ (taking into account that $q + p = 1$). So, in the case of the H-triplets alphabet the equation (4.34) is also true.

In the case of the H-tetraplets alphabet, its 16 members are components of the vector $[3, 2]^{(4)} = [3333, 3332, 3323, 3322, 3233, 3232, 3223, 3222, 2333, 2332, 2323, 2322, 2233, 2232, 2223, 2222]$ and their percentages are represented by appropriate components of the percent vector $[q, p]^{(4)} = [qqqq, qqqp, qqpq, qqpp, qpqq, qpqp, qppq, qppp, pqqq, pqqp, pqpq, pqpp, ppqq, ppqp, pppq, pppp]$. In this case, the sum of percentages of homo-rooted members 3333, 3332, 3323, 3322 is equal to q^2 again since $(qqqq + qqqp + qqpq + qqpp) = qqq(q + p) + qqp(q + p) = qqq + qqp = qq(q + p) = q^2$; the sum of percentages of homo-rooted members 2233, 2232, 2223, 2222 is equal to p^2 again since $(ppqq + ppqp + pppq + pppp) = ppq(q + p) + ppp(q + p) = ppq$

$+ppp = pp(q+p) = p^2$; the sum of percentages of all hetero-rooted members 3233, 3232, 3223, 3222, 2333, 2332, 2323, 2322 is equal to $2qp$ again since $(qpqq + qpqp + qppq + qppp + pqqq + pqqp + pqpq + pqpp) = qpq(q+p) + qpp(q+p) + pqq(q+p) + pqp(q+p) = qpq + qpp + pqq + pqp = qp(q+p) + pq(q+p) = 2qp$. So, in the case of the H-tetraplets alphabet the equation (4.34) is also true.

By analogy one can check that in other cases of H-n-plet alphabets with $n = 5, 6, 7, \ldots$ the equation (4.34) is also true. For all these cases, in the equation (4.34), the summand q^2 means the total percentage of all homo-rooted members relative to the H-monoplet 3; the summand p^2 means the total percentage of all homo-rooted members relative to the H-monoplet 2; the summand $2qp$ means the total percentage of all hetero-rooted members in any H-n-plets alphabet ($n = 2, 3, 4, \ldots$). On this basis, the following general and idealized rule in the frame of the described model approach is proposed (by analogy with the formulation of the Hardy-Weinberg law):

— Percentages of homo-rooted and hetero-rooted members of H-n-plet alphabets ($n = 2, 3, 4, \ldots$) in the H-n-plets sequence of any genomic DNA remain practically constant for H-n-plet alphabets under different n (in the framework of the formulated rule of stochastic organization of genomic H-n-plet sequences).

One can suppose that genetic phenomena described by the Hardy-Weinberg law have an evolutionary connection with the described phenomena of long DNA sequences of hydrogen bonds 3 and 2.

4.12.5. *Hydrogen n-plet sequences and formalisms of quantum informatics*

This section considers H-n-plet sequences in genomic DNAs as quantum-like systems, which can be characterized by formalisms and notions from quantum informatics. Mathematical symbols used here are taken from the book [15].

As is known, a quantum bit (or qubit) is a unit of quantum information. For two-level quantum systems used as qubits, the state |0> is identified with the vector [1, 0], and similarly the state |1> with the vector [0, 1]. Two possible states for a qubit are the states |0> and |1>, which correspond to the states of 0 and 1 for a classical bit. It is possible to form a linear combination of states (4.35):

$$|\psi > \sqrt{q}|0 > +\sqrt{p}|1 > \tag{4.35}$$

Here the symbol |ψ> means a state of a qubit. Values $\sqrt{q}$ and $\sqrt{p}$ are called amplitudes of probabilities q and p (the terms "probabilities" and "percentages" are synonyms in this section).

Put another way, the state of a qubit is a vector in a two-dimensional vector space. The standard notation for states in quantum mechanics is the Dirac notation "| >". The special states |0> and |1> are known as computational basis states, and form an orthonormal basis for this vector space [15, p. 13]. As is known, we cannot examine a qubit to determine its quantum state, that is, the values of q and p. Instead, quantum mechanics tells us that we can only acquire much more restricted information about the quantum state. When we measure a qubit, we get either the result 0, with probability q, or the result 1, with probability p. Naturally, the probabilities must sum to one: $q + p = 1$.

Geometrically, we can interpret this as the condition that the qubit's state be normalized to length 1. Thus, a qubit's state is a unit vector in a two-dimensional complex vector space. More generally, a system of "n" qubits is considered in quantum informatics. The computational basis states of this system are written in the form $|x_1x_2\ldots x_n>$; a quantum state of such a system is specified by 2^n amplitudes of probabilities [15, p. 17]. In our model approach we interpret different H-n-representations (described below) of DNA sequences as quantum-like systems of n-qubits.

In our model approach for genetic informatics, we represent a qubit as a two-level quantum-like system: in this system of a long DNA sequence one level corresponds to the indicator «3 hydrogen bonds» and the second level to the oppositional indicator «2 hydrogen bonds». In other words, such «genetic» qubit is represented by these oppositional indicators and the state of such qubit is a vector in its appropriate two-dimensional Hilbert space H. One can assume that the state |0> corresponds to the state «3 hydrogen bonds», and the state |1> to the state «2 hydrogen bonds». In this case, the state of a long DNA sequence as a quantum-like system in the form of a sequence of H-monoplets 3 and 2 is described by the expression (4.35) where q and p are probabilities (or percentages) of its nucleotides with 3 or 2 hydrogen bonds correspondingly.

The tensor product of the two-dimensional Hilbert space H ⊗ H gives one four-dimensional Hilbert space with the following separable pure state of a quantum 2-qubit system (4.36):

$$\begin{aligned}&|\psi>^{(2)} = |\psi> \otimes \ |\psi> = \left(\sqrt{q}\,|0> + \sqrt{p}\,|1>\right) \otimes \left(\sqrt{q}\,|0> + \sqrt{p}\,|1>\right) = \\ &\left(\sqrt{q}\times\sqrt{q}\right)|00> + \left(\sqrt{q}\times\sqrt{p}\right)|01> + \left(\sqrt{p}\times\sqrt{q}\right)|10> + \left(\sqrt{p}\times\sqrt{p}\right)|11> = \\ &= q\,|00> + \sqrt{qp}\,|01> + \sqrt{pq}\,|10> + p\,|11>\end{aligned} \tag{4.36}$$

Such a 2-qubit system (4.36), which is generated by the second tensor power $|\psi>^{(2)}$ of a state $|\psi>$ of a qubit (4.35), has 4 computational basis states denoted |00>, |01>, |10>, |11>. Probabilities (or percentages) of these states satisfy the normalization condition (4.37):

$$q^2 + qp + pq + p^2 = 1. \tag{4.37}$$

We note the following strong correspondence between the denotations of computational basis states |00>, |01>, |10>, |11> and appropriate values of probability amplitudes for them $\sqrt{q}\sqrt{q}$, $\sqrt{q}\sqrt{p}$, $\sqrt{p}\sqrt{q}$, $\sqrt{p}\sqrt{p}$: in any of the denotations of these computational basis states, replacements of the symbols 0 and 1 correspondingly by values $\sqrt{q}$ and $\sqrt{p}$ give the value of the probability amplitude of this state. For example, such replacement in the denotation of the state |01> gives its probability amplitude $\sqrt{q}\sqrt{p}$ in accordance with (4.36). Such strong correspondence between denotations of computational basis states and values of their probability amplitudes holds true in other cases of the tensor powers $|\psi>^{(n)}$ of a state $|\psi>$ of a qubit (4.35).

The third tensor power $|\psi>^{(3)}$ of a state $|\psi>$ of a qubit gives a 3-qubit system with 8 computational basis states denoted $|000>$, $|001>$, $|010>$, $|011>$, $|100>$, $|101>$, $|110>$, $|111>$. The same denotations of cells arise in the genetic matrix $[C, A; T, G]^{(3)}$ of 64 triplets and their H-triplet presentations (Figs. 4.12.1) when each cell is numerated by the result of modulo-2 addition of binary numberings of its row and column. Probabilities of these states are qqq, qqp, qpq, qpp, pqq, pqp, ppq, ppp correspondingly and they satisfy the normalization condition:

$$qqq + qqp + qpq + qpp + pqq + pqp + ppq + ppp = 1. \qquad (4.38)$$

The nth tensor power $|\psi>^{(n)}$ of a state $|\psi>$ of a qubit (4.35) gives an appropriate n-qubit system with 2^n computational basis states, each of which has its own value of probability amplitude (or square root of its percentage) calculated by analogy; formally it is possible to simply replace symbols 0 and 1 in the denotations of the computational basis states by values $\sqrt{q}$ and $\sqrt{p}$ correspondingly to get values of probabilities for members of the H-n-plets alphabet in this model of a genomic DNA text.

The states of the studied composite quantum-like sequences of H-plets, H-triplets, and so on, are described above as a tensor product of the state vectors of their components like $|\psi> \otimes |\psi> \otimes |\psi>$... In quantum informatics, a state described in this way is called pure separable quantum state [15, p. 96]. Correspondingly, from the formal point of view, the state of each of the genomic DNA sequences of H-duplets, H-triplets, etc., is a pure separable quantum state.

The described results give pieces of evidence about the important role of complexes of DNA H-n-plets in the algebraic structures and emergent properties of the genetic code system. We suppose that the genetic encoding is based on stochastic principles related to the rules of stochastic organization of genomic DNAs. The noted algebraic role of DNA H-n-plets is connected with multi-dimensional hypercomplex numbers, including 2^n-dimensional hyperbolic numbers. In particular, turning back to hyperbolic numbers in Figs. 4.12.2–4.12.4, one can suppose that in some genetic tasks their coefficients are percentages (probabilities) of DNA H-n-plets. These probabilities can be also used as coefficients in other multi-dimensional hypercomplex numbers noted in this book as related to the genetic code system, including Cockle's split-quaternions. We believe that further study of the role of stochastic rules in the organization of the genetic system and inherited physiological structures will lead to new important results in the future.

4.13. Some concluding remarks

The presented results of the study on the hidden rules in the distribution of the percentages of n-plets in long DNA texts of various organisms are consistent with Jordan's claim that life's missing laws are the rules of chance and probability of the quantum world [199]. The described results show the existence of previously unknown genetic regularities. The presented universal rules of genomic DNAs testify to the coexistence of stochastic and deterministic traits in genomes or, in other words, testify about stochastic-deterministic character of genomic DNAs. The stochastic-deterministic properties of genomic DNAs exist in parallel with the long-known facts of the coexistence of stochastic and deterministic features in inherited biological macrostructures, which are considered in the next chapter.

The above-described generalization of Chargaff's second rule — for representations of every tested genomic DNA as a sequence of n-plets — was empirically tested and validated for up to about $n = 50$ in the work [42], where comparison analysis between real DNA sequences and appropriate random sequences was also conducted.

Considering the views of Jordan and Schrödinger about the dictatorial role of structured informatics of genetic molecules for the whole organism [199], it is natural to think that the structural features of DNA informatics leave their marks on all genetically inherited biological systems and phenomena. This is consistent with the fact that all physiological systems must be structurally aligned with genetic coding to be transmitted in genetically encoded form to offspring. This is also consistent with the point of view that the main task of living organisms is to transfer genetic information along the chain of generations.

Schrodinger noted: "*From all we have learned about the structure of living matter, we must be prepared to find it working in a manner that cannot be reduced to the ordinary laws of physics...because the construction is different from anything we have yet tested in the physical laboratory*" [121]. For comparison, the enzymes in biological organisms work a million times more effectively than catalysts in the laboratory. What takes the enzyme in the body to make in 1 second, a catalyst in the laboratory can make only in 100 thousand years. We believe that such ultra-efficiency of enzymes in biological bodies is defined not only by laws of physics, but also by quantum-logical relations in the genetic system, and therefore — in line with Schrodinger — this ultra-efficiency cannot be reduced to the ordinary laws of physics. As far as we understand, the found Gestalt rules of "dictatorial" DNA texts are not derived from the known laws of physics, and therefore refer to the special laws of the structuring of living things that Schrödinger spoke about.

The obtained results can be used, in particular, for developing our knowledge about principles of brain activities and about the relation between living and non-living matter. These themes are actively discussed in the scientific community [209, 210].

The results described in this chapter additionally show that the genetic code structures are connected with a whole multi-level system of interconnected orthogonal matrices of different orders woven together and form tensor families of orthogonal matrices. These orthogonal matrices can be hypothetically considered as quantum gates for quantum computing inside organisms.

The presented universal rules of stochastic organization of n-plets of hydrogen bonds in H-n-plet sequences of genomic DNAs attract additional attention to hydrogen bonds as one of the most important components of life. The demonstrated connection between the probabilities of the systems of n-plets of hydrogen bonds in DNA sequences and some algebras of 2^n-dimensional hypercomplex numbers allows a comparison of different genomic DNA sequences by means of their personal systems of 2^n-dimensional hyperbolic numbers («hypercomplex passports» of genomic DNA sequences) for studies in the field of evolutionary biology, medicine and biotechnology. It is essential that these rules show their connections with Pascal's triangle and the Hardy-Weinberg law, which is the main law of population genetics. It opens some new avenues of research in comparison genetics.

The mentioned algebras of hypercomplex numbers and their algebraic extensions, which are associated with the considered genetic structures, possess appropriate metrics which can be used for modeling metric properties of inherited biological phenomena. Regarding the metric features of biological phenomena, it can be noted that thoughts about metric spaces are spread by some authors even to mathematics as a high form of intellectual activity. For example, the book [211] notes that a mathematician feels that in mathematics there is a certain metric that unites ideas — that all mathematics is a network of results that are interconnected by a huge number of connections; had we been able to introduce this highly developed sense of mathematical closeness — the mental metric of a mathematician — into the program, we could create a primitive artificial mathematician.

By these statements of Hofstadter, the book [212, p. 115] emphasizes: "*In other words, artificial intelligence could be brought closer to mathematical thinking, if it were possible to realize the metric properties of the human thinking space... We are ready to go further and say that consciousness itself is geometrically structured: existentially, a person is geometric... In our minds, when constructing texts through which we perceive the World, something very similar to what happens in morphogenesis happens. We are ready to see in the depths of consciousness the same geometric images that are revealed in morphogenesis*".

The theme of stochastics of genomic DNAs will be continued in subsequent chapters, including the chapter on quantum interrelations in connection with Frohlich's theory of long-range action.

Appendix 4.A. An example of nucleotide *n*-plet percentages in one of the bacterial genomes

This appendix presents data about percentages of nucleotide *n*-plets in the rhizobacteria *Bradyrhizobium japonicum* strain E109, complete genome, 9224208 bp (initial data were taken from https://www.ncbi.nlm.nih.gov/nuccore/CP010313.1?report=genbank; the length of this sequence is more than 20 times shorter than the length of the human chromosome №1). Figs. A4.I.1–A4.I.5 show matrices of phenomenological percentages of 4 nucleotides, 16 duplets, 64 triplets, and 256 tetraplets in corresponding layers of this genomic DNA. These percentage matrices for the bacteria and the percentage matrices (Figs. 4.1.1–4.1.3) for human chromosome №1 have similar properties for *m*-positional tetra-groupings of *n*-plets, which are expressed by the equalities (4.1) and (4.2).

Regarding projector tetra-groupings of *n*-plets in different layers of the genomic DNA of *Bradyrhizobium japonicum* strain E109, it was noted above

%C	%A		0.3184	0.1819
%T	%G	=	0.1815	0.3182

Fig. 4.A.1. The transformation of the symbolic matrix of 4 nucleotides into the numeric matrix of nucleotide percentages in the case of the genomic DNA of the *Bradyrhizobium japonicum* strain E109.

%CC	%CA	%AC	%AA		0.0781	0.0587	0.0467	0.0367
%CT	%CG	%AT	%AG		0.0515	0.1302	0.0469	0.0517
%TC	%TA	%GC	%GA	=	0.0720	0.0145	0.1216	0.0719
%TT	%TG	%GT	%GG		0.0365	0.0584	0.0468	0.0779

Fig. 4.A.2. The matrix of phenomenological percentages of the 16 duplets in the genomic DNA sequence of duplets in the *Bradyrhizobium japonicum* strain E109 (compare with Fig. 4.1.1 for human chromosome №1).

%CCC	%CCA	%CAC	%CAA	%ACC	%ACA	%AAC	%AAA
%CCT	%CCG	%CAT	%CAG	%ACT	%ACG	%AAT	%AAG
%CTC	%CTA	%CGC	%CGA	%ATC	%ATA	%AGC	%AGA
%CTT	%CTG	%CGT	%CGG	%ATT	%ATG	%AGT	%AGG
%TCC	%TCA	%TAC	%TAA	%GCC	%GCA	%GAC	%GAA
%TCT	%TCG	%TAT	%TAG	%GCT	%GCG	%GAT	%GAG
%TTC	%TTA	%TGC	%TGA	%GTC	%GTA	%GGC	%GGA
%TTT	%TTG	%TGT	%TGG	%GTT	%GTG	%GGT	%GGG

0.0131	0.0154	0.0136	0.0129	0.0145	0.0083	0.0097	0.0068
0.0148	0.0344	0.0145	0.0185	0.0053	0.0182	0.0075	0.0126
0.0174	0.0040	0.0446	0.0328	0.0205	0.0043	0.0205	0.0106
0.0126	0.0183	0.0185	0.0343	0.0075	0.0144	0.0052	0.0146
0.0149	0.0133	0.0044	0.0021	0.0353	0.0210	0.0195	0.0153
0.0105	0.0323	0.0043	0.0039	0.0210	0.0443	0.0209	0.0171
0.0151	0.0020	0.0206	0.0133	0.0195	0.0044	0.0356	0.0150
0.0069	0.0125	0.0082	0.0153	0.0098	0.0133	0.0148	0.0133

Fig. 4.A.3. The matrix of phenomenological percentages of the 64 triplets in the genomic DNA sequence of triplets in the *Bradyrhizobium japonicum* strain E109 (compare with Fig. 4.1.2).

.0023	.0025	.0033	.0030	.0052	.0017	.0035	.0021	.0022	.0037	.0018	.0019	.0023	.0023	.0016	.0015
.0026	.0057	.0038	.0055	.0014	.0051	.0027	.0045	.0029	.0058	.0024	.0022	.0015	.0036	.0019	.0018
.0048	.0011	.0119	.0082	.0066	.0011	.0081	.0030	.0014	.0006	.0057	.0050	.0025	.0010	.0045	.0028
.0039	.0052	.0048	.0096	.0021	.0044	.0018	.0052	.0014	.0019	.0029	.0047	.0018	.0022	.0015	.0039
.0036	.0024	.0016	.0004	.0141	.0072	.0094	.0057	.0042	.0044	.0010	.0006	.0053	.0043	.0026	.0026
.0020	.0090	.0015	.0004	.0073	.0162	.0085	.0091	.0035	.0085	.0012	.0016	.0039	.0072	.0035	.0019
.0056	.0005	.0068	.0037	.0082	.0016	.0173	.0055	.0024	.0005	.0048	.0034	.0018	.0006	.0058	.0036
.0019	.0043	.0022	.0055	.0035	.0051	.0059	.0056	.0019	.0027	.0024	.0039	.0015	.0013	.0030	.0027
.0027	.0031	.0035	.0031	.0012	.0009	.0006	.0005	.0059	.0062	.0048	.0047	.0058	.0036	.0042	.0029
.0035	.0057	.0034	.0036	.0006	.0016	.0005	.0005	.0059	.0173	.0049	.0067	.0018	.0081	.0025	.0057
.0040	.0010	.0103	.0092	.0017	.0004	.0013	.0010	.0069	.0013	.0167	.0102	.0098	.0017	.0070	.0040
.0027	.0030	.0050	.0082	.0011	.0011	.0006	.0010	.0045	.0080	.0058	.0118	.0024	.0066	.0014	.0047
.0036	.0031	.0007	.0004	.0065	.0033	.0035	.0031	.0037	.0036	.0010	.0007	.0094	.0065	.0037	.0035
.0026	.0058	.0006	.0004	.0043	.0069	.0044	.0024	.0026	.0095	.0010	.0016	.0053	.0141	.0041	.0036
.0028	.0005	.0047	.0030	.0036	.0009	.0061	.0032	.0040	.0005	.0048	.0034	.0059	.0011	.0059	.0027
.0015	.0021	.0020	.0029	.0024	.0016	.0036	.0026	.0016	.0034	.0018	.0033	.0023	.0052	.0021	.0024

Fig. 4.A.4. The matrix of phenomenological percentages of the 256 tetraplets in the genomic DNA sequence of tetraplets in the *Brudyrhizobium japonicum* strain E109 (compare with Fig. 4.1.3).

in the expressions (4.13) and (4.14) that the following equalities are fulfilled in all 7 mosaic matrices of duplets and triplets in Fig. 4.5.3:

$$\Sigma\%\heartsuit \approx \Sigma\%\diamondsuit \approx 0.277, \qquad \Sigma\%\spadesuit \approx \Sigma\%\clubsuit \approx 0.223$$

%C ≈ 0.3184	%G ≈ 0.3182	%A ≈ 0.1819	%T ≈ 0.1815
Σ%CN ≈ 0.3185	Σ%GN ≈ 0.3182	Σ%AN ≈ 0.1820	Σ%TN ≈ 0.1814
Σ%NC ≈ 0.3183	Σ%NG ≈ 0.3182	Σ%NA ≈ 0.1818	Σ%NT ≈ 0.1816
Σ%CNN ≈ 0.3197	Σ%GNN ≈ 0.3201	Σ%ANN ≈ 0.1806	Σ%TNN ≈ 0.1797
Σ%NCN ≈ 0.3166	Σ%NGN ≈ 0.3173	Σ%NAN ≈ 0.1836	Σ%NTN ≈ 0.1825
Σ%NNC ≈ 0.3189	Σ%NNG ≈ 0.3173	Σ%NNA ≈ 0.1815	Σ%NNT ≈ 0.1824
Σ%CNNN ≈ 0.314	Σ%GNNN ≈ 0.3183	Σ%ANNN ≈ 0.1816	Σ%TNNN ≈ 0.1817
Σ%NCNN ≈ 0.3184	Σ%NGNN ≈ 0.3185	Σ%NANN ≈ 0.1816	Σ%NTNN ≈ 0.1815
Σ%NNCN ≈ 0.3185	Σ%NNGN ≈ 0.3181	Σ%NNAN ≈ 0.1823	Σ%NNTN ≈ 0.1811
Σ%NNNC ≈ 0.3182	Σ%NNNG ≈ 0.3180	Σ%NNNA ≈ 0.1820	Σ%NNNT ≈ 0.1818

Fig. 4.A.5. Percentages of nucleotides C, G, A, T, and the sums of percentages of *n*-plets with these nucleotides at their attributive positions in corresponding tetra-groupings of *n*-plets in the case of the genomic DNA text of the *Bradyrhizobium japonicum* strain E109 (compare with Fig. 4.2.1).

For example, in the case of the mosaic (4 × 4)-matrix with the tetra-grouping №1 for 16 duplets (see Fig. 4.5.3), the following percentage sums appear for this genomic DNA based on percentages of duplets in Fig. 4.A.2:

$$\Sigma\%\heartsuit \approx \%CA + \%CG + \%AA + \%AG$$
$$\approx 0.0587 + 0.1302 + 0.0367 + 0.0517$$
$$\approx 0.277;$$
$$\Sigma\%\diamondsuit \approx \%TC + \%TT + \%GC + \%GT$$
$$\approx 0.0720 + 0.0365 + 0.1216 + 0.0468$$
$$\approx 0.277;$$
$$\Sigma\%\spadesuit \approx \%CC + \%CT + \%AC + \%AT$$
$$\approx 0.0781 + 0.0515 + 0.0467 + 0.0469$$
$$\approx 0.223;$$
$$\Sigma\%\clubsuit \approx \%TA + \%TG + \%GA + \%GG$$
$$\approx 0.0145 + 0.0584 + 0.0719 + 0.0779$$
$$\approx 0.223.$$

Chapter 5

Biological Dualism "Stochastics-Determinism", Phenomena of Gestalt Biology, and Bio-Symmetries

The previous chapter presented the universal rules of stochastic-deterministic organization for genomic DNAs of higher and lower organisms. The stochastic-deterministic properties of genomic DNAs exist in parallel with the long-known facts of the coexistence of stochastic and deterministic features in inherited biological macrostructures, which are considered here. This chapter connects the universal rules of stochastic-deterministic organization of genomic DNAs with the interrelationship of random and determined in biological bodies. Analysis and modeling of this interrelationship by formalisms of matrix algebra and matrix genetics will be important for new understandings of the role of random and deterministic in biological self-organization and also for generating new approaches to build bridges between biology, informatics, and mathematical natural science.

5.1. Dualism "stochastics-determinism" in genetics and phenomena of Gestalt biology

Genetics as a science began with Mendel's discovery of the stochastic rules of inheritance of traits in experiments on the crossing of organisms. Many processes in living bodies are of stochastic nature. The well-known expressions "gene noise" or "cell noise" reflect the fact that even genetically identical cells within the same tissue exhibit different levels of protein expression, sizes, and structures due to the stochastic nature of interactions among individual molecules in cells.

The stochastic nature of genetic processes in the "small" does not interfere with the inheritance of the traits determined in the "big" (so-called Gestalt phenomena). For example, fingerprints are different for all people, although fingers on the whole have a typical shape and composition (three phalanges, nails, etc.). According to Mendel's law of independent inheritance of traits, information from the level of DNA molecules dictates the macrostructure of living bodies through many independent channels, despite strong noises. Thus, hair, eye and skin colors are inherited independently of each other. Accordingly, each organism is a machine of multi-channel noise immunity encoding.

All physiological systems are genetically inherited through their genetic coding. Therefore, for the origins of the phenomena "stochastics-determinism" in inherited macrostructures, one should look in the genetic system of DNA informatics. We believe that the mentioned rules of stochastic-deterministic features of genomic DNAs are useful for modeling and understanding such inherited phenomena.

The discovered genomic DNA phenomena of the relative independence of the sums of percentages of *n*-plets from the values of individual percentages in these sums (see Chapter 4) resemble the phenomenon of perceiving a musical melody: it can be reproduced by different musical instruments and in different frequency ranges, that is, by significantly changing the frequency of each of its note elements, but despite these changes, the melody remains generally recognizable.

Many such phenomena of perception, in which there is relative independence of the integral form from its constituent individual components, are studied in Gestalt psychology. The described universal regularities in the preservation of total percentages in tetra-groupings of *n*-plets relatively regardless of the percentage of individual *n*-plets in genomic multilayer DNA texts allow the development of Gestalt genetics [70]. It seems natural to think that Gestalt genetics is interrelated with Gestalt psychology, which studies some genetically inherited properties of our brain regarding the perception of the environment.

The phenomena of perception of visual, auditory, and other images, studied by Gestalt psychology, reflect the fundamental inherited property of the psyche — to seek in a disparate whole. Thanks to the ability to think in Gestalts, you can understand the sentence, even if you change the order of the letters in each word and leave only the beginning and end in place. For example, you can easily understand the phrase, strongly "mutated" by local permutations: "Aoccdrnig to a rscheearch at Cmabrigde Uinervtisy, it deosn't mttaer in waht oredr the ltteers in a wrod are» (this "mutated" phrase example is taken from https://www.dictionary.com/e/typoglycemia/).

Gestalt genetics studies holistic genetic patterns that are relatively independent of particular components. Gestalt genetics comes in contact with the teachings of the creator of analytical psychology C. Jung and his associate Nobel Laureate in Physics W. Pauli about the archetypes of the unconscious; in particular, they linked these archetypes with the Yin-Yang schemes of the ancient Chinese book *I-Ching* and its table of 64 hexagrams, which have deep structural analogies with the DNA alphabets (these analogies were noted by some reputable geneticists — G. S. Stent [213], and Nobel Laureate in Medicine F. Jacob [214]; extended information on these analogies is in publications [12, 13, 157]). In our opinion, the origins of the genetically inherited ability of the brain to work with Gestalt images should be sought in Gestalt genetics. In particular, Gestalt genetics is capable of providing new approaches to understanding the noise immunity of genetic information under mutations of DNA texts.

In addition to the phenomena of Gestalt psychology, in living organisms there are many genetically inherited physiological phenomena, in which the same whole pattern is realized in conditions of a wide variety of constituent elements, and which can be attributed to Gestalt biology (this name is proposed as uniting genetically inherited Gestalt-like phenomena of different types). For example, Gestalt biology includes some genetically inherited phenomena of morphogenesis (laws of phyllotaxis, spiralization of biological structures at various levels and branches of biological evolution), as well as some functional phenomena (homeostasis at different stages of ontogenesis;

the processing of sensory information from different sense organs according to main psychophysical law of Weber-Fechner; the biomechanical phenomenon, known as Bernstein's problem, that the general target task of body movement is performed exactly regardless of the inaccuracies of its constituent motor subtasks). In particular, Gestalt genetics includes a classic observation in embryology from K. Baer that chick embryos are vastly different, while the resulting adult organisms are remarkably similar. One can think that various genetically inherited phenomena of Gestalt biology are based on Gestalt genetics.

We should also recall that the molecular composition of a living body is constantly changing while the shape of the body is maintained. Our body's proteins are involved in continuous life-death cycles of assembling and disassembling into amino acids. For example, the half-life of the hormone insulin is 6–9 minutes. In other words, genetically inherited parts of our body are constantly dying and being reborn. Considering such phenomena, the renowned physiologist A. G. Gurvich claims: "*The main problem in biology is maintaining shape while constantly renewing the substrate*" [215]. In our opinion, the described phenomena of Gestalt genetics with its Gestalt rules of percentages in genomic multilayer DNA texts are directly related to this fundamental problem of biology.

Gestalt genetics provides new approaches to understanding ontogeny. Using the terminology of Gestalt psychology and Gestalt therapy, it was suggested to interpret ontogenetic processes as the stepwise processes of "closing" certain genetic gestalts [70]. One should note that all physiological systems are forced to bear the structural stamp of the genetic code since they should be genetically encoded for transmission to descendants and survival.

Gestalt genetics also concerns the phenomenon of the biomechanics of movements, described by N. A. Bernstein: the general target task of the movement is performed exactly regardless of the inaccuracies of its constituent motor subtasks [216]. For example, when repeating an exact hit with a hammer on a nail, a person each time uses different trajectories, speeds, and accelerations of body parts with changes in both flexions in the

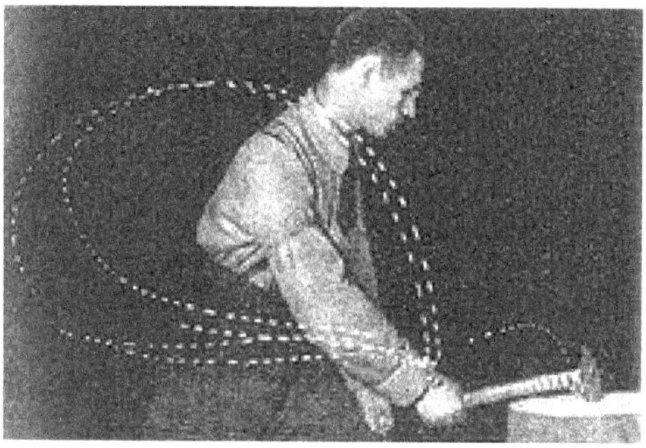

Fig. 5.1.1. Illustration to the Bernstein problem: a cyclogram of cutting metal with a chisel and hammer (from [217]), https://en.wikipedia.org/wiki/Nikolai_Bernstein.

joints, the activity of many muscles of each joint, and many motoneurons of each muscle. This question of how the central nervous system is capable of adequately controlling the many degrees of freedom of the musculoskeletal system was first addressed by Bernstein and is now known usually as the «Bernstein problem». The degrees of freedom problem in motor control states that there are multiple ways for humans or animals to perform a movement to achieve the same goal using redundant neurophysiological degrees of freedom. How the nervous system "chooses" a subset of these near-infinite degrees of freedom is an overarching difficulty in understanding motor control and motor learning. In other words, under normal circumstances, no simple one-to-one correspondence exists between a motor problem (or task) and a motor solution.

Gestalt genetics provides new opportunities for understanding the known physiological phenomena of our body. For example, mature people with a congenital absence of limbs feel a phantom limb's existence with a feeling of pain in them, although they have no experience of using them [218, 219].

Gestalt psychology meets the well-known phenomena of perception constancy, reflecting the ability of the brain to stably recognize the shapes of objects and other structures of the external world under a significant change in the conditions of their presentation to the sense organs. For example, a change in the color illumination of a room does not prevent a person from recognizing the shapes and colors of objects in it, although other light frequencies from objects come to the retina.

5.2. Biological symmetries, geometric transformations, and Gestalt biology

The inherited symmetry phenomena in the forms of biological bodies have long aroused the interest of natural scientists as one of the most remarkable and mysterious natural phenomena. The very concept of symmetry arose from the observations of the ancients over the shape of living bodies. One can believe that the ancient principle "*like begets like*" appeared in connection with observing phenomena of bio-symmetries. The same types of symmetry are realized at different lines and levels of biological evolution: shells of mollusks and foraminifera, bodies of annelids and centipedes, fish fins, bird feathers, flowers, leaves and shoots of plants, etc. The works of many modern scientists are devoted to bio-symmetries (see, for example, [220, 221]). Problems of biological symmetries at the macromolecular level were considered at a special Nobel Symposium [222]. They are covered in specialized journals and discussed at many international conferences, systematically held, for example, by the International Symmetry Association (https://isa.symmetry.hu/). The study of bio-symmetries is an integral part of the problems of Gestalt biology and biological dualism "stochastics-determinism". In modern science, the concept of symmetry acts as one of the most important and deeply developed methodological principles for constructing scientific theories. Many impressive successes in mechanics, physics, and chemistry were the result of understanding natural phenomena from the standpoint of invariant group analysis. H. Weyl emphasized that the method of symmetry is the guiding principle of modern mathematics and its applications.

The study of symmetries is associated with the famous Erlangen program of F. Klein in 1872, which approved a fundamentally new, invariant-group view of geometry: geometry began to be understood as the science of invariants of transformation groups. From these positions, Euclidean geometry appeared as just one of many possible geometries, each of which is based on its own group of transformations: Euclidean geometry — on the group of similarity transformations (which includes transformations of mirror reflection, rotation, translation, and scaling), affine — on the group of affine transformations, conformal — on the group of Möbius (or circular) transformations, projective — on the group of projective transformations, etc. In the light of the Erlangen program, speaking of the geometry and symmetries of living bodies, we will have in mind, first of all, those geometric groups of transformations, in accordance with the properties of which the interconnection of the constituent parts into a single whole in supramolecular biological structures is built.

One can note that from a geometric (i.e., invariant-group) point of view, all classical bio-morphology is essentially based on a group of similarity transformations. Examples of these are morphological studies on mirror symmetry and asymmetry of bio-bodies, multi-block bio-forms with similarity symmetries according to A. V. Shubnikov [221], scale volumetric growth, dwarfs and giants in organisms of the same species, etc. Non-Euclidean symmetries, in relation to which Euclidean symmetries are a special case, are also widely realized in bio-bodies; these bio-symmetries were revealed and systematized in the doctoral dissertation [155].

Let's first focus on multi-block structures, the components of which are conjugated into a single whole based on certain rules that are repeated at different levels and branches of biological evolution. We consider supramolecular biostructures, having the form of chains or manifolds ($S = S_1 \cup S_2 \cup S_3 \ldots \cup S_n \ldots$) with division into commensurate and regularly arranged elements (or motivic units) S_n. Fig. 5.2.1 shows examples of such manifolds seen in the extensive literature on bio-symmetries.

The general rule of such partitioned varieties is that the previous motivic unit is translated into the next by some fixed similarity transformation g, i.e., neighboring motivic units S_k are mutually conjugate based on the iterative algorithm

$$S_{k+1} = g \times S_k \tag{5.1}$$

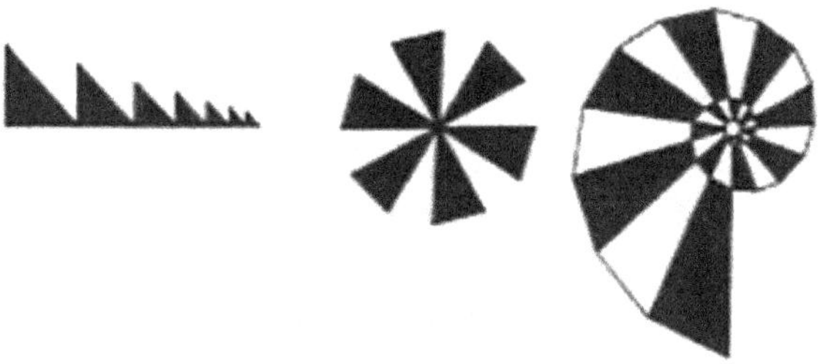

Fig. 5.2.1. Geometric examples of similarity cyclomerias or multi-block configurations with similarity automorphisms (from [221]).

This can be seen as a specific geometric implementation of the principle "like begets like". Accordingly, by repeatedly applying m times the begetting (or generating) transformation g to the motivic unit S_k, we obtain the block S_{k+m}. Speaking in the language of mathematics, a cyclic group (or semigroup) of transformations G acts on the set S_k (the finite number of motivic units in a biological object is not considered here). In other words, the membered varieties organized in this way have a cyclic automorphism group, and their motivic units line up along the orbit of the corresponding cyclic group. For brevity, such configurations are called "cyclomerias", regardless of whether the generating transformation g in (5.1) is Euclidean or non-Euclidean. In the history of the study of cyclomerias with similarity symmetries, there were important works by Shubnikov [221], in connection with which they are often called configurations with Shubnikov's similarity symmetries.

Similarity transformations in biology are also known in connection with the scale type of volumetric growth observed in many animals and plants over long periods of ontogeny. It is accompanied by a mutually consistent growth behavior of the set of small zones distributed over the volume of the body, which is geometrically described by a scaling transformation. In this case, knowing the transformation of only three points of the growing configuration, one can judge the transformation of the set of its remaining points.

It turns out that Shubnikov's similarity symmetries and the scale form of volumetric growth do not at all exhaust the geometrically regular types of mutual conjugation of parts in the structure and growth transformations of living bodies. The next section presents material on this.

5.3. Principle "like begets like", local-similarity transformations, and cyclomerias

In previous chapters, we met several times with the Poincare conformal disk model of hyperbolic geometry, which has some connections with phenomenological structures of genetic matrices, algebraic holography, the principle "like begets like", etc. It was also noted in Chapter 1 that the geometry of the space of visual perception is associated with the conformal Poincaré model of hyperbolic geometry [31]. But do conformal transformations and conformal symmetries have any relation to the processes of morphogenesis and biological symmetries in the inherited configurations of living bodies? Yes, they do. This paragraph continues the algebraic-biological theme of geometric conformal transformations linked with inherited phenomena of bio-symmetries in living bodies.

Transformations in which each local zone of the body undergoes a similarity transformation (the form is preserved in the "small") are called conformal. It was from the standpoint of local similarity that K. Gauss, who created the doctrine of conformal transformations, approached conformal transformations. For the case of a three-dimensional space, to which we restrict ourselves here, all conformal transformations are exhausted, as is known, by the group of Möbius transformations. This group in the literature is sometimes also called conformal, circular, analagmatic, inverse radii, Möbius, Liouville, Kelvin (these scientists made a significant contribution to the study of its mathematical properties and the development of its physical applications). Recall the basic information about this group.

By definition, the Möbius transformation group is the group of point transformations that preserves the magnitudes of the angles and transforms spheres into spheres. By Liouville's theorem, in the case of a three-dimensional space, this group coincides with the group of conformal transformations

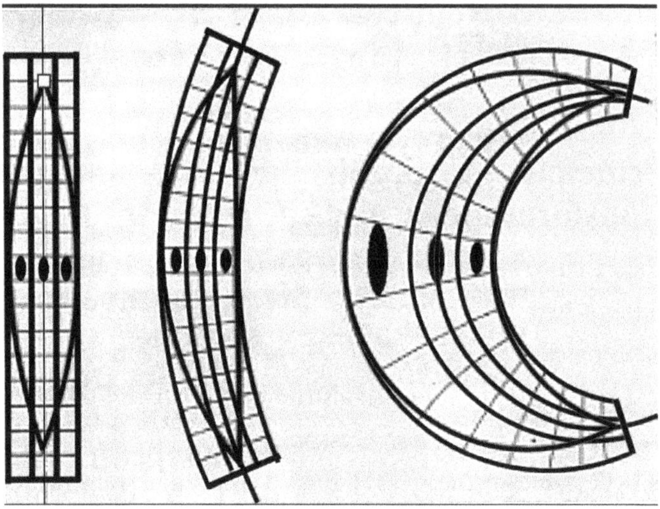

Fig. 5.3.1. An example of a Möbius transformation of figures that preserves local similarity when changing the shape as a whole (from [157]).

(in the two-dimensional case, the group of conformal transformations is wider than the Möbius group). If a small neighborhood of each point of a three-dimensional body undergoes a similarity transformation of each local zone, then the shape of the body in the "whole" can change dramatically from the usual point of view, since similarity in the "small" is not accompanied by similarity in the "whole" in the general case when similarity transformations are not the same in different local areas (Fig. 5.3.1 shows such an example). Correspondingly, in the general case of Möbius (that is, local similarity) transformations, the principle "like begets like" is geometrically realized in the Euclidean sense only for transformations of local zones: under conformal transformation, each local small zone begets a new local zone having a similar configuration.

If the neighborhoods of all points are scaled equally, then the whole body experiences a similarity transformation, which is a special case of Möbius transformations. Möbius transformations of three-dimensional images have their own characteristic set of invariants: lines of curvature of surfaces and isogonals to them, umbilical points, etc. The group of these transformations divides the entire set of figures into closed classes of Möbius-equivalent figures, which are somewhat wider (due to the additional three parameters of the group), than the usual classes of equivalent figures of the similarity group. The totality of all Möbius transformations of three-dimensional space forms a 10-parametric group, whose analytical expression in canonical pentaspherical coordinates coincides with the group of Lorentz transformations of five variables. Any Möbius transformation in n-dimensional space can be represented by a superposition of one similarity transformation and one transformation of inversion with respect to the sphere of unit radius.

Conformal symmetries have a long history of applications in physics. Back in the 19th century, W. Thompson (Lord Kelvin) developed a method for solving the electrostatic Laplace equation using the Möbius inversion transformation with respect to the sphere. This method is known as the method of "electrical" or "mirror" images and is widely used in mathematical physics. In 1909, it was discovered that Maxwell's equations of electrodynamics are invariant under the Möbius transformation group. The mathematical apparatus of quantum mechanics turned out to be closely related to conformal symmetries, due to how quantum mechanics can be represented as a theory of broken conformal symmetry. In physics, several variants of conformal field theory have been developed, a kinematic cosmology based on conformal symmetry has been created, etc. For more information on conformal symmetries in physics, including known conformal conservation laws, see the article [32]. Now, after this brief digression into the history of conformal symmetries in mathematics and physics, let's move on to the topic of conformal (Möbius) symmetries in biology, starting with the inherited phenomenology of mutual conjugations of individual blocks in multi-block bio-bodies.

So, similarity cyclomerias of Shubnikov are known in biology (Fig. 5.2.1). And what configurations are formed if the Möbius transformation acts as the generating transformation g from (5.1)? And do the Möbiusian (or conformal) cyclomerias that arise in this way have analogies in biological configurations? The analysis shows that the variety of cyclomeria in this

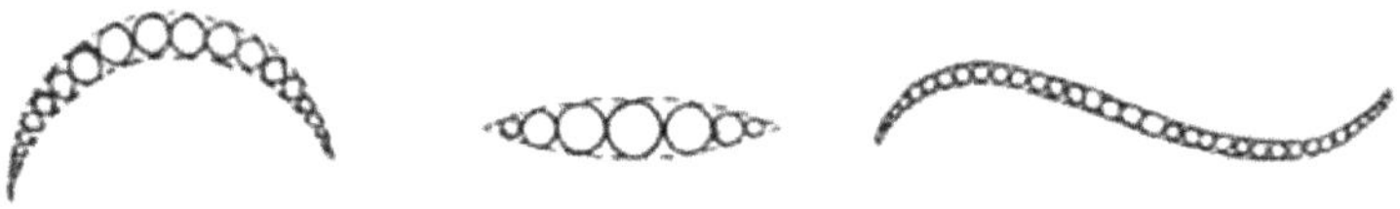

Fig. 5.3.2. Examples of configurations of Möbius cyclomerias (from [157]).

case expands noticeably and includes more complicated configurations: lyre-shaped, crescent-shaped, bud-shaped, etc. (Fig. 5.3.2), which are absent in similarity cyclomerias of Shubnikov.

In these Möbius cyclomerias, the motivic units already differ from each other in their form, and the change of these units along the cyclomeria can be essentially non-monotone. It is important that Möbius cyclomerias are clearly realized in bio-bodies at various lines and levels of biological evolution in parallel with Euclidean cyclomerias, but until recently they were not studied in the science on bio-symmetries. Let's illustrate this with specific examples.

As is known, the horns of many animals have spiral or rectilinear conical shapes, described by the Euclidean cyclomerias. At the same time, in other animals, the horns have significantly different configurations of non-Euclidean cyclomerias. For example, Fig. 5.3.3 (left) shows a segmented horn of *Pantholops hodgsoni orongo* and its model as a cyclomeria with a Möbius generating transformation (so-called loxodromic type).

Continuing the list of examples, let's name the antennae of insects in which, as in the horns of animals, not only Euclidean but also Möbius cyclomerias are observed (Fig. 5.3.3, right). The same is true for the human body: Fig. 5.3.4 shows how the sequence of vertebral discs of the human thoracic spine is reproduced by the Möbius cyclomeria (loxodrome type), in contrast to the configurations of the spine of many animals (for example, lizards), which are described by Euclidean cyclomerias. Note that by modifying the generating Möbius transformation and thus distorting the "normal" configuration of the spinal cyclomeria, it is possible to obtain models of morphogenetic anomalies of the spinal column. This is consistent with our other data on the realization of Euclidean and non-Euclidean cyclomeria in humans, animals, and plants not only in normal but also in pathological morphogenesis.

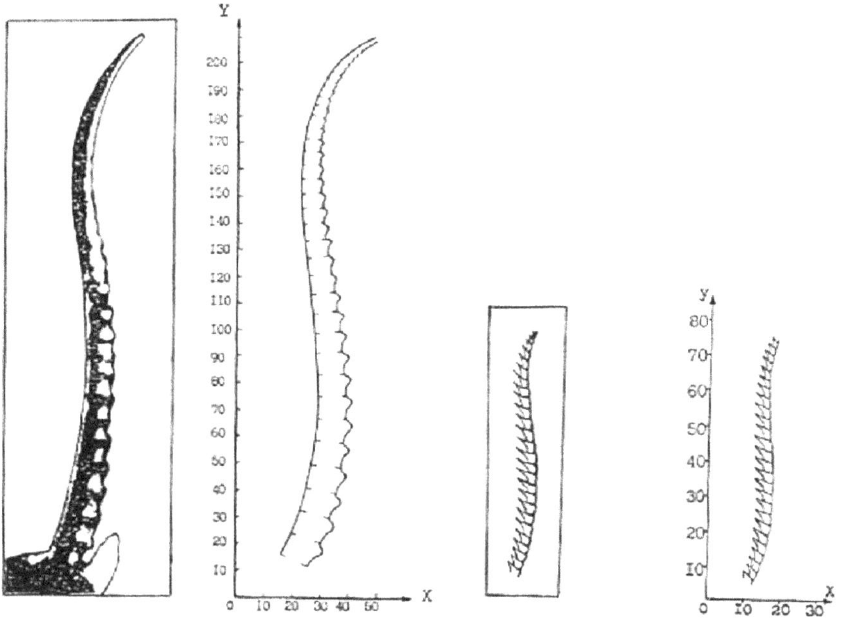

Fig. 5.3.3. Left: a comb-like antenna of an insect and its model as a Möbius cyclomeria (from [157]). Right: a segmented horn of the *Pantholops hodgsoni orongo* [223] and its modeling by a Möbius cyclomeria.

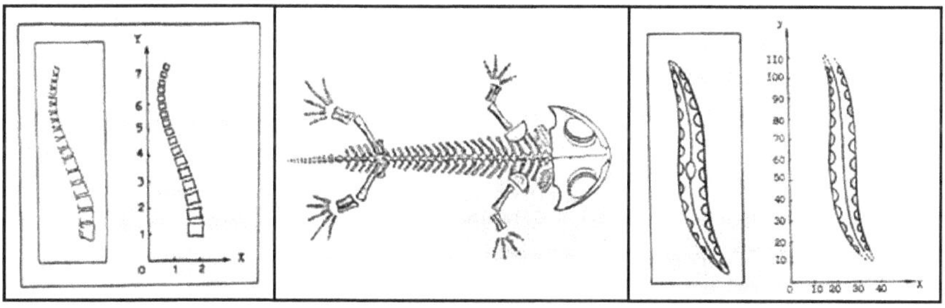

Fig. 5.3.4. Examples of Möbius cyclomerias. Left: vertebra of the torso part of the human spine (from [224]) and its model as a Möbius cyclomeria. Middle: for comparison, a similarity cyclomeria in the structure of the torso part of the spine is shown (from [224]). Right: the shell (from [153]) and its model as a Möbius cyclomeria.

Möbius cyclomerias, along with Euclidean ones, are also realized in the configurations of bird feathers. For example, in addition to feathers of a straight or spiral shape, there are lyre-shaped feathers, e.g., in lyrebirds, which got their name because of the characteristic shape of the tail feathers. Möbius

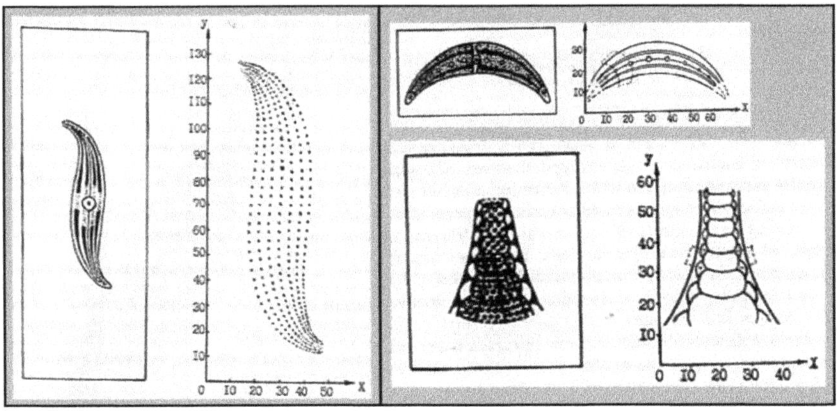

Fig. 5.3.5. Examples of Möbius cyclomerias. Left: the unicellular organism *Naviculla hippocampus* (from [224]) and its model as a series of Möbius cyclomerias. Top right: the desmidean *Glosterium lebleinii* (from [225]) and its model as a series of Möbius cyclomerias. Bottom right: the skeleton of *Stichocapsinae radiolaria* (from [226]) and its model as a Möbius cyclomeria of the elliptical type.

cyclomeria of the lyre-shaped (or loxodromic) type are also observed in the structures on the shells of sea urchins, in the generatrix lines of the spiral shells of some mollusks, the structural lines on the body of the seahorse, etc. Fig. 5.3.5 shows additional biological examples of Möbius cyclomerias.

The use of iterative algorithms in the formation of multi-block biostructures is functionally justified due to the obvious convolution of information necessary for their genetic coding and morphogenetic implementation. Euclidean and non-Euclidean cyclomerias are related not only to static biological configurations, but also to the organization of growth movements and motor patterns, including features of physiological normal postures in animal organisms; these movements can often be represented as a wave process of restructuring of cyclomerias due to a change in the type of begetting transformation in cyclomerias [32].

5.4. Principle "like begets like", local-similarity transformations, and volumetric growth

Now let's turn to the mysterious phenomenon of volumetric growth of living bodies. This biological growth is fundamentally different from the surface growth of crystals, which occurs due to the addition of a substance on the surface of the crystal and does not affect its internal regions. The volumetric growth of living bodies is an example of the regular cooperative behavior of many of their constituent parts. It is accompanied by a changing of the internal regions of the body. An example is the well-known scale type of volumetric growth, which is observed in plants and animals (the growth of leaves and flowers of some plants, insect larvae, mature fish, etc.) and is expressed in a proportional increase in all parts of the body. Knowledge of transforming the arrangement of only three points of a body subjected to scale transformation, one can determine the transformation of the entire set of its points. But along with the scale type of volumetric growth in living bodies, a local-similar (that is, conformal or Möbius) type of volume growth is also realized, in relation to which scale growth is its special case.

With the conformal type of volumetric growth, only small zones of the body are scaled, and the shape of the entire body as a whole can change non-linearly (see the example in Fig. 5.3.1). Inherited scale and conformal types of volumetric growth can be considered as a geometric implementation of the principle "like creates like" in relation to small areas of the body. Let us give specific examples of conformal volumetric growth. At the same time, we emphasize that, as in the case of the cyclomeria described above, scale and conformal types of volumetric growth are realized in similar biological bodies. For example, the caps of the fruiting bodies for many species of cap mushrooms are scaled in the course of individual development. At the same time, there are species of fungi in which the growth transformation of the cap of fruiting bodies over a long period of ontogenesis is described by a conformal transformation (see the example in Fig. 3.10.2 in Chapter 3). Scaled and Möbius kinds of three-dimensional growth are also observed in the individual development of composite flowers, fruits, etc.

Biological bodies can be constructed based on multi-stage symmetrical units, when the body made of symmetrical blocks of the first order is combined into symmetrical blocks of the second order, etc. In particular, this scheme manifests itself in the kinematics of the human body, where the mirror symmetry of the two halves of the body, which acts as second-order blocks, is supplemented by an approximate Möbius symmetry of the longitudinal proportions of the three-component kinematic blocks which comprise this structure, viz. the phalanges of fingers, the three-membered extremities (shoulder-forearm-wrist and hip-shin-foot) and the three-membered body (in anthropology the body is subdivided into the upper, torso, and lower parts). Let us explain that an approximate Möbius symmetry exists among all these three-membered blocks. The anthropometric data about these three-membered blocks are in [227, 228]. In the unidimensional case, the groups of Möbius and projective transformations coincide and the so-called cross-ratio (or the double ratio, the anharmonic ratio, or the wurf) is the main invariant of both of these groups [229].

The value of the cross-ratio of four points A, B, C, and D, lying on the same line, is calculated by the following known formula:

$$W = (C - A)(D - B)/\{(C - B)(D - A)\} \tag{5.2}$$

The expressions in parentheses denote the lengths of corresponding segments between noted points of the line. One can consider these four points as ends of corresponding three members of each three-membered block. Under the considered nonlinear transformations in one-dimensional space, the lengths of segments, as well as the ratio between the lengths of two segments, can change, but the value of the cross-ratio (5.2) always remains unchanged.

Calculation of the cross-ratio value for each of the named three-member blocks of the human body in different ages along its life gives for all of them almost the same value $P \approx 1.31$ as it was published on the basis of known anthropometric data in [32, 143]. Table 5.4.1 shows this fact for the middle finger in men of different ages. This is especially interesting because the growth of the human body is essentially non-linear (Fig. 5.4.1): for instance, its upper part grows 2.4-fold, the torso 2.8-fold, and the lower part 3.8-fold.

There are some arguments that the value $P \approx 1.31$ of the cross-ratios in the human body blocks is connected with the morphogenetic phyllotaxis laws and the so-called golden cross-ratio (or the golden wurf) $W = \varphi^2/2 = 1.309\ldots$, where $\varphi = (1 + 5^{0.5})/2 = 1.618\ldots$ is the golden ratio [32, 143, 157].

Table 5.4.1. The cross-ratio W of the middle finger in men of different ages. The lengths of phalanges (in cm) are taken from [228]. AB — the basal phalanx; BC — the medium phalanx; CD — the end phalanx.

Age (years)	AB	BC	CD	W
4	2.42	1.43	0.86	1.31
6	2.64	1.65	1.02	1.31
8	3.00	1.88	1.19	1.31
10	3.10	1.96	1.25	1.31
12	3.34	2.13	1.37	1.31
14	3.56	2.27	1.46	1.31
16	4.08	2.57	1.64	1.31
18	4.19	2.65	1.69	1.31
21	4.41	2.78	1.76	1.31

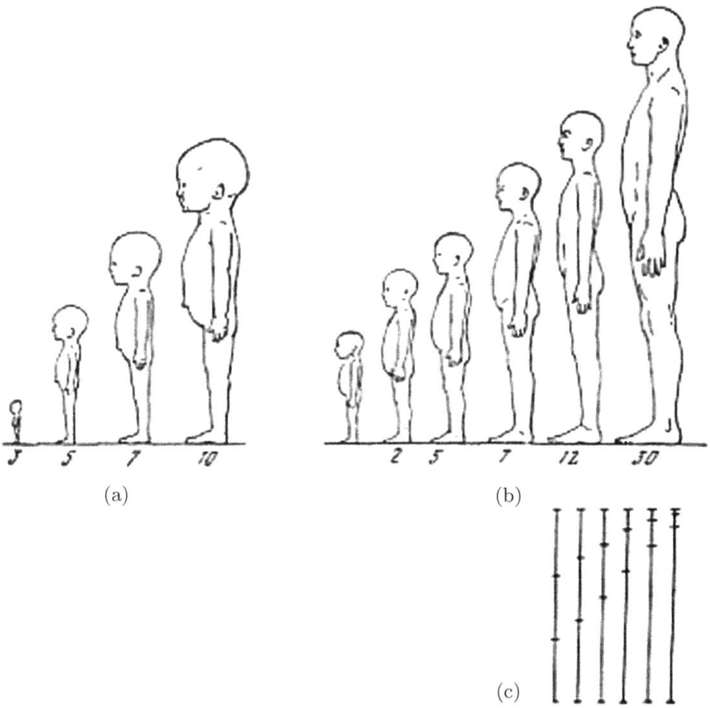

Fig. 5.4.1. Changes of the human body with age (from [157]). (a, b) Antenatal and postnatal stages: (a) in lunar months, (b) in years — the first on the left is a newborn. (c) Three-stretch parts whose wurfs are equal to 1.31.

For example, this golden cross-ratio $\varphi^2/2$ arises if the general expression (5.2) of the cross-ratio is calculated for the case of lengths $AB = \varphi^2$, $BC = \varphi$, and $CD = 1$.

The presented results show that all the three-membered blocks of the human kinematics in the straight posture are practically Möbius equivalent and Möbius invariable during the lifetime. This is largely true for dwarfs and giants and a broad range of highly organized animals. These findings are discussed in more detail in [143].

5.5. Some concluding remarks

The biological dualism of stochastics-determinism is one of the most mysterious and important phenomena that require intensive study and algebraic modeling. It seems that progress in its study can give a lot to modern science and biological technologies. The solution of many problems of genetic inheritance, morphogenesis, bio-symmetries, biological self-organization, non-local bioinformatics, and Gestalt biology is connected with the study of this phenomenon.

A number of research results in the field of matrix genetics described in this book indicate some connection of this phenomenon with the mathematics of vector spaces, the geometric features of which correspond to the conformal Poincaré model of hyperbolic geometry of a two-dimensional space and its extensions to the case of multi-dimensional spaces. This geometry has special properties which can be useful for understanding inherited biological structures. It is possible that the dualism of stochastics-determinism is due to a certain extent to the principles of informatics and the peculiarities of the propagation of information signals in living matter. When analyzing these issues in the future, it seems useful to take into account works on relativistic approaches to information signals, which can be of a quasi-stochastic nature, but at the same time carry specific information [231].

Additional materials on the topic of biological dualism "stochastics-determinism" will be presented in subsequent chapters of the book.

Chapter 6

The Doctrine of Energy-Informational Evolution on Bio-Antenna Arrays

This chapter is devoted to the possibilities of considering the evolution of biological systems in connection with the unique emergent properties of antenna arrays, that is, systems of mutually matched antennas widely used in technology. Materials are presented for the proposition that the evolution of biosystems can be formally considered as the evolution of systems of bio-antenna arrays and their energy-information wave activity, which participates in biological computation and contributes to the unification of body parts into a coherent whole. The use of digital antenna arrays in technology is based on their tensor-matrix theory. A structural analogy of this theory with the tensor-matrix features of genetic coding systems is described, as well as algebraic modeling of the universal rules for the stochastic DNA organization of the genomes of higher and lower organisms. This analogy is just one of the facts presented in the chapter in support of the usefulness of borrowing knowledge from modern antenna technology to consider the evolution of biosystems.

This chapter continues a description of results of studying universal rules and symmetries in structures of long single-stranded DNA sequences in eukaryotic and prokaryotic genomes. The previous chapters presented the tensor family of genetic matrices $[C, A; T, G]^{(n)}$, each of which combines all members of one of the n-plet alphabets of DNA: the alphabet of 4 nucleotides C, A, T, G; the alphabet of 16 duplets; the alphabets of 64 triplets, etc. (Figs. 1.2.1 and 1.4.2).

To show new results in this chapter, let us return to matrices of percentages of n-plets in different n-texts of DNA of the human chromosome №1 (Figs. 4.1.1–4.1.3), in which each of the n-plets from symbolic matrices in

Figs. 1.2.1 and 1.4.2 is replaced by its percentage value in the corresponding n-text of this DNA. At first glance, the set of values in the percentage matrices (Figs. 4.1.1–4.1.3) is quite chaotic. It has the following features regarding the percentage of separate n-plets:

- Percentage of presented n-plets significantly depends on the order of letters in them. For example, the percentage of duplets CG and GC, having the same letter composition, differ several times: %CG = 0.0103, and %GC = 0.0440;
- The numerical percentage matrices for duplets, triplets, and tetraplets (Figs. 4.1.1–4.1.3) are not tensor powers of the nucleotide percent (2×2)-matrix (Fig. 4.1.1).

But in fact, these matrices are naturally interconnected, which will be discussed below.

The main aim of this chapter is the following:

1) Describe revealed system analogies between matrices, which are used in the theory of digital antenna arrays, and percentage matrices of universal stochastic organization of DNA sequences in eukaryotic and prokaryotic genomes from publications [70, 116, 232] (as in Figs. 4.1.1–4.1.3);
2) Present the first application of algebraic matrix operations from the theory of digital antenna arrays to model numeric interconnections inside families of mentioned percentage matrices of n-texts of genomic DNAs;
3) Show the first materials of the doctrine about the key role of mathematics of the digital bio-antenna arrays and their wave functions (including radiation patterns) for understanding the evolutionary phenomena of self-organization of living bodies, which are based on wave energy-information intercommunication among their components [233–235].

With the presented approach, biological evolution is considered as the evolution of systems of digital bio-antenna arrays (or gratings) and their interrelated wave multi-path radiations, providing energy-information communication of body parts with each other and with the outside world. From the point of view of this doctrine, energy-information communication orientation appears in evolution, which can be considered as an important system-forming factor of biological essences.

6.1. System analogies between genomic percentage matrices and matrices in the theory of digital antenna arrays

Is there a meaningful algebraic-matrix relationship between the phenomenological probability matrices $\mathbf{P}_1$, $\mathbf{P}_2$, $\mathbf{P}_3$, and $\mathbf{P}_4$ (Figs. 4.1.1–4.1.3)? Yes, such a relationship exists. Its study shows important probability properties of stochastic organization of genomic DNAs, which lead to perspective ideas about a connection of genomic DNAs with the theory of digital antenna arrays and also with formalisms of quantum informatics. Let us explain this.

The number of members in an n-plets alphabet of DNA is 4^n. Accordingly, when passing from the n-plets alphabet to the $(n+1)$-plets alphabet, which is represented in the same tensor family of matrices $[C, A; T, G]^{(n)}$ (Figs. 1.2.1 and 1.4.2), the number of alphabetic members becomes 4 times more. This tensor multiplication of any n-plet by the (2×2)-matrix [C, A; T, G] associates with each n-plet from the first alphabet four $n+1$-plets of the second alphabet, that is, it is an operation of tetra-multiplicating (or tetra-breeding) of alphabetic members in this transition to the next alphabet.

To simplify the explanation, let us consider in more detail the case of the relationship between the percentage of 4 monoplets in the 1-text representation of genomic DNA and the percentage of 16 duplets in its 2-text representation. In this case, for example, each of the 4 monoplets C, A, T, G turns out to be associated — through the tensor product of matrices — immediately with four corresponding duplets shown on the right side of expressions (6.1):

$$[C] \otimes [C, A; T, G] = [CC, CA; CT, CG],$$
$$[A] \otimes [C, A; T, G] = [AC, AA; AT, AG],$$
$$[T] \otimes [C, A; T, G] = [TC, TA; TT, TG],$$
$$[G] \otimes [C, A; T, G] = [GC, GA; GT, GG]. \quad (6.1)$$

Concerning the relationships between the named percentages of 4 monoplets and 16 duplets, we accept by analogy the following: the percentage matrix of each quadruple of duplets indicated in (6.1) is the tensor product of the percentage matrices of monoplets [%C], [%A], [%T], [%G] by the corresponding (2×2)-matrices with unknown real coefficients following expressions (6.2) and (6.3):

$$[\%C] \otimes [B_C] = [\%CC, \%CA; \%CT, \%CG]),$$
$$[\%A] \otimes [B_A] = [\%AC, \%AA; \%AT, \%AG]),$$
$$[\%T] \otimes [B_T] = [\%TC, \%TA; \%TT, \%TG]),$$
$$[\%G] \otimes [B_G] = [\%GC, \%GA; \%GT, \%GG]) \quad (6.2)$$

where

$$B_C = [b_{C00}, b_{C01}; b_{C10}, b_{C11}], \quad B_A = [b_{A00}, b_{A01}; b_{A10}, b_{A11}],$$
$$B_T = [b_{T00}, b_{T01}; b_{T10}, b_{T11}], \quad B_G = [b_{G00}, b_{G01}; b_{G19}, b_{G11}] \quad (6.3)$$

These matrices B_C, B_A, B_T, B_G (6.3) map the percentage of each nucleotide to the percentage of four duplets, and therefore they are conditionally called tetra-multiplicating matrices. From (6.2) and (6.3), one can derive the expressions for the named 16 coefficients in terms of the ratios of the percentages of the corresponding duplets and monoplets:

$$[\%C \times b_{C00}, \%C \times b_{C01}; \%C \times b_{C10}, \%C \times b_{C11}] = [\%CC, \%CA; \%CT, \%CG]) \rightarrow$$
$$\rightarrow b_{C00} = \%CC / \%C, b_{C01} = \%CA / \%C, b_{C10} = \%CT / \%C, b_{C11} = \%CG / \%C;$$
$$[\%A \times b_{A00}, \%A \times b_{A01}; \%A \times b_{A10}, \%A \times b_{A11}] = [\%AC, \%AA; \%AT, \%AG]) \rightarrow$$

$$\rightarrow b_{A00} = \%AC / \%A,\ b_{A01} = \%AA / \%A,\ b_{A10} = \%AT / \%A,\ b_{A11} = \%AG / \%A;$$
$$[\%T \times b_{T00}, \%T \times b_{T01}; \%T \times b_{T10}, \%T \times b_{T11}] = [\%TC, \%TA; \%TT, \%TG]) \rightarrow$$
$$\rightarrow b_{T00} = \%TC / \%T,\ b_{T01} = \%TA / \%T,\ b_{T10} = \%TT / \%T,\ b_{T11} = \%TG / \%T;$$
$$[\%G \times b_{G00}, \%G \times b_{G01}; \%G \times b_{G10}, \%G \times b_{G11}] = [\%GC, \%GA; \%GT, \%GG]) \rightarrow$$
$$\rightarrow b_{G00} = \%GC / \%G,\ b_{G01} = \%GA / \%G,\ b_{G10} = \%GT / \%G,\ b_{G11} = \%GG/\%G \quad (6.4)$$

We need to algebraically express the system connection between the entire percentage (2 × 2)-matrix of 4 monoplets [%C, %A; %T, %G] and the percentage (4 × 4)-matrix of 16 duplets (see the symbolic matrix in Fig. 4.1.1 on the left). This can be done using an algebraic operation called the Hadamard product for matrices, which is indirectly related to the tensor product. The Hadamard product is known in the theory of digital antenna arrays, artificial intelligence, and many other areas of matrix analysis of systems. We will use the Hadamard product due to the convenience of expressing the desired matrix relationships with it, as well as the usefulness of emerging associations between the matrix representations of genetic systems and other systems, primarily in the theory of digital antenna arrays.

The Hadamard product (also known as the element-wise product, entry-wise product, or Schur product [236]) is a binary operation that takes two matrices of the same dimensions and produces another matrix of the same dimension as the operands, where each element *i, j* is the product of elements *i, j* of the original two matrices. In other words, for two matrices *A* and *B* of the same dimension $m \times n$, the Hadamard product $A \bigcirc B$ is a matrix of the same dimension as the operands, with elements given by $(A \bigcirc B)_{ij} = (A)_{ij}(B)_{ij}$. The symbol $\bigcirc$ denotes the Hadamard product. The expression (6.5) shows an example of the Hadamard product for two (2 × 2)-matrices:

$$\begin{vmatrix} \mathbf{a}_{00}, \mathbf{a}_{01} \\ \mathbf{a}_{10}, \mathbf{a}_{11} \end{vmatrix} \bigcirc \begin{vmatrix} \mathbf{b}_{00}, \mathbf{b}_{01} \\ \mathbf{b}_{10}, \mathbf{b}_{11} \end{vmatrix} = \begin{vmatrix} \mathbf{a}_{00}\mathbf{b}_{00}, \mathbf{a}_{01}\mathbf{b}_{01} \\ \mathbf{a}_{10}\mathbf{b}_{10}, \mathbf{a}_{11}\mathbf{b}_{11} \end{vmatrix} \quad (6.5)$$

$$\begin{vmatrix} \%C, \%A \\ \%T, \%G \end{vmatrix} \circ \begin{vmatrix} B_C, B_A \\ B_T, B_G \end{vmatrix} = \begin{vmatrix} \%C\times B_C, \%A\times B_A \\ \%T\times B_T, \\ \%G\times B_G \end{vmatrix} =$$

%CC	%CA	%AC	%AA
%CT	%CG	%AT	%AG
%TC	%TA	%GC	%GA
%TT	%TG	%GT	%GG

Fig. 6.1.1. Representation of the 16-duplet percentage matrix (shown at the bottom) via the Hadamard product of the monoplet percentage (2 × 2)-matrix with the block (2 × 2)-matrix, in which the entries B_C, B_A, B_T, B_G are (2 × 2)-matrices from (6.3) with the coefficients defined in the expression (6.4).

The Hadamard product is a principal submatrix of the tensor (or Kronecker) product [237].

Fig. 6.1.1 shows the Hadamard product of two (2 × 2)-matrices, in the second of which the entries B_C, B_A, B_T, B_G are tetra-multiplicating (2 × 2)-matrices with real coefficients from expressions (6.3). Overall, the second matrix appears as a colony or a union of 4 tetra-multiplicating matrices. This Hadamard product of two matrices (shown in Fig. 6.1.1, top) defines a percentage matrix of 16 duplets (Fig. 6.1.1, bottom).

Now let's move on from the general symbolic expression of the relationship of percentage matrices of 4 monoplets and 16 duplets (Fig. 6.1.1) to the analysis of the numeric relationship of phenomenological percentages in the concrete case of DNA of human chromosome № 1 (from Fig. 4.1.1). In this case, symbolic expressions (6.3) and (6.4) give the numeric tetra-multiplicating matrices shown in Fig. 6.1.2. Each of them contains 4 real numbers, which can be interpreted as percentages in fractions of one (their values are rounded here to 6 decimal places).

Fig. 6.1.2 shows that the tetra-multiplicating matrices B_C, B_A, B_T, and B_G differ significantly from each other in their sets of numerical entries. But at the same time, as it turns out, they obey the general rule: the sum of all entries in each of these matrices is equal to 1 (accurate to the third decimal place) (Fig. 6.1.3). This rule adds to the list of genetic gestalt rules, in which

B_C =

0.054090 / 0.208499	0.072739 / 0.208499
0.071341 / 0.208499	0.010305 / 0.208499

=

0.259426	0.348870
0.342165	0.049425

B_A =

0.050327 / 0.291001	0.095037 / 0.291001
0.074290 / 0.291001	0.071368 / 0.291001

=

0.172944	0.326587
0.255291	0.245250

B_T =

0.060080 / 0.291756	0.063123 / 0.291756
0.095680 / 0.291756	0.072859 / 0.291756

=

0.205925	0.216355
0.327945	0.249726

B_G =

0.044024 / 0.208744	0.060083 / 0.208744
0.050459 / 0.208744	0.054195 / 0.208744

=

0.210899	0.287831
0.241727	0.259624

Fig. 6.1.2. Tetra-multiplicating matrices from expressions (6.3) and (6.4) in the case of DNA of human chromosome №1.

Matrix	Sum of matrix entries
B_C	0.259426 + 0.348870 + 0.342165 + 0.049425 = **0.999886**
B_A	0.172944 + 0.326587 + 0.255291 + 0.245250 = **1.000072**
B_T	0.205925 + 0.216355 + 0.327945 + 0.249726 = **0.999951**
B_G	0.210899 + 0.287831 + 0.241727 + 0.259624 = **1.000081**

B_C	B_A
B_T	B_G

○

0.999886	1.000072
0.999951	1.000081

Fig. 6.1.3. Equality to the unity of the sum of all 4 entries in each of the tetra-multiplicating matrices B_C, B_A, B_T, B_G in the case of DNA of human chromosome №1. Top: different summands give practically the same sums in each of these matrices. Bottom: an additional visual presentation of this colony of matrices B_C, B_A, B_T, B_G, and their corresponding sums of entries.

the value of the sum practically does not change with a wide variability of the summand values [70, 117].

In the DNA of different genomes, the sets of entries in the tetra-multiplicating matrices B_C, B_A, B_T, and B_G can be significantly different. But the unit sum rule for all four entries in different tetra-multiplicating matrices B_C, B_A, B_T, and B_G (represented by a particular example in Fig. 6.1.3) is universal and holds for the entire wide range of eukaryotic and prokaryotic genomes according to [70, 116, 232, 238–241].

Considering the equalities (6.4) and the unit sum rule (Fig. 6.1.3), calculations of sums of 4 entries in each of the (2 × 2)-quadrants of the (4 × 4)-matrix give the following:

$$\%CC + \%CA + \%CT + \%CG = \%C \times (b_{C00} + b_{C01} + b_{C10} + b_{C11}) \approx \%C,$$
$$\%AC + \%AA + \%AT + \%AG = \%A \times (b_{A00} + b_{A01} + b_{A10} + b_{A11}) \approx \%A,$$
$$\%TC + \%TA + \%TT + \%TG = \%T \times (b_{T00} + b_{T01} + b_{T10} + b_{T11}) \approx \%T,$$
$$\%GC + \%GA + \%GT + \%GG = \%G \times (b_{G00} + b_{G01} + b_{G10} + b_{G11}) \approx \%G \quad (6.6)$$

The expressions (6.6) correspond to the universal phenomenological rule of genomic DNA [70, 116, 232, 239–241] and are represented by expressions: %C ≈ Σ %CN, %A ≈ Σ %AN, %T ≈ Σ %TN, %G ≈ Σ %GN, where N denotes any of the nucleotides C, A, T, G.

In connection with publications on the relationship of DNA alphabets with the formalisms of quantum informatics [70, 242], we especially note the following. The four entries of each tetra-multiplicating matrix B_C, B_A, B_T, and B_G, giving the unit sum, can be considered as probabilities in a two-qubit quantum system, which is traditionally expressed as follows in Dirac notations [15, p. 16]:

$$|\psi_{12}> = \alpha_{00}\,|00> + \alpha_{01}\,|01> + \alpha_{10}\,|10> + \alpha_{11}\,|11> \quad (6.7)$$

In (6.7), the coefficients α_{00}, α_{01}, α_{10}, α_{11} are called the probability amplitudes of the corresponding computational basis states |00>, |01>, |10>, |11>. They satisfy the normalization condition for probabilities (6.8):

$$\alpha_{00}^2 + \alpha_{01}^2 + \alpha_{10}^2 + \alpha_{11}^2 = 1, \quad (6.8)$$

which is associated with the indicated unit sum rule of genomes (Fig. 6.1.3). Some interesting connections between the stochastic organization of genomic DNAs and formalisms of quantum informatics are later described in this book.

One can also note that — in the case of the considered genomic DNA — numeric tetra-multiplicating matrices B_C, B_A, B_T, and B_G are not completely independent of each other but their entries are interrelated. Fig. 6.1.4 shows the colonial union of these 4 matrices into a single (4×4)-matrix using data from Fig. 6.1.2. In this colonial (4×4)-matrix, the sums of the entries in rows

$\begin{bmatrix} B_C & B_A \\ B_T & B_G \end{bmatrix} =$

0.259427	0.348870	0.172945	0.326585	**1.107827**
0.342167	0.049426	0.255290	0.245251	**0.892135**
0.205927	0.216354	0.210899	0.287829	**0.921009**
0.327944	0.249727	0.241725	0.259624	**1.079021**
1.135465	**0.864378**	**0.880859**	**1.119288**	

Fig. 6.1.4. The colonial (4×4)-matrix combines 4 tetra-multiplicating (2×2)-matrices B_C, B_A, B_T, B_G. The numeric data taken from Fig. 6.1.2 in the case of DNA of human chromosome №1 are shown. Sums of entries in each of the columns and the rows are marked in bold.

%CC	%CA	%AC	%AA
%CT	%CG	%AC	%AG
%TC	%TA	%GC	%GA
%TT	%TG	%GT	%GG

∘

B_{CC}	B_{CA}	B_{AC}	B_{AA}
B_{CT}	B_{CG}	B_{AT}	B_{AG}
B_{TC}	B_{TA}	B_{GC}	B_{GA}
B_{TT}	B_{TG}	B_{GT}	B_{GG}

=

%CCC	%CCA	%CAC	%CAA	%ACC	%ACA	%AAC	%AAA
%CCT	%CCG	%CAT	%CAG	%ACT	%ACG	%AAT	%AAG
%CTC	%CTA	%CGC	%CGA	%ATC	%ATA	%AGC	%AGA
%CTT	%CTG	%CGT	%CGG	%ATT	%ATG	%AGT	%AGG
%TCC	%TCA	%TAC	%TAA	%GCC	%GCA	%GAC	%GAA
%TCT	%TCG	%TAT	%TAG	%GCT	%GCG	%GAT	%GAG
%TTC	%TTA	%TGC	%TGA	%GTC	%GTA	%GGC	%GGA
%TTT	%TTG	%TGT	%TGG	%GTT	%GTG	%GGT	%GGG

Fig. 6.1.5. Representation of the 64-triplet percentage matrix (shown at bottom) via the Hadamard product of the duplet percentage (4 × 4)-matrix with the block (4 × 4)-matrix, whose 16 entries B_{CC}, B_{CA}, B_{CG}, B_{CG}, B_{AC}, B_{AA}, B_{AT}, B_{AG}, B_{CC}, B_{CA}, B_{CT}, B_{CG}, B_{GC}, B_{GA}, B_{GT}, B_{GG} are tetra-multiplicating (2 × 2)-matrices from Fig. 6.1.6.

(as well as in columns), which are located symmetrically, are approximately equal.

Let us now turn to the algebraic relationship between the percentage (4×4)-matrix of 16 duplets and the percentage (8×8)-matrix of 64 triplets (Figs. 4.1.2 and 4.1.3). Fig. 6.1.5 shows this relationship via the Hadamard product by analogy with Fig. 6.1.1.

Tetra-multiplicating matrices in Fig. 6.1.5 are defined by analogy with the previous case shown in expression (6.4). Fig. 6.1.6 presents these 16 tetra-multiplicating matrices, whose entries are equal to ratios of percentage of corresponding triplets and duplets shown in Figs. 4.1.1 and 4.1.2.

In the concrete case of percentages in 2-texts and 3-texts of DNA of human chromosome №1 (represented in Figs. 4.1.1 and 4.1.2), the

B_{CC}=	%CCC / %CC	%CCA / %CC	;	B_{CA}=	%CAC / %CA	%CAA / %CA
	%CCT / %CC	%CCG / %CC			%CAT / %CA	%CAG / %CA
B_{AC} =	%ACC /%AC	%ACA / %AC	;	B_{AA} =	%AAC /%AA	%AAA / %AA
	%ACT /%AC	%ACG / %AC			%AAT /%AA	%AAG / %AA
B_{CT} =	%CTC / %CT	%CTA / %CT	;	B_{CG} =	%CGC / %CG	%CGA / %CG
	%CTT / %CT	%CTG / %CT			%CGT / %CG	%CGG / %CG
B_{AT} =	%ATC / %AT	%ATA / %AT	;	B_{AG} =	%AGC / %AG	%AGA / %AG
	%ATT / %AT	%ATG / %AT			%AGT / %AG	%AGG / %AG
B_{TC} =	%TCC / %TC	%TCA / %TC	;	B_{TA} =	%TAC / %TA	%TAA / %TA
	%TCT / %TC	%TCG / %TC			%TAT / %TA	%TAG / %TA
B_{GC} =	%GCC / %GC	%GCA / %GC	;	B_{GA} =	%GAC / %GA	%GAA / %GA
	%GCT / %GC	%GCG / %GC			%GAT / %GA	%GAG / %GA
B_{TT} =	%TTC / %TT	%TTA / %TT	;	B_{TG} =	%TGC / %TG	%TGA / %TG
	%TTT / %TT	%TTG / %TT			%TGT / %TG	%TGG / %TG
B_{GT} =	%GTC / %GT	%GTA / %GT	;	B_{GG} =	%GGC / %GG	%GGA / %GG
	%GTT / %GT	%GTG / %GT			%GGT / %GG	%GGG / %GG

Fig. 6.1.6. 16 symbolic tetra-multiplicating matrices noted in Fig. 6.1.5, whose entries are equal to ratios of percentage of corresponding triplets and duplets shown in Figs. 4.1.1 and 4.1.2.

tetra-multiplicating matrices from Fig. 2.6 get the numeric forms shown in Fig. 6.1.7.

The sum of all 4 entries in each of these 16 tetra-multiplicating matrices in Figs. 6.1.5–6.1.7 is equal to 1 (accurate to the second decimal place) (Fig. 6.1.8).

It is this unit sum rule that determines, in particular, the universal phenomenological rule of genomic DNA previously described in [70, 116, 232, 238–241] and represented in expression (2.8) by some analogy with expression (6.7):

B_{CC} =

0.255962	0.347236
0.342614	0.053725

; B_{CA} =

0.209464	0.255911
0.245935	0.289207

B_{AC} =

0.234964	0.392793
0.322369	0.050406

; B_{AA} =

0.152210	0.388563
0.249937	0.209134

B_{CT} =

0.246475	0.178701
0.281582	0.292675

; B_{CG} =

0.243414	0.220494
0.251465	0.282438

B_{AT} =

0.177344	0.261409
0.321454	0.239696

; B_{AG} =

0.201857	0.313426
0.226093	0.258937

B_{TC} =

0.264259	0.326923
0.370469	0.038782

; B_{TA} =

0.174753	0.314608
0.307214	0.203470

B_{GC} =

0.284965	0.330797
0.326511	0.057502

; B_{GA} =

0.160038	0.326148
0.220872	0.292257

B_{TT} =

0.206149	0.207045
0.389283	0.196943

; B_{TG} =

0.200019	0.267209
0.272874	0.260144

B_{GT} =

0.189482	0.221009
0.286418	0.303943

; B_{GG} =

0.231754	0.295197
0.218745	0.255051

Fig. 6.1.7. Tetra-multiplicating matrices from Fig. 6.1.6 in the case of numeric data of DNA of human chromosome №1.

B_{CC}	B_{CA}	B_{AC}	B_{AA}
B_{CT}	B_{CG}	B_{AT}	B_{AG}
B_{TC}	B_{TA}	B_{GC}	B_{GA}
B_{TT}	B_{TG}	B_{GT}	B_{GG}

○

0.9995	1.0005	1.0005	0.9998
0.9994	0.9978	0.9999	1.0003
1.0004	1.0000	0.9998	0.9993
0.9994	1.0002	1.0009	1.0007

Fig. 6.1.8. Left: symbols of the 16 tetra-multiplicating (2 × 2)-matrices from Figs. 6.1.5–6.1.7. Right: a matrix of corresponding sums of 4 entries in these (2 × 2)-matrices; each of these sums is equal to 1 with high-level precision.

$$\%CC \approx \Sigma\, \%CCN,\ \%CA \approx \Sigma\, \%CAN,\ \%CT \approx \Sigma\, \%CTN,\ \%CG \approx \Sigma\, \%CGN,$$
$$\%AC \approx \Sigma\, \%ACN,\ \%AA \approx \Sigma\, \%AAN,\ \%AT \approx \Sigma\, \%ATN,\ \%AG \approx \Sigma\, \%AGN,$$
$$\%TC \approx \Sigma\, \%TCN,\ \%TA \approx \Sigma\, \%TAN,\ \%TT \approx \Sigma\, \%TTN,\ \%TG \approx \Sigma\, \%TGN,$$
$$\%GC \approx \Sigma\, \%GCN,\ \%GA \approx \Sigma\, \%GAN,\ \%GT \approx \Sigma\, \%GTN,\ \%GG \approx \Sigma\, \%GGN, \tag{6.9}$$

where N denotes any of the nucleotides C, A, T, and G.

In the case of the considered genomic DNA, 16 numeric tetra-multiplicating matrices from Fig. 6.1.2 are not completely independent of each other but their entries are interrelated by some additional relations. Fig. 6.1.9 shows the colonial union of these 16 matrices into a single (8 × 8)-matrix using data from Fig. 6.1.7. In this multi-block or colonial (8 × 8)-matrix, the sums of the entries in rows (as well as in columns), which are located mirror-symmetrically, are approximately equal (by analogy with the previous case in Fig. 6.1.4).

Such series of algebraic relationships — using the Hadamard product like in Figs. 6.1.1 and 6.1.5 — between the percentage ($2^n \times 2^n$)-matrices of 4^n n-plets and the percentage ($2^{n+1} \times 2^{n+1}$)-matrices of 4^{n+1} (n+1)-plets can be

B_{CC}	B_{CA}	B_{AC}	B_{AA}	
B_{CT}	B_{CG}	B_{AT}	B_{AG}	
B_{TC}	B_{TA}	B_{GC}	B_{GA}	=
B_{TT}	B_{TG}	B_{GT}	B_{GG}	

0.255962	0.347227	0.209464	0.255911	0.234964	0.392793	0.152210	0.388563	**2.237094**
0.342613	0.053731	0.245935	0.289207	0.322369	0.050406	0.249937	0.209134	**1.763331**
0.246475	0.178701	0.243414	0.220494	0.177344	0.261409	0.201857	0.313426	**1.843119**
0.281582	0.292675	0.251465	0.282438	0.321454	0.239696	0.226093	0.258937	**2.154341**
0.264259	0.326923	0.174753	0.314608	0.284965	0.330797	0.160038	0.326148	**2.182491**
0.370469	0.038782	0.307214	0.203470	0.326511	0.057502	0.220872	0.292257	**1.817077**
0.206149	0.207045	0.200019	0.267209	0.189482	0.221009	0.231754	0.295197	**1.817864**
0.389283	0.196943	0.272874	0.260144	0.286418	0.303943	0.218745	0.255051	**2.183400**
2.356791	**1.642028**	**1.905139**	**2.093480**	**2.143507**	**1.857555**	**1.661505**	**2.338713**	

Fig. 6.1.9. The colonial (8 × 8)-matrix, which combines 16 tetra-multiplicating (2 × 2)-matrices from Figs. 6.1.5 and 6.1.7. The numeric data taken from Fig. 6.1.7 in the case of DNA of human chromosome №1 are shown. Sums of entries in each of the columns and the rows are marked in bold.

continued by analogy for cases $n = 4, 5, 6, \ldots$ (but n should be much smaller than the length of genomic DNA). It should be additionally noted that the same matrix representation based on the Hadamard product of matrices (like in Figs. 6.1.1 and 6.1.5) is appropriate for the stochastic organization of n-texts in epi-chains of genomic DNAs (the notion of "epi-chains" was described in Chapter 4).

The matrix-tensor results described in this book lead to new approaches in two important scientific fields:

1) usage of analogies between matrix genetics [12, 116, 238, 239] and tensor-matrix theory of digital antenna arrays, where using the Hadamard product for matrices provides a huge profit in calculations and synthesis of highly effective antenna arrays (or grids) [243; 244, p. 442].
   In this field, the doctrine of energy-information biological evolution is developed based on multi-level interrelated bio-antenna arrays providing multi-beam radiation and absorption of wave energy and information, and also possessing properties of self-organization and development in ontogenesis and phylogenesis;
2) modeling molecular-genetic structures and universal stochastic rules of genomic DNAs based on formalisms of quantum informatics [70, 242, etc.]. The next sections describe these approaches in more detail.

6.2. The doctrine of energy-information evolution based on bio-antenna arrays and their wave functioning

The method of revealing analogies is one of the basic methods in the development of sciences, where it plays an important heuristic role, as is known at least from the time of B. Bolzano. Planck wrote: "*We thus find that it is a characteristic of every new idea occurring in science that it combines in a certain original manner two distinct series of facts*" [245].

Overall, this section presents the doctrine of energy-information evolution based on multi-level interrelated bio-antenna arrays and their wave functioning. The doctrine was generated, firstly, by non-trivial analogies between tensor-matrix descriptions in matrix genetics and the theory of digital antenna arrays and, secondly, by many facts on the importance of emission and absorption of wave energy and signals by using bio-antenna arrays in genetically inherited physiological systems. From the point of view of this doctrine, DNA informatics uses molecular bio-antenna arrays, whose wave functioning is coordinated with other genetically inherited macro-physiological bio-antenna arrays providing in living bodies many hereditary physiological structures and processes related to emission and absorption of wave energy and signals: compound faceted eyes in insects; inherited photonic crystal arrays determining species patterns on butterfly wings and other animal bodies; the innate ability to echolocation in animals; and others. Some of these inherited phenomena relate to intelligent actions of living bodies; for this reason, the proposed doctrine includes the proposition that inherited intelligent abilities of organisms are closely related to the functioning of multi-level bio-antenna arrays, the studying of which can lead to new approaches in the creation of artificial intelligence. Below some of these inherited phenomena and their relations to engineering technologies of digital antenna arrays are presented in more detail.

In his famous book *What is Life?*, Schrodinger made the auspicious proposal that genetic material is an "aperiodic crystal" [121, 246]. The described doctrine of energy-informational evolution continues this line of thought but significantly supplements it with the provision of the key role of bio-antenna arrays and their wave functioning. The doctrine proposes to consider DNA and RNA as a germinal aperiodic crystal of bio-antenna arrays, which serves as a seed for the aperiodic crystallization of the entire developing organism as a huge growing set of bio-antenna arrays interrelated with each other. The processes of this aperiodic crystallization of the inherited body can be modified under the influence of the environment, nutritional conditions, stages of onto- and phylogenetic development, etc. The functioning of these developing bio-antenna arrays is accompanied by exchanges of wave energy and signals among physiological subsystems and external sources and also by related processes of biological self-organization with the participation of resonance mechanisms and principles of energy minimization. The doctrine considers bio-antenna arrays and their wave functioning as the key element of energy-information evolution, which deserve special attention and study (for brevity, this doctrine can also be called the doctrine of bio-antenna evolution).

This section represents a set of facts and arguments supporting the described doctrine. To start this presentation, let us briefly recall basic facts about digital antenna arrays and their tensor-matrix theory in engineering technologies.

6.2.1. *Regarding antenna arrays and matrix approaches in their theory*

Modern technology is saturated with antenna arrays of various types, including medical ultrasound scanning technology (on multi-channel platforms with digital emitter arrays), sonar systems, radio relay stations, radio astronomic devices, etc. Their importance has led to the intensive development of the mathematical theory of transmitting and receiving antenna arrays of various types, which is presented in many publications. This developing mathematical theory provides new, previously unknown means of matrix-tensor analysis not only for antenna technology but also for mathematical natural science [244]. In particular, one can productively use elements of this matrix-tensor analysis (including the Hadamard product of matrices) in genetic research.

Antenna arrays consist of many separate antennas, each of which can emit or absorb waves of a specific frequency range (Fig. 6.2.1). These sets of antennas in antenna arrays can be located in space in a variety of ways on cylindrical, conical, spiral, and other surfaces periodically, non-equidistantly, quasi-randomly concerning each other, etc. The dimensions of engineering antenna arrays can vary greatly depending on the purpose: from giant arrays in radio astronomy and radio relay stations to nanoantenna arrays, with which modern science associates the development of new types of computer

Fig. 6.2.1. Examples of engineering antenna arrays (taken from https://commons.wikimedia.org/wiki/File:PAK_FA_AESA_maks2009.jpg and https://commons.wikimedia.org/wiki/File:Traqueur_acquisition.JPG, permission granted under the GNU Free Documentation License).

technology (photonics) and energy. The design of nanoantenna arrays can be associated with photonic crystals and liquid crystals, which are actively used in modern technology and which are realized for millions of years in genetically inherited physiological structures, will be discussed in more detail below. The mathematical description of the operation of engineering antenna arrays to emit or receive waves is almost the same.

Antenna arrays are used to provide such kinds of radiation patterns (or beamforming, or spatial filtering), which a single antenna element is unable to (these are emergence properties of antenna arrays). Let's list some of them in line with the book [247]. The directivity of the action of the simplest separate antenna — a symmetrical vibrator — is low. To increase the direction of action, already at the first stages of the development of antenna technology, they began to use a system of vibrators — antenna arrays. Currently, antenna arrays are the most common class of antennas due to the following reasons, the knowledge of which is important for understanding and studying the role of bio-antenna arrays in living organisms:

- An array of N elements allows one to increase approximately N times the directivity and, accordingly, the gain of the antenna compared to a single radiator, as well as to narrow the beam to improve the accuracy of determining the angular coordinates of the radiation source in navigation, radar, and other radio systems. With the help of antenna arrays, it is possible to raise the dielectric strength of the antenna and increase the level of radiated (or received) power;

- Antenna arrays are a unique tool for providing communication noise immunity and extracting a weak signal from a strong noise. For this reason, they are used, for example, in satellite communications technology in navigation satellite systems GPS, GLONASS, and Galileo, where the corresponding unique property of antenna arrays to overcome natural noises or special jamming is characterized by the following citation: *"Low power of the received satellite navigation signals leads to situations when jamming of natural or artificial origin, commensurable in power with the wanted signal, may present a danger for navigation communications integrity. In the case of receiving navigation signals against the background of powerful electromagnetic jamming, a substantial increase of signal-to-jamming ratio can only be achieved by way of spatial filtration of signals through the application of antenna arrays"* [248]. This special property of antenna arrays is especially important for understanding the well-known phenomenal ability of organisms to work with weak information signals against a background of strong noise, providing communication noise immunity. The same property of antenna arrays allows us to rethink the amazing noise immunity of genetic coding, as well as the multi-channel nature of genetic inheritance, expressed by Mendel's law of independent inheritance of traits *(vide infra);*
- An important advantage of antenna arrays is the ability to quickly (inertialess) survey space by swinging the beam or antenna via electrical methods (electrical scanning), as well as the possibility of using many beams in parallel with them operating at different frequencies and other characteristics. The antenna system is a necessary link to ensure the noise immunity of the radio system, taking into account adaptation to the interference environment.

Fig. 6.2.2. A conditional radiation pattern (beamforming) of a multi-beam digital antenna array.

Given these unique emergent properties of antenna arrays, it can be expected that organisms are forced to use them in their life activity. The doctrine stated in this article confirms this expectation on a set of examples of the structure and functioning of inherited physiological systems based on antenna arrays and their energy wave. We continue the theme of the features of antenna arrays.

An important class of antenna arrays is phased antenna arrays capable of operating with multi-beam radiation patterns (Fig. 6.2.2). A phased array usually means an electronically scanned and computer-controlled array of antennas, which creates a beam of radio waves that can be electronically steered to point in different directions without moving the antennas [249, 250].

In simple antenna arrays, the radiofrequency current from the transmitter is fed to the individual antenna elements with a differential phase relationship so that the energy from the separate elements adds together to increase the far-field power in the desired direction and suppress radiation in undesired directions (due to wave interference). In other words, the specificity of a phased array antenna is that the amplitude-phase distribution among elements is not fixed, but can be adjusted (changed in a controlled manner) during operation. Thanks to this, it is possible to move the beams of the antenna array in certain sectors of space without mechanical rotations of the antenna. Since the size of an antenna array must extend many wavelengths to achieve high gain, phased arrays are mainly practical at the high-frequency end of the radio spectrum (usually between 300 MHz and 300 GHz). Phased arrays have spread to civilian applications such as 5G MIMO for cell phones. The phased array principle is also used in acoustics for medical ultrasound imaging scanners, oil and gas prospecting, etc.

The operation of antenna arrays is associated not only with the transmission and reception of information signals but also with the reception and transmission of energy. For example, there are rectennas (or rectifying antennas, https://en.wikipedia.org/wiki/Rectenna), which are non-linear antennas designed to convert the field energy of an incident electromagnetic wave into direct current energy. The simplest design option can be a half-wave vibrator, between the arms of which a device with one-way conduction (for example, a diode) is installed. To increase the gain, such engineering devices are combined into multi-element arrays. In engineering, rectennas are considered promising for energy transmission over long distances, for example, when transporting energy to Earth from solar power plants in space. In living bodies, to pump the energy of electromagnetic waves through possible biological rectennas, semiconductors are needed, and they are present in us: "*... almost all types of living organisms are capable of biosynthesis of inorganic compounds, which by their nature are inorganic semiconductors and exhibit photochemical activity*" [251].

The defining feature of the new generation of radio engineering systems is the implementation of their antenna devices using the technology of digital antenna arrays (phased and others), which are often called Smart Antenna or Intelligent Antenna [243, 244]. These arrays additionally use analog-to-digital and digital-to-analog conversions to digitize the wave messages emitted or received by the antenna array to computer process these digital representations of antenna information with various mathematical algorithms to achieve certain goals. Such antenna arrays can operate in a multi-path mode with several partial radiation patterns almost simultaneously existing. In particular, smart antennas are used in numerous systems of radio holography, which is an analog of optical holography and uses mathematical matrices for calculating and designing antenna arrays or holographic antennas (see, for example, [166, 167, 247]). In the active digital antenna arrays, the radar transmitter is distributed ("smeared") over the antenna array and is one of the most reliable elements: if several solid-state modules break down, there is no significant reduction in performance (previously for radars with a passive array, if the transmitter elements came out of the ensemble, the system is "blind").

It is in Smart Antennas that matrix methods are a key tool for the rational synthesis of antenna arrays, as well as the development of effective algorithms for computer processing of digital antenna information and digital beamforming with high noise immunity.

The problems arising here about Smart Antennas, sometimes containing hundreds and thousands of information processing channels, have led to the development of new tools for matrix analysis that have not previously been encountered in other areas of science and technology. First of all, we are talking about the tensor-matrix theory of antenna arrays by the Ukrainian scientist V. Slyusar, who proposed some new operations with matrices (a list of his thematic publications is on https://en.wikipedia.org/wiki/Vadym_Slyusar). These new operations in the theory of digital antenna arrays (Smart Antennas) are closely related to the Hadamard product of matrices, which, as shown above, turned out to be adequate for describing the universal rules for the stochastic organization of genomic DNA (Figs. 6.1.1 and 6.1.5). In technics, the apparatus of the tensor-matrix theory of antenna arrays makes it possible to simplify computational processes "*by replacing the inversion of the multi-dimensional matrix product (P × P)-1 with an identical procedure for more simply formed Hadamard products…*" [244, pp. 441–442]. The huge savings in computational costs when using the Hadamard product for matrices in the theory of digital antenna arrays is illustrated by the following citation:

"*Calculating the quadratic form by identity is reduced to the Hadamard product. As a result, for a 32 × 32 antenna array in each channel with 32 synthesized frequency filters in 32 distance intervals, we can decrease the amount of multiplication by 8,004 times and the number of summations by 8,456 times concerning the initial notation. In this case, the number of multiplication operations is decreased by more than 268.845 billion, as compared to a four-coordinate model based on a traditional matrix product"* [243, p. 10; 244, p. 442].

Tensor-matrix theory allows creating digital twins of multi-coordinate measuring systems and neural hypernetworks.

Similar matrix methods associated with the Hadamard product for matrices are used in artificial intelligence systems, machine learning, reducing the number of calculations when implementing the tensor sketch method to reduce the data dimension, etc.

Regarding fractal-like structures in biological bodies, one can consider fractal antennas, which are actively used in modern technologies (see, for example, [252–254]).

6.2.2 *Inherited phenomena and the genetic doctrine of energy-informational evolution based on bio-antenna arrays*

The theory of antenna arrays says the following. If we have a problem of extracting a weak signal from strong noise, then we are forced to use the antenna lattices with their unique properties that individual antennas do not have. This makes one suspect that the information systems of living organisms, with their high-level noise immunity and ability to extract weak signals against a background of strong noise, use the unique advantages of antenna arrays.

As noted above, the method of analogies is one of the main methods for the development of science. It was also shown above that the Hadamard product for matrices, which is effectively used in the tensor-matrix theory of digital antenna arrays, unexpectedly showed its involvement in matrix representations of the universal rules for the stochastic organization of genomic DNAs. In short, the tensor-matrix formalisms in DNA informatics turned out to be similar to the formalisms of the tensor-matrix theory of digital antenna arrays.

This gave rise to the initial ideas of the stated doctrine of evolution based on bio-antenna arrays and their wave functioning. In this case, we are talking primarily about electromagnetic radiation. But mechanical and electrical oscillations in living bodies are closely connected because many tissues are piezoelectric (nucleic acids, actin, dentin, tendons, bone, etc.). The mathematics of mechanical and electrical oscillations is analogical (so-called "electro-mechanical analogies" are well known). The articles [36, 118] describe some aspects of this theme related to resonance phenomena.

Let us describe some of the inherited physiological phenomena that are included in this doctrine as its natural components due to the following:

- these phenomena are associated with the mechanisms of generation and reception of multicomponent wave radiation, whose parameters are specific to different types of organisms and are inherited from generation to generation;
- they can be explained naturally based on the idea of the energy-information functioning of bio-antenna arrays with their multi-beam radiation patterns.

6.2.2.1. *Complex faceted eyes of insects as receiving bio-antenna arrays*

Insects, crustaceans, and some other invertebrates receive visual information about the world around them, which is necessary for solving many vital tasks of an intellectual nature (search for food, escape from predators, etc.), through complex faceted eyes. These eyes serve as bio-antenna arrays (grids, or lattices) for receiving electromagnetic waves of the corresponding frequency range. They are formed by special structural units — ommatidia, the corneal lens of which has the form of a convex hexagon — facets. The compound eyes of insects are motionless, located on the sides of the head, and can occupy almost the entire surface of the head. The image perceived by such eyes is "recalculated" from the numerous ommatidia, which point in slightly different directions. The eyes of various insect species consist of

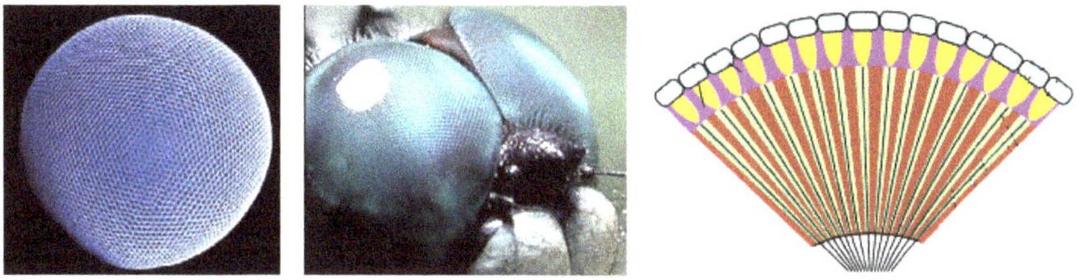

Fig. 6.2.3. Examples of complex faceted eyes. Left: compound eye of Antarctic krill as imaged by an electron microscope (from https://en.wikipedia.org/wiki/Compound_eye, photo by Gerd Alberti and Uwe Kils, permission granted under the GNU Free Documentation License). Middle: Dragonfly compound eyes (from https://en.wikipedia.org/wiki/File:Dragonfly_eye_3811.jpg, photo by David L. Green, permission granted under the Creative Commons Attribution-Share Alike 3.0 Unported license). Right: insect compound eye diagram (from https://tftwiki.ru/wiki/Arthropod_eye, permission granted under the GNU Free Documentation License, Version 1.2, or any later version published by the Free Software Foundation).

many ommatidia: a worker ant has about 100, a housefly has about 4,000, a worker bee 5,000, butterflies have up to 17,000, and dragonflies up to 30,000. Fig. 6.2.3 shows examples of complex faceted eyes, which serve to receive electromagnetic waves and have structures like engineering antenna arrays.

Compound eyes provide a large angular field of vision since the small "elementary eyes" — ommatidia — are directed in different directions and can cover a huge angular field up to a full sphere. The compound eye can quickly detect movement since its temporal inertia is 15 times less than that of the human eye. The uniformity of sensitivity, the great depth of the depicted space, the stereoscopic vision, the sensitivity to the polarization of radiation and, finally, the amazing miniature of the compound eyes should also be noted. Due to these properties, compound eyes are biological prototypes of many antenna arrays in engineering technologies.

6.2.2.2. *Inherited phenomena of biological echolocation*

It is known that many organisms have an innate ability for echolocation, which is important for their life and survival (https://en.wikipedia.org/

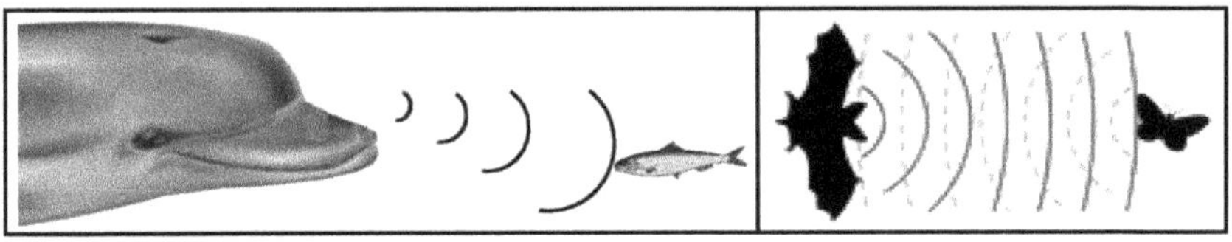

Fig. 6.2.4. Examples of biological echolocation: echolocation in dolphins (left) and bats (from https://commons.wikimedia.org/wiki/File:Chiroptera_echolocation.svg, permission granted under the Creative Commons CC0 License).

wiki/Animal_echolocation). At the same time, due to the mechanisms of echolocation, for example, dolphins and bats can recognize not only the distance to the target but also the dimensions and shape of the objects they track, by analogy with how active digital antenna arrays (Smart Antennas) make it possible in technology (Fig. 6.2.4).

For example, the dolphin's receiving-radiating hydroacoustic system allows it to locate a pellet that has fallen into the water at a distance of 15 meters; distinguish between the material and dimensions of objects of the same shape, differing by a few percent; distinguish, like a tomograph, details of the internal structure and shape of objects in water or a layer of silt; detect edible fish at a distance of three kilometers and distinguish it from those that are not food. There is evidence that dolphins can send images of objects to each other using their sonar systems [171].

Sonar of dolphins uses the generation of ultrasonic vibrations by special membranes, which form a directed ultrasonic beam, the frequency and directivity pattern of which can change. It also uses many hydroacoustic receptors distributed at a high density on the front side and a lower density over the surface of the entire body. These receptors "*form a multi-element broadband hydroacoustic receiving antenna with a circular pattern. This holographic reception subsystem provides illumination of the underwater environment, operating both in active and passive modes... Dolphin has several hydroacoustic information systems that partially overlap with each other and work in parallel*" [255]. Recall that mechanical and electromagnetic oscillations in organisms are interrelated since many biological tissues are piezoelectric.

In wildlife, electrolocation is also widespread with the generation and reception of electric fields for solving by organisms many vital tasks of a search, evaluation, and communication nature. In active electrolocation, the animal senses its surrounding environment by generating electric fields and detecting distortions in these fields using electroreceptor organs [256]. This electric field is generated using a specialized electric organ consisting of modified muscle or nerves. This field may be modulated so that its frequency and waveform are unique to the species and, sometimes, the individual.

The perception of electric fields and their changes is usually carried out by many electroreceptors that are distributed throughout the body, forming bio-antenna arrays, the cumulative readings of which are processed by the body. For example, the platypus possesses receiving bio-antenna arrays having almost 40,000 electroreceptors arranged in a series of stripes along with the bill. These same abilities of electrolocation are used by organisms for electrical communication with other organisms. For example, weakly electric fish can also communicate by modulating the electrical waveform they generate, an ability known as electrocommunication [257]. They may use this for mate attraction and territorial displays.

In particular, electrolocation is used for jamming avoidance response, which is a behavior performed by some species of weakly electric fish. It occurs when two electric fish with wave discharges meet — if their discharge frequencies are very similar, each fish will shift its discharge frequency to increase the difference between the two fish's discharge frequencies. By doing this, both fish prevent jamming of their sense of electroreception [258]. Lists of literature and illustrations on electrolocation, electroreception, and electrical communications are available on the websites https://en.wikipedia.org/wiki/Electroreception, https://en.wikipedia.org/wiki/Electrocommunication, and https://en.wikipedia.org/wiki/Jamming_avoidance_response, respectively.

One of the biological examples of directed electromagnetic radiation is mitogenetic radiation, discovered in 1923 by A. G. Gurwitsch in experiments on cell proliferation of an onion [259–262]. This radiation is associated with electromagnetic radiation in the ultraviolet range [263]. It is characterized by a high orientation, discovered by Gurwitsch himself and confirmed by the experiments of other authors [264].

One can note that the detection of objects and their geometric characteristics based on biological echolocation is an act of intellectual activity. Therefore, the phenomenon of biological echolocation is an evidence in favor of the fact that the innate ability for intellectual activity in living organisms is based on inherited systems of bio-antenna arrays with their wave activity (smart bio-antennas, by analogy with Smart Antennas in technology). The development of this topic of "system-antenna intelligence" leads to the understanding that bio-antenna arrays with their wave energy and self-organization mechanisms are not just transmitting and receiving devices, but are the key participants of intellectual activity in the living.

We also note that researchers at Durham University (England) taught methods of biological echolocation to a group of blind and sighted people using a 10-week training program. Their study showed that the human body can learn the skills of biological click-based echolocation regardless of age and degree of vision loss. The results of this work indicate the promise of teaching biological echolocation to people who are blind [265].

6.2.2.3. *Inherited biophotonic crystals*

This section presents information about the inherited sets of interconnected bio-photonic crystals that form bio-antenna array systems.

Modern engineering technologies actively use so-called photonic crystals to control the spatial distribution of photon beams [266–268]. The technologies of antennas and antenna arrays based on photonic crystals are being intensively developed [269]. A photonic crystal is a periodic optical nanostructure that affects the motion of photons. Photonic crystals contain regularly repeating regions of high and low dielectric constants. Photons (behaving as waves) either propagate through this structure or not, depending on their wavelength. This gives rise to distinct optical phenomena, such as inhibition of spontaneous emission, high-reflecting omnidirectional mirrors, and low-loss waveguiding. The periodicity of the photonic crystal structure must be around half the wavelength of the electromagnetic waves to be diffracted. One should note that, as is known, living bodies possess inherited opportunities to manage photonic beams using physical principles of photonic crystals with their properties of photon gratings, etc. Many inherited biological phenomena of structural coloration and animal reflectors are built on this, including the beautiful coloring of butterfly wings, peacock feathers, etc. (see details and lists of references in https://en.wikipedia.org/wiki/Photonic_crystal, https://en.wikipedia.org/wiki/Animal_reflectors, https://en.wikipedia.org/wiki/Structural_coloration).

Figs. 6.2.5 and 6.2.6 show some examples of inherited interrelated regular manifolds of bio-photonic crystals forming biological analogies of technical antenna arrays. These examples give pieces of evidence that living bodies with their genetic system skillfully encode and inherit entire manifolds of bio-antenna arrays.

The amazing coloring of peacock feathers is due to the play of light on the photonic crystals in them [270]. The chameleon also changes the color of its body reversibly, using multiple photonic crystals on its surface and optical interference on them. This was recently established in studies by

(a) (b) (c)

Fig. 6.2.5. Photonic crystals form heritable species patterns on butterfly wings, which allow differentiation between different species of butterflies. (a) Rhetenor Blue Morpho (from https://commons.wikimedia.org/wiki/File:Morpho_rhetenor_rhetenor_MHNT_dos.jpg, permission granted under the Creative Commons CC0 License); (b) European peacock (from https://commons.wikimedia.org/wiki/File:Paon-du-jour_MHNT_CUT_2013_3_14_Cahors_Dos.jpg, permission granted under the Creative Commons CC0 License); (c) regular sets of photonic crystal scales on a wing of a peacock butterfly (*Aglais io*) under the microscope (from https://ru.wikipedia.org/wiki/Чешуекрылые, permission granted under the Creative Commons CC0 License).

Fig. 6.2.6. Peacock feather coloring and body-color changes in chameleons are linked to ordered inherited arrays of bio-photonic crystals (from https://en.wikipedia.org/wiki/Indian_peafowl and https://en.wikipedia.org/wiki/Chameleon, permission granted under the Creative Commons Attribution 3.0 Unported license and the Creative Commons Attribution-Share Alike 4.0 International license, respectively).

Swiss scientists who revealed the existence in the skin of chameleons of two layers of cells with highly ordered structures of small (~130 nm) photonic nanocrystals of guanine, which serves as one of the nitrogenous bases in DNA [271]. Tension or relaxation in the system of these cells helps animals to quickly change color by changing the ability of the skin to reflect light.

Many animals in nature such as fish or beetles employ responsive photonic crystals, which form matrix structures, in their intelligent-like activities for camouflage, signaling, or baiting their prey [272] (https://en.wikipedia.org/wiki/Photonic_crystal_sensor).

It is natural to assume that the genetic transfer of inherited properties of photonic crystals in biological bodies is built on the fact that the molecular genetic structures themselves possess the properties of photonic crystals. One can recall here Schrödinger's definition of chromosomes as aperiodic crystals [121].

We believe that spatial characteristics of ensembles of genetic and other biological molecules, which form complex diffraction structures, play a managing role of photonic crystals in the problem of controlling photon beams that are generated and absorbed by these molecules (the range of photon frequencies in living bodies can be very wide, far beyond the optical range).

In our view, the inherited morphogenetic processes in living bodies are also determined to a large extent by biological photon beams, the course of which is not accidental, but is strictly organized by a system of spatial characteristics of ensembles of genetic and other biological molecules as photonic crystals. In the course of ontogenesis, based on electromagnetic (photonic) interactions, new molecular materials are involved in a naturally growing biological body, which leads to the appropriate growth of the managing system of bio-photonic crystals and the growth of numbers of photon beams. Of course, quantum-mechanic laws of resonances in molecular photonic interactions play an important role. Photonic crystals are related to the topic of nanoantenna arrays in two ways [269, 273, 274]. They are used in optical nanoantenna and microwave technologies. Usually, a nanoantenna is understood as a miniature antenna, the dimensions of which do not exceed hundreds of microns, and one of the dimensions is 100 or fewer nanometers. Nanoantennae are also often referred to as nanodevices that allow the scattering of radiation in a given direction [268]. Photonic crystals and their lattices are such devices; the location of many photonic crystals as a screen next to the nanoantenna can change its functional characteristics. In particular, the fluorescence of a single molecule that enters the nanoantenna gap can be increased by a factor of 10 [268].

In engineering technics, photonic crystals are composed of periodic dielectric, metallo-dielectric or even superconductor microstructures or nanostructures. As is known, active biological systems have extraordinarily high dielectric properties [275]. For this reason, dielectric photonic crystals, which have wide application in technics and are actively analyzed in the theory of photonic crystals and their applications in antenna arrays, are the most interesting for the theme of bio-antenna arrays.

One should add that the theme of nanoantennae, which are used in living nature for millions of years, is considered in modern science as very important: scientists believe that progress in technologies of nanoantenna production can lead to revolutions in computers (using photonic principles) and energetics (effective usage of solar energy) [268, 269]. Nanoantennae are one of the elements needed to create quantum computers. The presented doctrine on the evolutionary role of bio-antenna arrays may be useful for understanding the mechanisms of photosynthesis in plant leaves.

6.2.2.4. *Liquid biocrystals and heritable bio-antenna arrays*

DNA and RNA are bioliquid crystal structures, like other components of the living. The set of photonic crystals, particular cases of which are described in the previous section, contains a subclass of liquid crystal photonic crystals. Representatives of this subclass are among the most scientifically interesting and promising for applications of bio-photonic structures.

Liquid crystal is a state of matter that has properties between those of conventional liquids and those of solid crystals [276]. For instance, a liquid crystal may flow like a liquid, but its molecules may be oriented in a crystal-like way. There are many different types of liquid crystal phases, which can be distinguished by their different optical properties. Examples of liquid crystals can be found both in the natural world and in technological applications. Widespread liquid crystal displays use liquid crystals. To date, tens of thousands of types of multicomponent liquid crystal compounds have been synthesized for technical purposes around the world.

Due to the anisotropy of their structure, liquid crystals can scatter electromagnetic waves incident on them in selected directions. If a liquid crystal object has dimensions characteristic of nanoantenna arrays, then due to this anisotropic ability, it can serve as a nanoantenna array (as was mentioned for nanoantennae in the previous section). Emission at a photonic bandgap created by the periodic dielectric structure of the liquid crystal gives a low-threshold high-output device with stable monochromatic emission [277, 278].

Lyotropic liquid crystalline phases are abundant in living systems, the study of which is referred to as lipid polymorphism. Accordingly, lyotropic liquid crystals attract particular attention in the field of biomimetic chemistry. In particular, biological membranes and cell membranes are a form of liquid crystals. These liquid crystal membrane phases can also host important proteins such as receptors freely "floating" inside, or partly outside, the membrane. Many other biological structures exhibit liquid crystal behavior. For instance, the concentrated protein solution that is extruded by a spider to generate silk is, in fact, a liquid crystal phase. DNA and many polypeptides, including actively driven cytoskeletal filaments, can also form liquid crystal phases [279]. Together, these biological applications of liquid crystals form an important part of current academic research. The piezoelectric effect exists in liquid crystals; for this reason, interrelated electromagnetic and vibrational phenomena can coexist in them [280].

As noted above, antenna arrays are capable of generating narrow and precision directional electromagnetic beams of multiply enhanced intensity. Getting in a targeted way on the piezoelectric areas in living tissues, such electromagnetic influences cause mechanical vibrations in the corresponding small (almost point) areas. Therefore, the wave activity of bio-antenna arrays can lead to a very heterogeneous distribution of piezoelectric vibrations in cells and other tissue structures. These heterogeneous influences from one of the bio-antenna arrays on biological tissue can also lead to spatial-temporal reorganizing of other bio-antenna arrays in this tissue. Such interrelated wave influences of different bio-antenna arrays on each other, which reorganize their interconnected antenna systems, are participants of self-organization in living bodies.

Of particular interest for biological and applied research are liquid crystals called chiral phases or twisted nematic. For these crystals, the direction of the orientation of molecules in successive layers changes in a spiral (Fig. 6.2.7).

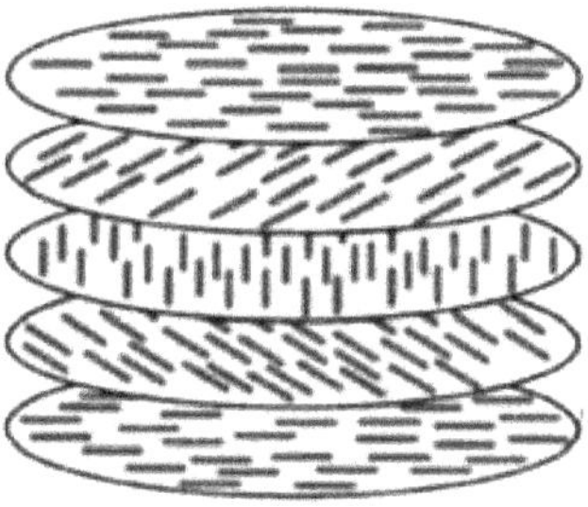

Fig. 6.2.7. Chiral nematic phase (from https://en.wikipedia.org/wiki/Liquid_crystal, permission granted under the Creative Commons Attribution-Share Alike 3.0 Unported license).

As a rule, chiral nematic crystals are obtained from compounds with pronounced anisotropy — unequal properties in different directions. It is these properties that, as is known, DNA and RNA, which are naturally twisted into a double helix, possess. It can be said that the desire for such helical chiral packing is an "inherent" property of their double-stranded molecules.

Since their nitrogenous bases absorb ultraviolet, these chiral nematic crystals, when obtained in the laboratory, have a peculiar "color", which manifests itself in the form of an abnormally high optical activity (the ability to rotate the radiation polarization plane), which is tens and hundreds of times higher than the optical activity of single molecules [281]. This fact emphasizes the need to take into account the emergent properties of multi-element nanoantenna arrays and their wave activity, which are fundamentally different from the properties of a single nanoantenna.

The chiral twisting that occurs in chiral liquid crystal phases also makes the system respond differently to right- and left-handed circularly polarized light. These materials can thus be used as polarization filters [282]. The theme of circularly polarized light from biological chiral liquid crystals is connected with the fundamental problem of biomolecular asymmetry by Pasteur [283]. Sets of helical chiral biomolecules can serve as helical antenna arrays, which radiate and absorb electromagnetic waves of the corresponding circular polarization. This provides the opportunity for helical chiral biomolecules to exchange radio waves of the corresponding circular polarization selectively with helical biomolecules of the same kind of chirality. In radio transmission technologies, circular polarization is often used where the relative orientation of the transmitting and receiving antennas cannot be easily controlled, such as in animal tracking and spacecraft communications (that is, for example, a spacecraft rotation does not influence the communication). In other words, regarding these helical antenna arrays, the factor of chirality (left or right polarization) is very important for communication. We add that the directional properties of a helical axial radiation antenna can be determined by considering the spiral as a linear antenna array consisting of some emitters — turns [247]. The principle of the chiral stereochemical organization of biomolecules in living nature is deeply related to the informational principle of communication among biomolecules based on bio-antenna arrays working with electromagnetic waves of appropriate circular polarization. Helices were termed long ago as «*curves of life*» due to the multiple implementations of inherited helical structures and processes in living bodies on various lines and levels of biological evolution [180, 181]. For example, in the human body, helical structures are genetically inherited from generation to generation in muscles, heart, vessels, bones, tendons, ligaments, nerves, organs of hearing (cochlea, ear), etc.

An analog of a spiral antenna in the form of left-handed helices was recently discovered by Swedish scientists in the tail of a spermatozoon. These authors suggest that these helical structures in particular can "*play a role in controlling the swimming direction of spermatozoa*" [284].

The ability of biological molecules for such spontaneous formation of ordered ensembles is important not only for the life and development of organisms but also for the technical nano-design of new molecular structures and their ensembles to create new drugs, sensors, nano-robots, etc. Researchers note the chemical diversity of biological "building blocks" (amino acids, lipids, nucleotides) for such nano-constructions, as well as their tendency for the spontaneous formation of complex spatial structures [281].

For example, it has been observed that concentrated solutions of short DNA oligomers develop liquid crystal ordering as the result of a hierarchically structured supramolecular self-assembly; this finding suggests a novel scenario for the abiotic origin of nucleic acids [285]. One can mention that the phenomena of self-organization of nanostructures are intensively studied in modern science [286]. From the point of view of convenience for nano-assembly technologies, nucleic acids have the best properties [281].

Nanoantennae based on DNA are already used in scientific technologies: scientists at the University of Montreal in Canada have created glowing nanoantennae from DNA molecules to track the relationships within proteins. These nanoantennae are capable of fluorescence and can absorb radiation at one wavelength and emit light at a different frequency depending on the molecular environment [287].

The structural properties of liquid crystals combine probabilistic and deterministic characteristics. The liquid crystal in the simplest case is a structure consisting of ordered molecular layers, which still retain some diffusion degrees of freedom characteristic of the behavior of molecules in a liquid solution. There is a well-known comparison of liquid crystals with a stream of logs floating down a river: in general, they are all lined up in the same direction, downstream, although each log floats on its own [288]. Thus, in the properties of liquid crystals, the "probability-determinism" dualism is represented at the molecular level, which is characteristic of biological phenomena of various levels and considered in Gestalt genetics and Gestalt biology [25, 70, 241, 289]. We also note that the chemical composition of liquid crystals can be very different under the external similarity of their integral configuration.

6.2.2.5. *Bio-antenna arrays, resonances, code biology, and morphogenesis*

Antennas are resonance devices. The described doctrine of bio-antenna arrays attracts additional attention to the important role of resonances in living bodies. This role of resonances in biological structures was presented in Chapter 3. It is no coincidence that many structural features of a molecular genetic system turn out to be consistent with the mathematical formalisms of the theory of resonances of oscillatory systems with many degrees of freedom [118, 131, 133].

One of the developing areas of modern theoretical biology is the direction of "code biology", which studies the coordinated interaction between various subsystems of the body based on certain codes [290]. In our opinion, at least many biological codes are arranged on the principles of resonances in some analogy with how a car door can be opened remotely using an electronic key. The antenna of such a key sends electromagnetic waves carrying the desired code to the car. Different cars need different wave codes but all of them are based on resonance interrelations.

Concerning morphogenesis, it should be remembered that it proceeds in the environment of polymer matrices with dispersed liquid crystals, which are DNA and RNA, some proteins, membranes and cytoskeleton of cells [291]. It is known that the optical and other properties of these polymeric media are regulated by the action of an electric field. In this case, various non-trivial morphological structures arise; under certain conditions, such hetero-phase polymers cure [292]. Wave activities of bio-antenna arrays can influence such hetero-phase polymers in living bodies and their morphological forms at different ontogenetic stages.

6.3. Regarding regeneration phenomena, genetics, and antenna arrays

The cells of a multicellular organism must exchange information for their coordinated behavior in the processes of ontogenesis and life activity. This implies the presence of a system of long-distance information links between many cells of the body. Violation of these connections is fraught with oncology, morphological deformities, and other negative consequences. The activity of bio-antenna arrays that emit and receive electromagnetic waves can be one of the types of long-range information links between cells and other structures of the body. At the same time, in biological piezoelectric media, electromagnetic waves generate mechanical vibrations and waves.

It has already been noted above that many living organisms have an innate ability for electrolocation and electrical communication. It is natural to believe that these abilities for remote wave electrical signal communication, inherited from generation to generation, did not arise from scratch, but are a continuation of the forms of electromagnetic wave communication between elements within the body itself. These electromagnetic forms exist alongside well-known biochemical communications using the exchange of molecular elements, as well as communications based on the action of mechanical forces. The neurons of the nervous system, with their action potentials, are a well-known example of remote signaling based on bioelectricity. Modern data from some authors suggest that the bioelectrical connection between cells of other types that are not neurons is also used as a means of remote communication (see, for example, [293]).

In connection with the topic of long-range operational communication between elements of the body, let us recall the above-mentioned remarkable emergent properties of antenna arrays, which are ensured precisely by the coordinated combination of many individual antennas into a single system and which are used in bio-antenna arrays inside the body:

1) the possibility of generating and receiving precision-directed beams (including narrow beams for point zones) with multiple increases in the level of radiated and received power;
2) the possibility of emission and reception of many such beams at once, having different frequencies and other characteristics that are independent of each other;
3) providing communication noise immunity and extracting a weak signal from a strong noise;
4) ensuring the noise immunity of information messages with the possibility of adapting to the interference environment.

For the design and calculation of digital antenna array systems in technology, their tensor-matrix theory was developed [243] which, in particular, allows you to create new approaches to artificial intelligence systems. Its algebraic formalisms turned out to be related to matrix-tensor genetics [12, 13, 70, 79, 238, 239]. This gives an additional reason to think that the intellectual properties of living organisms and their subsystems can be based on the energy-information activity of bio-antenna array systems. As noted above, the functioning of bio-antenna arrays in physiological systems, for example, in echolocation and faceted organs of vision, is directly related to the performance of intellectual functions of searching and evaluating objects of the external world.

The problem of inherited intellectual abilities has acquired particular relevance in connection with the creation of artificial intelligence systems. In this case, the following point of view is known: "*In fact, intelligence — a purposeful response to available information, often anticipating the future — is not restricted to the minds of some privileged species. It is distributed throughout biology, at many different spatial and temporal scales. There are not just intelligent people, mammals, birds, and cephalopods. Intelligent, purposeful problem-solving behavior can be found in parts of all living things: single cells and tissues, individual neurons and networks of neurons, viruses, ribosomes, and RNA fragments, down to motor proteins and molecular networks... Modern biology faces a fundamental knowledge gap when trying to explain meaningful, intelligent behavior. How can a system composed of cells and electrical signals generate a well-adapted body with behavior and mental states? If cells are not intelligent, how can intelligent behavior emerge from a distributed system composed of them? This fundamental mystery permeates biology. All biological phenomena are, in a sense, "group decisions" because organisms are made of individual parts — organs, tissues, cells, organelles, molecules. What properties of living systems enable components to work together toward higher-level goals?*" [294]. The authors of the cited article argue that the architecture of modular systems underlies evolutionary and organismal intelligence.

They emphasize the basic meaning of the following evolutionary principles represented in biological bodies:

1) the principle of modularity;
2) the principle of a hierarchy of modules: "*lower-level modules combine to form increasingly sophisticated higher-levels modules, which then become new building blocks for even higher-level modules, and so on. Similarly, the local metabolic and signaling goals of the cells integrate toward a*

morphogenetic outcome of building and repairing complex organs. Thus, increasingly sophisticated intelligence emerges from hierarchies of modules" [ibid].

But these principles are actively used in the engineering construction of modular digital antenna arrays (smart antenna) where lower-level antenna arrays of modular type are combined to form increasingly sophisticated higher-level modules, which then become new building blocks for even higher-level modules, and so on (systems of antenna arrays-subarrays-subsubarrays-...). It is one of the additional arguments in favor of a basic meaning of bio-antenna arrays for the development and provision of biological intelligence.

Cells are one of the important types of biological modules from which a living organism is built. Many of the key elements of a cell, such as its DNA, membranes, many polypeptides, and cytoskeleton, are liquid crystals [295, 296]. These crystals, as mentioned above, can serve as liquid crystal nanoantenna arrays. At the same time, they can participate in the intellectual activity at the cellular level. We note in passing that the cytoskeleton of neurons has long been considered by some authors as a computing environment where nerve cells perform their tasks [297]. The activity of transceiver liquid crystal bio-antenna arrays can be considered a special chapter of bioelectric physiology. They have connections with endogenous bioelectric fields and have been shown to regulate many patterning, morphological, and regenerative processes [293, 298].

It is natural to believe that the electromagnetic wave activity of liquid crystals and other nanoantenna arrays of cells is involved, among other things, in the distribution of rest potentials on cell membranes. This applies, in particular, to the cells in the initial stages of formation of a multicellular organism. In turn, an artificial change in electrical potentials on cellular liquid crystal membranes under the influence of external factors (for example, under the influence of ionophores) can cause rearrangements in the organism's system of bio-antenna arrays and their functioning.

The location in the cell of DNA molecules and cytoskeleton elements involved in the activity of transmitting and receiving bio-nanoantenna arrays is characterized by a certain asymmetry and polarity, which may be involved in the phenomenon of the polarity of the distribution of potentials on the cell membrane. We emphasize this in connection with the fact that the polarity of the distribution of potentials on cell membranes is important for morphogenetic processes and the regeneration of organs and tissues. This is shown by the impressive results of the experimental works of M. Levin and his colleagues at Tufts University, USA (https://ase.tufts.edu/biology/labs/levin/publications/, https://www.youtube.com/watch?v=XheAMrS8Q1c).

These works were carried out on flatworms — planarians, which have a head, a true brain, and many other organs, including a tail. Planarians are champions in organ and tissue regeneration and, in this regard, have been studied in biology, including bioelectric physiology, for over 100 years. If you cut its body into pieces — for example, into 200 pieces — every piece will rebuild exactly what's needed to make a perfect little worm. This is a system where every single piece knows exactly what a correct planarian looks like and builds the right organs in the right places and then stops. In the works of Levin and his colleagues, it was shown that the determination of the correct position of the organ in the body of the worm occurs due to bioelectric potentials and is controlled by the distribution of membrane potentials in not-yet-differentiated cells. Differences between the cells of the head and tail appear in the first three hours after amputation. In these works, it was shown that an artificial change (with the help of ionophores without application of external electrical influence) of the typical distribution of the membrane potential on an amputated fragment of the worm's body leads to the formation of planarians with two heads and sometimes an unusual body shape [299].

It is especially remarkable that if both heads are cut off from the resulting two-headed worm, then a two-headed worm is again formed from the remaining middle fragment. And this procedure can be repeated many times, each time receiving the regeneration of a two-headed worm. This means that the memory of the correct structure of the body, which should be formed after regeneration, turns out to be radically changed, although the genome of this organism did not change. Therefore, the memory that tells the worm about how many heads it should have is not contained in the genome at all!

These experimental facts have led Levin to the important idea that emergent phenomena of an ensemble of cells are key participants of morphological and some other processes. More precisely, the association of cells into a multicellular ensemble leads to the manifestation of emergent properties in the emerging supracellular system: such multicellular ensembles can determine the type of morphological patterns formed, largely independently of the information in DNA, which is essential for the coding of proteins. This new point of view, justified by Levin, is very different from the widespread idea that the information sequences of DNA and RNA determine everything in a developing organism. It is quite possible that each higher level of combination of cells and supracellular structures into ensembles exhibits its emergent properties; that is, the development and behavior of complex multicellular organisms are determined by a multi-level system of emergent properties.

Considering the mentioned experimental results, Levin puts forward the concept that "*the electrical blueprints orchestrate life*" [300]. More precisely, he believes that all cells in our body communicate with each other using electrical signals. Corresponding electrical networks of cells (much like networks on the brain) are formed, which process information including pattern memories in the same way that brains hold other kinds of memory and learning. Besides this, Levin writes that these electrical signals have a bioelectrical code to control growth and form. A feature article on the bioelectric code highlights the importance of cracking it: "*We review emerging progress in reading and re-writing anatomical information encoded in bioelectrical states... Cracking the bioelectric code will enable much-improved control over biological patterning, advancing basic evolutionary developmental biology as well as enabling numerous applications in regenerative medicine and synthetic bioengineering*" [293].

The idea that cell systems combined into an ensemble have emergent properties led Levin to fruitful experiments on combining genetically unchanged cells into an ensemble. These experiments showed that from the union of such cells, each of which has its evolutionary history, organisms that have never existed before can arise. For example, Levin showed that the combination of cells from the skin of a frog into a single ensemble leads to the appearance of a tiny body of a completely new design. Such bodies, which are biological robots called xenobots, can move, navigate a labyrinth, explore the environment, feed and heal themselves, and exhibit emergent group behavior, although they do not have a nervous system, do not have a brain, and are simply collections of skin cells. These xenobots can live 10–14 days [300].

Our doctrine of energy-informational evolution based on bio-antenna arrays is also based on the idea of the emergence of system properties: the functional capabilities of bio-antenna arrays that combine many antenna elements radically exceed the capabilities of individual antenna elements. But this doctrine is broader than Levin's concept that "the electrical blueprints orchestrate life" and differs significantly from it. These differences are characterized by the following main statements of our doctrine which may be, in particular, useful for understanding the results of the experiments of Levin and his colleagues:

1. In the biological connection between the parts of each bio-body, an important role is played by electromagnetic waves of bio-antenna arrays (in addition to bioelectric communication based on distributed electrical potentials and the exchange of electrical signals). Thus, the participants in this connection are also the magnetic components of these waves and not just electrical potentials;
2. Electromagnetic waves of bio-antenna arrays cause mechanical vibrations in a variety of piezoelectric biological media; these vibrations can generate a whole class of additional mechanical phenomena in bio-bodies,

sometimes striking in their unusualness and unexpectedness, as noted above;

3. The emergent property of bio-antenna arrays to operate with narrowly directed beams of electromagnetic waves (including operating simultaneously with many independent beams having a different frequency, phase, polarization, and other characteristics) provides targeted and selective connections between many different elements of the body, including remote from each other;
4. The emergent property of phased antenna arrays to change the direction and intensity of rays of produced and received wave radiation due to a change in the phase relationships in the operation of their individual antennas (without moving the antennas themselves) gives a new approach to the theme of cyclic processes in biological bodies: corresponding changes in the phase ratios of the elements of bio-antenna arrays make it possible to rebuild the spatial-temporal characteristics of wave energy rays, cyclically redirecting and redistributing internal energy flows in space and time to provide cyclic activations and deactivations of biological components;
5. The types of polarization of electromagnetic waves of bio-antenna arrays must be taken into account when considering the relationships between the elements of the bio-body since they are associated with the problems of biological dissymmetry and molecular chirality, which have been considered in science since the time of Pasteur (many organisms are known to have polarization vision and can distinguish between types of polarized light);
6. The unique ability of antenna arrays to provide noise-resistant multi-channel operation with the extraction of weak signals against the background of strong noise allows us to rethink the phenomenal ability of organisms to work with weak information signals against the background of strong noise in many parallel channels to provide information noise immunity. According to Mendel's law of independent inheritance of traits, information from the level of DNA molecules dictates the macrostructure of living bodies through many independent channels, despite strong

noises. For example, hair, eye, and skin colors are inherited independently of each other. Accordingly, each organism is a machine of multi-channel noise immunity encoding;
7. Electromagnetic waves from bio-nanoantenna arrays are photon fluxes, which in general are the subject of quantum mechanics and quantum informatics (in contrast to the problem of the distribution of electrical potentials on surfaces);
8. Memory and intellectual abilities are connected with systems of bio-antenna arrays and their wave-coordinated activity;
9. Electromagnetic waves of bio-antenna arrays are involved in the operational transfer and redistribution of energy between the elements of the body.

The algebraic analogies between the matrix-tensor representations of the universal rules of the stochastic organization of genomic DNAs and the tensor-matrix theory of digital antenna arrays, also based on the Hadamard product for matrices, additionally lead to the following new thoughts:

- Stochastic organization of information sequences of genomic DNAs is a special case of biological self-organization based on the wave activity of bio-antenna arrays, examples of which are presented at all levels of the body;
- The genetic code itself is one of the offspring of wave activity and self-organization of bio-antenna arrays, and this code interrelates to other inherited physiological offspring of bio-antenna arrays in the organism. The secret of the structural organization and origin of the genetic code, as well as the origin of organisms, must be sought not in the random combination of the molecular elements of the genetic code, but in the emergent properties of self-organizing systems of bio-antenna arrays with their wave energy activity.

Accordingly, it is not genes that are the dictators of all life activity, since they themselves are built into the information-energy coherences of bio-antenna arrays and are produced by them. Dialects of the genetic codes arise in the course of the evolution of bio-bodies, when new living conditions appear and, if necessary, the dynamic systems of bio-antenna arrays associated with quantum mechanics are corrected. The genomes themselves and the genetic system are an integral part of the wave activity of ensembles of bio-antenna arrays and can be considered biological smart-antennas.

Previously, the following short definition of life was known: "*Life is a partnership between genes and mathematics*" [6]. But in the light of the above facts, a new short definition is proposed: "*Life is a partnership of organisms with bio-antenna arrays and their wave energy*" [233]. Of course, the role of mathematics in revealing the content of this definition, which is implicitly related to the algebraic theories of smart antennas, resonances, quantum mechanics, and quantum informatics is exceptionally great.

6.4. Some concluding remarks

The scientific community has long discussed the relationship between innate knowledge and knowledge acquired during life. The extreme point of view is formulated in Plato's famous statement that to know means to remember, awakening, as it were, from sleep. In particular, Kepler was a supporter of this point of view; it is described in detail in the article by the Nobel Laureate in Physics W. Pauli [301]. Close to this is the widespread opinion that our body, with its nervous system and everything else, already carries in hidden forms the fullness of knowledge, whose parts come into our consciousness when they are insistently requested. But our body grows from a single fertilized cell carrying a genome with DNA and RNA molecules. It is obvious that this original cell, together with its genome, cannot contain all the named completeness of knowledge. But then additional knowledge should come gradually in the course of ontogeny to the developing multicellular organism from the outside. In this case, it is the bio-antenna arrays that can serve as the system through which cosmic and planetary wave energy-informational influences enter the body from the outside world; these external influences are necessary for the adequate development and replenishment of the body's knowledge.

From the point of view of the stated doctrine, biological evolution can be formally considered as the evolution of digital bio-antenna arrays to increase the efficiency of communications between the parts of the body and the outside world. At the same time, communication orientation appears in evolution. The factor of increasing the possibilities of such communication can be considered a system-forming factor in biology according to the theory of P. K. Anokhin [302]. It is possible to develop the concept of "digital antenna biology and biological evolution" with access to a more general concept of the "antenna universe", which understands the world as a system of digital antennas distributed randomly and non-randomly.

Antenna arrays can be made from different chemical materials but perform similar functions. The radiations of antenna arrays can be the same even if the antennas are made of different materials. It is just as the same codes can work with transceivers made of different materials or as the same theory of resonances is applied to different antenna systems.

The topic of antenna arrays turns out to be cross-cutting for all levels of biological organization — from the genomes of higher and lower organisms to inherited sensory and other life support systems, including those associated with intellectual activity. Because of this, the presented doctrine is useful as a unifying platform for considering biological phenomena of different levels from a unified point of view, the development of algebraic and quantum biology, as well as for applied work on organ regeneration, medical diagnostics, artificial intelligence, etc.

In relation to biology, the ideology of antenna arrays draws attention to the fact that the collective use of an electromagnetic (or other) field by bio-antenna arrays can give very special properties of communication and impact on the surrounding world which, in principle, cannot be given by the use of the same field by one element. The properties of biophenomena are the emergent properties of the collective (coherent) use of physical fields (electromagnetic, sound, etc.) by a system of many elements. In contrast, in biology they sometimes try to explain biological phenomena by the action of unknown fields, but do not discuss that these phenomena can be caused by known physical fields generated and perceived by bio-antenna arrays with amazing efficiency, unattainable for a single antenna.

At present, humanity is intensively encircling the Earth with technical antenna arrays on land, underwater, in the air, and space. From the point of view of the stated doctrine of bio-antenna arrays, this activity can be considered as a system modification of the Earth according to the bio-antenna arrays system image and likeness of a living organism.

Chapter 7 Genetic Code Systems and Quantum Biology

One of the creators of quantum mechanics, P. Jordan, credited with authoring the first work on quantum biology, postulated the following. The mechanisms of living organisms are associated with what he referred to as his 'amplifier theory', based on Bohr's notion of the 'irreversible act of amplification', required to bring the fuzzy quantum reality into sharp focus by 'observing' it [19]. Jordan claimed that "*life's missing laws were the rules of chance and probability (the indeterminism) of the quantum world that were somehow scaled up inside living organisms*" [199]. It is these laws of chance and probability, postulated by Jordan, that we are looking for in the probabilistic characteristics of long DNA sequences of hydrogen bonds and nitrogenous bases [80, 153–155, 238]. Long DNA sequences are very important and convenient for such computerized research of possible biological rules of probabilities, which in Jordan's day was impossible. The main idea of all quantum mechanics is that everything is described only as probabilistic. Correspondingly the author believes that the described study systems of probabilities in structural organization of long DNA sequences is important.

Many authors have long suggested the quantum-informational nature of living bodies. For example, R. Penrose, treating the body as a quantum computer, appeals to tubulin proteins that can switch from one state to another by analogy with triggers [303]. The articles [304–306] suppose that the genetic code is related to the quantum Grover's algorithm. One can show that the analysis of living bodies at a deeper — genomic — level leads to more convincing and fruitful pieces of evidence about organisms as quantum information entities (or as quantum-like entities whose modeling can use the formalisms of quantum informatics).

This chapter describes the theme of the affinity between the doctrine of bio-antenna evolution, presented in Chapter 6, and the formalisms of quantum informatics and quantum mechanics. Briefly speaking, the doctrine links together quantum informatics and antenna array theory based on genetic and physiological phenomena that testify to the expediency of considering such a connection.

Many authors have supposed that living organisms use principles of quantum informatics, for example, [36, 42, 116, 133, 150, 242, 289, 303–306, 307–320]. The below-presented results support the thoughts of many authors about quantum computing in biological organisms.

7.1. DNA alphabets, 2*n*-qubit systems, and quantum-like entanglement

As shown in Chapter 1, the set of DNA alphabets of 4 nucleotides, 16 duplets, 64 triplets, and, in general, 4^n n-plets is presented in the form of a single tensor family of matrices $[C, A; T, G]^{(n)}$ where (n) is the tensor power (Figs. 1.2.1 and 1.4.2). The tensor product gives a way of putting separate vector spaces together to form larger vector spaces and it is one of the basic instruments in quantum informatics: "*This construction is crucial to understanding the quantum mechanics of multiparticle systems*" [15]. Thus, DNA informatics through tensor interrelations in the system of DNA alphabets is connected with the formalisms of quantum informatics already at the level of the DNA alphabets.

It is presented below that the quantitative characteristics of the stochastic organization of genomic DNAs (see examples in Figs. 6.1.1–6.1.9 in Chapter 6) show its connection with quantum informatics and the notion of "qubits" and, therefore, also the connection of this evolutionary doctrine with quantum informatics. In short, this doctrine, which includes the topic of nano-antenna arrays, includes also formalisms of quantum informatics connected with qubits, tensor products, Hilbert spaces, etc. To our best knowledge, quantitative relationships related to qubits are absent in the engineering systems of antenna arrays, and therefore they are a distinctive feature of this biological doctrine (correspondingly bio-antenna arrays should be considered as a very particular case of all sets of possible antenna arrays).

Besides, the above-noted operation of the Hadamard product of matrices, which made it possible to algebraically express the phenomenological relationship of genomic probability matrices (Figs. 6.1.1 and 6.1.5), is also used in quantum informatics; this fact allows us to take a fresh look at the stochastic organization of genomic DNAs and this evolutionary doctrine in general from a quantum-like point of view. Let's explain this in more detail.

In this chapter, we will rely on the terminology of the classic book [15] and use the generally accepted Dirac notations:

— $|\psi>$ means a vector, which is known also as a ket-vector;
— $<\psi|$ means a vector, which is dual to $|\psi>$ and known as a bra-vector;
— $<\varphi|\psi>$ means scalar (or inner) product between the vectors $<\varphi|$ and $|\psi>$;
— $|\varphi> \otimes |\psi>$ means the tensor product of $|\varphi>$ and $|\psi>$.

In quantum informatics, those vector spaces H are considered, which are equipped with the scalar (or inner) product (so-called Hilbert spaces). Let H_1 and H_2 be quantum mechanical state spaces, that is, finite-dimensional Hilbert spaces with orthonormal basis states $|\alpha_i>$ and $|\beta_j>$, where $i = 1, \ldots, n$ and $j = 1, \ldots, m$. By a postulate of quantum mechanics, the state space of the composite system is given by the tensor product $H_1 \otimes H_2$ with basis states $\{|\alpha_i> \otimes |\beta_j>\}$, or in more compact notation, $\{|\alpha_i\beta_j>\}$. "*The state space of a composite physical system is the tensor product of the state spaces of the component physical systems*" [15, p. 10]. If a quantum state can be represented as a vector of a Hilbert space, such a state is called a pure quantum state. If a pure state $|\psi> \in H_1 \otimes H_2$ can be written in the form $|\psi> = |\psi_1> \otimes |\psi_2>$, where $|\psi_i>$ is a pure state of the ith subsystem, it is said to be separable. Otherwise, it is called entangled or non-separable [15, p. 96].

As is known, a quantum bit (or qubit) is a unit of quantum information. For two-level quantum systems used as qubits, the state $|0>$ is identified with the vector [1, 0], and the state $|1>$ with the orthogonal vector [0, 1].

Two possible states for a qubit are the states $|0>$ and $|1>$, which correspond to the states of 0 and 1 for a classical bit. The difference between bits and qubits is that a qubit can be in a state other than $|0>$ or $|1>$. It is possible to form linear combinations of states, often called superpositions (7.1):

$$|\psi> = \alpha|0> + \beta|1> \tag{7.1}$$

The symbol $|\psi>$ means a state of a qubit. The numbers α and β can be complex numbers but in our case, it is enough to think of them as real numbers. Put another way, the state of a qubit is a vector in a two-dimensional vector space. The standard notation for states in quantum mechanics is the Dirac notation "$|\ >$". The special states $|0>$ and $|1>$ are known as computational basis states and form an orthonormal basis for this vector space [15, p. 13]. As is known, we cannot examine a qubit to determine its quantum state, that is, the values of α and β. Instead, quantum mechanics tells us that we can only acquire much more restricted information about the quantum state. When we measure a qubit, we get either the result 0, with probability $|\alpha|^2$, or the result 1, with probability $|\beta|^2$. Naturally, $|\alpha|^2 + |\beta|^2 = 1$, since the probabilities must sum to one. Geometrically, we can interpret this as the condition that the qubit's state is normalized to length 1. Values α and β are called amplitudes of probabilities. Thus, in general, a qubit's state is a unit vector in a two-dimensional complex vector space. Let us emphasize again that when a qubit is measured, it only ever gives "0" or "1" as the measurement result — probabilistically.

In the more general case, a system of "*n*" qubits is considered in quantum informatics. The computational basis states of this system are written in the form $|x_1x_2...x_n>$; a quantum state of such a system is specified by 2^n amplitudes [15, p. 17]. In our model approach, we in general interpret each of the considered *n*-plet texts of genomic DNAs as a multi-qubit quantum system.

In technical devices of quantum informatics, a qubit can be represented in many ways based on different pairs of binary-oppositional indicators: for example, by two electronic levels of an atom; by two kinds of polarization of a single photon (vertical polarization and horizontal polarization), etc. In our model approach for genetic informatics, we represent qubits based on different pairs of binary-oppositional indicators of adenine A, guanine G, cytosine C, and thymine T, which were shown in Table 1.2.1. As we noted, each of these DNA bases can be uniquely defined by two kinds of mentioned binary-oppositional indicators (see § 1.2 of Chapter 1). More precisely, to characterize each of the DNA letters C, A, T, G we will use the binary-oppositional indicators "amino-or-keto" (C = A = 0, T = G = 1) and the binary-oppositional indicators "pyrimidine-or-purine" (C = T = 0, A = G = 1). Based on each of these pairs of binary-oppositional indicators, two options of qubits can be considered under a model representation of stochastic genomic DNA sequences as quantum-like systems (those qubits, which are introduced in genetic informatics based on binary-oppositional indicators of genetic molecules, can be conditionally called «genetic qubits» or briefly «g-qubits»).

Let us introduce, firstly, the notion of a genetic qubit as a two-level quantum-like system, which represents a genomic DNA based on the indicators «amino-or-keto»: in this quantum system, one level corresponds to the indicator «amino» and the second level to the oppositional indicator «keto». In other words, such a genetic qubit is represented by these oppositional indicators, and the state of such a qubit is a vector in its appropriate two-dimensional Hilbert space H_1. One can assume that the state $|0>$ corresponds to the state «amino», and the state $|1>$ to the state «keto» (we mark these states in red to emphasize their relations to the red binary numeration of rows of matrices in Fig. 1.2.1). By analogy with the expression (7.1), a state of such genetic qubit can be shown by the expression (7.2), where α_0 and α_1 are amplitudes of probabilities of these computational basis states "amino" and "keto":

$$|\psi_1> = \alpha_0\,|0> + \alpha_1\,|1>, \qquad \alpha_0^2 + \alpha_1^2 = 1 \tag{7.2}$$

The qubit (7.2) corresponds to a pure state of a quantum system in the form of a DNA stochastic sequence of amino and keto molecules.

Secondly, let us introduce the notion of another genetic qubit as a two-level quantum system, which represents a genomic DNA based on the indicators «pyrimidine-or-purine»: in this quantum system, one level corresponds to the indicator «pyrimidine» and the second level to the indicator «purine». In other words, this genetic qubit is represented by these two indicators, and the state of such qubit is a vector in its appropriate two-dimensional Hilbert space H_2. One can assume that the state $|0>$ corresponds to the state «pyrimidine», and the state $|1>$ to the state «purine» (we mark these states in blue to emphasize their relations to the blue binary numeration of columns of matrices in Fig. 1.2.1). By analogy with the expression (7.1), a state of such a genetic qubit can be expressed by the expression (7.3), where β_0 and β_1 are amplitudes of probabilities of these computational basis states:

$$|\psi_2> = \beta_0\,|0> + \beta_1\,|1>, \qquad \beta_0^2 + \beta_1^2 = 1 \tag{7.3}$$

The qubit (7.3) corresponds to a pure state of a quantum system in the form of a sequence which consists of pyrimidines and purines.

So, we have two different two-dimensional Hilbert spaces H_1 and H_2 to which pure states of genetic qubits (7.2) and (7.3) belong correspondingly. In our genetic case, the tensor product of the two-dimensional Hilbert space gives one four-dimensional Hilbert space $H_{12} = H_1 \otimes H_2$, whose vectors correspond to appropriate 2-qubit systems. Its four computational basis states are expressed by the tensor product of the computational basis states of its subspaces H_1 and H_2 and have the following forms, corresponding to 4 nitrogenous bases C, A, T, and G:

— $|00>$ corresponds to C defined by the pair of indicators "amino and pyrimidine";
— $|01>$ corresponds to A defined by the pair of indicators "amino and purine";

— |10> corresponds to T defined by the pair of indicators "keto and pyrimidine";

— |11> corresponds to G defined by the pair of indicators "keto and purine".

These four computational basis states |00>, |01>, |10>, |11> symbolize 4 orthonormal basis vectors [1, 0, 0, 0], [0, 1, 0, 0], [0, 0, 1, 0], [0, 0, 0, 1]. The computational basis states can be also denoted as |C>, |A>, |T>, |G>, and correspondingly their amplitudes of probabilities can be denoted γ_C, γ_A, γ_T, γ_G in general.

In general, corresponding 2-qubit systems are presented by the expression (7.4):

$$|\psi_{12}> = \gamma_{00}\,|00> + \gamma_{01}\,|01> + \gamma_{10}\,|10> + \gamma_{11}\,|11> \tag{7.4}$$

where the normalization condition should be fulfilled:

$$(\gamma_{00})^2 + (\gamma_{01})^2 + (\gamma_{10})^2 + (\gamma_{11})^2 = 1 \tag{7.5}$$

There are two following options for the interrelation between amplitudes of probabilities γ_{00}, γ_{01}, γ_{10}, γ_{11} of the 2-qubit state $|\psi_{12}>$ and amplitudes of probabilities α_0, α_1, β_0, β_1 of the qubit's states $|\psi_1>$ and $|\psi_2>$.

The first option is if these amplitudes are interrelated by the equalities (7.6):

$$\gamma_{00} = \alpha_0\beta_0, \quad \gamma_{01} = \alpha_0\beta_1, \quad \gamma_{10} = \alpha_1\beta_0, \quad \gamma_{11} = \alpha_1\beta_1 \tag{7.6}$$

It is realized if the state $|\psi_{12}>$ can be expressed by the tensor product of the states $|\psi_1>$ and $|\psi_2>$:

$$\begin{aligned}|\psi_{12}> &= |\psi_1> \otimes |\psi_2>\\ &= (\alpha_0\,|0> + \alpha_1\,|1>) \otimes (\beta_0\,|0> + \beta_1\,|1>)\\ &= \alpha_0\beta_0\,|00> + \alpha_0\beta_1\,|01> + \alpha_1\beta_0\,|10> + \alpha_1\beta_1\,|11>\end{aligned} \tag{7.7}$$

In quantum informatics, such state $|\psi_{12}>$ is called a pure separable state. The sign of the tensor product here indicates that these subsystems are completely independent of each other.

The second option is if the state $|\psi_{12}>$ cannot be expressed by the tensor product of the states $|\psi_1>$ and $|\psi_2>$. In this case the state $|\psi_{12}>$ is called non-separable or entangled, and you have inequalities such as in (7.8):

$$|\psi_{12}> \neq |\psi_1> \otimes |\psi_2>, \quad \gamma_{00} \neq \alpha_0\beta_0, \quad \gamma_{01} \neq \alpha_0\beta_1, \quad \gamma_{10} \neq \alpha_1\beta_0, \quad \gamma_{11} \neq \alpha_1\beta_1 \quad (7.8)$$

Let's study which of these two options is realized by nature, for example, in the DNA of human chromosome № 1. In this DNA, the percentage α_0^2 of amino molecules (C and A) is equal to 0.499500, and the percentage α_1^2 of keto molecules (T and G) is equal to 0.500500 (rounding to the sixth decimal place). In other words, $\alpha_0^2 \approx \alpha_1^2$ by Chargaff's second parity rule, which states that %C ≈ %G and %A ≈ %T in long enough single-stranded DNAs (correspondingly you have %C + %A ≈ %T + %G).

In the considered DNA, the percentage β_0^2 of pyrimidines (C and T) is equal to 0.500255, and the percentage β_1^2 of purines (A and G) is equal to 0.499745 (rounding to the sixth decimal place). In other words, $\beta_0^2 \approx \beta_1^2$ by Chargaff's second parity rule as well. Using these phenomenological data, you have the following amplitudes of probabilities for computational basis states $|00>$, $|01>$, $|10>$, $|11>$ in a supposed case of the pure separable states (7.7):

$$\begin{aligned}
\alpha_0\beta_0 &= 0.499500^{0.5} \times 0.500255^{0.5} = 0.499877,\\
\alpha_0\beta_1 &= 0.499500^{0.5} \times 0.499745^{0.5} = 0.499622,\\
\alpha_1\beta_0 &= 0.500500^{0.5} \times 0.500255^{0.5} = 0.500377,\\
\alpha_1\beta_1 &= 0.500500^{0.5} \times 0.499745^{0.5} = 0.500122. \qquad (7.9)
\end{aligned}$$

Resuming, in the supposed case of pure separable states (7.7) of the considered chromosomal DNA, you should have approximate equalities of all amplitudes of probabilities:

$$\alpha_0\beta_0 \approx \alpha_0\beta_1 \approx \alpha_1\beta_0 \approx \alpha_1\beta_1 \qquad (7.10)$$

But in reality, the considered DNA shows the completely different situation of pronounced inequality in the amplitudes of the probabilities of the nitrogenous bases C, A, T, G. These probabilities were shown in

Fig. 4.1.1 of Chapter 4: %C = 0.208499, %A = 0.291001, %T = 0.291756, %G = 0.208744. One can show corresponding amplitudes of the probabilities using their denotations from (7.4):

$$\begin{aligned}\gamma_{00} &= (\%C)^{0.5} = 0.208499^{0.5} = 0.456617,\\ \gamma_{01} &= (\%A)^{0.5} = 0.291001^{0.5} = 0.539445,\\ \gamma_{10} &= (\%T)^{0.5} = 0.291756^{0.5} = 0.540144,\\ \gamma_{11} &= (\%G)^{0.5} = 0.208744^{0.5} = 0.456885. \end{aligned} \tag{7.11}$$

Amplitudes of probabilities in the abstract case (7.9) and in reality (7.11) are significantly different with respect to the criterion (7.8) of the second option about quantum entanglement. This fact is evidence that, in the human chromosomal DNA, its 2-qubit state $|\psi_{12}>$ is entangled. It means that the 2-qubit state $|\psi_{12}>$ of the composite quantum system cannot be divided into two separate, completely independent parts corresponding to two sequences of "amino- and keto-molecules" and "pyrimidine and purine molecules".

The result obtained on the quantum entanglement of the stochastic nucleotide sequence of the chromosomal DNA is important due to the importance of the concept of entanglement for quantum informatics. The classic book [15, p. xxiii] emphasizes: "*Entanglement is a key element in effects such as quantum teleportation, fast quantum algorithms, and quantum error-correction. It is, in short, a resource of great utility in quantum computation and quantum information. There is a thriving research community currently fleshing out the notion of entanglement as a new type of physical resource, finding principles which govern its manipulation and utilization*". Usually, in speeches about quantum entanglements, researchers focus their attention on the entanglement of atoms and other elementary particles or the entanglement of spins. But this section draws attention to a question about quantum entanglement (or quantum-like entanglement) in quite different objects: stochastic information sequences of genomic DNAs.

Similar results about the existence of such entanglement were obtained for many other genomic DNAs. These obtained results about the stochastic properties of genomic DNAs allow believing that molecular-genetic systems widely use quantum entanglement (or quantum-like entanglement) to provide their biological functions and that the topic of entanglement should be included in the future development of algebraic biology and the doctrine of evolution based on bio-antenna arrays.

But what about a criterion like (7.8) of quantum entanglement in presentations of genomic DNAs as sequences of duplets, triplets, tetraplets, etc.? It turns out that in these cases you again meet similar quantum-like entanglement but for $2n$-qubit quantum systems ($n = 2, 3, 4, \ldots$). In other words, these $2n$-qubit states of the composite quantum systems cannot be divided into separate, completely independent parts corresponding to two sequences of "amino- and keto-molecules" and "pyrimidine and purine molecules". It can be shown by analogy with the described case of the 1-textual presentations of single-stranded genomic DNA. Such quantum-like entanglements, realized in 1-, 2-, 3-, 4-textual representations of genomic DNAs sequences, have regular interrelations each with other, which are reflected in universal rules of stochastic organization of genomic DNAs and in a special family of algebraic equations using the Hadamard product (see their examples in Chapter 6, Figs. 6.1.1 and 6.1.5).

In addition, we briefly note the following. We have just considered genetic qubits (7.2) and (7.3) built on the binary-opposition pairs "amino-keto" (C+A vs. G+T) and "pyrimidines-purines" (T+C vs. A+G). But the molecular DNA alphabet of four nucleotides is arranged by nature in such a way that it is endowed with another pair of binary-oppositional features associated with the complementarity of nitrogenous bases: in the DNA double strand, the bases C and G are connected by 3 hydrogen bonds ("strong link"), and A and T by 2 hydrogen bonds ("weak link"). Accordingly, it is also possible to consider genetic qubits built on this pair of oppositions "strong link-weak link". This leads to the possibility of a similar analysis of the four-dimensional Hilbert space and 2-qubit systems arising on the following pairs of genetic qubits: 1) "amino-keto" and "strong bond-weak bond"; 2) "pyrimidine-purine" and "strong bond-weak bond". It turns out that in these 2-qubit representations of stochastic sequences of genomic DNAs, you also have appropriate quantum entanglements.

Now let's consider interrelations among the following $2n$-qubit states ($n = 1, 2, 3$), which are represented by sets of probability amplitudes for 4 monoplets, 16 duplets, and 64 triplets in the DNA sequence of human chromosome №1, whose probabilities are shown in Chapter 4, Figs. 4.1.1–4.1.3 (we denote these $2n$-qubit states by symbols $|\psi_4>$, $|\psi_{16}>$, and $|\psi_{64}>$ correspondingly; values are rounded to 4 decimal places):

$$|\psi_4> = 0.2085^{0.5}\,|00> + 0.2910^{0.5}\,|01> + 0.2918^{0.5}\,|10> + 0.2087^{0.5}\,|11>, \tag{7.12}$$

$$\begin{aligned}|\psi_{16}> = {} & 0.0541^{0.5}\,|0000> + 0.0727^{0.5}\,|0001> + 0.0503^{0.5}\,|0010> + 0.0950^{0.5} \\ & |0011> + 0.0713^{0.5}\,|0100> + 0.0103^{0.5}\,|0101> + 0.0743^{0.5}\,|0110> \\ & + 0.0714^{0.5}\,|0111> + 0.0601^{0.5}\,|1000> + 0.0631^{0.5}\,|1001> + 0.0440^{0.5} \\ & |1010> + 0.0601^{0.5}\,|1011> + 0.0957^{0.5}\,|1100> + 0.0729^{0.5}\,|1101> \\ & + 0.0505^{0.5}\,|1110> + 0.0542^{0.5}\,|1111>\end{aligned} \tag{7.13}$$

$$\begin{aligned}|\psi_{64}> = {} & 0.0138^{0.5}|000000> + 0.0188^{0.5}|000001> + 0.0152^{0.5}|000010> \\ & + 0.0186^{0.5}|000011> + 0.0118^{0.5}\,|000100> + 0.0198^{0.5}\,|000101> \\ & + 0.0145^{0.5}\,|000110> + 0.0369^{0.5}\,|000111> + 0.0185^{0.5}\,|001000> \\ & + 0.0029^{0.5}\,|001001> + 0.0179^{0.5}\,|001010> + 0.0210^{0.5}\,|001011> \\ & + 0.0162^{0.5}\,|001100> + 0.0025^{0.5}|001101> + 0.0238^{0.5}|001110> \\ & + 0.0199^{0.5}\,|001111> + 0.0176^{0.5}|010000> + 0.0127^{0.5}|010001> \\ & + 0.0025^{0.5}|010010> + 0.0023^{0.5}|010011> + 0.0132^{0.5}|010100> \\ & + 0.0194^{0.5}|010101> + 0.0144^{0.5}|010110> + 0.0224^{0.5}|010111> \\ & + 0.0201^{0.5}|011000> + 0.0209^{0.5}|011001> + 0.0026^{0.5}|011010> \\ & + 0.0029^{0.5}|011011> + 0.0239^{0.5}|011100> + 0.0178^{0.5}|011101> \\ & + 0.0161^{0.5}|011110> + 0.0185^{0.5}|011111> + 0.0159^{0.5}|100000> \\ & + 0.0196^{0.5}|100001> + 0.0110^{0.5}|100010> + 0.0199^{0.5}|100011>\end{aligned}$$

$$
\begin{aligned}
&+ 0.0125^{0.5}|100100> + 0.0146^{0.5}|100101> + 0.0096^{0.5}|100110> \\
&+ 0.0196^{0.5}|100111> + 0.0223^{0.5}|101000> + 0.0023^{0.5}|101001> \\
&+ 0.0194^{0.5}|101010> + 0.0128^{0.5}|101011> + 0.0144^{0.5}|101100> \\
&+ 0.0025^{0.5}|101101> + 0.0133^{0.5}|101110> + 0.0176^{0.5}|101111> \\
&+ 0.0197^{0.5}|110000> + 0.0198^{0.5}|110001> + 0.0146^{0.5}|110010> \\
&+ 0.0195^{0.5}|110011> + 0.0096^{0.5}|110100> + 0.0112^{0.5}|110101> \\
&+ 0.0126^{0.5}|110110> + 0.0160^{0.5}|110111> + 0.0372^{0.5}|111000> \\
&+ 0.0188^{0.5}|111001> + 0.0199^{0.5}|111010> + 0.0190^{0.5}|111011> \\
&+ 0.0145^{0.5}\,|111100> + 0.0153^{0.5}|111101> + 0.0119^{0.5}|111110> \\
&+ 0.0138^{0.5}\,|111111> \qquad (7.14)
\end{aligned}
$$

One can calculate that the following inequalities hold for these $2n$-qubit states (7.12)–(7.14):

$$|\psi_4> \otimes |\psi_4> \neq |\psi_{16}>; \qquad (7.15)$$

$$|\psi_4> \otimes |\psi_{16}> \neq |\psi_{64}>; \qquad (7.16)$$

$$|\psi_4> \otimes |\psi_4> \otimes |\psi_4> \neq |\psi_{64}>. \qquad (7.17)$$

The inequality (7.15) testifies that the 4-qubit state $|\psi_{16}>$ cannot be expressed by the tensor product $|\psi_4> \otimes |\psi_4>$. The inequality (7.16) testifies that the 6-qubit state $|\psi_{64}>$ cannot be expressed by the tensor product $|\psi_4> \otimes |\psi_4> \otimes |\psi_4>$. The inequality (7.17) testifies that the 6-qubit state $|\psi_{64}>$ cannot be expressed by the tensor product $|\psi_4> \otimes |\psi_{16}>$.

Similar algebraic numerical data hold for different genomic DNAs. These results give materials for a possible consideration of genomic DNAs as interconnected quantum-like systems having multi-level networks of quantum-like entangled states.

7.2. The Hadamard product, nonlinear transformations in quantum computation, and the stochastics of genomic DNAs

Quantum mechanics is fundamentally linear. Quantum computers are naturally well suited to implementing linear operations. One of the important tasks for developing quantum computing and its applications is implementing nonlinear transformations of quantum states. This problem is considered in a series of modern works by different authors, for example, [321, 322]. These works noted that nonlinear subroutines may prove key to a range of applications of quantum computing from solving nonlinear equations to data processing and quantum machine learning.

We showed above that the universal rules of stochastic organization of genomic DNAs are algebraically expressed by using the Hadamard product, which is used in the tensor-matrix theory of digital antenna arrays (Figs. 6.1.1 and 6.1.5). But the Hadamard product is also used in quantum computing tasks to create algorithms implementing basic nonlinear transformations of quantum states. The work [321] introduces a notion of the "quantum Hadamard product", which is used to generate new sets of quantum states.

Let us return to the unexpected arising of the Hadamard product in algebraic expressions of universal interconnections between genetic probability matrices, whose sums of entries are equal to 1 (see examples in Figs. 6.1.1 and 6.1.5 in Chapter 6). Note formal connections of these matrices with quantum $2n$-qubits. For example, Fig. 6.1.1 shows the (2×2)-matrix [%C, %A; %T, %G] of probabilities of 4 nucleotides in the 1-textual presentation of genomic DNAs; 4 entries of this matrix can be interpreted as squares of probability amplitudes of the following 2-qubit system since their sum is equal to 1:

$$|\psi_{2q}> = (\%C)^{0.5}\,|00> + (\%A)^{0.5}\,|01> + (\%T)^{0.5}\,|10> + (\%G)^{0.5}\,|11> \quad (7.18)$$

where %C + %A + %T + %G = 1.

Also Fig. 6.1.1 shows the (4 × 4)-matrix of probabilities of 16 duplets in the 2-textual presentation of genomic DNAs:

[%CC, %CA, %AC, %AA; %CT, %CG, %AT, %AG;
%TC, %TA, %GC, %GA; %TT, %TG, %GT, %GG],

whose 16 entries can be interpreted as squares of probability amplitudes of the following 4-qubit system since their sum is equal to 1:

$$\begin{aligned}|\psi_{4q}\rangle = {} & (\%CC)^{0.5}\,|0000\rangle + (\%CA)^{0.5}\,|0001\rangle + (\%AC)^{0.5}|0010\rangle + (\%AA)^{0.5} \\ & |0011\rangle + (\%CT)^{0.5}\,|0100\rangle + (\%CG)^{0.5}\,|0101\rangle + (\%AT)^{0.5}\,|0110\rangle \\ & + (\%AG)^{0.5}\,|0111\rangle + (\%TC)^{0.5}\,|1000\rangle + (\%TA)^{0.5}\,|1001\rangle \\ & + (\%GC)^{0.5}\,|1010\rangle + (\%GA)^{0.5}\,|1011\rangle + (\%TT)^{0.5}\,|1100\rangle \\ & + (\%TG)^{0.5}\,|1101\rangle + (\%GT)^{0.5}\,|1110\rangle + (\%GC)^{0.5}\,|1111\rangle,\end{aligned}$$

where

$$\begin{aligned}&\%CC + \%CA + \%AC + \%AA + \%CT + \%CG + \%AT + \%AG + \%TC + \%TA \\ &+ \%GC + \%GA + \%TT + \%TG + \%GT + \%GG = 1. \qquad (7.19)\end{aligned}$$

By analogy, the (8 × 8)-matrix (Fig. 6.1.5) of probabilities of 64 triplets in the 3-textual presentation of genomic DNAs can be interpreted as a matrix representation of the 6-qubit system.

In any of the tested genomic DNAs, these 2-qubit, 4-qubit, 6-qubit states turn out to be algebraically connected by using the Hadamard product of matrices (Figs. 6.1.1 and 6.1.5). Correspondingly, the genetic (2 × 2)-matrix of probabilities, representing a 2-qubit state, is nonlinearly transformed into the genetic (4 × 4)-matrix of probabilities, representing an appropriate 4-qubit state. So, the described analysis of universal rules of stochastic organization of genomic DNAs reveals the unexpected way of nonlinear transformations of quantum states, which — as far as we can judge — is unknown in works on quantum computing but fundamental for living nature; this way connects quantum states of 2*n*-qubit systems in Hilbert spaces of different 2*n*-dimensions [235]. These nonlinear transformations are particular cases of the notion of tensor-unitary (or tensor-orthogonal) transformations described later in the book.

One can suppose that a discovery of this "genetic" way of such transformations of quantum states opens new perspectives for developing quantum computing in close connection with bio-information patents of living nature. One can also suppose that this way can lead, for example, to more biomorphic options of quantum machine learning.

7.3. Universal rules of stochastics of genomic DNAs and quantum informatics

The matrix equalities in Figs. 6.1.1 and 6.1.5 can be represented in equivalent linear forms based on combinations of genetic 2-qubit states. It is the phenomenological fact that the sum of all 4 entries in each of the mentioned tetra-multiplicating (2 × 2)-matrices (B_C, B_A, B_T, B_G, B_{CC}, B_{CA}, B_{CT}, B_{CG}, B_{AC}, B_{AA}, B_{AT}, B_{AG}, B_{TC}, B_{TA}, B_{TT}, B_{TG}, B_{GC}, B_{GA}, B_{GT}, B_{GG}) is equal to 1 with high accuracy as it was shown in Figs. 6.1.3 and 6.1.8 for the particular example of the DNA of human chromosome №1 (similar equalities hold for other tested genomic DNAs). Correspondingly, each of the noted genetic (2 × 2)-matrices can be interpreted as a matrix representation of an appropriate 2-qubit system. One can explain this by the following examples for the numeric matrices from Figs. 6.1.3 and 6.1.8 and equations (6.2)–(6.5) from Chapter 6 (numeric values are shown below for the DNA of human chromosome №1, rounded to 4 decimal places):

$$B_C = [\%CC / \%C, \%CA / \%C; \%CT / \%C, \%CG / \%C]$$
$$= [0.2594, 0.3489; 0.3422, 0.0494], \text{ which represents the following 2-qubit state:}$$

$$|\psi_{BC}> = (\%CC / \%C)^{0.5}\,|00> + (\%CA / (\%C)^{0.5}\,|01> + (\%CT / \%C)^{0.5}\,|10> + (\%CG / \%C)^{0.5}\,|11>$$
$$= 0.2594^{0.5}\,|00> + 0.3489^{0.5}\,|01> + 0.3422^{0.5}\,|10> + 0.0494^{0.5}\,|11>; \quad (7.20)$$

$$B_A = [\%AC / \%A, \%AA / \%A; \%AT / \%A, \%AG / \%A] =$$
$$= [0.1729, 0.3266; 0.2553, 0.2452]$$, which represents the following 2-qubit state:

$$|\psi_{BA}> = (\%AC / \%A)^{0.5} |00> + (\%AA / (\%A)^{0.5} |01> + (\%AT / \%A)^{0.5} |10> + (\%AG / \%A)^{0.5} |11>$$
$$= 0.1729^{0.5} |00> + 0.3266^{0.5} |01> + 0.2553^{0.5} |10> + 0.2452^{0.5} |11>; \quad (7.21)$$

$$B_T = [\%TC / \%T, \%TA / \%T; \%TT / \%T, \%TG / \%T]$$
$$= [0.2059, 0.2164; 0.3279, 0.2497]$$, which represents the following 2-qubit state:

$$|\psi_{BT}> = (\%TC / \%T)^{0.5}|00> + (\%TA / (\%T)^{0.5}|01> + (\%TT / \%T)^{0.5}|10> + (\%TG / \%T)^{0.5} |11>$$
$$= 0.2059^{0.5} |00> + 0.2164^{0.5} |01> + 0.3279^{0.5} |10> + 0.2497^{0.5} |11>; \quad (7.22)$$

$$B_G = [\%GC / \%G, \%GA / \%G; \%GT / \%G, \%GG / \%G]$$
$$= [0.2109, 0.2878; 0.2417, 0.2596]$$, which represents the following 2-qubit state:

$$|\psi_{BG}> = (\%GC / \%G)^{0.5}|00> + (\%GA / (\%G)^{0.5}|01> + (\%GT / \%G)^{0.5}|10> + (\%GG / \%G)^{0.5} |11>$$
$$= 0.2109^{0.5} |00> + 0.2878^{0.5} |01> + 0.2417^{0.5} |10> + 0.2596^{0.5} |11>; \quad (7.23)$$

After the explanation of representations of the tetra-multiplicating (2 × 2)-matrices B_C, B_A, B_T, B_G by the corresponding 2-qubits (7.20)–(7.23), similar representations can be done by analogy regarding the 16 tetra-multiplicating (2 × 2)-matrices B_{CC}, B_{CA}, B_{CT}, B_{CG}, B_{AC}, B_{AA}, B_{AT}, B_{AG}, B_{TC}, B_{TA}, B_{TT}, B_{TG}, B_{GC}, B_{GA}, B_{GT}, B_{GG} (from Figs. 6.1.6–6.1.8 in Chapter 6). We show such representation only for the matrix B_{TG} for brevity:

$$B_{TG} = [\%TGC / \%TG, \%TGA / \%TG; \%TGT / \%TG, \%TGG / \%TG]$$
$$= [0.2000, 0.2672; 0.2729, 0.2601]$$, which represents the following 2-qubit state:

$$|\psi_{BTG}> = (\%TGC / \%TG)^{0.5} |00> + (\%TGA / \%TG)^{0.5} |01> + (\%TGT / \%TG)^{0.5} |10> + (\%TGG / \%TG)^{0.5} |11>$$
$$= 0.2000^{0.5} |00> + 0.2672^{0.5} |01> + 0.2729^{0.5} |10> + 0.2601^{0.5} |11>. \quad (7.24)$$

The lengths of all corresponding four-dimensional vectors whose components are 4 probability amplitudes in such genetic 2-qubits are equal to 1, that is, they are equal to each other. It means that these vectors form a single class of unitarily equivalent vectors: each of these vectors can be transformed into another vector of this class by unitary transformations (or orthogonal transformations in the case of real components) since unitary transformations of vectors preserve their lengths. It is interesting since all calculations in quantum computers are based on precisely unitary transformations. Accordingly, the revealed multifaceted relationship between the features of the stochastic organization of genomic DNAs and unitary (orthogonal) transformations is another evidence of the deep conjugation of molecular genetics with the formalisms of quantum mechanics and quantum informatics.

The matrix of probabilities of nucleotides in 1-textual presentation of the considered DNA (Fig. 4.1.1 from Chapter 4) represents the following 2-qubit state $|\psi_{CATG}>$ as well:

$$[\%C, \%A; \%T, \%G] = [0.208499, 0.291001; 0.291756, 0.208744],$$
$$|\psi_{CATG}> = (\%C)^{0.5} |00> + (\%A)^{0.5} |01> + (\%T)^{0.5} |10> + (\%G)^{0.5} |11>$$
$$= 0.208499^{0.5} |00> + 0.291001^{0.5} |01> + 0.291756^{0.5} |10> + 0.208744^{0.5} |11> \quad (7.25)$$

Now let us express the matrix equality from Fig. 6.1.1 in Chapter 6 by an equivalent linear form based on combinations of the described genetic 2-qubits using the tensor product. Let's start with the 2-qubit $|\psi_{CATG}>$ (7.25) of the probability amplitudes of the 4 nucleotides in a 1-textual representation of a long single-stranded DNA:

$$|\psi_{CATG}> = (\%C)^{0.5}\ |00> + (\%A)^{0.5}\ |01> + (\%T)^{0.5}\ |10> + (\%G)^{0.5}\ |11> \quad (7.26)$$

Then tensor multiply each of the 4 terms of this 2-qubit $|\psi_{CATG}>$ by the corresponding 2-qubit from the family of four 2-qubits $|\psi_{BC}>$, $|\psi_{BA}>$, $|\psi_{BT}>$ and $|\psi_{BG}>$ (7.20)–(7.23) as shown in (7.27):

$$\begin{aligned}
&(\%C)^{0.5}\ |00>\otimes\ |\psi_{BC}> + (\%A)^{0.5}\ |01> \otimes\ |\psi_{BA}> + (\%T)^{0.5}\ |10> \otimes\ |\psi_{BT}> \\
&+ (\%G)^{0.5}\ |11> \otimes\ |\psi_{BG}> \\
&= (\%C)^{0.5}|00> \otimes \{(\%CC / \%C)^{0.5}|00> + (\%CA / (\%C)^{0.5}|01> + (\%CT / \\
&\quad \%C)^{0.5}|10> + (\%CG / \%C)^{0.5}\ |11>\} + (\%A)^{0.5}|01> \otimes \{(\%AC / \%A)^{0.5}|00> \\
&\quad + (\%AA / (\%A)^{0.5}|01> + (\%AT / \%A)^{0.5}|10> + (\%AG / \%A)^{0.5}\ |11>\} \\
&\quad + (\%T)^{0.5}|10> \otimes \{(\%TC / \%T)^{0.5}|00> + (\%TA / (\%T)^{0.5}|01> + (\%TT / \\
&\quad \%T)^{0.5}|10> + (\%TG / \%T)^{0.5}\ |11>\} + (\%G)^{0.5}|11> \otimes \{(\%GC / \%G)^{0.5}|00> \\
&\quad + (\%GA / (\%G)^{0.5}|01> + (\%GT / \%G)^{0.5}|10> + (\%GG / \%G)^{0.5}\ |11>\} \\
&= (\%CC^{0.5}\ |0000> + \%CA^{0.5}\ |0001> + \%CT^{0.5}\ |0010> + \%CG^{0.5}\ |0011>) \\
&\quad + (\%AC^{0.5}\ |0100> + \%AA^{0.5}\ |0101> + \%AT^{0.5}\ |0110> + \%AG^{0.5}\ |0111>) \\
&\quad + (\%TC^{0.5}\ |1000> + \%TA^{0.5}\ |1001> + \%TT^{0.5}\ |1010> + \%TG^{0.5}\ |1011>) \\
&\quad + (\%GC^{0.5}|1100> + \%GA^{0.5}\ |1101> + \%GT^{0.5}|1110> + \%GG^{0.5}\ |1111>)
\end{aligned} \quad (7.27)$$

Let us verify that the use of component-wise tensor multiplication by 2-qubits $|\psi_{BC}>$, $|\psi_{BA}>$, $|\psi_{BT}>$, and $|\psi_{BG}>$ in (7.27) automatically ensures that the phenomenological rules (the expression (6.6) in Chapter 6) hold for genomic DNAs. We show this execution for equality in the first line in (6.6): $\%C \approx \%CC + \%CA + \%CT + \%CG$. Using (7.27), corresponding to the 16-dimensional Hilbert space, we calculate in its four-dimensional subspace the square of the length of the vector corresponding to the expression $\%CC^{0.5}\ |0000> + \%CA^{0.5}\ |0001> + \%CT^{0.5}\ |0010> + \%CG^{0.5}\ |0011>$. This square of length L^2 is equal to the sum of the squares of the coordinates of the given vector:

$$\begin{aligned}
L^2 &= \%CC + \%CA + \%CT + \%CG \\
&= \%C \times \{(\%CC / \%C) + (\%CA / \%C) + (\%CT / \%C) \\
&\quad + (\%CG / \%C)\} \\
&= \%C
\end{aligned} \quad (7.28)$$

In expression (7.28), the sum of the terms in curly brackets is equal to 1, since it represents the sum of the squared probability amplitudes from the 2-qubit $|\psi_{BC}>$ (7.20). Hence, as a result of the described tensor operation, we obtain, in the language of formalisms of quantum informatics, the verifiable equality %C = %CC + %CA + %CT + %CG, which now represents the metric relation of equality for the lengths of the considered vectors in four-dimensional and 16-dimensional Hilbert spaces.

Thus, expressions (7.26) and (7.27), referring respectively to vectors in four-dimensional and 16-dimensional Hilbert spaces, show a numerical relationship between the sets of phenomenological probabilities of 4 monoplets and 16 duplets in the representation of genomic DNAs. We emphasize that in this case, we expressed the desired relationship between the two described sets of probabilities 4 monoplets and 16 duplets in genomic DNAs without using the Hadamard product operation for matrices, replacing it with a specific component-by-component tensor multiplication by 2-qubits. This makes it possible to model in the language of metric relations and quantum information formalisms the universal rules of the stochastic organization of genomic DNAs described earlier, examples of which are given by expressions (6.6) and (6.9) in Chapter 6 and which are reduced to algorithmic tensor combinations of 2-qubit systems.

The set of 16 computational basic states |0000>, |0001>, …, |1111> specified in (7.27) defines a (4 × 4)-table in which the rows are binary numbered by the first two digits of these expressions, and the columns are binary numbered by their last two digits, as shown in Fig. 7.3.1.

At the same time, the corresponding probability amplitudes or the probabilities of these computational basic states are uniquely entered into each cell of the table. The resulting table of probabilities for these 16 computational base states of a 4-qubit is structurally similar to the DNA alphabet matrix of 16 duplets and the probability matrix of 16 duplets shown in Figs. 4.1.1 and 4.1.2 in Chapter 4.

Similarly, using this 4-qubit containing the probability amplitudes of 16 duplets (7.27), and component-by-component tensor-multiplying its 16 computational base states by 2-qubit representations of the corresponding

	00	01	10	11
00	%CC	%CA	%AC	%AA
01	%CT	%CG	%AT	%AG
10	%TC	%TA	%GC	%GA
11	%TT	%TG	%GT	%GG

Fig. 7.3.1. The table of probabilities of 16 computational basis states of the 4-qubit system (7.27) corresponds to percentage values of 16 duplets in the 2-textual representation of any genomic DNA nucleotide sequence.

16 tetra-multiplicating (2 × 2)-matrices B_{CC}, B_{CA}, B_{CT}, B_{CG}, B_{AC}, B_{AA}, B_{AT}, B_{AG}, B_{TC}, B_{TA}, B_{TT}, B_{TG}, B_{GC}, B_{GA}, B_{GT}, B_{GG} (from Chapter 6, Figs. 6.1.6–6.1.8), you end up with a 64-qubit; raising to the second power the amplitudes of the probabilities of its 64 computational basis states, you get the phenomenological values of the probabilities of 64 triplets in a 3-text representation of genomic DNAs.

In the general case, the relationship between the probabilities of n-plets and $(n+1)$-plets in n-text and $(n+1)$-text representations of genomic DNAs is expressed for each concrete genomic DNA by the following general algorithm based on quantum information formalisms and generalizing expression (7.27):

— using a $2n$-qubit representing the probability amplitudes of n-plets in the genomic DNA, and applying the component-by-component tensor product of each of its 4^n computational basis states by 2-qubits corresponding to 4^n tetra-multiplicating (2×2)-matrices of probabilities, you get a $2(n+1)$-qubit representing the probability amplitudes of $(n+1)$-plets in that DNA (n = 1, 2, 3, 4, …, but much smaller than the DNA length).

These multiple uses of 2-qubits, which have four computational basis states, models the phenomenological tetra-structuring relations, according to which, in n-textual and $(n+1)$-textual representations of genomic DNAs, the probability of any n-plet is equal to the sum of the probabilities of four $(n+1)$-plets [25, 70, 82, 116] (here n = 1, 2, 3, 4, ... is relatively small). Examples of these tetra-relations are the following equalities, the fulfillment of which for the case of DNA of human chromosome №1 can be verified using the probability values from Figs. 4.1.1b–4.1.3 in Chapter 4:

$$\%C \approx \%CC + \%CA + \%CT + \%CG;$$
$$\%CA \approx \%CAC + \%CAA + \%CAT + \%CAG;$$
$$\%CAC \approx \%CACC + \%CACA + \%CACT + \%CACG; \text{ etc.} \quad (7.29)$$

Similar tetra-relations, penetrating the stochastic organization of genomic DNAs and associated with 2-qubits of quantum informatics, are conveniently visualized in graphic forms of multitiered mandalas, described below.

In the next chapter, we describe tensor-unitary operations, whose idea was originated from these phenomenological relationships (like the expression (7.27)) between the probabilities of n-plets and $(n+1)$-plets in n-text and $(n+1)$-text representations of genomic DNAs.

7.4. Mandalas and universals of the stochastic organization of genomic DNAs

Many authors have long drawn attention to the fact that the family of DNA alphabets, which contain 4 nucleotides, 4^2 duplets, 4^3 triplets, etc., is conveniently visualized in 4-section pie charts like the one shown in Fig. 7.4.1. Similar 4-section pie charts are called sometimes mandalas since they resemble ancient mandala images, which have been used for thousands of years. One can see examples of such tiered mandalas of the family of DNA alphabets on websites https://www.researchgate.net/figure/Genetic-code-Wikipedia_fig1_338865739, https://www.pinterest.com/pin/350647520964620594/, http://www.kozlenkoa.narod.ru/gencode1.htm, etc. In particular, tiered mandalas of DNA alphabets are connected with Yin-Yang schemes of the ancient Chinese book *I-Ching* (https://wiki.deldebbio.com.br/index.php?title=File:Knowledge-iching-mandala.jpg).

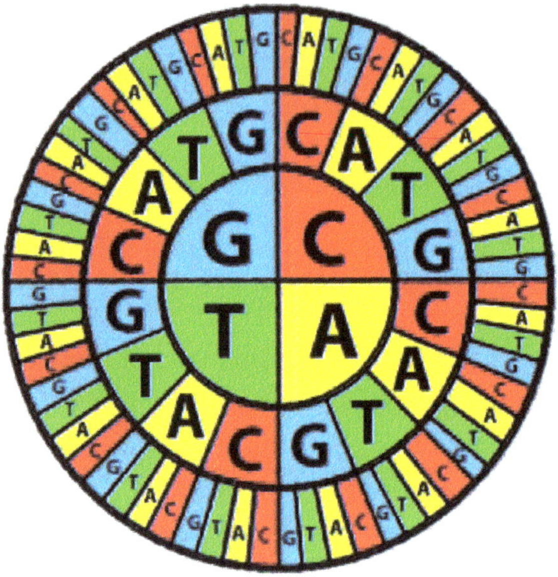

Fig. 7.4.1. A graphic representation of the family of interrelated DNA alphabets of 4 nucleotides, 4^2 duplets, and 4^3 triplets as a tiered pie chart having 4 sectors. The first tier contains 4 nucleotides C, A, T, and G. The second tier contains 16 duplets whose second letters are shown. The third tier contains 64 triplets whose third letters are shown.

By analogy let us create mandalas not for DNA alphabets but for genomic probabilities of n-plets in n-textual representations of genomic DNAs. In such a mandala, each of its cellules contains a percentage of an appropriate n-plet in the DNA (instead of the n-plet symbol shown in Fig. 7.4.1). More precisely, the first tier contains percentages %C, %A, %T, %G of 4 nucleotides in the 1-textual representation of considered DNA; the second tier contains percentages of 16 duplets (%CC, %CA, %CT, %CG, ...) in the 2-textual representation of the DNA; the third tier contains percentages of 64 triplets (%CCC, %CCA, %CCT, ...) in the 3-textual representation of the DNA (this section describes only a case of genomic probability mandalas for values $n = 1, 2, 3$ though cases of higher values n can be considered as well). In these mandalas, the angular width of each cellule is proportional to the percentage value of the corresponding n-plet, that is, on each tier, its cellules differ in angular width (in contrast to the DNA alphabets mandala in Fig. 7.4.1, in which all cellular of one tier are the same in angular width).

Families of probabilities of 4 nucleotides, 16 duplets, and 64 triplets in genomic DNAs are connected correspondingly with 2-qubit, 4-qubit, and 6-qubit states $|\psi_4>$, $|\psi_{16}>$, $|\psi_{64}>$ as noted in expressions (7.12)–(7.14):

$$|\psi_4> = \sqrt{\%C}|00> + \sqrt{\%A}|01> + \sqrt{\%T}|10> + \sqrt{\%G}|11> \tag{7.30}$$

$$\begin{aligned}|\psi_{16}> = {} & \sqrt{\%CC}|0000> + \sqrt{\%CA}|0001> + \sqrt{\%CT}|0010> \\ & + \sqrt{\%CG}|0011> + \sqrt{\%AC}|0100> + \sqrt{\%AA}|0101> \\ & + \sqrt{\%AT}|0110> + \sqrt{\%AG}|0111> + \sqrt{\%TC}|1000> \\ & + \sqrt{\%TA}|1001> + \sqrt{\%TT}|1010> + \sqrt{\%TG}|1011> \\ & + \sqrt{\%GC}|1100> + \sqrt{\%GA}|1101> + \sqrt{\%GT}|1110> \\ & + 0\sqrt{\%GG}|1111>\end{aligned} \tag{7.31}$$

$$\begin{aligned}|\psi_{64}> = {} & \sqrt{\%CCC}|000000> + \sqrt{\%CCA}|000001> + \sqrt{\%CCT}|00010> \\ & + \sqrt{\%CCG}|000011> + \sqrt{\%CAC}|000100> + \sqrt{\%CAA}|000101> \\ & + \sqrt{\%CAT}|000110> + \sqrt{\%CAG}|000111> + \sqrt{\%CTC}|001000> \\ & + \sqrt{\%CTA}|001001> + \sqrt{\%CTT}|001010> + \sqrt{\%CTG}|001011> \\ & + \sqrt{\%CGC}|001100> + \sqrt{\%CGA}|001101> + \sqrt{\%CGT}|001110> \\ & + \sqrt{\%CGG}|001111> + \sqrt{\%ACC}|010000> + \sqrt{\%ACA}|010001> \\ & + \sqrt{\%ACT}|010010> + \sqrt{\%ACG}|010011> + \sqrt{\%AAC}|010100> \\ & + \sqrt{\%AAA}|010101> + \sqrt{\%AAT}|010110> + \sqrt{\%AAG}|010111> \\ & + \sqrt{\%ATC}|011000> + \sqrt{\%ATA}|011001> + \sqrt{\%ATT}|011010> \\ & + \sqrt{\%ATG}|011011> + \sqrt{\%AGC}|011100> + \sqrt{\%AGA}|011101> \\ & + \sqrt{\%AGT}|011110> + \sqrt{\%AGG}|011111> + \sqrt{\%TCC}|100000> \\ & + \sqrt{\%TCA}|100001> + \sqrt{\%TCT}|100010> + \sqrt{\%TCG}|100011> \\ & + \sqrt{\%TAC}|100100> + \sqrt{\%TAA}|100101> + \sqrt{\%TAT}|100110> \\ & + \sqrt{\%TAG}|100111> + \sqrt{\%TTC}|101000> + \sqrt{\%TTA}|101001> \\ & + \sqrt{\%TTT}|101010> + \sqrt{\%TTG}|101011> + \sqrt{\%TGC}|101100>\end{aligned}$$

$$+\sqrt{\%TGA}\,|101101\rangle + \sqrt{\%TGT}\,|101110\rangle + \sqrt{\%TGG}\,|101111\rangle$$
$$+\sqrt{\%GCC}\,|110000\rangle + \sqrt{\%GCA}\,|110001\rangle + \sqrt{\%GCT}\,|10010\rangle$$
$$+\sqrt{\%GCG}\,|110011\rangle + \sqrt{\%GAC}\,|110100\rangle + \sqrt{\%GAA}\,|110101\rangle$$
$$+\sqrt{\%GAT}\,|110110\rangle + \sqrt{\%GAG}\,|110111\rangle + \sqrt{\%GTC}\,|111000\rangle$$
$$+\sqrt{\%GTA}\,|111001\rangle + \sqrt{\%GTT}\,|111010\rangle + \sqrt{\%GTG}\,|111011\rangle$$
$$+\sqrt{\%GGC}\,|111100\rangle + \sqrt{\%GGA}\,|111101\rangle + \sqrt{\%GGT}\,|111110\rangle$$
$$+\sqrt{\%GGG}\,|111111\rangle \qquad (7.32)$$

From the standpoint of $2n$-qubit states of genomic DNAs (7.30)–(7.32), the genomic probability mandalas can be called "quantum information mandalas" presenting genomic stochastics. Fig. 7.4.2 represents the quantum information mandala of DNA of human chromosome №1 whose n-plet percentages and corresponding $2n$-qubit states are shown above in expressions (7.12)–(7.14).

One should note the following features of the quantum information mandala in Fig. 7.4.2:

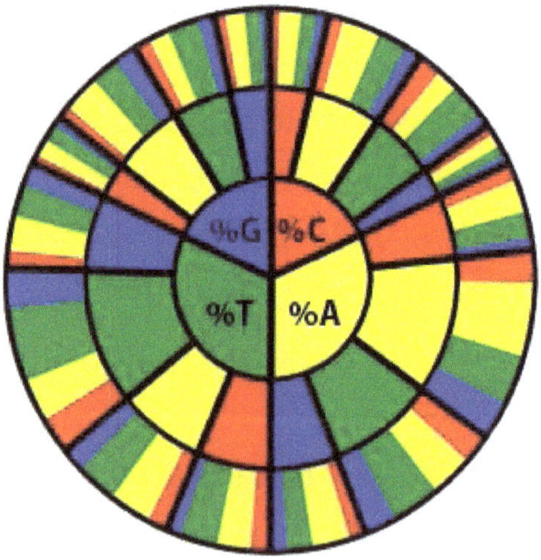

Fig. 7.4.2. The quantum information mandala of DNA of human chromosome №1 (from [234]).

1) An angular width of any cellule of the first tier is practically equal to the sum of angular widths of 4 appropriate cellules of the second tier since $\%N \approx \%NC + \%NA + \%NT + \%NG$, where N refers to any of nucleotides C, A, T, G;
2) An angular width of any cellule of the second tier is practically equal to the sum of angular widths of 4 appropriate cellules of the third tier in accordance with the described phenomenological rules of genomic stochastics [80, 116, 155, 234, 239].

One can expand this genomic mandala (Fig. 7.4.2) by adding fourth, fifth, and sixth tiers to them, but for pairs of adjacent tiers in these expanded mandalas, the same rules will again be fulfilled: the angular width of any cell of the previous tier will be practically equal to the sum of the angular widths of the four corresponding cells of the next tier (in accordance with the universal rules of stochastic organization of genomic DNAs).

Various genomes can significantly differ from each other by n-plet probabilities in their genomic DNAs. Correspondingly their quantum information mandalas differ from each other. Each genomic DNA is characterized by its individual quantum information mandala. Fig. 7.4.3 shows examples of quantum information mandalas of some eukaryotic and prokaryotic genomic DNA.

But the comparative analysis of different chromosomes of eukaryotic genomes reveals that, in each of the tested genomes, quantum information mandalas of its chromosomes are very similar to each other. Fig. 7.4.2 and 7.4.4 confirm this observation by examples of similarities of quantum information mandalas of human somatic and sex chromosomes.

We briefly note that the phenomenological rules of stochastic organization of genomic DNAs provide an opportunity to build for each of the 4 nucleotides, 16 duplets, 64 triplets, and so on its individual mandala with analogical features since the stochastic organization of genomic DNAs possesses fractal-like features. For example, Fig. 7.4.5 shows individual stochastic mandalas for each of the 4 cellules of the first tier in the DNA of

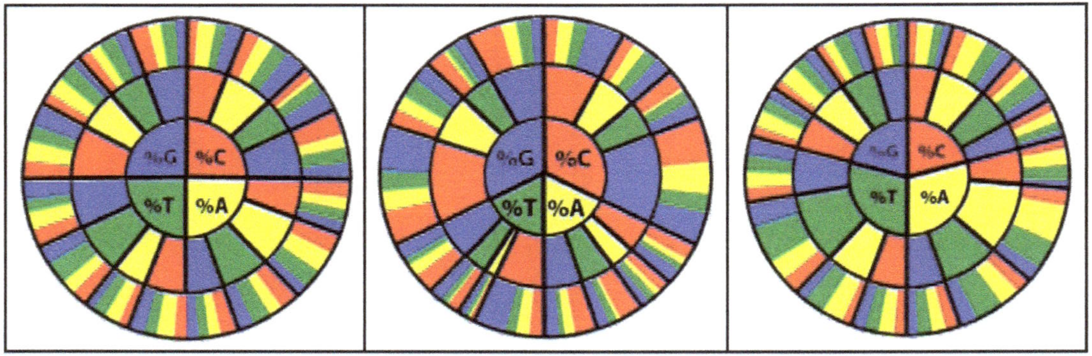

Fig. 7.4.3. Examples of quantum information mandalas of eukaryotic and prokaryotic genomic DNAs. Left: bacteria *Escherichia coli* CFT073, complete genome, GenBank: AE014075.1, 5231428 bp (initial data taken from https://www.ncbi.nlm.nih.gov/nuccore/AE014075.1?report=genbank). Middle: rhizobacteria *Bradyrhizobium japonicum* strain E109, complete genome, 9224208 bp (initial data taken from https://www.ncbi.nlm.nih.gov/nuccore/CP010313.1?report=genbank). Right: fruit fly *Drosophila melanogaster* chromosome 2L, NCBI Reference Sequence: NT_033779.5 LOCUS NT_033779, 23513712 bp (initial data taken from https://www.ncbi.nlm.nih.gov/nuccore/NT_033779.5).

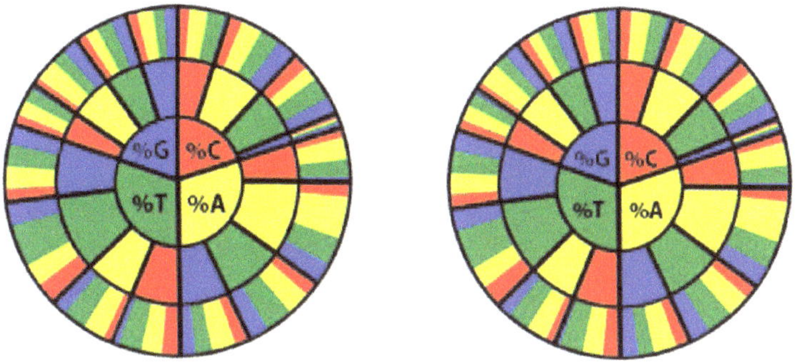

Fig. 7.4.4. Quantum information mandalas of stochastic organization of human sex chromosomes X (at left) and Y (at right).

human chromosome №1 (Fig. 7.4.2). Correspondingly, each genomic DNA can be considered as a fractal-like system of many interrelated quantum information mandalas.

Each of the individual stochastic mandalas of separate cellules at different tiers in such quantum information mandalas has a fractal-like structure as well.

We suppose that the mandala-like organization of eukaryotic and prokaryotic genomes relates to the ancient psychological theme of mandalas, which were used as meditation tools to achieve "enlightenment" and healing by millions of Buddhists, Hindus, and other believers. The creator of analytical psychology C. Jung and his collaborator Nobel Laureate in Physics W. Pauli considered the mandala to be an innate archetype of the unconscious and the conjugation of a cosmogram with a psychogram, capable of harmonizing the psyche. It can be thought that the ancient ideas of believers about the significance of mandalas reflect the feelings of enlightened sages about the genetic foundations of living organisms, represented in the stochastic organization of genomic DNAs. That the ancient sages also felt the importance of the stochastic (probabilistic) principles of nature organization is reflected in the well-known fact that the *I-Ching* was used in ancient China as a divination book with procedures for determining probabilities in a series of repeated tests.

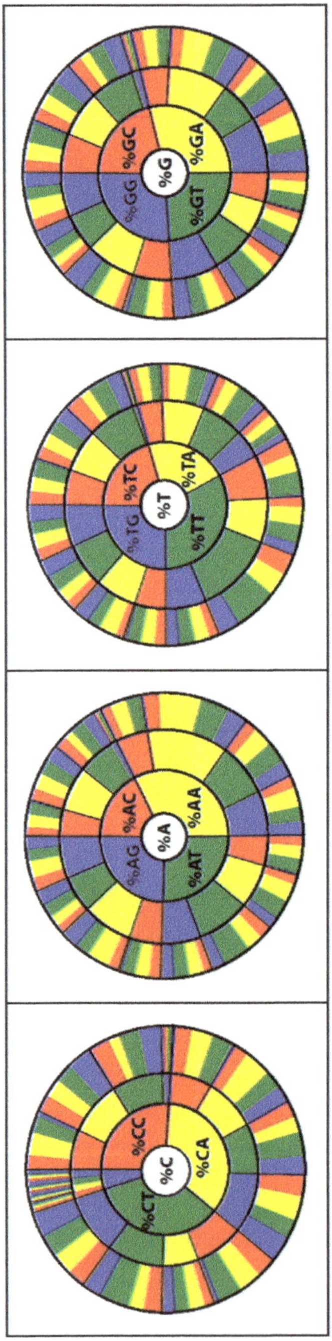

Fig. 7.4.5. Individual stochastic mandalas for each of the 4 cellules of the first tier in the DNA of human chromosome №1.

In our time some kinds of music are used in biotechnological processes (for example, in wine production and stimulation of plant growth), musical therapy, etc. Techniques for musical dubbing of such quantum information mandalas of various genomes are being developed to get "genomic quantum information music", which can be applied in different fields: biotechnology, regenerative medicine, musical therapy, and improvements of ancient technologies of mandala-musical meditations for enlightenment and healing. The affinity of the genetic harmony of organisms with musical harmony are studied in the special "Center for Interdisciplinary Research of Musical Creativity" of the Moscow State Tchaikovsky Conservatory (existing since 2012, http://www.mosconsv.ru/ru/groups.aspx?id=135683).

7.5. Genetic matrices and quantum gates

Let us additionally consider a connection of the mentioned genetic matrices with quantum gates of quantum informatics. In the previous chapters, we repeatedly noted the connection between genetic matrices — under their decomposition into sparse matrices — with unitary matrices having real entries (see Figs. 1.5.2, 1.5.3, 1.6.3, 1.6.4, 2.3.1, 2.3.2, 4.12.2–4.12.4). Unitary matrices with real entries are used in quantum informatics as quantum logic elements (quantum gates) for performing quantum computations on their basis [15]. Each unitary matrix can be used as a quantum gate. In the case of multi-qubit systems, the operation of the tensor product of matrices is of key importance in connection with the postulate of quantum mechanics: the state space of a composite system is the tensor product of the state spaces of its components. The tensor product of two unitary matrices always generates a unitary matrix [22, p. 38].

In light of this, it is especially interesting that the entire genetic Rademacher (4×4)-matrix R_4 (Figs. 1.4.3 and 1.5.3) is the sum of the tensor products of four orthogonal (2×2)-matrices U_0, U_1, U_2, and U_3 (Fig. 7.5.1). The set of these 4 matrices is closed concerning multiplication and defines the multiplication table of Cockle's split-quaternions, which was met many

$$R_4 = \begin{vmatrix} 1, -1, & 1, -1 \\ 1, & 1, -1, -1 \\ 1, -1, & 1, -1 \\ -1, -1, & 1, & 1 \end{vmatrix} = U_0 \otimes U_0 + U_0 \otimes U_1 + U_3 \otimes U_2 + U_3 \otimes (-U_3)$$

$$U_0 = \begin{vmatrix} 1, 0 \\ 0, 1 \end{vmatrix}; \quad U_1 = \begin{vmatrix} 0, -1 \\ 1, \ 0 \end{vmatrix}; \quad U_2 = \begin{vmatrix} -1, 0 \\ 0, 1 \end{vmatrix}; \quad U_3 = \begin{vmatrix} 0, -1 \\ -1, 0 \end{vmatrix}$$

*	U_0	U_1	U_2	U_3
U_0	U_0	U_1	U_2	U_3
U_1	U_1	$-U_0$	U_3	$-U_2$
U_2	U_2	$-U_3$	U_0	$-U_1$
U_3	U_3	U_2	U_1	U_0

Fig. 7.5.1. The genetic Rademacher (4×4)-matrix of 16 duplets (from Figs. 1.4.3 and 1.5.2) is the sum of the tensor products of four orthogonal (2×2)-matrices U_0, U_1, U_2, and U_3. The set of these 4 matrices defines the shown multiplication table of Cockle's split-quaternions.

times in the previous chapters. It should be noted that unitary matrices U_0, U_1, U_2 and U_3 (Fig. 7.5.1) have relations to quantum gates used widely in quantum computing [15, p. xxx]. One can suppose that these 4 unitary matrices are used as quantum gates in quantum computing in organisms.

Exponentiation of orthogonal matrices U_1, U_2 and U_3 into ordinary integer powers $n = 2, 3, 4, \ldots$ gives cyclic groups of matrices with the following periods: $U_1^n = U_1^{n+4}$, $U_2^n = U_2^{n+2}$, $U_3^n = U_3^{n+2}$. Here we especially note a connection of cyclic groups with algebraic properties of genetic orthogonal matrices since such cyclic groups can be useful for modeling many inherited cyclic processes in physiology of organisms.

Considering the quantum-informational character of molecular-genetic systems and also the important role of unitary matrices in quantum computing, it is interesting that the genetic matrices R_4 and R_8 (Figs. 1.4.3, 1.5.2, and 1.5.3) are connected with inter-complementary sparse unitary matrices described below.

Let us begin with a consideration of the genetic matrix R_4 of 16 doublets. We reveal that its cross-like character relates to a pair of two sparse unitary matrices, which are mirror-anti-symmetric to each other. Fig. 7.5.2 shows that the genetic cross-like matrix R_4 is the sum of two sparse matrices R_{4R} and R_{4L}: $R_4 = R_{4R} + R_{4L}$. These two sparse matrices are inter-complementary in an algebraic sense since they are mirror-antisymmetric to each other concerning their positive and negative entries and they jointly form the non-sparse matrix R_4. Using the analogy with our stereoscopic vision by means of two — left and right — eyes, we conditionally call the pair of complementary matrices R_{4R} and R_{4L} as the stereoscopic pair (or, briefly, the stereo-pair) where the matrix R_{4R} is called the right stereo-matrix and the matrix R_{4L} is called the left stereo-matrix.

Taking the factor $2^{-0.5}$, each of these stereo-matrices R_{4R} and R_{4L} is the orthogonal matrix:

$$(2^{-0.5} \times R_{4R}) \times (2^{-0.5} \times R_{4R})^T = I; \quad (2^{-0.5} \times R_{4L}) \times (2^{-0.5} \times R_{4L})^T = I. \quad (7.33)$$

In each of the orthogonal matrices $2^{-0.5} \times R_{4R}$ and $2^{-0.5} \times R_{4L}$, the set of its rows contains an orthogonal system of functions. The possibilities of using these matrices in the spectral analysis of vectors are described in [206].

One should note that each of the two stereo-matrices R_{4R} and R_{4L} (Fig. 7.5.2) is the sum of two orthogonal matrices shown in Fig. 7.5.3: $R_{4R} = K_0 + K_1$, $R_{4L} = K_2 + K_3$. Really, $K_0 \times K_0^T = I_4$, $K_1 \times K_1^T = I_4$, $K_2 \times K_2^T = I_4$ and $K_3 \times K_3^T = I_4$, where I_4 is the identity matrix of order 4 and the matrix K_0 resembles the well-known quantum gate "controlled-NOT": [1, 0, 0, 0; 0, 1, 0, 0; 0, 0, 0, 1; 0, 0, 1, 0] [15, p. xxxi]. Rows and columns of sparse matrices K_0, K_1, K_2 and K_3 correspond to orthogonal systems of functions.

$$\begin{bmatrix} 1 & -1 & 1 & -1 \\ 1 & 1 & -1 & -1 \\ 1 & -1 & 1 & -1 \\ -1 & -1 & 1 & 1 \end{bmatrix} = \begin{bmatrix} 1 & -1 & 0 & 0 \\ 1 & 1 & 0 & 0 \\ 0 & 0 & 1 & -1 \\ 0 & 0 & 1 & 1 \end{bmatrix} + \begin{bmatrix} 0 & 0 & 1 & -1 \\ 0 & 0 & -1 & -1 \\ 1 & -1 & 0 & 0 \\ -1 & -1 & 0 & 0 \end{bmatrix}$$

Fig. 7.5.2. The cross-like matrix R_4 is the sum of two sparse matrices R_{4R} and R_{4L}, non-zero entries of which coincide with non-zero entries in corresponding quadrants along the main diagonal and the secondary diagonal of the matrix R_4.

$$R_{4R} = \begin{bmatrix} 1 & -1 & 0 & 0 \\ 1 & 1 & 0 & 0 \\ 0 & 0 & 1 & -1 \\ 0 & 0 & 1 & 1 \end{bmatrix} = \begin{bmatrix} 1 & 0 & 0 & 0 \\ 0 & 1 & 0 & 0 \\ 0 & 0 & 0 & -1 \\ 0 & 0 & 1 & 0 \end{bmatrix} + \begin{bmatrix} 0 & -1 & 0 & 0 \\ 1 & 0 & 0 & 0 \\ 0 & 0 & 1 & 0 \\ 0 & 0 & 0 & 1 \end{bmatrix} = K_0 + K_1$$

$$R_{4L} = \begin{bmatrix} 0 & 0 & 1 & -1 \\ 0 & 0 & -1 & -1 \\ 1 & -1 & 0 & 0 \\ -1 & -1 & 0 & 0 \end{bmatrix} = \begin{bmatrix} 0 & 0 & 1 & 0 \\ 0 & 0 & 0 & -1 \\ 0 & -1 & 0 & 0 \\ -1 & 0 & 0 & 0 \end{bmatrix} + \begin{bmatrix} 0 & 0 & 0 & -1 \\ 0 & 0 & -1 & 0 \\ 1 & 0 & 0 & 0 \\ 0 & -1 & 0 & 0 \end{bmatrix} = K_2 + K_3$$

Fig. 7.5.3. Each of the stereo-matrices R_{4R} and R_{4L} is the sum of two unitary matrices: $R_{4R} = K_0 + K_1$, $R_{4L} = K_2 + K_3$, where K_0, K_1, K_2 and K_3 are unitary matrices.

The genetic (8 × 8)-matrix R_8 can be also represented via orthogonal genetic (2 × 2)-matrices U_0, U_1, U_2 and U_3 from Fig. 7.5.1:

$$R_8 = U_0 \otimes U_0 \otimes U_0 + U_0 \otimes U_0 \otimes (-U_3) + U_0 \otimes U_1 \otimes U_0 + U_0 \otimes U_1 \otimes (-U_3) + U_3 \otimes U_2 \otimes U_0 + U_3 \otimes U_2 \otimes (-U_3) + U_3 \otimes (-U_3) \otimes U_0 + U_3 \otimes (-U_3) \otimes (-U_3) \quad (7.34)$$

A detailed analysis of this case concerning the matrix R_8, including the study of relationship between triplets and the amino acids encoded by them, is presented in [206].

One can note that the presented study belongs to the field of applications of algebraic methods and notions of quantum mechanics for modeling molecular-genetic structures where works of many authors exist, for example, the following [11, 22, 23, 32, 38, 182, 189, 303, 310, 317, 318, 323, 324, 326–331, 333, 335–345, 344, 346–349].

7.6. Frohlich's theory about collective quantum effects in biological systems

The above-described results on universal rules of the stochastic-deterministic organization of genomic DNAs indicate the presence of long-range relationships in the composition of genomic DNA texts. These results are associated with the well-known vibration-resonance theory of H. Frohlich about long-range coherence in biological systems, that is, about collective quantum effects in biological systems [275, 350–354]. This theory testifies in favor of the fact that living bodies are a kind of analog of Bose-Einstein condensate, which plays a prominent role in the theory of the physical vacuum. This theory is supported by many authors. Let us consider this association.

Nobel Laureate R. Penrose uses Frohlich's theory to substantiate his ideas about the brain as a quantum computer and writes about it as follows: "*The distinguished physicist Herbert Frohlich (who, in the 1930s, had made one of the fundamental breakthroughs in the understanding of 'ordinary' low-temperature superconductivity) suggested a possible role for collective quantum effects in biological systems. ... Frohlich was led to propose, in 1968, that there should be vibrational effects within active cells, which would resonate with microwave electromagnetic radiation, at 10^{11} Hz, as a result of a biological quantum coherence phenomenon. Instead of needing a low temperature, the effects arise from the existence of a large energy of metabolic drive. There is now some respectable observational evidence, in many biological systems, for precisely the kind of effect that Frohlich had predicted in 1968. ... He argued that so long as the energy of metabolic drive is large enough, and the dielectric properties of the materials concerned are sufficiently extreme, then there is the possibility of large-scale quantum coherence similar to that which occurs in the phenomena of superconductivity and superfluidity — sometimes referred to as Bose-Einstein condensation — even at the relatively high temperatures that are present in biological systems... In a Bose-Einstein condensate (which occurs also in the action of a laser), large numbers of particles participate collectively in a single quantum state. There is a wavefunction, for this state, of the kind that would be appropriate for a single particle — but now it applies all at once to the entire collection of particles that are participating in the state. ... With a Bose-Einstein condensate, it is as though the entire system containing a large number of particles behaves as a whole very much as the quantum state of a single particle would, except that everything is scaled up appropriately. There is a coherence on a large scale, where many of the strange features of quantum wavefunctions hold at a macroscopic level*" [303, p. 352 and 367].

F. Fröhlich (the son of Herbert Fröhlich) wrote the article "Genetic code as language" [355] regarding quantum coherence states and long-range communication in genomes. He noted: "*Beyond the chromosome, the genome as a whole must contain some sort of long-distance communication in order not to produce on one section antigens produced on other parts. So there is a complexity of entities which use genetic language in different ways. There might be said to be a logic of cells*" [355, p. 194]. H. Fröhlich's hypothesis "*suggests that long-range coherent vibrations will lead to resonance between a differentiated cell with its own characteristic vibrations and the chromosome such that the chromosome-particular region responding to this characteristic frequency will be activated or opened up so it can produce the appropriate proteins. Such a resonance could transport the embryological, already partially induced cells to their target and there they would be further fixed into producing the correct proteins for this organ through superimposed resonance and why if it were too mature, it would not so adapt. There might be degrees of resonance*" [ibid, p. 199]. "*This have been done more specifically in the case of cell division during mitosis, where it has been hypothesized that corresponding chromosomes line up through resonance having the same frequency, finding each other from the apparent confusion of the genome by long-range communication [Holland, 1972]. Resonance oscillations draw like to like. More generally there might be some kind of coherent long-range interaction among the members of the genome creating the self-marking. If this broke down through subsequent mutations, auto-immune diseases would arise*" [355, p. 203].

Fröhlich's synchronous large-scale collective oscillations imply intercellular microwave emissions which would constitute a non-chemical and non-thermal interaction between cells [356]. Some evidence of a non-thermal influence of coherent microwave radiation on the genome conformational state in *E. coli* has been reported [357], which may indicate that "*chromosomal DNA could be the target of mm microwave irradiation within this system. Also low intensity microwave irradiation of leukocytes results in a significant increase in bio-photon emission in the optical range, the origin of which is thought to involve DNA. Also it is worth noticing the possible influence of the concept of bio-coherence on the very particular dipolar system, which is water.*

It can be considered the possibility that biological water might itself support coherent dipolar excitations extending over mesoscopic regions; thus water, instead of passive space-filling solvent, would be raised to an important singular position whose full significance has yet to be elucidated" [356].

Living organisms on earth are exposed to a whole spectrum of electromagnetic radiation modalities, related to the physical constitution of Earth and our planetary position in the solar and galaxy systems. The article [209] presents results of an extensive literature survey, comprising 175 articles from 1950 to 2015, dealing with effects of electromagnetic radiation on *in vitro* and *in vivo* life systems, indicating that typical discrete coherent frequencies of electromagnetic waves are able to stabilize cells, whereas others cause a clear destabilization. The authors wrote: "*We find support for the hypothesis of H. Fröhlich, that a driven set of oscillators condenses in a broad energy range, may activate a vibrational mode in life organisms at room temperature. Taking into account the life sustaining frequencies, as extracted from literature, an algorithm of coherent frequencies of standing waves for the stability of biological order was inferred. Interestingly, we found that the origin of the particular biological algorithm can be mathematically approached by a selected "tempered Pythagorean" reference acoustic scale. The algorithm expresses one-dimensional wave equations known for vibrating strings*". These results, which are connected with standing waves, can be considered as having close correlations with results about the connection of the harmonic progression with genomes of higher and lower organisms, which is described in Chapter 11.

Beside, the article [209] reminds us that Fröhlich claimed oscillating charges in a thermal bath, in which a large number of quanta may condense into a single state known as a Bose condensate, which could constitute a physical and non-thermal interaction between cells. These authors note that a relation between 'living' and non-living matter has also been made by W. Pauli: he stated that the mental and material domains are governed by common ordering principles, and can be understood as "*complementary aspects of the same reality*". In other words, the analogy with Bose-Einstein condensate can be distributed also on the mental domain.

Frohlich's quantum concept allows consideration of biological structures as a kind of analog of the quantum Bose-Einstein condensate associated with the concept of a physical vacuum. An important feature of the physical vacuum is its binary-oppositional (Yin-Yang) property: entities generated from it appear in the form of a mutually opposite pair, for example, particles — antiparticles, waves — anti-waves, etc. When viewed from the perspective of Frohlich's concept of biological structures as special analogs of Bose-Einstein condensates that do not require ultra-low temperatures, the following thought suggests itself: this vacuum-like quantum state of organisms is responsible for many of the known binary-oppositional features of genetic and genetically inherited biological structures.

Let us recall that, according to modern concepts, the physical vacuum is not a void with invariable properties, but a complex hierarchical system in space-time capable of dynamic evolution. The structure of the vacuum seems to be very complex, consisting of zero-point oscillations (fluctuations of quantized force fields). Fluctuations are continuously born and die, but in each area of space and at each moment in time, their average number remains unchanged. These states of the physical vacuum are called vacuum condensates. The properties of matter are determined by the properties of vacuum condensates. One of the important types of these vacuum condensates is Bose-Einstein condensate. It represents the state of aggregation of matter which is based on bosons (particles with integer spin) cooled to temperatures close to absolute zero. In such a strongly cooled state, a sufficiently large number of boson atoms are in their minimum possible quantum states and quantum effects begin to manifest themselves at the macroscopic level. Bose condensate was predicted by Einstein in 1925. 70 years later, in 1995, the first Bose condensate was obtained in the United States from a supercooled gas of rubidium atoms (this work was marked by the 2001 Nobel Prize). A quantum system consisting of an arbitrary number of bosons and an even number of fermions (particles with half-integer spin) is itself a composite boson. Currently, the Bose-Einstein condensate is being studied by physicists in different countries for new technologies and theoretical concepts based on it, as well as the possibility of obtaining this condensate at room temperature (see, for example, [358, 359]).

Frohlich's theory attracts many scientists developing concepts of cooperative interrelations in biomolecular ensembles and of the important role of resonances in functions of these ensembles including DNA (see, for example, [360]).

7.7. Some concluding remarks

A great discovery of twentieth-century physics was the probabilistic nature of physical phenomena at atomic scales, described by quantum mechanics. Informational molecules of heredity DNA and RNA belong to the microworld of quantum mechanics and they should obey the principles of quantum mechanics. Many authors have supposed that living organisms use principles of quantum informatics and quantum mechanics. This chapter is devoted to additional opportunities to introduce formalisms of quantum mechanics and quantum informatics on the basis of symmetric peculiarities of DNA alphabets for modeling some genetic phenomena including the universal rules of stochastic-deterministic organization of genomic DNAs. The presented results support the thoughts of many authors about quantum computing in biological organisms.

The doctrine of energy-informational evolution based on bio-antenna arrays has quantum mechanical aspects. The Hadamard product is appropriate not only for modeling the universal rules of stochastic-deterministic organization of genomic DNAs but also for nonlinear transformations in quantum computation. The discovered universal rules in the cooperative organization of genomes can be explained from Frohlich's theory about collective quantum effects in biological systems by analogy with Bose-Einstein condensates.

The revealed new connection of genetics with quantum informatics opens up the possibility of introducing formalisms of quantum mechanics and quantum informatics into algebraic biology for the mutual enrichment of these sciences and the inclusion of biology in the field of developed mathematical natural science. There are about 100 trillion cells in the human body, forming a single system. The formalisms of quantum informatics and Gestalt genetics can help in understanding such coherence phenomena. The work [361] contains a review about quantum computing and the problems of artificial intelligence.

Chapter 8 Genetic Code Systems and Mathematical Variety

In this book, the algebra-genetic matrices are applied to separate subsystems of the genetic code system having amazing emergence properties. Presented results of this approach relate equally to alphabets of DNA and RNA nucleobases, which are separate parts of the holistic genetic code system having emergence properties. In this chapter the symbol of uracil U from the RNA alphabet is used instead thymine T from DNA, which is used in other chapters. As described in Chapter 1, the genetic four-letter alphabet (A, C, G, T/U) bears three types of binary-oppositional traits (1. A = C, G = T/U; 2. A = G, C = T/U; 3. A = T/U, C = G). In this chapter, we continue an investigation of these three kinds of binary oppositions by introducing three special (2 × 2)-matrices associated with them. This investigation additionally shows a mathematical variety of algebraic tools and notions which are connected with non-trivial structures of molecular ensembles of the genetic coding system and are used to reveal the emergent properties of the genetic system. This chapter presents algebraic connections between the numeric properties of nucleobase alphabets with metric tensors, special numeric values (such as the ring ratio and the golden ratio), doubly stochastic matrices, and unitary matrices. Described results are useful for revealing bio-informational patents of living nature and new genetic symmetries, and also for developing algebraic biology.

8.1. Symmetric matrices, metric tensors, and three kinds of binary oppositions

The four letters (or the four nitrogenous bases A, C, G, T/U) of the genetic alphabet represent specific poly-nuclear constructions with special biochemical properties. The set of these four constructions is not absolutely heterogeneous, but it bears the substantial symmetric system of distinctive-uniting attributes (or, more precisely, pairs of an "attribute and anti-attribute"):

1) A = C & G = U (according to the attributes: amino-mutating or non-amino-mutating under the action of nitrous acid HNO_2 [362, 363]; the same division is given by the attributes "keto" or "amino" [107]);
2) C = U & A = G (according to the binary-opposite attributes: "pyrimidine" or "non-pyrimidine", that is, purine);
3) C = G & A = U (according to the attributes: three or two hydrogen bonds are materialized in these complementary pairs).

Between 1999–2001, Petoukhov proposed to use these three kinds of biomolecular binary oppositions to construct square genetic matrices where genetic *n*-plets are disposed and enumerated in strict orders [145, 329–331]. Based on these three kinds of genetic molecular oppositions, three mapping relations from R = {A, C, G, U} to N = {0, 1, 2, 3} were defined in [364, 365] as follows:

- α: R → {0, 1} with α (G) = α (U) = 0, α (A) = α (C) = 1;
- β: R → {1, 2} with β (C) = β (U) = 1, β (A) = β (G) = 2;
- γ: R → {2, 3} with γ (A) = γ (U) = 2, γ (C) = γ (G) = 3.

These three mappings were applied to the standard genetic code in [364, 365] and derived three (8×8)-matrices associated with the genetic code. Their basic symmetrical properties were investigated in [364–366].

These three kinds of binary opposition in the RNA alphabet are represented by three (2×2) square matrices:

- The first kind of binary opposition: G = U = 0, A = C = 1, amino-mutating absent/present (0, 1)-combination, this kind can be represented as a symbolic (2×2)-matrix of nucleobases and corresponding numerical mutating values of 0 or 1 as follows:

$$H_\alpha = \begin{array}{|c|c|}\hline A & G \\ \hline U & C \\ \hline \end{array} \quad H_\alpha(0,1) = \begin{array}{|c|c|}\hline 1 & 0 \\ \hline 0 & 1 \\ \hline \end{array}$$

- The second kind of binary opposition: C = U = 1, A = G = 2, pyrimidine/purine ring-based (1, 2)-combination, this equivalence can be represented as a symbolic (2 × 2)-matrix of nucleobases and corresponding numerical ring values of 1 or 2 as follows:

$$H_\beta = \begin{array}{|c|c|}\hline A & C \\ \hline U & G \\ \hline \end{array} \quad H_\beta(1,2) = \begin{array}{|c|c|}\hline 2 & 1 \\ \hline 1 & 2 \\ \hline \end{array}$$

- The third kind of binary opposition: A = U = 2, C = G = 3, hydrogen bond-based (2, 3)-combination, this kind can be represented as a symbolic (2 × 2)-matrix of nucleobases and corresponding numerical bond values of 2 or 3 as follows:

$$H_\gamma = \begin{array}{|c|c|}\hline C & A \\ \hline U & G \\ \hline \end{array} \quad H_\gamma(2,3) = \begin{array}{|c|c|}\hline 3 & 2 \\ \hline 2 & 3 \\ \hline \end{array}$$

These (2 × 2)-matrices provide building blocks for possible construction — by using their tensor powers — of matrices of higher dimensions. In this chapter, we study these three pairs of basic (2 × 2)-matrices [H_α, $H_\alpha(0, 1)$], [H_β, $H_\beta(1, 2)$], [H_γ, $H_\gamma(2, 3)$] to establish their mathematical features.

One can see that each of the matrices $H_\alpha(0,1)$, $H_\beta(1,2)$, and $H_\gamma(2,3)$ is diagonal, that is, both entries along every diagonal are identical. Correspondingly, such type of diagonal (2 × 2)-matrices is represented in a general form by the following matrix H:

$$H = \begin{array}{|c|c|}\hline a & b \\ \hline b & a \\ \hline \end{array}$$

As described in Chapter 3, such symmetric matrices are connected with the theory of resonances and the notion of metric tensors, which are essential for developing the proposed concept of the morpho-resonance field. By definition, the metric tensor in an n-dimensional affine space with the introduced scalar multiplication operation is given by a nondegenerate symmetric matrix $||g_{ij}||$, $g_{ij} = g_{ji}$ [151, p. 157], which is satisfied by the structure of the symmetric real (2×2)-matrices of vibro-systems considered in Chapter 3. The coordinates g_{ij} of the metric tensor are pairwise scalar products of the vectors of the frame, on which it is built. If a square root is extracted from a symmetric matrix of a metric tensor, then a new symmetric matrix is obtained, the columns of which are the vectors of this frame or the basic repers (and which, in turn, can be treated as a new metric tensor).

Let us study the mentioned symmetric nondegenerate matrices $H_\alpha(0,1) = [1, 0; 0, 1]$, $H_\beta(1,2) = [2, 1; 1, 2]$, and $H_\gamma(2,3) = [3, 2; 2, 3]$ from the standpoint of their interpretation as metric tensors. For this, one should calculate their square roots. The first matrix $H_\alpha(0,1) = [1, 0; 0, 1]$ is the identity matrix and not interesting for our analysis. We start from the second matrix $H_\beta(1,2) = [2, 1; 1, 2]$, the square root from which is a symmetric matrix:

$$\begin{aligned} H_\beta(1,2)^{0.5} &= [2, 1; 1, 2]^{0.5} \\ &= [1.36602540378\ldots, 0.36602540378\ldots; \\ &\quad 0.36602540378\ldots, 1.36602540378\ldots] \end{aligned} \tag{8.1}$$

The value 1.36602540378... is called the "ring ratio" here. It has interesting properties in cooperation with the second entry 0.36602540378... in the matrix (8.1). Firstly, all decimal places for both numbers coincide, by analogy just as the decimal places for the golden ratio $\varphi = 1.61803398875...$ and its reciprocal value $\varphi^{-1} = 1.61803398875...$ coincide. Secondly, both of these numbers in (8.1) have apparently an infinite number of non-zero decimal places, but their product (8.2) gives a finite number of 0.5. This is analogous to how the product of the golden ratio φ and its reciprocal (which have an infinite number of non-zero decimal places) gives the integer 1.

$$1.36602540378\ldots \times 0.36602540378\ldots = 0.5 \tag{8.2}$$

The columns of the obtained matrix (8.1) define two vectors of the frame, on which the metric tensor [2, 1; 1, 2] is built: $\bar{X}_1 = [1.36602540378\ldots, 0.36602540378\ldots]$ and $\bar{X}_2 = [0.36602540378\ldots, 1.36602540378\ldots]$. Pairwise scalar products of these vectors of the frame gives entries of the considered metric tensor [2, 1; 1, 2]: $\bar{X}_1 \times \bar{X}_1 = 2$, $\bar{X}_1 \times \bar{X}_2 = \bar{X}_2 \times \bar{X}_1 = 1$, $\bar{X}_2 \times \bar{X}_2 = 2$.

The symmetric matrix (8.1), containing the ring ratio, in turn can be considered as a metric tensor. The expression (8.3) shows its square root, which is a new symmetric matrix.

$$([2, 1; 1, 2]^{0.5})^{0.5} = [1.15803700647..., 0.15803700647...; 0.15803700647..., 1.15803700647...] \tag{8.3}$$

Again, all decimal places in its two entries coincide. The product of these two entries gives a value 0.183012701892219... (8.4). This value is exactly half of the smaller value 0.36602540378… in the previous metric tensor (8.1).

$$1.15803700647\ldots \times 0.15803700647\ldots = 0.183012701892\ldots \tag{8.4}$$

The symmetric matrix (8.3), in turn, can be considered as a metric tensor. The expression (8.5) shows its square root, which is a new symmetric matrix.

$$(([2, 1; 1, 2]^{0.5})^{0.5})^{0.5} = [1.07360134521..., 0.07360134521...; 0.07360134521..., 1.07360134521...] \tag{8.5}$$

Again, all decimal places in its two entries coincide. The product of these two entries gives a value 0.07901850322... (8.6), which is exactly half of the smaller value 0.15803700645... in the previous metric tensor (8.3).

$$1.07360134521... \times 0.07360134521... = 0.07901850322... \tag{8.6}$$

The examples (8.1)–(8.6) show that the recurrent use of the algorithm for taking the square root of the symmetric matrices of the metric tensor generates an arbitrarily long hierarchical sequence of metric tensors (or a sequence of coordinate repers), as if nested into each other.

Now let us return to the symmetric matrix $H_\gamma(2,3) = [3, 2; 2, 3]$. Its square root generates a symmetric matrix (8.7), which is directly connected with the golden ratio $\varphi = (1 + 5^{0.5})/2 = 1.61803...$ and its reversal value $\varphi^{-1} = (1 - 5^{0.5})/2 = 0.61803...$

$$(H_\gamma(2,3))^{0.5} = [3, 2; 2, 3]^{0.5} = [\varphi, \varphi^{-1}; \varphi^{-1}, \varphi] \tag{8.7}$$

This new matrix is called the golden matrix or the golden metric tensor [12, 13]. The square root of the golden matrix (8.7) gives a symmetric matrix (8.8), which can be also considered as a metric tensor.

$$\begin{aligned}((H_\gamma(2,3))^{0.5})^{0.5} &= ([3, 2; 2, 3]^{0.5})^{0.5} \\ &= [1.24767439061..., 0.24767439061...; \\ &\quad 0.24767439061..., 1.24767439061...]\end{aligned} \tag{8.8}$$

In both kinds of entries in the generated matrix (8.8), all decimal places coincide. The product of these two kinds of entries (8.9) gives the value 0.30901699437..., which is exactly half of the value φ^{-1} in the previous golden metric tensor (8.7).

$$0.24767439061... \times 1.24767439061... = 0.30901699437... = \varphi^{-1}/2 \tag{8.9}$$

The square root from the metric tensor (8.8) gives a new metric tensor (8.10):

$$\begin{aligned}(([3, 2; 2, 3]^{0.5})^{0.5})^{0.5} &= [1.11142227249..., 0.11142227249...; \\ &\quad 0.11142227249..., 1.11142227249...]\end{aligned} \tag{8.10}$$

In both kinds of entries in the metric tensor (8.10), all decimal places coincide. The product of these two kinds of entries (8.10) gives the value

0.12383719530..., which is exactly half of the smaller value 0.24767439061... in the previous metric tensor (8.8). This recurrent algorithm for taking the square root in such a sequence of metric tensors, which includes the golden metric tensor, can be applied again and again with similar results.

We consider (2 × 2)-matrices $H_\alpha(0,1) = [1, 0; 0, 1]$, $H_\beta(1,2) = [2, 1; 1, 2]$, $H_\gamma(2,3) = [3, 2; 2, 3]$ and sequences of square roots from them. But in a similar way one can analyze the tensor (Kronecker) powers of these three matrices, which have the form of symmetric nondegenerate matrices of the 4th, 8th, and generally 2^n orders and which can be treated as metric tensors of vector spaces of the corresponding dimensionalities. Below in Chapter 13, the tensor powers of the matrix [3, 2; 2, 3] and the golden matrix $[\varphi, \varphi^{-1}; \varphi^{-1}, \varphi]$ are specially considered in connection with the theme of affinity between the parametric organization of molecular genetic system and ratios of musical harmony associated with the theory of resonances.

8.2. Symmetric matrices, hyperbolic numbers, and three kinds of binary oppositions

As is known [58], the noted matrix H = [a, b; b, a] is the matrix representation of hyperbolic number a + *j*b, where *j* is the imaginary unit of hyperbolic numbers and satisfies the condition $j^2 = +1$ (not to be confused with complex numbers whose matrix representation is [a, b; −b, a] and whose imaginary unit *i* satisfies the condition $i^2 = -1$). Correspondingly, symmetric matrices $H_\alpha(0,1) = [1, 0; 0, 1]$, $H_\beta(1,2) = [2, 1; 1, 2]$, and $H_\gamma(2,3) = [3, 2; 2, 3]$ present appropriate hyperbolic numbers:

- the matrix $H_\alpha(0,1) = [1, 0; 0, 1]$ presents hyperbolic number h = 1 + 0j;
- the matrix $H_\beta(1,2) = [2, 1; 1, 2]$ presents hyperbolic number h = 2 + j;
- the matrix $H_\gamma(2,3) = [3, 2; 2, 3]$ presents hyperbolic number h = 3 + 2j;
- the golden matrix $[\varphi, \varphi^{-1}; \varphi^{-1}, \varphi]$ presents hyperbolic numbers $h = \varphi + \varphi^{-1}j$;
- all metric tensors, noted above in the preliminary section, also present corresponding hyperbolic numbers.

As described in Chapters 1–3, hyperbolic numbers are connected in many ways with the structures of genetic coding, as well as with the theory of resonances of oscillatory systems with many degrees of freedom. The theory of resonances pays special attention to the eigenvectors and eigenvalues of the matrices describing the oscillatory system. Symmetric matrices presenting hyperbolic numbers always have real eigenvalues, which is one of the possible reasons for the connections of the genetic coding system with hyperbolic numbers. And what eigenvectors and eigenvalues do the considered symmetric matrices have? Let's describe them.

The matrix $H_\beta(1,2) = [2, 1; 1, 2]$ has eigenvalues 1 and 3. The matrix $H_\gamma(2,3) = [3, 2; 2, 3]$ has eigenvalues 1 and 5. The golden matrix $[\varphi, \varphi^{-1}; \varphi^{-1}, \varphi]$ has eigenvalues 1 and $5^{0.5}$. All these matrices have identical eigenvectors $[-0.5^{0.5}, 0.5^{0.5}]$.

Tensor powers of considered matrices, for example, $H_\beta(1,2) = [2, 1; 1, 2]$ and $H_\gamma(2,3) = [3, 2; 2, 3]$, can be interpreted as matrices presenting multi-dimensional numbers from those algebras, which are extensions of two-dimensional algebra of hyperbolic numbers. For instance, Fig. 8.2.1 shows the second tensor power of the matrix $H_\beta(1,2) = [2, 1; 1, 2]$, which presents the four-dimensional number $h_4 = 4 + 2e_1 + 2e_2 + e_3$ from the algebra, whose multiplication table of basic elements is also shown.

Fig. 8.2.2 shows the third tensor power of the same matrix, which presents an eight-dimensional number from the corresponding algebra, which is the extension of the two-dimensional algebra of hyperbolic numbers.

Chapter 13, which is devoted to the theme of musical harmony in ensembles of molecular parameters of the genetic code system, contains analogous data about the second and third tensor powers of the matrix

$$\begin{vmatrix} 4 & 2 & 2 & 1 \\ 2 & 4 & 1 & 2 \\ 2 & 1 & 4 & 2 \\ 1 & 2 & 2 & 4 \end{vmatrix}$$

×	e_0	e_1	e_2	e_3
e_0	e_0	e_1	e_2	e_3
e_1	e_1	e_0	e_3	e_2
e_2	e_2	e_3	e_0	e_1
e_3	e_3	e_2	e_1	e_0

Fig. 8.2.1. The second tensor power of the matrix $H_\beta(1,2) = [2, 1; 1, 2]$, which presents four-dimensional number $h_4 = 4 + 2e_1 + 2e_2 + e_3$ from the algebra, whose multiplication table is shown on the right.

8	4	4	2	4	2	2	1
4	8	2	4	2	4	1	2
4	2	8	4	2	1	4	2
2	4	4	8	1	2	2	4
4	2	2	1	8	4	4	2
2	4	1	2	4	8	2	4
2	1	4	2	4	2	8	4
1	2	2	4	2	4	4	8

×	e_0	e_1	e_2	e_3	e_4	e_5	e_6	e_7
e_0	e_0	e_1	e_2	e_3	e_4	e_5	e_6	e_7
e_1	e_1	e_0	e_3	e_2	e_5	e_4	e_7	e_6
e_2	e_2	e_3	e_0	e_1	e_6	e_7	e_4	e_5
e_3	e_3	e_2	e_1	e_0	e_7	e_6	e_5	e_4
e_4	e_4	e_5	e_6	e_7	e_0	e_1	e_2	e_3
e_5	e_5	e_4	e_7	e_6	e_1	e_0	e_3	e_2
e_6	e_6	e_7	e_4	e_5	e_2	e_3	e_0	e_1
e_7	e_7	e_6	e_5	e_4	e_3	e_2	e_1	e_0

Fig. 8.2.2. The third tensor power of the matrix $H_\beta(1,2) = [2, 1; 1, 2]$, which presents the eight-dimensional number $h_8 = 8 + 4e_1 + 4e_2 + 2e_3 + 4e_4 + 2e_5 + 2e_6 + e_7$ from the algebra, whose multiplication table is shown on the right.

$H_\gamma(2,3) = [3, 2; 2, 3]$ and the golden matrix $[\varphi, \varphi^{-1}; \varphi^{-1}, \varphi]$, which present appropriate four- and eight-dimensional numbers from the same algebras.

8.3. Symmetric genetic matrices and doubly stochastic matrices

The tensor families of matrices $H_\beta(1,2)^{(n)} = [2, 1; 1, 2]^{(n)}$ and $H_\gamma(2,3) = [3, 2; 2, 3]^{(n)}$, $[\varphi, \varphi^{-1}; \varphi^{-1}, \varphi]^{(n)}$, whose initial members are shown in Figs. 8.4.1, 8.4.2, 13.2.1, and 13.3.1, have the following mutual feature: in each of the matrices, sums of numeric entries in each of its rows and in each of its columns are identical. For example, in each of the matrices $[2, 1; 1, 2]^{(n)}$, all sums in each row and column are equal to 3^n. In each of the matrices $[3, 2; 2, 3]^{(n)}$, all sums in each of its rows and columns are equal to 5^n, and so on. If in each of these matrices all its members are divided by the sum of each of its rows or columns, then a matrix is formed in which the sum of the members in each row and column is 1. In mathematics, such matrices are called doubly stochastic. They have essential applications in many scientific fields: probability, combinatorics, linear programming and planning, theory of games and optimization, forming of coalitions, models for oncology study, economy, etc.

Doubly stochastic matrices are used for modeling stochastic phenomena in genetics. For example, the Hardy-Weinberg law, which is called in biological literature as the basis of mathematical constructions in population genetics and contemporary evolutionary theory, is simulated on the basis of the tensor family $[p, q; q, p]^{(n)}$ of bisymmetric doubly stochastic matrices where $p + q = 1$ [206].

Doubly stochastic matrices are related to the structure of the DNA double helix, in which the fractional content of complementary nucleobases is: %C = %G and %A = %T. Accordingly, the matrix 2[%C, %A; %T, %G] is doubly stochastic because the sums of the percentages in each row and column are 1.

Doubly stochastic matrices relate to unitary matrices as the following theorem claims [367]:

- if a square $(n \times n)$-matrix $\mathrm{M} = |\mathrm{m}_{ij}|$ is unitary then a $(n \times n)$-matrix $\mathrm{B} = |\mathrm{b}_{ij}|$, where $\mathrm{b}_{ij} = |\mathrm{m}_{ij}|^2$, is doubly stochastic.

In line with this theorem, Fig. 8.3.1 shows an interrelation between four unitary (2×2)-matrices and a doubly stochastic (2×2)-matrix of percentages of nucleotides in DNA double helices where %C = %G and %A = %T. Two of these four unitary matrices present transformations of mirror reflections. The other two unitary matrices are transformations of rotations of vectors clockwise and counterclockwise and simultaneously they are matrix presentations of complex number $z = \sqrt{2\%C} + i \times \sqrt{2\%A}$, where $i^2 = -1$ is the imaginary unit of complex numbers.

This interrelation of the percentage structural features of DNA double helices with unitary matrices of rotations clockwise and counterclockwise can be used for algebraic modeling of well-known Gestalt morphogenetic phenomena of left and right spirals realized in a great number of genetically

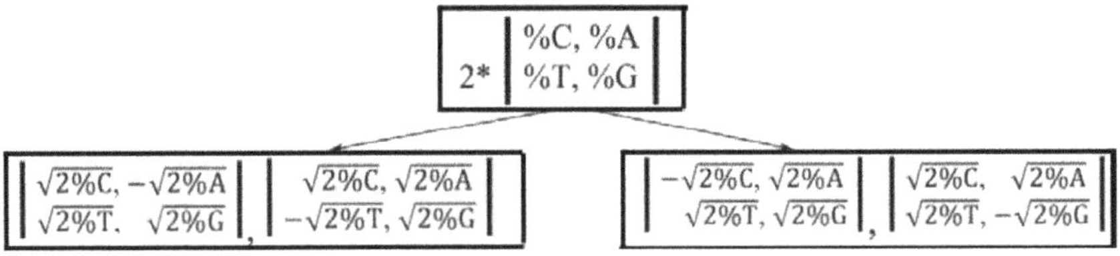

Fig. 8.3.1. Relationship between the doubly stochastic matrix of nucleotide fractions in the DNA double helix (upper row) and four unitary matrices (lower row). The two unitary matrices on the left represent the rotation transformations, and the other two unitary matrices on the right represent the mirror image transformations.

inherited biological structures. The described connection of matrices of nucleotide percentage in DNA double helices with complex numbers, which are presented by noted unitary matrices, is important for algebraic and quantum mechanical modeling of genetic phenomena since quantum mechanics is closely related to complex numbers.

Such interrelations of (2×2)-matrices, shown in Fig. 8.4.1, can be generalized for $(2^n \times 2^n)$-matrices produced by tensor powers of the initial unitary (2×2)-matrix and the doubly stochastic (2×2)-matrix (as it is known, tensor powers of unitary matrices generate new unitary matrices of increased orders). Tensor powers of (2×2)-matrices, which present complex numbers, produce $(2^n \times 2^n)$-matrices, which present algebraic extensions of complex numbers.

8.4. Some concluding remarks

The genetic code systems contain many non-trivial regular structures, that are affinity to a variety of algebraic tools and notions. These tools and notions can be explored to effectively simulate them for including the genetic code system into the field of developed mathematical natural science. Analysis of the molecular ensemble of binary-oppositional traits of the genetic code system allow us to reveal this affinity and propose a variety of new algebraic ways to study the genetic code system and its emergence properties.

The discovered connections between phenomenological features of the genetic alphabets, which are presented in appropriate matrix forms, and important notions of matrix and tensor analysis include wide topics of metric tensors, doubly stochastic matrices, unitary matrices, hyperbolic numbers, and their algebraic extensions. The obtained results are useful for deeper understanding the genetic code system, creating effective algorithms of informational processing, developing the theory of morpho-resonance field and, perhaps, for progress in quantum computing. The next chapters will continue to study the intersections between the worlds of genetic structures and algebraic formalisms.

Chapter 9

Algebraic Modeling of Biological Dualism "Stochastics-Determinism" and Gestalt Phenomena

It is not enough just to say about the stochastics-determinism dualism in biosystems. It is desirable to indicate the mathematical operations that allow modeling of this dualism. The chapter is devoted to the algebraic modeling of interrelations between stochastic molecular processes and inherited deterministic structures in biological bodies. Studying universal rules of stochastic organization of DNAs in genomes of higher and lower organisms, it was revealed that structural features of these genomic stochastics are related to algebraic operations called "tensor-unitary transformations". These operations allow algebraic modeling of the biological dualism "stochastics-determinism", physiological Gestalt phenomena, and other biological structures. In this chapter we present features of tensor-unitary transformations and their possible applications in algebraic biology and understanding of genetic intelligence. It is supposed that they are useful for the development of quantum biomechanics to model biosimilar quantum systems capable of tensor self-development with the inheritance of memory of past states.

9.1. Tensor-unitary transformations and modeling dualism "stochastics-determinism"

It is natural to model the above-presented universal rules of the stochastic-deterministic organization of genomic DNAs in the language of quantum mechanics and quantum informatics, which are based on the concept of probability. It can be noted here that the concept of a quantum computer first appeared in 1980 in the book of Yu. I. Manin precisely in connection with the analysis of high-speed processing of DNA information in chromosomes by "*genetic automata*". This book predicted the important role of unitary transformations and tensor products: "*A quantum automaton must be abstract: its mathematical model must use only the most general quantum principles, without prejudging physical realizations. Then the model of evolution is a unitary rotation in a finite-dimensional Hilbert space, and the model of virtual division into subsystems corresponds to the decomposition of space into a tensor product. Somewhere in this picture, an interaction should find its place, which is traditionally described by Hermitian operators and probabilities*" [368, p. 15]. Thus, the very birth of quantum informatics was due to the desire to understand the features of genetic informatics and its stochastics!

The universal rules of stochastic organization of genomic DNAs, which are presented in our work as multilayer texts written in the languages of tensor interconnected alphabets of DNA n-plets, correspond to this prediction by Manin. Modeling of their interconnections led to the introduction of a new concept of "tensor-unitary transformations". One can recall here that in quantum mechanics and quantum informatics, unitary transformations have fundamental meaning: the evolution of closed quantum systems is described by unitary transformations, and all calculations in quantum informatics are based on unitary operators acting as quantum gates. Unitary transformations (they are orthogonal transformations in the case of real components) of vectors preserve their lengths and the inner products.

Tensor-unitary transformations are transformations that preserve the norms (lengths) of vectors when they are tensor transformed into vectors of a space of increased dimension [369]. Unlike unitary transformations, which transform the original n-dimensional vector into an n-dimensional vector of the same dimension, tensor-unitary transformations transform the original ("mother") n-dimensional vector into a "daughter" m-dimensional vector of increased dimension ($m > n$) while preserving the length of the mother vector. One can say that they provide the "inheritance" of the vector length into the length of a new tensor-transformed vector belonging to a new space of tensorial increased dimension; this new space can be interpreted as the configuration space of a multi-parameter system whose number of parameters increases in a course of its development (by analogy with biosystems under their ontogenetic growth).

Applying these tensor-unitary transformations to vectors uses a two-step process. First, the original n-dimensional vector, called the mother vector, is represented as the sum of all its basis vectors with their coordinate weights (for example, the vector [x, y] is represented as [x, 0] + [0, y], that is, x[1, 0] + y[0,1]). Then each of these weight basis vectors is tensor-multiplied by the so-called "normvector" (qubit-like vector) ensuring the preservation of the length of the mother vector in the daughter vector of the tensor-increased dimension. By definition, a normvector is a k-dimensional vector $[\alpha_0, \alpha_1, \alpha_2, \ldots, \alpha k_{-1}]$, whose sum of squares of coordinates is equal to one: $\alpha_0^2 + \alpha_1^2 + \alpha_2^2 + \cdots + \alpha_{k-1}^2 = 1$. Its coordinates can be considered as probability amplitudes.

One can briefly mention that the tensor-unitary transformations can be represented as a very special case of the Hadamard products for block vectors or matrices when it comes to component-wise multiplication of a vector just by qubit-like normvectors, the set of components of which coincides with the set of probability amplitudes of the corresponding qubit or polyqubit. For the operation of tensor-unitary transformations a special symbol ⍝ can be proposed (its Unicode 0 × 235d, https://unicodemap.org/details/0x235D/index.html). This symbol resembles the shape of a thumb with a nail, which is one of the representatives for the biological gestalt phenomena of the inheritance of deterministic configurations under conditions of stochastic interactions of individual molecules in cells.

Why do universal stochastic rules of genomic DNAs relate to tensor-unitary transformations? What mathematical features are provided by these transformations? Could they be used for modeling the biological gestalt phenomena and the dualism "stochastics-and-determinism"? Why does living nature prefers the usage of tensor-unitary transformations in genetic systems? The main aim of this chapter is to study these questions and present some results and thoughts about appropriate algebraic tools for modeling important biological phenomena, which were not algebraically modeled early but now have been included in the field of mathematical natural science.

The "qubit-like" property of normvectors introduces an element of probability (stochasticity) into the values of individual coordinates of the daughter multi-dimensional vector, whose special groupings of coordinates turn out to be deterministic memory carriers of the coordinates of the mother vector. Thus, tensor-unitary transformations, being stochastic-deterministic transformations, generate stochastic-deterministic vectors and make it possible to model the biological dualism "stochastics-determinism". Let us explain this with simple examples.

One can consider the simplest mother vector function [x(t), y(t)] = [x(t), 0] + [0, y(t)] and tensor multiply its first weighted basis vector [x(t), 0] with the normvector $[\alpha_0, \alpha_1]$, and its second weighted basis vector [0, y(t)] with the normvector $[\beta_0, \beta_1]$ (here $\alpha_0^2 + \alpha_1^2 = 1$, and $\beta_0^2 + \beta_1^2 = 1$):

$$\begin{aligned} \vec{D}_1 &= [x(t), 0] \otimes [\alpha_0, \alpha_1] + [0, y(t)] \otimes [\beta_0, \beta_1] \\ &= [x(t)\alpha_0, x(t)\alpha_1, y(t)\beta_0, y(t)\beta_1] \end{aligned} \tag{9.1}$$

Using the mentioned symbol $\circledcirc$ of the tensor-unitary transformation, this expression can be written also in the following form:

$$[x(t), y(t)] \circledcirc \{[\alpha_0, \alpha_1], [\beta_0, \beta_1]\} = [x(t)\alpha_0, x(t)\alpha_1, y(t)\beta_0, y(t)\beta_1] \tag{9.2}$$

The values of the individual coordinates of the daughter vector function $\vec{D}_1$ (9.1) are stochastic since each of them contains one of the probability amplitudes α_0, α_1, β_0, β_1. There exist infinitely many options for the values of α_0, α_1, β_0, and β_1, and their random selection changes the values of these individual coordinates, but the lengths of the daughter and mother vectors will be equal under all these selections and under each fixed value of the variable t. In other words, using different sets of the values α_0, α_1, β_0, β_1, you get many daughter vector functions $\vec{D}_1$, which differ from each other by their coordinates but all of them have the same length under any fixed value of the parameter t. This is confirmed by calculating the length $||\vec{D}_1||$ of the vector function (9.1) under each fixed value of a variable t:

$$||\mathbf{D}_1|| = \{x^2\alpha_0^2 + x^2\alpha_1^2 + y^2\beta_0^2 + y^2\beta_1^2\}^{0.5}$$
$$= \{x^2(\alpha_0^2 + \alpha_1^2) + y^2(\beta_0^2 + \beta_1^2)\}^{0.5} = (x(t)^2 + y(t)^2)^{0.5} \tag{9.3}$$

This length (9.3) of the daughter four-dimensional vector $\vec{D}_1$ is equal to the length of the mother two-dimensional vector function [x(t), y(t)] at any fixed time t. In other words, the daughter vector $\vec{D}_1$, having stochastic coordinates and the increased dimensionality, carries an exact "memory" of the length of the mother vector function at any fixed moment in time, regardless of the stochasticity of its individual coordinates, and in this respect is a deterministic entity. One can say that here you have stochasticity with hidden determinism. Now let us note other deterministic features of this daughter's four-dimensional vector $\vec{D}_1$, which are connected with its projections on coordinate planes.

A four-dimensional vector space with a Cartesian system of numbered coordinates $[X_0, X_1, X_2, X_3]$ contains 6 coordinate planes:

$$(X_0, X_1), (X_2, X_3), (X_0, X_2), (X_0, X_3), (X_1, X_2), (X_1, X_3) \tag{9.4}$$

In the plane (X_0, X_1), the daughter vector $\vec{D}_1$ (9.1) is represented by its projection vector $\vec{M}_{01} = [x(t)\alpha_0, x(t)\alpha_1, 0, 0]$. Its length $||\mathbf{M}_{01}||$ (9.5) is equal to the first coordinate of the mother vector [x(t), y(t)] at any fixed time t:

$$||\mathbf{M}_{01}|| = (a(t)^2\alpha_0^2 + a(t)^2\alpha_1^2)^{0.5} = \{a(t)^2(\alpha_0^2 + \alpha_1^2)\}^{0.5} = a(t) \tag{9.5}$$

In the plane (X_2, X_3), the daughter vector $\vec{D}_1$ is represented by its projection vector $\vec{M}_{23} = [0, 0, y(t)\beta_0, y(t)\beta_1]$. Its length $||\mathbf{M}_{23}||$ (9.6) is equal to the second coordinate of the mother vector [x(t), y(t)] at any fixed time t:

$$||\mathbf{M}_{23}|| = \{y(t)^2\beta_0^2 + y(t)^2\beta_1^2\}^{0.5} = \{y(t)^2(\beta_0^2 + \beta_1^2)\}^{0.5} = y(t) \tag{9.6}$$

Expressions (9.3), (9.5), and (9.6) indicate that the daughter four-dimensional vector $\vec{D}_1$, having stochastic coordinates, contains in the groupings of these coordinates the exact "memory" of all coordinates and length of the mother two-dimensional vector, that is, in this respect it is a deterministic entity.

Thus, tensor-unitary transformations and families of daughter vectors, generated by them, make it possible to model the biological dualism "stochastics-determinism", and the phenomena of Gestalt biology, in which only relations in groupings of elements are important and meaningful, but not individual elements in themselves. For example, we recognize a musical melody even when it is performed on different instruments and in different frequency ranges with a change in the sound frequencies of individual notes, which turn out to be insignificant, in contrast to the ratios in the groupings of musical frequencies. Since tensor-unitary transformations make it possible to model biological Gestalt phenomena, they can also be conditionally called Gestalt transformations, and the stochastic-deterministic vectors (or vector functions) generated by them can be called Gestalt vectors.

One should emphasize that usually, unitary (or orthogonal) operators are understood as unitary square matrices, but in the case of tensor-unitary transformations, the component-wise tensor product of a mother vector and corresponding number of normvectors is used to get a stochastic-deterministic object (number of normvectors is equal to the dimensionality of the mother vector, whose components should each be multiplied by an individual normvector).

It should be noted that the daughter four-dimensional vector $\vec{D}_1$ (9.1) is not only the carrier of the exact memory of the mother two-dimensional vector coordinates but also the carrier of new information in the planes (X_0, X_2), (X_0, X_3), (X_1, X_2), (X_1, X_3) of four-dimensional space. In these planes, the daughter vector $\vec{D}_1$ is represented by its projection vectors $\vec{M}_{02} = [x(t)\alpha_0, 0, y(t)\beta_0, 0]$, $\vec{M}_{03} = [x(t)\alpha_0, 0, 0, y(t)\beta_1]$, $\vec{M}_{12} = [0, x(t)\alpha_1, y(t)\beta_0, 0]$, $\vec{M}_{13} = [0, x(t)\alpha_1, 0, y(t)\beta_1]$ whose lengths are not equal to the coordinates of the mother vector. The values of their lengths carry new information, which is entered into them by the stochastic coordinates of the normvectors α_0, α_1, β_0, β_1. These stochastic components can reflect, for example, the specifics of the stage of ontogenesis or the impacts on the organism from the external environment. Thus, the **tensor-unitary transformation is a memory expansion operation** (while maintaining all the previous memory) for a tensor-extended multi-parameter system which, along with the memory of the "ancestor" parameters, contains many new parameters or information. Correspondingly, tensor-unitary transformations can be used for modeling growing stochastic-deterministic biosystems of morphogenetic, biorhythmic, and other kinds where the number of parameters and degrees of freedom are increased step by step.

The values of the dot products of two mother vectors — for example, the vectors [x, y] and [s, r] — and of two of their daughter vectors are equal only when tensor-unitary transformations of both mother vectors use the same set of normvectors, for example, $[\alpha_0, \alpha_1]$ and $[\beta_0, \beta_1]$. To illustrate this, one can take two mother vectors [x, y] and [s, r], whose dot product is equal to xs + yr. Now let us tensor-unitary transform both mother vectors into their daughter vectors $\vec{P}$ and $\vec{V}$ by using these normvectors $[\alpha_0, \alpha_1]$ and $[\beta_0, \beta_1]$:

$$\begin{aligned} \vec{P} &= [x, y] \otimes \{[\alpha_0, \alpha_1], [\beta_0, \beta_1]\} = [x\alpha_0, x\alpha_1, y\beta_0, y\beta_1] \\ \vec{V} &= [s, r] \otimes \{[\alpha_0, \alpha_1], [\beta_0, \beta_1]\} = [s\alpha_0, s\alpha_1, r\beta_0, r\beta_1] \end{aligned} \tag{9.7}$$

Calculating the dot product of these two daughter vectors $\vec{P}$ and $\vec{V}$ gives the same value as the dot product of their mother vectors:

$$\begin{aligned} \vec{P} \bullet \vec{V} &= x\alpha_0 s\alpha_0 + x\alpha_1 s\alpha_1 + y\beta_0 r\beta_0 + y\beta_1 r\beta_1 \\ &= xs(\alpha_0^2 + \alpha_1^2) + yr(\beta_0^2 + \beta_1^2) = xs + yr \end{aligned} \tag{9.8}$$

One can easily check that if components of two mother vectors are tensor multiplied by different normvectors (for example, the mother vector [x, y] is tensor-unitary transformed by normvectors $[\alpha_0, \alpha_1]$ and $[\beta_0, \beta_1]$, and the mother vector [s, r] is tensor-unitary transformed by normvectors $[\gamma_0, \gamma_1]$

and $[\delta_0, \delta_1]$), then the dot product of the daughter vectors are not equal to the dot product of the mother vectors.

Tensor-unitary transformations can be repeatedly applied to generate or develop from a mother vector more and more complex multi-parameter daughter vectors with a step-by-step increase in the dimension of their configuration spaces. In these growing daughter vectors, the connections of parameters from mother vectors and also from previous daughter vectors will be preserved simultaneously with arising of new parameters and relationships. Tensor-unitary transformations make it possible to record a step-by-step history of the development of a tensor-growing multi-parametric stochastic-deterministic system in a form of sequences (or a set of the sequences) of daughter vectors corresponding to an initial mother vector (or a set of the mother vectors).

Regarding this step-by-step history of development by repeating of tensor-unitary transformations, let us consider an example where the above four-dimensional daughter vector function $\vec{D}_1$ (9.1) is tensor-unitarily transformed into an eight-dimensional daughter vector function $\vec{D}_2$. This example will show that all data about lengths and components of the daughter vector function $\vec{D}_1$ are preserved in certain groupings of coordinates of the daughter vector function $\vec{D}_2$, that is in projections of $\vec{D}_2$ into appropriate subspaces of eight-dimensional space with Cartesian coordinate system.

Regarding this step-by-step history of development of multi-parametric systems by repeating of tensor-unitary transformations, let us consider an example when the above four-dimensional daughter vector function $\vec{D}_1$ (9.1) is tensor-unitarily transformed into an eight-dimensional daughter vector function $\vec{D}_2$. This example will show that all data about lengths and components of the daughter vector function $\vec{D}_1$ are preserved in certain groupings of stochastic coordinates of the daughter vector function $\vec{D}_2$, that is, in projections of $\vec{D}_2$ into appropriate subspaces of eight-dimensional space with Cartesian coordinate system. To simplify the notation, we will not constantly indicate that the coordinates of the considered vector functions are functions of time.

To construct $\vec{D}_2$, by analogy with the above, we represent the vector $\vec{D}_1$ as the sum of its weighted basis vectors (9.9):

$$\vec{D}_1 = [x\alpha_0, 0, 0, 0] + [0, x\alpha_1, 0, 0] + [0, 0, y\beta_0, 0] + [0, 0, 0, y\beta_1] \quad (9.9)$$

Then we tensor multiply each of these four weight basis vectors in (9.9) by one of the 4 normvectors (9.10):

$$[c_0, c_1], [d_0, d_1], [p_0, p_1], [r_0, r_1],$$
$$\text{where } c_0^2 + c_1^2 = 1, d_0^2 + d_1^2 = 1, p_0^2 + p_1^2 = 1, r_0^2 + r_1^2 = 1 \quad (9.10)$$

The resulting eight-dimensional daughter vector $\vec{D}_2$ has the form (9.11):

$$\begin{aligned}\vec{D}_2 &= [x\alpha_0, 0, 0, 0] \otimes [c_0, c_1] + [0, x\alpha_1, 0, 0] \otimes [d_0, d_1] + [0, 0, y\beta_0, 0] \\ &\quad \otimes [p_0, p_1] + [0, 0, 0, y\beta_1] \otimes [r_0, r_1] \\ &= [x\alpha_0 c_0, x\alpha_0 c_1, 0, 0, 0, 0, 0, 0] + [0, 0, x\alpha_1 d_0, x\alpha_1 d_1, 0, 0, 0, 0] \\ &\quad + [0, 0, 0, 0, y\beta_0 p_0, y\beta_0 p_1, 0, 0] + [0, 0, 0, 0, 0, 0, y\beta_1 r_0, y\beta_1 r_1] \\ &= [x\alpha_0 c_0, x\alpha_0 c_1, x\alpha_1 d_0, x\alpha_1 d_1, y\beta_0 p_0, y\beta_0 p_1, y\beta_1 r_0, y\beta_1 r_1] \quad (9.11)\end{aligned}$$

The values of the individual coordinates of this daughter vector $\vec{D}_2$ are also probabilistic in nature, since each of them contains the amplitude of the probabilities, that is, in this aspect this vector is stochastic.

Calculation of the length $||D_2||$ of the vector $\vec{D}_2$ (9.11), considering the relations (9.10), gives the following:

$$\begin{aligned}||D_2|| &= \{(x\alpha_0 c_0)^2 + (x\alpha_0 c_1)^2 + (x\alpha_1 d_0)^2 + (x\alpha_1 d_1)^2 + (y\beta_0 p_0)^2 + (y\beta_0 p_1)^2 \\ &\quad + (y\beta_1 r_0)^2 + (y\beta_1 r_1)^2\}^{0.5} \\ &= \{(x\alpha_0)^2(c_0^2 + c_1^2) + (x\alpha_1)^2(d_0^2 + d_1^2) + (y\beta_0)^2(p_0^2 + p_1^2) \\ &\quad + (y\beta_1)^2(r_0^2 + r_1^2)\}^{0.5} \\ &= \{(x^2(\alpha_0^2 + \alpha_1^2) + y^2(\beta_0^2 + \beta_1^2)\}^{0.5} = (x^2 + y^2)^{0.5} \quad (9.12)\end{aligned}$$

At any time, this length (9.12) of the daughter eight-dimensional vector function $\vec{D}_2$ is equal to the length of the mother two-dimensional vector function [x(t), y(t)] and the length (9.3) of the previous daughter four-dimensional vector function $\vec{D}_1$. In other words, at any fixed point in time, in the grouping of its coordinates, the daughter vector function $\vec{D}_2$ carries an exact "memory" of the length of the mother vector function, as well as the length of the daughter vector function $\vec{D}_1$, regardless of the stochasticity of its individual coordinates. Accordingly, the stochastic vector function $\vec{D}_2$ in this aspect is a deterministic entity and belongs to the class of stochastically deterministic vector functions (or vectors). We now also note other deterministic features of this daughter vector function $\vec{D}_2$ generated by the tensor-unitary transformation.

We enumerate through X_i the order of coordinates in the eight-dimensional vector space with the Cartesian coordinate system: $[X_0, X_1, X_2, X_3, X_4, X_5, X_6, X_7]$. Let us show that the groupings of 8 weight basis vector functions of $\vec{D}_2$ (9.11) contain exact information (memory) about each coordinate of the mother vector function [x(t), y(t)] and each coordinate of the previous daughter four-dimensional vector function $\vec{D}_1$ at any fixed time.

In the subspace (X_0, X_1, X_2, X_3) the eight-dimensional vector function $\vec{D}_2$ (9.11) is represented by the vector function $\vec{V}_{0123} = [x\alpha_0 c_0, x\alpha_0 c_1, x\alpha_1 d_0, x\alpha_1 d_1]$. At any time, its length (9.13) is equal to the length and first coordinate of the mother weight basis vector [x, 0] and the length of the two-dimensional vector $\vec{M}_{01}$ (9.5), which is the projection of the daughter four-dimensional vector of the previous generation $\vec{D}_1$ (9.1):

$$\begin{aligned} ||\vec{V}_{0123}|| &= (x^2\alpha_0^2 c_0^2 + x^2\alpha_0^2 c_1^2 + x^2\alpha_1^2 d_0^2 + x^2\alpha_1^2 d_1^2)^{0.5} \\ &= \{x^2\alpha_0^2(c_0^2 + c_1^2) + x^2\alpha_1^2(d_0^2 + d_1^2)\}^{0.5} \\ &= \{x^2\alpha_0^2 + x^2\alpha_1^2\}^{0.5} \\ &= x\{\alpha_0^2 + \alpha_1^2\}^{0.5} \\ &= x(t) \end{aligned} \tag{9.13}$$

In the subspace (X_4, X_5, X_6, X_7) the eight-dimensional vector function $\vec{D}_2$ (9.11) is represented by the vector function $\vec{V}_{4567} = [y\beta_0 p_0, y\beta_0 p_1, y\beta_1 r_0, y\beta_1 r_1]$. At any time, its length (9.14) is equal to the length and second coordinate of the mother vector function [x(t), b(t)], as well as the length of the two-dimensional vector $\vec{M}_{01}$ (9.5), which is the projection of the daughter four-dimensional vector of the previous generation $\vec{D}_1$ (9.1):

$$\begin{aligned}||\mathbf{V}_{4567}|| &= \{y^2\beta_0^2p_0^2 + y^2\beta_0 p_1^2 + y^2\beta_1^2r_0^2 + y^2\beta_1^2r_1^2\}^{0.5}\\ &= \{y^2\beta_0^2\,(p_0^2 + p_1^2) + y^2\beta_1^2\,(r_0^2 + r_1^2)\}^{0.5}\\ &= \{y^2\beta_0^2 + y^2\beta_1^2\}^{0.5}\\ &= y\{\beta_0^2 + \beta_1^2\}^{0.5}\\ &= b(t)\end{aligned} \qquad (9.14)$$

Let us now show that the groupings of coordinates of the eight-dimensional daughter vector function $\vec{D}_2$ (9.11) also carry the exact memory of all four coordinates of the four-dimensional daughter vector function of the previous generation $\vec{D}_1$ (9.1).

In the plane (X_0, X_1), the eight-dimensional vector function $\vec{D}_2$ (9.11) is represented by the vector function $\vec{V}_{01} = [x\alpha_0 c_0, x\alpha_0 c_1]$. Its length (9.15) is equal to the first coordinate of the daughter vector function $\vec{D}_1$ (9.1):

$$||\mathbf{V}_{01}|| = (x^2\alpha_0^2c_0^2 + x^2\alpha_0^2c_1^2)^{0.5} = x\alpha_0(c_0^2 + c_1^2)^{0.5} = x\alpha_0 \qquad (9.15)$$

In the plane (X_2, X_3), the eight-dimensional vector function $\vec{D}_2$ (9.11) is represented by the vector function $\vec{V}_{23} = [x\alpha_1 d_0, x\alpha_1 d_1]$. Its length (9.16) is equal to the second coordinate $x\alpha_1$ of the daughter vector $\vec{D}_1$ (9.1):

$$||\mathbf{V}_{23}|| = (x^2\alpha_1^2d_0^2 + x^2\alpha_1^2d_1^2)^{0.5} = x\alpha_1(d_0^2 + d_1^2)^{0.5} = x\alpha_1 \qquad (9.16)$$

In the plane (X_4, X_5), the eight-dimensional vector function $\vec{D}_2$ (9.11) is represented by the vector function $\vec{V}_{45} = [y\beta_0 p_0, y\beta_0 p_1]$. Its length (9.17) is equal to the third coordinate $y\beta_0$ of the daughter vector function $\vec{D}_1$ (9.1):

$$||\mathbf{V}_{45}|| = (y^2\beta_0^2p_0^2 + y^2\beta_0^2p_1^2)^{0.5} = y\beta_0(p_0^2 + p_1^2)^{0.5} = y\beta_0 \qquad (9.17)$$

In the plane (X_6, X_7), the eight-dimensional vector function $\vec{D}_2$ is represented by the vector function $\vec{V}_{67} = [y\beta_1 r_0, y\beta_1 r_1]$. Its length (9.18) is equal to the fourth coordinate $y\beta_1$ of the daughter vector function $\vec{D}_1$ (9.1):

$$||\mathbf{V}_{67}|| = (y^2\beta_1^2r_0^2 + y^2\beta_1^2r_1^2)^{0.5} = y\beta_1(r_0^2 + r_1^2)^{0.5} = y\beta_1 \qquad (9.18)$$

Thus, indeed, the groupings of the coordinates of the daughter eight-dimensional vector function $\vec{D}_2$, presented in the expressions (9.13)–(9.18), carry the exact memory of all the coordinates of the mother vector function [x(t), y(t)] and the daughter four-dimensional vector function of the previous generation $\vec{D}_1$. The set of other groupings of coordinates of the daughter eight-dimensional vector $\vec{D}_2$, for example, in the subspaces (X_0, X_2), (X_1, X_7), (X_1, X_3, X_4, X_5), etc., correspond to vectors whose lengths are not equal to the coordinates of the mother vector and the previous daughter vector $\vec{D}_1$, and can carry new information introduced by the parameters of the normvectors used, which can reflect various factors that affect the complexity of the organism as a developing multi-parametric system.

In addition, the following case should be noted. Coordinates of normvectors can be not only fixed numeric values but also normalized functions of time or other variables. The expression (9.19) shows an example of the two-dimensional normvectors $[\alpha_0, \alpha_1]$ whose coordinates are functions of variable t:

$$\overrightarrow{F(t)} = [\alpha_0(\mathrm{t}), \alpha_1(\mathrm{t})]/\{\alpha_0(\mathrm{t})^2 + \alpha_1(\mathrm{t})^2\}^{0.5} \tag{9.19}$$

At any value t, the sum of squares of its coordinates is equal to 1 (9.20), that is, the vector function $\overrightarrow{F(t)}$ satisfies the definition of normvectors and can be used in tensor-unitary transformations:

$$\begin{aligned}&\{\alpha_0(\mathrm{t})/(\alpha_0(\mathrm{t})^2 + \alpha_1(\mathrm{t})^2)^{0.5}\}^2 + \{\alpha_1(\mathrm{t})/(\alpha_0(\mathrm{t})^2 + \alpha_1(\mathrm{t})^2)^{0.5}\}^2\\ &= \{\alpha_0(\mathrm{t})^2 + \alpha_1(\mathrm{t})^2\}/\{\alpha_0(\mathrm{t})^2 + \alpha_1(\mathrm{t})^2\}\\ &= 1\end{aligned} \tag{9.20}$$

In particular, in this case, when the coordinates of the normvectors are normalized functions of time or another variable, the above expressions (9.3), (9.5), (9.6), (9.8), (9.12)–(9.14) about the memory of "ancestors" in vector lengths are not changed but the expressions (9.15)–(9.18) about new information acquire forms of essentially new vector functions, which depend on vector functions in used normvectors (9.21):

$$
\begin{aligned}
||\mathbf{V}_{01}|| &= (x^2\alpha_0^2c_0^2 + x^2\alpha_0^2c_1^2)^{0.5} = x\alpha_0(c_0^2 + c_1^2)^{0.5} = x(t) \times \alpha_0(t),\\
||\mathbf{V}_{23}|| &= (x^2\alpha_1^2d_0^2 + x^2\alpha_1^2d_1^2)^{0.5} = x\alpha_1(d_0^2 + d_1^2)^{0.5} = x(t) \times \alpha_1(t),\\
||\mathbf{V}_{45}|| &= (y^2\beta_0^2p_0^2 + y^2\beta_0^2p_1^2)^{0.5} = y\beta_0(p_0^2 + p_1^2)^{0.5} = y(t) \times \beta_0(t),\\
||\mathbf{V}_{67}|| &= (y^2\beta_1^2r_0^2 + y^2\beta_1^2r_1^2)^{0.5} = y\beta_1(r_0^2 + r_1^2)^{0.5} = y(t) \times \beta_1(t)
\end{aligned}
\tag{9.21}
$$

This case of using normvectors, whose coordinates are normalized functions of variables, gives new abilities to model some biological phenomena, for example, cellular differentiations when the cells change to specialized types.

It should be emphasized that it is the lengths of daughter vectors and their projections into subspaces of the multi-dimensional configuration space of a multi-parameter system that carry the memory of past and new information. Therefore, in the analysis of genetic diseases, possibly associated with disturbances in the stochastic characteristics of gene sequences, it is necessary to focus on a comparative analysis of the lengths of stochastic vectors and their projections. It is important that in shortened DNA sequences (for example, sequences of separate genes or their combinations) modeling will be based on component-by-component tensor multiplication not on qubit-like normvectors, but on some other vectors; the use of these new vectors generates daughter vectors of a different length than the mother vector. The obtained new daughter vectors, in the case of shortened DNA sequences of genes, can be compared with genomic DNA daughter vectors, paying attention not only to length differences, but also to the values of the dot products between daughter vectors of genes and of genomes.

9.2. Tensor-unitary transformations and multi-reproductions of configurations

Tensor-unitary transformations make it possible to simulate multi-reproductions (multi-replications) of geometric configurations. It can be illustrated by the following example. Let the variable coordinates of the

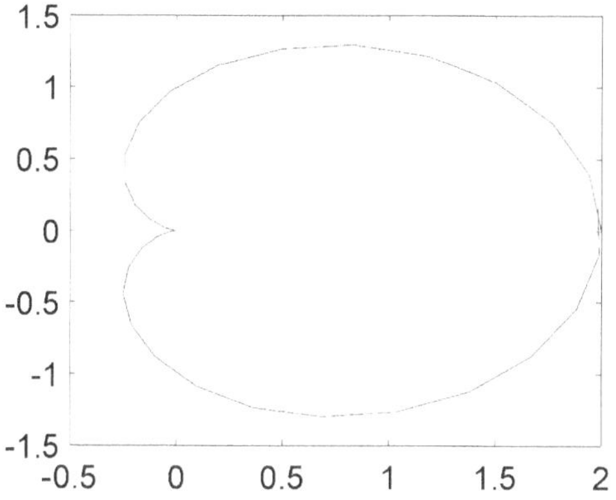

Fig. 9.2.1. Mother cardioid, parametrically specified by the mother vector [x(t), y(t)] where x(t) and y(t) are presented in the expression (9.22).

mother vector function [x(t), y(t)] correspond to the functions of the parametric definition of the curve (Fig. 9.2.1), called the cardioid (we will call it the mother cardioid):

$$x(t) = \cos(t) \times (1 + \cos(t)); \qquad y(t) = \sin(t) \times (1 + \cos(t)) \qquad (9.22)$$

For simplicity, consider the tensor-unitary transformation of the mother vector function [x(t), y(t)] (where x(t) and y(t) are presented by the expression (9.22)), in which both coordinates are tensor multiplied by the same normvector $\overrightarrow{W} = [0.5^{0.5}, 0.5^{0.5}]$, giving a daughter vector function $\vec{R}_4(t)$ (9.23):

$$\begin{aligned}\vec{R}_4(t) &= [x(t), 0] \otimes \overrightarrow{W} + [0, y(t)] \otimes \overrightarrow{W} \\ &= [x(t) \times 0.5^{0.5}, x(t) \times 0.5^{0.5}, y(t) \times 0.5^{0.5}, y(t) \times 0.5^{0.5}] \qquad (9.23)\end{aligned}$$

In the four planes (X_0, X_2), (X_1, X_3), (X_0, X_3) and (X_1, X_2) of the Cartesian coordinate system, this daughter vector function $\vec{R}_4$ is represented by the same projection vector function $[x(t) \times 0.5^{0.5}, y(t) \times 0.5^{0.5}]$, whose coordinates correspond to the parametric description of the cardioid (Fig. 9.2.2). Thus, in these four different two-dimensional subspaces of the four-dimensional

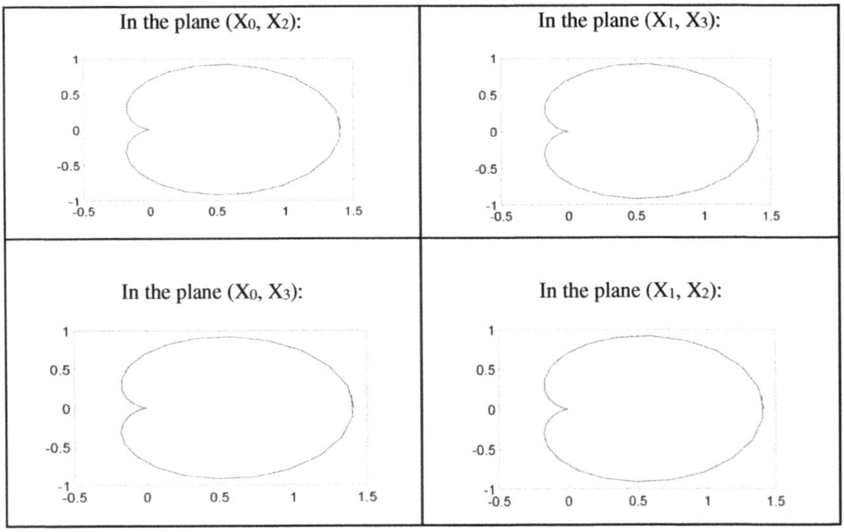

Fig. 9.2.2. Cardioids described parametrically by the projections of the daughter four-dimensional vector function $\vec{R}_4(t)$ (9.23) in the planes (X_0, X_2), (X_0, X_3), (X_1, X_2) and (X_1, X_3) of the Cartesian system coordinates of four-dimensional space.

configuration space of the multi-parameter system, four identical daughter cardioids are generated, which are slightly reduced copies of the mother cardioid. This fact illustrates the general proposition that **tensor-unitary transformations can serve as multi-reproduction operations for structures** given by the mother vector. In other words, these tensor-unitary operations, prompted by the universal rules of the stochastic-deterministic organization of genomic DNAs, make it possible to model in the matrix-tensor language not only the biological dualism "stochastics-determinism" with its gestalt phenomena, but also, for example, the phenomena of cell division under processes of mitosis and meiosis.

If the named normvector has different signs for its coordinates of the same magnitude, for example, $[0.5^{0.5}, -0.5^{0.5}]$, then under the tensor-unitary transformation of the same parent vector, instead of the daughter vector function (9.23), another four-dimensional daughter vector function $\vec{S}_4(t)$ (9.24) arises:

$$\vec{S}_4(t) = [x(t), 0] \otimes [0.5^{0.5}, -0.5^{0.5}] + [0, y(t)] \otimes [0.5^{0.5}, -0.5^{0.5}]$$
$$= [x(t) \times 0.5^{0.5}, -x(t) \times 0.5^{0.5}, y(t) \times 0.5^{0.5}, -y(t) \times 0.5^{0.5}] \quad (9.24)$$

In the plane (X_0, X_2), from this vector function $\vec{S}_4(t)$ there is a vector projection $[x(t) \times 0.5^{0.5}, y(t) \times 0.5^{0.5}]$. In the plane ($X_0$, X_3), from the vector function $\vec{S}_4(t)$ there is a vector projection $[x(t) \times 0.5^{0.5}, -y(t) \times 0.5^{0.5}]$. In the plane ($X_1$, X_2), from the vector function $\vec{S}_4(t)$ there is a vector projection $[-x(t) \times 0.5^{0.5}, y(t) \times 0.5^{0.5}]$. In the plane ($X_1$, X_3), from the vector function $\vec{S}_4(t)$, there is a vector projection $[-x(t) \times 0.5^{0.5}, -y(t) \times 0.5^{0.5}]$.

Fig. 9.2.3 shows the parametrically specified cardioids corresponding to these projections. It shows that this change in the sign of one of the coordinates of the normvector leads to the fact that the set of the four planes (X_0, X_2), (X_1, X_3), (X_0, X_3) and (X_1, X_2) of the four-dimensional space splits into pairs of planes, in which there are mirrored cardioids (in contrast to identical cardioids in Fig. 9.2.2). This is one example of the fact

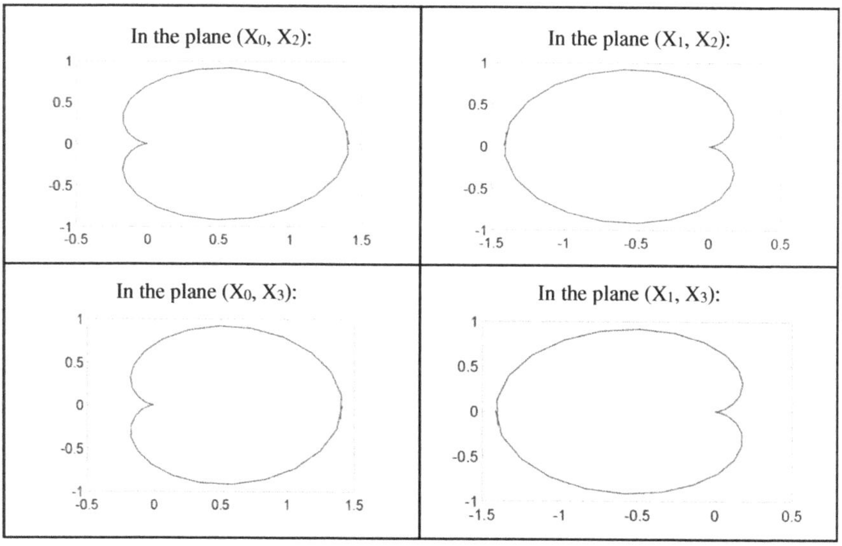

Fig. 9.2.3. Mirrored pairs of cardioids parametrically described by vector projections of the daughter four-dimensional vector function $\vec{S}_4(t)$ (9.24) in the planes (X_0, X_2), (X_0, X_3), (X_1, X_2), and (X_1, X_3) of the Cartesian coordinate system of four-dimensional space.

that in algebraic biology, tensor-unitary transformations make it possible to model the appearance of mirror-symmetric structures in the described reproduction of images from the mother vector function.

Let us now consider an example of the case when, under a tensor-unitary transformation, each coordinate of the mother two-dimensional vector function [x(t), y(t)] is tensor multiplied by an individual two-dimensional normvector with unequal coordinates. For example, let the first coordinate of this mother vector function be tensor multiplied by the normvector $[\alpha_0, \alpha_1]$, $\alpha_0 \neq \alpha_1$, $\alpha_0^2 + \alpha_1^2 = 1$, and the second coordinate by the normvector $[\beta_0, \beta_1]$, $\beta_0 \neq \beta_1$, $\beta_0^2 + \beta_1^2 = 1$. The tensor-unitary transformation of the mother vector function [x(t), y(t)] using such normvectors generates a four-dimensional daughter vector function $\overrightarrow{W_4}(t)$ (9.25):

$$\begin{aligned}\overrightarrow{W_4}(t) &= [x(t), 0] \otimes [\alpha_0, \alpha_1] + [0, y(t)] \otimes [\beta_0, \beta_1] \\ &= [x(t)\alpha_0, x(t)\alpha_1, y(t)\beta_0, y(t)\beta_1] \end{aligned} \tag{9.25}$$

In a four-dimensional space with a Cartesian coordinate system (X_0, X_1, X_2, X_3) in the planes (X_0, X_2), (X_1, X_3), (X_0, X_3), and (X_1, X_2) the following vector projections of this daughter vector function $\overrightarrow{W_4}$ exist:

$$[x(t)\alpha_0, y(t)\beta_0], [x(t)\alpha_1, y(t)\beta_1], [x(t)\alpha_0, y(t)\beta_1], [x(t)\alpha_1, y(t)\beta_0] \tag{9.26}$$

Since the mother vector [x(t), y(t)] parametrically defines some "mother" curve (for example, a cardioid), then each of these four vectors (9.26) also defines a certain parametrically defined curve in its plane. Each of such curves is a repetition of the mother curve, but with deformation due to the scaling of the coordinate axes.

One can turn now to the phenomena of coordinated bio-cycles. The organism is a huge chorus of coordinated cyclic processes, whose number increases as it develops ontogenetically from the embryonic to the mature state with step-by-step increase of new degrees of freedom, which are coordinated with those already existing. For example, in an adult human body, this number reaches enormous values, since it contains approximately 100 trillion cells participating in these sets of coordinated cycles. Moreover, all these cyclic processes occur through stochastic interactions between individual molecules in cells. According to the provisions of ancient chrono-medicine, all our diseases are the result of violations of this coherence.

But how to simulate such development of huge sets of coordinated cyclic processes under conditions of stochasticity of individual parameters of an organism and a huge step-by-step increase in the dimension of its configuration space? Tensor-unitary transformations allow us to offer a model answer to this question since they generate stochastic-deterministic vector functions; in these vector functions, with a purely stochastic nature of the values of individual coordinates, sets of coordinated stable cycles on certain groupings of coordinates can be represented, i.e., on certain groupings of subspaces of the configuration space of a developing multi-parametric organism. As is known, in mathematics, the tensor product of matrices is used to increase the dimension of a vector space. The genetic coding system, together with the family of DNA alphabets, is structured specifically for the tensor product of matrices which links, for example, matrix representations of different DNA alphabets into a single tensor family of matrices [12, 13].

One can add that tensor-unitary transformation can be applied not only in the case of the algebra of real numbers but also in cases of different systems of multi-dimensional numbers: complex numbers, hyperbolic numbers, quaternions by Hamilton, split-quaternions by Cockle, etc.

In quantum mechanics, unitary transformations describe the evolution of closed quantum systems. The author believes that tensor-unitary transformations are useful for the development of quantum mechanics of non-closed quantum systems that develop with an increase in their multicomponent composition, like biological bodies that develop in the course of onto- and phylogenesis.

Speaking of modeling a set of coordinated cyclic processes in an organism, one should also recall the connection between structured DNA alphabets and a family of matrices (9.27):

$$[1, 1; 1, 1]^{(n)} \otimes [1, 1, -1, -1; 1, 1, -1, -1; 1, -1, 1, -1; -1, 1, -1, 1] \quad (9.27)$$

There are interesting actions of such matrices (9.27) to voluntary 2^n-dimensional vectors having arbitrary coordinates $a_i(t)$ which can be, for example, cyclic functions of time. In a general case, actions of $(2^n \times 2^n)$-matrices (9.27) to a voluntary 2^n-dimensional vector with coordinates unrelated to each other gives a new block-structured vector with interrelated coordinates combined in pairs by a binary-oppositional principle. If coordinates of an initial vector of a system are cyclic or wave functions, then the obtained new vector have block-structured sets of coordinates whose blocks are equal to each other up to the sign, and their cyclic changes are synchronized in time (that is, cyclic behavior of the considered system in its different configuration subspaces becomes strongly interrelated). Figuratively speaking, these matrix operators create order out of chaos.

9.3. Gestalt vectors and projection operators based on genetic matrices

In accordance with §9.1 of this chapter, we call gestalt vector functions (or gestalt vectors) those daughter stochastic-deterministic vector functions that result from tensor-unitary transformations of mother vector functions. For example, the above daughter vector functions $\vec{D}_1$ (9.1) and $\vec{D}_2$ (9.11) are examples of gestalt vectors obtained by tensor-unitary transformations from the mother vector function [x(t), y(t)]. This name is due to the fact that, as described above, the gestalt vectors model the features of biological gestalt phenomena, the content of which is determined by the relationship between the groupings of their elements, while the meanings of individual elements are not significant (by analogy with the gestalt phenomena of the perception of a musical melody, recognizable when performed in different frequency ranges, that is, perceived largely independently of the frequency value of each individual musical note).

As described in §9.1 using the examples of gestalt vector functions $\vec{D}_1$ (9.1) and $\vec{D}_2$ (9.11), part of the set of possible groupings of coordinates in gestalt vectors carries the memory of the past states of the developing multicomponent system (that is, knowledge of the coordinates of the parent vector and child gestalt vectors of previous generations), and the other part of the groupings of coordinates in gestalt vectors carries new information that appeared only at this stage in the development of the system and of its multi-dimensional configuration space. For algebraic selection of certain groupings of coordinates in genetic gestalt vectors, projection operators are needed. Are there any projection operators in the structures of the genetic coding system?

It turns out that there is an extensive set of tensorial interconnected projection operators based on matrix representations of structured DNA alphabets. This set of mutually related genetic matrices that meet the criteria of projection operators and are associated with universal projection rules for the stochastic organization of genomic DNAs is described in [70, Figs. 9.1–9.7].

In particular, in Chapter 4, Fig. 4.6.1 the (4 × 4)-matrices of oblique projection operators are described: $\mathbf{P}_0$ = [1 0 1 0; 1 0 1 0; 0 0 0 0], $\mathbf{P}_1$ = [0 0 0 0; 0 0 0 0; 0 1 0 1; 0 1 0 1], $\mathbf{P}_2$ = [0 1 0 1; 0 1 0 1; 0 0 0 0; 0 0 0 0]; $\mathbf{P}_3$ = [0 0 0 0; 0 0 0 0; 1 0 1 0; 1 0 1 0]. What are the results of the action of these genetic projection operators on the gestalt vectors of the corresponding dimension? Let us give an example of their action on the four-dimensional gestalt vector $\vec{D}_1 = [x(t)\alpha_0, x(t)\alpha_1, y(t)\beta_0, y(t)\beta_1]$, described above and shown in the expression (9.1):

$$\vec{D}_1 \times \mathbf{P}_0 = [x(t)\alpha_0(t) + y(t)\beta_0(t), 0, x(t)\alpha_0(t) + y(t)\beta_0(t), 0] \quad (9.28)$$

The new vector function (9.28) formed as a result of this action has an interesting property: its non-zero first and third coordinates are identical, and therefore synchronized in time (if the parameter t denotes time). This means that the resulting vector function consists of two synchronized parts, regardless of what variables or constant values $\alpha_0(t)$ and $\beta_0(t)$ are used in the coordinates of the normvector of the tensor-unitary transformation. In many respects, similar results are obtained when the same vector function $\vec{D}_1$ is subjected to skew genetic projections that give the expressions $\vec{D}_1 \times \mathbf{P}_1$, $\vec{D}_1 \times \mathbf{P}_2$, $\vec{D}_1 \times \mathbf{P}_3$. This synchronization of parts of each vector function thus formed is useful for modeling a number of developing biosystems containing synchronized subsystems.

For the case of eight-dimensional Gestalt vectors, for example, $\vec{D}_2$ (9.11), there is a whole family of genetic (8 × 8)-matrices of oblique projection operators shown in Chapter 4, Figs. 4.6.3–4.6.7. An example from this family is the matrix $\mathbf{P}_8$ = [1 0 1 0 1 0 1 0; 1 0 1 0 1 0 1 0; 0 0 0 0 0 0 0 0; 0 0 0 0 0 0 0 0; 0 0 0 0 0 0 0 0; 0 0 0 0 0 0 0 0; 0 0 0 0 0 0 0 0; 0 0 0 0 0 0 0 0]. Its action on the eight-dimensional vector $\vec{D}_2 = [x\alpha_0c_0, x\alpha_0c_1, x\alpha_1d_0, x\alpha_1d_1, y\beta_0p_0, y\beta_0p_1, y\beta_1r_0, y\beta_1r_1]$ (9.11) generates the eight-dimensional vector function (9.29):

$$\vec{D}_2 \times \mathbf{P}_8 = [x(t)\alpha_0c_0 + x(t)\alpha_0c_1, 0, x(t)\alpha_0c_0 + x(t)\alpha_0c_1, 0, x(t)\alpha_0c_0 + x(t)\alpha_0c_1, 0, x(t)\alpha_0c_0 + x(t)\alpha_0c_1, 0] \quad (9.29)$$

In this vector function, the non-zero first, third, fifth, and seventh coordinates are identical and synchronized in time (if the parameter t denotes time). All other coordinates are equal to zero. This means that the generated vector function consists of four synchronized parts, regardless of what variables or constant values are used in the coordinates of the normvectors of the tensor-unitary transformation. In many respects, similar results are obtained when the same vector function $\vec{D}_2$ is affected by other genetic oblique projectors from Chapter 4, Figs. 4.6.3–4.6.7.

Passing to the cases of Gestalt vectors of even higher dimension and acting on them with the corresponding genetic matrices of oblique projections, we obtain vector functions with an increasing number of non-zero identical coordinates (all other coordinates are equal to zero).

In connection with the presented doctrine of evolution based on bio-antenna arrays, it should be noted that the theory of antenna arrays in technology is associated with projection operators. Correspondingly, for the search words "projection operators and antenna arrays", the browser gives many publications. Therefore, the presence of projection operators in the above-mentioned genetic matrices, as well as the existence of universal projection rules in the stochastic organization of genomic DNAs, can be considered as additional evidences in favor of the adequacy of this doctrine.

9.4. Stochastic determinism as an antipode to deterministic chaos

Our world is full of random events. The life and development of genetically inherited bio-bodies are inextricably linked with random interactions of molecules, against which the inherited deterministic macrostructures with the parental traits are realized. So, hidden determinism and the corresponding laws of stochastic determinism are hidden behind these accidents. The important role of probabilities in Nature is reflected in quantum mechanics based on the concept of probabilities. At the same time, as the Nobel Laureate in Physics R. Feynman stated, "*nobody really understands quantum mechanics*". The studies of the biological dualism "stochastics-determinism" and the universal rules of the stochastic organization of genomic DNAs, presented in this chapter, draw attention to the need to develop the theory of "stochastic determinism" (or "chaotic determinism") as an antipode to the well-known theory of "deterministic chaos". One can believe that in the future theory of stochastic determinism, a prominent place will be occupied by the tensor-unitary transformations described above, which model the named dualism, biological gestalt phenomena, and tensor reproduction of biostructures. Let's look at the differences between these two theories.

The theory of deterministic chaos (or dynamic chaos) has been developed by the works of a large number of mathematicians and physicists. It has been the subject of a large number of publications (see, for example, [370, 371]). Deterministic chaos is a phenomenon and a part in the theory of dynamical systems, in which the behavior of a nonlinear system looks random, despite the fact that it is determined by deterministic laws. The reason for this phenomenon is instability (sensitivity) with respect to the initial conditions and parameters: a small change in the initial condition over time leads to arbitrarily large changes in the dynamics of the system. Dynamics that is sensitive to the slightest changes in the initial conditions of the system, from which its development begins, and in which these slightest deviations multiply many times over time, making it difficult to predict the future states of the system, is often called chaotic. A well-known example of a system of deterministic chaos is Sinai billiards (https://mathworld.wolfram.com/SinaiBilliards.html).

What is the difference between stochastic determinism, represented in the phenomena of biological dualism "stochastics-determinism", and deterministic chaos? Table 9.4.1 shows the main differences.

Table 9.4.1. The main differences between stochastic determinism and deterministic chaos.

Deterministic chaos	**Stochastic determinism**
The birth of the random from the non-random is observed. A completely deterministic system generates unpredictable states that have the properties of a random process.	The birth of the deterministic (non-random) from the random is observed. A system characterized by random interactions of molecules at the cellular level gives rise to deterministic biological forms.
A small change in the initial condition leads to arbitrarily large changes in the dynamics of the system over time (it is instability or sensitivity of the system with respect to the initial conditions).	Random interactions among molecules at the beginning of the biosystem development have little effect on the final deterministic result of its development (stability or insensitivity with respect to initial conditions).
In systems of deterministic chaos, it is customary to consider multi-parameter systems with a fixed dimension of their configuration spaces (for example, Sinai billiards).	In the stochastic determinism of biological systems, systems are considered with tensor development of the dimension of their configuration spaces (for example, when sets of interconnected billiards of Sinai are born tensorially in the course of development).

9.5. Some concluding remarks

The stochastic organization of genomic DNAs must give pieces of evidence about special interconnections between formalisms of antenna arrays and of quantum informatics described above. Correspondingly the doctrine of evolution on bio-antenna arrays is connected with formalisms of quantum informatics and quantum mechanics. The described tensor-unitary transformations are useful for modeling dualism "stochastics-determinism" and it can be useful for creation of quantum mechanics of tensorial development of multi-parametric systems such as biological ones.

The theory of stochastic determinism as an antipode to deterministic chaos should be developed. Vibrational synchronization and vibrational mechanics can be one of the participants in ensuring the biological dualism "stochastics-determinism". The theory of vibrational mechanics itself should be further developed to cover vibrational phenomena in tensor developing multi-parameter systems. The described algebraic tools and results are useful for development of our knowledge about "genetic intelligence" whose notion was considered in Chapter 2.

Chapter 10

Harmonic Progression in the Stochastic Organization of Genomic DNAs

Living bodies are huge sets of various molecules, which have an amazing ability to inherit biological traits of organisms to the next generations. G. Mendel, in his experiments with plant hybrids, found that the transmission of traits under the crossing of organisms occurs by certain algebraic rules, despite the colossal heterogeneity and complexity of the molecular structures of their bodies.

This chapter represents the results of studying hidden algebraic rules regarding quantitative characteristics of *n*-plet groupings in information structures of DNAs. This study uses the so-called oligomer sums method [198, 232]. The method allows study of the oligomer cooperative organization by comparative analysis of total amounts of all *n*-plets (having fixed length *n*) from the certain equivalence classes of oligomers. The term "oligomer" here means a chain of a relatively small number of nucleotides and is analogous to the term "*n*-plet". One can note that regarding the quantitative analysis of DNA sequences, researchers usually study quantities and percentages of separate *n*-plets (that is, separate oligomers, having their length *n*) but not *n*-plet groupings. For example, Chargaff's second rule is based on such a study of the quantities of separate nucleotides A, T, C, and G. The work [191] studies quantities of separate *n*-plets.

Below the oligomer sums method and the results of its application to many genomes and separate nucleotide sequences are represented. In addition, it is noted that this method can be applied to the analysis not only of genomic sequences of nucleotides but also to the analysis of the amino acid sequences of long proteins (the example of the protein Titin is presented).

10.1. The hyperbolic rule in the oligomer cooperative organization of all human nuclear chromosomes

The term "oligomer" refers to a molecular complex of chemicals that consists of a few repeating units. Nucleobases — adenine A, thymine T, cytosine C, and guanine G — serve as such repeated units in DNA oligomers, which can have different lengths, and which are also called *n*-plets, where *n* refers to the oligomer length. Each of the nucleotide sequences in eukaryotic and prokaryotic genomes can be considered as a sequence of monomers (like A-C-A-T-G-T-...), duplets (like AC-AT-GT-GG-...), or triplets (like ACA-TGT-GGA-...), etc. The chapter describes the numerical analysis of sets of *n*-plets, which belong to the equivalence classes (or cooperative groupings) of A_1-oligomers, T_1-oligomers, C_1-oligomers, or G_1-oligomers correspondingly (their index 1 indicates that all oligomers of each class start with the same nucleotide A, T, C, or G). For example, the class of the A_1-oligomers contains the following *n*-plets: 4 duplets AA, AT, AC, and AG; 16 triplets AAA, AAT, AAC, AAG, ATA, ..., AGG; etc. The total amount of different kinds of *n*-plets, which start with the same nucleotide, under fixed *n* is equal to 4^{n-1}.

To simplify a theoretical explanation, let us consider the example of an analysis of the oligomer cooperative organization of human chromosome №1 by the oligomer sums (OS) method. The totality of data obtained by analyzing a nucleotide sequence by the OS method is called its OS representations. This method gives numeric sequences called OS sequences.

The application of the OS method to the analysis of the human chromosome №1 includes the following steps, which are typical also for the analysis of other DNA and RNA sequences:

- Firstly, one should calculate phenomenological quantities S_A, S_T, S_C, and S_G of monomers A, T, C, and G correspondingly in the considered nucleotide sequence. In the human chromosome № 1, the following quantities exist: S_A = 67070277, S_T = 67244164, S_C = 48055043, S_G = 48111528;

- Secondly, to construct the OS sequences, one should calculate the total amounts $\Sigma_{A,n,1}$, $\Sigma_{T,n,1}$, $\Sigma_{C,n,1}$, and $\Sigma_{G,n,1}$ of n-plets in equivalence classes of A_1-oligomers, T_1-oligomers, C_1-oligomers, and G_1-oligomers under $n = 1, 2, 3, 4, \ldots$ (here, for example, the symbol $\Sigma_{A,3,1}$ refers to the total amount of triplets which start with the nucleotide A). These total amounts regarding each of the classes are members of the appropriate OS sequence of the class. For analysis of human chromosomes and various eukaryotic and prokaryotic genomes, it was usually taken as $n = 1, 2, 3, \ldots, 19, 20$ or, in special cases, $n = 1, 2, 3, \ldots, 99, 100$ [198].

One can recall here that genomic sequences in the GenBank sites usually contain some letters N, indicating that there can be any nucleotide in this place (https://www.ncbi.nlm.nih.gov/books/NBK21136/). For this reason, the total amount of all monomers A, T, C, and G (that is, the sum $S_A + S_T + S_C + S_G$), calculated for the sequence from the GenBank, is slightly less than the complete length of the DNA sequence. But practically this is not essential for the results of the application of the OS method to analyze genomic sequences.

For human chromosome №1, phenomenological values of the total amounts of n-plets from the class of A_1-oligomers are shown in the graphical form for $n = 1, 2, 3, \ldots, 20$ in Fig. 10.1.1, left (in blue). Here the abscissa represents the values of n, and the ordinate represents the values of the total amounts $\Sigma_{A,n,1}$ of n-plets, which start with the nucleotide A. The amazing result is that all 20 phenomenological points $[n, \Sigma_{A,n,1}]$ lie — with a high level of accuracy — along with the hyperbola $H_{A,1} = S_A/n = 67070277/n$ shown in red in Fig. 10.1.1, middle. Deviations of phenomenological quantities $\Sigma_{A,n,1}$ from model values S_A/n lie in the range −0.030%/0.024%, that is, they comprise only one-hundredth of a percent (Fig. 10.1.1, right). Initial data on this chromosome were taken from the GenBank: https://www.ncbi.nlm.nih.gov/nuccore/NC_000001.11.

This result is striking because it shows that knowing only the number of nucleotides A, that is, only one member of the number series shown in

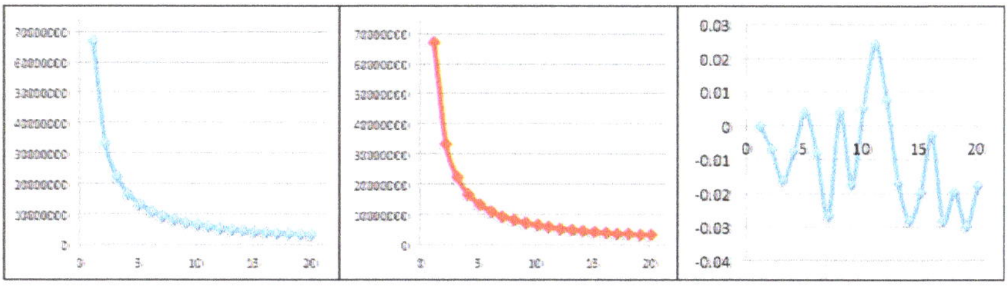

Fig. 10.1.1. The graphs of data for the OS sequences of n-plets from the class A_1-oligomers of the human chromosome №1. In these graphs, the abscissa represents the values $n = 1, 2, 3, \ldots, 20$. Left: the ordinate represents the set of phenomenological total amounts $\Sigma_{A,n,1}$ of n-plets beginning with the nucleotide A. Middle: the ordinate represents modeling values $S_A/n = 67070277/n$. The dots with coordinates $[n, S_A/n]$ belong to the shown hyperbola $H_{A,1} = S_A/n = 67070277/n$. Right: deviations of the real OS sequence $\Sigma_{A,n,1}$ from the model hyperbolic progression S_A/n in percentages.

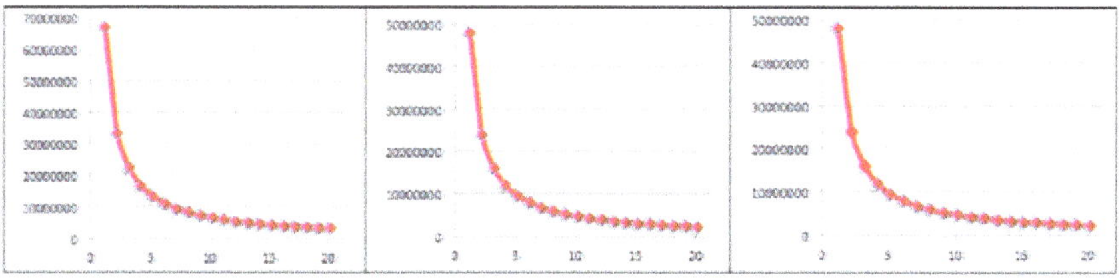

Fig. 10.1.2. Additional graph data to the OS representation of the human chromosome №1. The abscissa represents the values $n = 1, 2, 3, \ldots, 20$. The ordinate shows model values $H_{T,1}(n)$, $H_{C,1}(n)$, and $H_{G,1}(n)$ (in red) from (10.1), which practically coincide with phenomenological values $\Sigma_{T,n,1}$, $\Sigma_{C,n,1}$, and $\Sigma_{G,n,1}$ of the total amount of n-plets that start with the nucleotide T (left), C (middle), and G (right). The numerical data on this coincidence is shown below.

Fig. 10.1.1, left, one can predict with the high accuracy all other 19 members, each of which is a sum of 4^{n-1} possible kinds of n-plets. The number of possible kinds of n-plets in these sums is growing rapidly, becoming astronomically huge: 4, 16, 64, 256, 1024, ..., 4^{10}, ..., 4^{19}. Of course, in the human chromosome №1, for example, not all possible 4^{19} kinds of the mentioned 20-plets exist but the total amount of all those kinds of 20-plets, which exist in this chromosome, is practically equal to $S_A/20$ with a high level of accuracy shown below.

Similar results were obtained when studying in this chromosome the total amounts of n-plets, which start with the nucleotide T (Fig. 10.1.2, left),

C (Fig. 10.1.2, middle), and G (Fig. 10.1.2, right). The phenomenological values of the total amounts $\Sigma_{T,n,1}$, $\Sigma_{C,n,1}$, and $\Sigma_{G,n,1}$ of n-plets are also modeled effectively by appropriate hyperbolic progressions $H_{T,1}$, $H_{C,1}$, $H_{G,1}$ (10.1), which differ from each other only by their numerators S_T, S_C, and S_G:

$$H_{T,1} = S_T/n = 67244164/n,$$
$$H_{C,1} = S_C/n = 48055043/n,$$
$$H_{G,1} = S_G/n = 48111528/n \qquad (10.1)$$

Table 10.1.1 shows real and model values for the OS representation of the classes of A_1-, T_1-, C_1-, and G_1-oligomers of the human chromosome №1 for $n = 1, 2, 3, \ldots, 20$. The model values of the total amounts of n-plets, which start with a certain nucleotide (A, T, C, or G), are calculated correspondingly as values of the hyperbolic progressions $H_{A,1} = S_A/n = 67070277/n$, $H_{T,1} = S_T/n = 67244164/n$, $H_{C,1} = S_C/n = 48055043/n$, and $H_{G,1} = S_G/n = 48111528/n$. Deviations of real values from model values are also shown in percentages in accordance with the expression: 100(1 – (real value)/(model value)). One can see that these deviations are much less than 0.2% in all cases.

The model hyperbolic progressions $H_{A,1} = S_A/n$, $H_{T,1} = S_T/n$, $H_{C,1} = S_C/n$, and $H_{G,1} = S_G/n$ serve as mathematical standards for the described phenomenological facts. These hyperbolic progressions differ from each other only in the magnitude of numerators in their expressions, and therefore they can be specified by the general expression (10.2):

$$H_{N,1}(n) = S_N/n, \qquad (10.2)$$

where N refers to any of the nucleotides A, T, C, or G; and S_N refers to the number of corresponding monomers A, T, C, or G in the analyzed nucleotide sequence. If you know the total quantity S_N of the monomer N, you can predict — with a high level of accuracy — the total amounts of n-plets belonging to the class N_1-oligomers by using the general expression (10.2). These phenomenological facts testify in favor of the cooperative entity of the nucleotide sequence in the human chromosome №1.

Table 10.1.1. Real and model values to the OS representations of the classes of A_1-, T_1-, C_1-, and G_1-oligomers in human chromosome №1 are shown for $n = 1, 2, \ldots, 20$. The real total amounts of n-plets, which start with a certain nucleotide (A, T, C, or G), are indicated (in blue) jointly with their model values $H_{A,1}(n)$, $H_{T,1}(n)$, $H_{C,1}(n)$, and $H_{G,1}(n)$ from (10.1) (in red). Δ% refers to deviations of real values from model values in percentage (the model values are taken as 100%).

N	1	2	3	4	5	6	7	8	9	10
A										
Real	67070277	33537501	22360413	16768845	13413532	11179286	9584038	8383461	7453552	6706672
Model	67070277	33535139	22356759	16767569	13414055	11178380	9581468	8383785	7452253	6707028
Δ%A	0.000	−0.007	−0.016	−0.008	0.004	−0.008	−0.027	0.004	−0.017	0.005
T										
Real	67244164	33620498	22412993	16808862	13445360	11207274	9606748	8405040	7470145	6724359
Model	67244164	33622082	22414721	16811041	13448833	11207361	9606309	8405521	7471574	6724416
Δ%T	0.000	0.005	0.008	0.013	0.026	0.001	−0.005	0.006	0.019	0.001
C										
Real	48055043	24024903	16012711	12013624	9612227	8005708	6865944	6008215	5336968	4803919
Model	48055043	24027522	16018348	12013761	9611009	8009174	6865006	6006880	5339449	4805504
Δ%C	0.000	0.011	0.035	0.001	−0.013	0.043	−0.014	−0.022	0.046	0.033
G										
Real	48111528	24057606	16040889	12028924	9625086	8021235	6869132	6013412	5348337	4813156
Model	48111528	24055764	16037176	12027882	9622306	8018588	6873075	6013941	5345725	4811153
Δ%G	0.000	−0.008	−0.023	−0.009	−0.029	−0.033	0.057	0.009	−0.049	−0.042

(Continued)

Table 10.1.1. *(Continued)*

n	11	12	13	14	15	16	17	18	19	20
A										
Real	6095821	5588773	5160139	4792078	4472245	4192017	3946422	3726860	3531067	3354107
Model	6097298	5589190	5159252	4790734	4471352	4191892	3945310	3726127	3530015	3353514
Δ%A	0.024	0.007	−0.017	−0.028	−0.020	−0.003	−0.028	−0.020	−0.030	−0.18
T										
Real	6111970	5601854	5173904	4801395	4479492	4202773	3954021	3735327	3535288	3360459
Model	6113106	5603680	5172628	4803155	4482944	4202760	3955539	3735787	3539167	3362208
Δ%T	0.019	0.033	−0.025	0.037	0.077	0.000	0.038	0.012	0.110	0.052
C										
Real	4370502	4002753	3694018	3433636	3202830	3003511	2826568	2668499	2531448	2402186
Model	4368640	4004587	3696542	3432503	3203670	3003440	2826767	2669725	2529213	2402752
Δ%C	−0.043	0.046	0.068	−0.033	0.026	−0.002	0.007	0.046	−0.088	0.024
G										
Real	4374518	4013372	3701250	3435824	3210839	3006763	2830698	2673815	2532772	2407301
Model	4373775	4009294	3700887	3436538	3207435	3006971	2830090	2672863	2532186	2405576
Δ%G	−0.017	−0.102	−0.010	0.021	−0.106	0.007	−0.021	−0.036	−0.023	−0.072

By the corresponding compression of the ordinate in these Cartesian coordinate systems (that is, by appropriate scaling of numerators S_A, S_T, S_C, and S_G), each of these four hyperbolic sequences $H_{A,1} = S_A/n$, $H_{T,1} = S_T/n$, $H_{C,1} = S_C/n$, and $H_{G,1} = S_G/n$ reduces to the hyperbolic sequence (10.3):

$$y = 1/n, \tag{10.3}$$

which we call the canonic (or reference) hyperbolic sequence of OS representations (or the canonical OS sequence) of nucleotide sequences. In mathematics, the sequence (10.4)

$$1/1,\ 1/2,\ 1/3,\ 1/4,\ 1/5,\ \ldots,\ 1/n \tag{10.4}$$

is known long ago as a harmonic progression (or a harmonic sequence) where each term is the harmonic mean of the neighboring terms. For this reason, the revealed hyperbolic sequences in genomes can be also called genomic harmonic progressions and, in this mathematical sense, one can talk about the harmonic rules and the harmonious organization of genomes described below. The historically famous name "the harmonic progression" comes from the connection (10.1.4) with the series of harmonics in music. The sums of the first members of the harmonic progression (10.1.4) are called harmonic numbers. The rich centuries-old history of the study of harmonic progressions and harmonic series is associated with the names of Pythagoras, Orem (d'Oresme), Leibniz, Newton, Euler, Fourier, Dirichlet, Riemann, and other researchers. The generalization of the harmonic series is known as the Riemann zeta function. Using musical terminology, where the term "timbre" refers to the totality of the set of sound frequencies in a prolonged sound, one can conditionally say that the OS method represents the analyzed nucleotide sequence as some "oligomer timbre". The series of harmonic numbers serves as the discrete analog of the continuous function of natural logarithm ln n [372, p. 276]; this, in particular, connects the harmonic progression (10.1.4) with the Weber-Fechner logarithmic law, which is the main psychophysical law and dictates informatic peculiarities for all inherited sensory channels —

vision, hearing, smell, etc., whose organs (eyes, ears, nose, etc.) differ from each other in appearance very much. It testifies that genetic and different psychophysical levels of inherited biological informatics are structurally intercorrelated on the algebra-harmonic basis [36].

Given the relationship of the harmonic progression (10.1.4) with the four OS sequences for the four types of nucleotides A, T, C, and G, genomic sequences can be called tetra-harmonic sequences. Table 10.1.1 shows that the OS sequences of the total amounts of n-plets from the classes of A_1-oligomers and T_1-oligomers differ little from each other. The same is true for the OS sequences of the total amounts of n-plets from the classes of C_1- and G_1-oligomers. This fact is described by the expressions (10.5):

$$\Sigma_{A,n,1} \approx \Sigma_{T,n,1}, \quad \Sigma_{C,n,1} \approx \Sigma_{G,n,1} \tag{10.5}$$

In the particular case at $n = 1$, expressions (10.4) demonstrate the second Chargaff's rule on the approximate equality between the amounts of nucleotides A and T, as well as C and G in long DNA sequences. Correspondingly the phenomenological fact, described by expressions (10.4), is a certain generalization of Chargaff's second rule.

The results presented indicate, at least for the human chromosome №1, that there exist two general hyperbolic (or harmonic) rules regarding the total amounts of n-plets, which start with a certain nucleotide A, T, C, or G.

The first hyperbolic rule (about interrelations of oligomers in individual chromosomes):

- For any class of A_1-, T_1-, C_1-, or G_1-oligomers in individual chromosomes, the total amounts $\Sigma_{N,n,1}(n)$ of their n-plets, corresponding different n, are interrelated with each other through the general expression $\Sigma_{N,n,1} \approx S_N/n$ with a high level of accuracy (here N refers to any nucleotide A, T, C, or G; S_N refers to the number of monomers N; $n = 1, 2, 3, 4, \ldots$ is not too large compared to the full length of the nucleotide sequence). The phenomenological points with coordinates $[n, \Sigma_{N,n,1}]$ practically lie on the hyperbola having points $H_{N,1} = S_N/n$.

The second hyperbolic rule (about the similarity in the pairs of OS sequences):

- In individual chromosomes, two numeric OS sequences expressing the total amounts of *n*-plets, which start with nucleotides A and T, are approximately identical. The same is true for two numeric OS sequences expressing the total amounts of *n*-plets which start with nucleotides C and G (in accordance with the expressions (2.5)). Here $n = 1, 2, 3, 4, \ldots$ is not too large compared to the full length of the nucleotide sequence.

The obtained results of the hyperbolic (or harmonic) interrelationship of the amounts of *n*-plets, belonging to the indicated classes of oligomers, are not trivial. Theoretical counterexamples of artificial nucleotide sequences, which do not have such interrelation, can be indicated. For example, in the class of A_1-oligomers, one can mentally construct a long nucleotide sequence that contains many nucleotides A but does not have two adjacent nucleotides A, that is, does not contain a single AA duplet. Such a sequence does not have the hyperbolic interrelationship between the amounts of the nucleotide A and the total amounts of *n*-plets starting with A. It can be added that, in the same human chromosome № 1, the comparison of amounts of *n*-plets, consisting only of nucleotides of the same kind, for example, of the nucleotide A, shows an absence of the hyperbolic relationship between them. Really, in this case, the amount of the nucleotide A is equal to 67070277, the amount of the duplet AA — 10952057, triplet AAA — 2837038, tetraplet AAAA — 856207, and so on without their hyperbolic interrelation.

Let us continue the description of obtained results for the analysis of the human genome, which contains 22 autosomes and 2 sex chromosomes X and Y. These chromosomes are very different from each other in length, molecular weight, gene content, etc. Are there hyperbolic rules similar to formulated rules for the human chromosome №1? Yes, a positive answer to this question is obtained [198, 232]. For each of the 24 human chromosomes, knowing the quantity S_N of the monomer N (that is A, T, C, or G) allows you to calculate the total amounts of *n*-plets, which start with the oligomer N, with a high level of accuracy by using the general expression (10.2). Here $n = 1, 2, 3, \ldots$ but not very large in comparison with the length of the DNA sequence. Table 10.1.2 shows confirmational results of studying all 24 human chromosomes by the OS method under $n = 1, 2, 3, \ldots, 20$.

Table 10.1.2. Results of the analysis of all 24 human nuclear chromosomes by the OS method. For each of the chromosomes, quantities S_A, S_T, S_C, and S_G of monomers A, T, C, and G are shown to define the model hyperbolic progressions (10.2). The columns «Range %) show ranges of deviations for real OS series of corresponding n-plets ($n = 1, 2, \ldots, 20$) from their appropriate model values S_A/n, S_T/n, S_C/n, and S_G/n in percentages (in each case, an appropriate model value is taken as 100%). The left column shows chromosome numbers.

№	S_A	Range %	S_T	Range %	S_C	Range %	S_G	Range %
1	67070277	-0.030/0.024	67244164	-0.025/0.110	48055043	-0.088/0.068	48111528	-0.106/0.057
2	71791213	-0.079/0.087	71987932	-0.075/0.095	48318180	-0.097/0.072	48450903	-0.105/0.141
3	59689091	-0.021/0.045	59833302	-0.097/0.098	39233483	-0.130/0.081	39344259	-0.034/0.088
4	58561236	-0.065/0.044	58623430	-0.036/0.128	36236976	-0.039/0.127	36331025	-0.117/0.075
5	54699094	-0.052/0.040	54955010	-0.071/0.078	35731600	-0.012/0.132	35879674	-0.103/0.085
6	51160489	-0.039/0.057	51151754	-0.049/0.022	33520786	-0.092/0.061	33516767	-0.029/0.069
7	47058248	-0.104/0.040	47215040	-0.061/0.030	32317984	-0.086/0.091	32378859	-0.076/0.069
8	42641072	-0.061/0.068	42581941	-0.111/0.071	28600559	-0.110/0.069	28600963	-0.068/0.050
9	31752642	-0.134/0.090	31733822	-0.083/0.065	22487631	-0.099/0.141	22470915	-0.079/0.143
10	38875926	-0.081/0.052	39027555	-0.067/0.099	27639505	-0.058/0.085	27719976	-0.118/0.085
11	39286730	-0.032/0.084	39361954	-0.062/0.042	27903257	-0.139/0.056	27981801	-0.086/0.112
12	39370109	-0.096/0.056	39492225	-0.097/0.094	27092804	-0.076/0.078	27182678	-0.073/0.105
13	29224840	-0.067/0.077	29320872	-0.107/0.069	18341128	-0.107/0.141	18346620	-0.130/0.065
14	25606393	-0.109/0.100	25819249	-0.040/0.086	17733667	-0.137/0.077	17782016	-0.056/0.142
15	24508669	-0.085/0.179	24553812	-0.127/0.088	17752941	-0.090/0.162	17825903	-0.067/0.113
16	22558319	-0.122/0.080	22774906	-0.143/0.104	18172742	-0.146/0.074	18299976	-0.146/0.117
17	22639499	-0.141/0.105	22705261	-0.146/0.070	18723944	-0.134/0.072	18851500	-0.144/0.105
18	22087028	-0.160/0.071	22109347	-0.169/0.121	14574701	-0.090/0.134	14594335	-0.160/0.210
19	15142293	-0.160/0.024	15282753	-0.062/0.062	13954580	-0.103/0.097	14061132	-0.057/0.226
20	16455618	-0.106/0.129	16643030	-0.099/0.089	13037092	-0.062/0.116	13098788	-0.092/0.155
21	9943435	-0.161/0.083	9882679	-0.206/0.173	6864570	-0.134/0.277	6852178	-0.373/0.219
22	10382214	-0.175/0.084	10370725	-0.036/0.209	9160652	-0.258/0.155	9246186	-0.143/0.235
X	46754807	-0.078/0.084	46916701	-0.102/0.055	30523780	-0.116/0.179	30697741	-0.135/0.067
Y	7886192	-0.244/0.097	7956168	-0.063/0.185	5285789	-0.181/0.407	5286894	-0.247/0.142

These results demonstrate that both hyperbolic (or harmonic) rules №1 and №2 hold true for each of the human chromosomes with a high level of accuracy.

One can show that the obtained phenomenological data also leads to the third hyperbolic rule related to normalized versions of the OS sequences S_A/n, S_T/n, S_C/n, and S_G/n. Scaling the numerators S_A, S_T, S_C, and S_G by dividing by their total amount $S = S_A + S_T + S_C + S_G$, we obtain the corresponding scaling of all these OS sequences, which are termed as "normalized OS sequences" (10.6):

$$S_A/(nS),\ S_T/(nS),\ S_C/(nS),\ S_G/(nS) \tag{10.6}$$

It turns out that the normalized OS sequences of all human chromosomes are similar to each other with a high level of accuracy as Table 10.1.3 shows regarding the first main members S_A/S, S_T/S, S_C/S, and S_G/S of these hyperbolic sequences.

Table 10.1.3. Data for normalized OS sequences $S_A/(nS)$, $S_T/(nS)$, $S_C/(nS)$, and $S_G/(nS)$ of all human chromosomes are shown for comparison. Here $S = S_A + S_T + S_C + S_G$. Harmonic means of the values S_A/S, S_T/S, S_C/S, and S_G/S in each chromosome are also indicated.

Chromosome	S_A/S	S_T/S	S_C/S	S_G/S	Harmonic mean
1	0.2910	0.2918	0.2085	0.2087	0.243
2	0.2984	0.2993	0.2009	0.2014	0.241
3	0.3013	0.3020	0.1980	0.1986	0.239
4	0.3086	0.3089	0.1910	0.1915	0.236
5	0.3018	0.3032	0.1971	0.1979	0.239
6	0.3021	0.302	0.1979	0.197	0.239
7	0.2960	0.2970	0.2033	0.2037	0.241
8	0.2994	0.2990	0.2008	0.2008	0.240
9	0.2928	0.2926	0.2074	0.2072	0.243
10	0.2917	0.2929	0.2074	0.2080	0.243
11	0.2920	0.2926	0.2074	0.2080	0.243
12	0.2957	0.2966	0.2035	0.2042	0.242

(Continued)

Table 10.1.3. (*Continued*)

Chromosome	S_A/S	S_T/S	S_C/S	S_G/S	Harmonic mean
13	0.3069	0.3079	0.1926	0.1926	0.237
14	0.2945	0.2970	0.2040	0.2045	0.242
15	0.2896	0.2901	0.2097	0.2106	0.244
16	0.2758	0.2784	0.2221	0.2237	0.247
17	0.2730	0.2738	0.2258	0.2273	0.248
18	0.3011	0.3014	0.1987	0.1989	0.240
19	0.2591	0.2615	0.2388	0.2406	0.250
20	0.2778	0.2810	0.2201	0.2211	0.247
21	0.2964	0.2946	0.2047	0.2043	0.242
22	0.2651	0.2648	0.2339	0.2361	0.249
X	0.3019	0.3029	0.1971	0.1982	0.239
Y	0.2985	0.3012	0.2001	0.2001	0.240

The same results on the similarity of normalized OS sequences S_A/nS, S_T/nS, S_C/nS, and S_G/nS in all chromosomes of a particular genome were obtained when studying the genomes of a number of eukaryotes (until now, without a single exception in analyzed cases) [155, 198, 232]. Below, results for some eukaryotic genomes are shown. These results allow proposing the third hyperbolic (or harmonic) rule on the total amounts of n-plets, which start with A, T, C, or G.

The third hyperbolic rule (about the similarity of chromosomes):

- All chromosomes of any individual eukaryotic genome have approximately the same normalized OS sequences S_A/nS, S_T/nS, S_C/nS, and S_G/nS representing classes of A_1-, T_1-, C_1-, and G_1-oligomers ($n = 1, 2, 3, 4, \ldots$ is not too large compared to the full length of the nucleotide sequence).

These hyperbolic rules are candidates for the role of universal rules; the analysis of the widest variety of genomes is required to verify their universality.

Let us return to the harmonic progression (10.4) and recall its relation with the well-known concept of the harmonic mean. The harmonic means H of the positive real numbers $x_1, x_2, \ldots, x_n$ is defined by the expression (10.7):

$$H = \frac{n}{\frac{1}{x_1} + \frac{1}{x_2} + \cdots + \frac{1}{x_2}} \tag{10.7}$$

Knowing two neighboring members of the harmonic progression, one can calculate its next member. Here we can briefly mention that the harmonic mean is associated with the Pythagorean teaching on musical harmony and the aesthetics of proportions, presented in the famous numerical triangle published 2,000 years ago by Nicomachus of Gerasa in his book *Introduction into Arithmetic*. In accordance with this triangle, the Parthenon [373] and other great architectural objects were created because architecture was interpreted as movement music, and music as dynamic architecture (see more details in [12, 13, 374, 375]). Since the harmonic mean is related to the harmonic progression, the author indicates values of the harmonic mean in some figures of the article for comparison analysis of OS sequences in different nucleotide sequences (Table 10.1.3 and many others).

Each genomic DNA sequence with its total amount S of all nucleotides A, T, C, and G also contains total amounts S/n of n-plets (that is, S/2 duplets, S/3 triplets, etc.). These total amounts are members of the hyperbolic sequence S, S/2, S/3, …, S/n. Each member of this sequence is the sum of the four OS sequences S_A/n, S_T/n, S_C/n, and S_G/n (10.8):

$$S_A/n + S_T/n + S_C/n + S_G/n = S/n \quad \text{or} \quad S_A/nS + S_T/nS + S_C/nS + S_G/nS = 1/n \tag{10.8}$$

These linear superpositions are valid for a wide variety of genomes that differ only in individual coefficients S_A, S_T, S_C, and S_G.

Below we represent the results, which have been obtained on the basis of the analysis of very different genomes by the OS method and which testify in favor that the formulated hyperbolic rules have a general genomic significance.

10.2. The hyperbolic rules in all chromosomes of the fruit fly *Drosophila melanogaster*

This and upcoming sections are devoted to the analysis — by the OS method — of single-stranded DNA sequences from the complete sets of chromosomes of a few model eukaryotic organisms, which were used long ago in the study of genetics, development, and disease. Represented tabular data confirm that both hyperbolic (harmonic) rules regarding n-plets from the classes of A_1-, T_1-, C_1-, and G_1-oligomers hold for each described chromosome at $n = 1, 2, 3, 4, \ldots, 19, 20$ (although these rules are also satisfied for larger values of n, at least up to $n = 100$; the data tables for such large n are too cumbersome to include in the book).

Let us start with the fruit fly *Drosophila melanogaster*, which has been studied in biology labs for over eighty years. All initial data about its chromosomes were taken from GenBank (https://www.ncbi.nlm.nih.gov/genome/?term=drosophila+melanogaster). The resulting data in Table 10.2.1 confirm that — for all the chromosomes — the model hyperbolic progressions $H_{A,1}(n) = S_A/n$, $H_{T,1}(n) = S_T/n$, $H_{C,1}(n) = S_C/n$, and $H_{G,1}(n) = S_N/n$ from the expression (10.2) practically coincide with the real sequences of total amounts of n-plets from the classes A_1-, T_1-, C_1-, and G_1-oligomers at

Table 10.2.1. The results of the analysis of all chromosomes of *Drosophila melanogaster* by the OS method. The left column shows symbols of chromosomes. S_A, S_T, S_C, and S_G refer to the quantities of monomers A, T, C, and G in appropriate chromosomes. The columns "Range %" show maximal deviations of members of real OS sequences of n-plets from corresponding members of the model hyperbolic sequences $H_{A,1}(n) = S_A/n$, $H_{T,1}(n) = S_T/n$, $H_{C,1}(n) = S_C/n$, and $H_{G,1}(n) = S_G/n$ at $n = 1, 2, 3, \ldots, 20$ (the model values are taken as 100%).

№	S_A	Range %	S_T	Range %	S_C	Range %	S_G	Range %
X	6732793	−0.196/0.057	6774766	−0.125/0.090	4975870	−0.198/0.139	4992722	−0.148/0.213
2L	6853032	−0.259/0.128	6836080	−0.219/0.090	4912017	−0.239/0.313	4912383	−0.251/0.350
2R	7272860	−0.217/0.178	7235562	−0.144/0.304	5395216	−0.195/0.222	5376598	−0.222/0.323
3L	8143548	−0.142/0.196	8198331	−0.126/0.206	5825673	−0.211/0.108	5824515	−0.262/0.169
3R	9205526	−0.143/0.152	9197619	−0.145/0.132	6833716	−0.170/0.169	6817898	−0.231/0.192
4	425241	−1.759/0.488	436669	−0.423/0.744	232566	−1.463/1.299	236655	−0.855/1.369
Y	1056780	−0.494/0.314	1008635	−0.125/0.431	682725	−0.268/0.659	661579	−0.512/0.386

Table 10.2.2. Data of normalized OS sequences S_A/nS, S_T/nS, S_C/nS, and S_G/nS of all chromosomes of *Drosophila melanogaster* are shown for comparison. Here $S = S_A + S_T + S_C + S_G$. Harmonic means of the values S_A/S, S_T/S, S_C/S, and S_G/S in each chromosome are also indicated.

Chromosome	S_A/S	S_T/S	S_C/S	S_G/S	Harmonic mean
X	0.2868	0.2886	0.2120	0.2127	0.244
2L	0.2915	0.2907	0.2089	0.2089	0.243
2R	0.2877	0.2862	0.2134	0.2127	0.245
3L	0.2909	0.2929	0.2081	0.2081	0.243
3R	0.2872	0.2869	0.2132	0.2127	0.245
4	0.3195	0.3280	0.1747	0.1778	0.228
Y	0.3099	0.2958	0.2002	0.1940	0.239

$n = 1, 2, 3, \ldots, 20$. In all shown cases, the deviations of real sequences from model hyperbolic progressions are less than 1% as data in the tabular columns «Range %» indicate. This means that the formulated hyperbolic (harmonic) rules are fulfilled in the considered genome.

Table 10.2.2 shows data of normalized OS sequences for all chromosomes of *Drosophila melanogaster.*

10.3. The hyperbolic rules in all chromosomes of the nematode *Caenorhabditis elegans*

The section represents data about the analysis of single-stranded DNA sequences from the complete set of chromosomes of free-living soil nematode *Caenorhabditis elegans* by the OS method. This nematode has been widely used as a model organism in genetics for a long time. The *Caenorhabditis elegans* nuclear genome is approximately 100 Mb, distributed among six chromosomes. All initial data are taken from GenBank (https://www.ncbi.nlm.nih.gov/genome?term=caenorhabditis%20elegans).

The resulting data in Table 10.3.1 confirm that — for all the chromosomes — the model hyperbolic progressions $H_{A,1}(n) = S_A/n$, $H_{T,1}(n) = S_T/n$, $H_{C,1}(n) = S_C/n$, and $H_{G,1}(n) = S_N/n$ from the expression (10.2) practically coincide with

Table 10.3.1. The results of the analysis of all chromosomes of *Caenorhabditis elegans* by the OS method. The left column shows symbols of chromosomes. S_A, S_T, S_C, and S_G refer to the quantities of monomers A, T, C, and G in appropriate chromosomes. The columns "Range %" show maximal deviations of members of real OS sequences of n-plets from corresponding members of the model hyperbolic sequences $H_{A,1}(n) = S_A/n$, $H_{T,1}(n) = S_T/n$, $H_{C,1}(n) = S_C/n$, and $H_{G,1}(n) = S_G/n$ at $n = 1, 2, 3, \ldots, 20$ (the model values are taken as 100%).

№	S_A	Range %	S_T	Range %	S_C	Range %	S_G	Range %
1	4835939	-0.144/0.319	4848450	-0.160/0.294	2695890	-0.487/0.327	2692155	-0.498/0.218
2	4878209	-0.196/0.421	4869734	-0.229/0.109	2769232	-0.256/0.492	2762246	-0.253/0.257
3	4444681	-0.139/0.157	4423618	-0.269/0.156	2449158	-0.451/0.303	2466344	-0.173/0.362
4	5711043	-0.106/0.229	5730974	-0.253/0.177	3034784	-0.393/0.219	3017028	-0.199/0.414
5	6750403	-0.145/0.124	6760297	-0.164/0.203	3712075	-0.222/0.575	3701405	-0.418/0.286
X	5747200	-0.256/0.120	5734092	-0.166/0.167	3119741	-0.156/0.340	3117909	-0.272/0.256

Table 10.3.2. Data of normalized OS sequences S_A/nS, S_T/nS, S_C/nS, and S_G/nS of all chromosomes of *Caenorhabditis elegans* are shown for comparison. Here $S = S_A + S_T + S_C + S_G$. Harmonic means of the values S_A/S, S_T/S, S_C/S, and S_G/S in each chromosome are also indicated.

Chromosome	S_A/S	S_T/S	S_C/S	S_G/S	Harmonic mean
1	0.3208	0.3217	0.1789	0.1786	0.230
2	0.3193	0.3187	0.1812	0.1808	0.231
3	0.3225	0.3209	0.1777	0.1789	0.229
4	0.3265	0.3276	0.1735	0.1725	0.226
5	0.3226	0.3231	0.1774	0.1769	0.229
X	0.3244	0.3236	0.1761	0.1760	0.228

the real sequences of total amounts of n-plets from the classes A_1-, T_1-, C_1-, and G_1-oligomers at $n = 1, 2, 3, \ldots, 20$. In all shown cases, the deviations of real sequences from model hyperbolic (harmonic) progressions are less than 0.5% as data in the tabular columns «Range %» indicates. This means that the formulated hyperbolic (harmonic) rules are fulfilled in the considered genome.

Table 10.3.2 shows data of normalized OS sequences for all chromosomes of *Caenorhabditis elegans*.

10.4. The hyperbolic rules in all chromosomes of the house mouse *Mus musculus*

The section represents data about the analysis of single-stranded DNA sequences from the complete set of chromosomes of the laboratory mouse *Mus musculus*, which is a major model organism for basic mammalian biology, human disease, and genome evolution. All initial data are taken from the GenBank (https://www.ncbi.nlm.nih.gov/genome?term=mus%20 musculus).

The resulting data in Table 10.4.1 confirm that — for all the chromosomes — the model harmonic progressions $H_{A,1}(n) = S_A/n$, $H_{T,1}(n) = S_T/n$, $H_{C,1}(n) = S_C/n$, and $H_{G,1}(n) = S_N/n$ from the expression (10.2) practically coincide with the real sequences of total amounts of n-plets from the classes A_1-, T_1-, C_1-, and G_1-oligomers at $n = 1, 2, 3, \ldots, 20$. In all shown cases, the deviations of real sequences from model hyperbolic progressions are significantly less than 0.5% as data in the tabular columns «Range %» indicates. This means that the formulated hyperbolic (harmonic) rules are fulfilled in the considered genome.

Table 10.4.2 shows data of normalized OS sequences for all chromosomes of *Mus musculus*.

Table 10.4.1. The results of the analysis of all chromosomes of a house mouse *Mus musculus* by the OS method. The left column shows symbols of chromosomes. S_A, S_T, S_C, and S_G refer to the quantities of monomers A, T, C, and G in appropriate chromosomes. The columns "Range %" show maximal deviations of members of real OS sequences of n-plets from corresponding members of the model hyperbolic sequences $H_{A,1}(n) = S_A/n$, $H_{T,1}(n) = S_T/n$, $H_{C,1}(n) = S_C/n$, and $H_{G,1}(n) = S_G/n$ at $n = 1, 2, 3, \ldots, 20$ (the model values are taken as 100%).

№	S_A	Range %	S_T	Range %	S_C	Range %	S_G	Range %
1	56530182	-0.051/0.054	56416289	-0.126/0.067	39495313	-0.044/0.121	39467408	-0.150/0.107
2	51600126	-0.099/0.076	51679955	-0.060/0.063	37504114	-0.041/0.111	37542456	-0.110/0.036
3	46503996	-0.041/0.063	46631177	-0.057/0.092	31603703	-0.151/0.120	31659979	-0.063/0.075
4	43821952	-0.050/0.081	43922197	-0.056/0.076	32146231	-0.097/0.062	32165231	-0.050/0.090
5	42488105	-0.060/0.059	42515761	-0.063/0.050	31456650	-0.033/0.085	31459158	-0.218/0.052
6	42843713	-0.122/0.082	42886213	-0.107/0.048	30315703	-0.028/0.108	30290914	-0.083/0.094
7	40271749	-0.105/0.070	40509547	-0.110/0.079	30554235	-0.089/0.081	30519876	-0.019/0.125

(*Continued*)

Table 10.4.1. (*Continued*)

№	S_A	Range %	S_T	Range %	S_C	Range %	S_G	Range %
8	36224525	-0.043/0.083	36167473	-0.119/0.127	26616967	-0.147/0.104	26602467	-0.123/0.081
9	34722476	-0.079/0.066	34694585	-0.088/0.066	25880876	-0.154/0.185	25859081	-0.114/0.095
10	37185184	-0.066/0.068	37273294	-0.046/0.142	26277876	-0.099/0.096	26331308	-0.112/0.123
11	33401283	-0.069/0.049	33317397	-0.121/0.049	26022668	-0.057/0.177	26004597	-0.061/0.081
12	33897029	-0.111/0.069	34225639	-0.105/0.144	24374340	-0.061/0.094	24425412	-0.115/0.109
13	34255191	-0.119/0.088	34115119	-0.064/0.082	24377641	-0.079/0.119	24373242	-0.138/0.090
14	35695406	-0.135/0.027	35758968	-0.086/0.115	24980458	-0.057/0.135	25007278	-0.099/0.124
15	29177752	-0.135/0.069	29244798	-0.028/0.092	21121081	-0.050/0.180	21109684	-0.095/0.088
16	28035438	-0.054/0.115	28084677	-0.067/0.099	19439086	-0.125/0.023	19460557	-0.170/0.003
17	26251160	-0.075/0.097	26302830	-0.051/0.079	19586991	-0.085/0.060	19566481	-0.096/0.063
18	25615329	-0.064/0.075	25597990	-0.103/0.054	18095575	-0.099/0.159	18143740	-0.129/0.089
19	16732680	-0.097/0.094	16602953	-0.193/0.076	12449343	-0.181/0.201	12420880	-0.098/0.225
X	49660944	-0.069/0.049	49651848	-0.102/0.052	32081377	-0.049/0.093	32093826	-0.126/0.131
Y	26842991	-0.084/0.166	27013719	-0.072/0.107	17175367	-0.221/0.171	17092621	-0.163/0.171

Table 10.4.2. Data of normalized OS sequences S_A/nS, S_T/nS, S_C/nS, and S_G/nS of all chromosomes of a house mouse *Mus musculus* are shown for comparison. Here $S = S_A + S_T + S_C + S_G$. Harmonic means of the values S_A/S, S_T/S, S_C/S, and S_G/S in each chromosome are also indicated.

Chromosome	S_A/S	S_T/S	S_C/S	S_G/S	Harmonic mean
1	0.2946	0.2940	0.2058	0.2057	0.242
2	0.2894	0.2898	0.2103	0.2105	0.244
3	0.2973	0.2982	0.2021	0.2024	0.241
4	0.2882	0.2889	0.2114	0.2115	0.244
5	0.2872	0.2874	0.2127	0.2127	0.244
6	0.2928	0.2931	0.2072	0.2070	0.243
7	0.2839	0.2856	0.2154	0.2151	0.245
8	0.2884	0.2879	0.2119	0.2118	0.244
9	0.2866	0.2864	0.2136	0.2134	0.245
10	0.2926	0.2933	0.2068	0.2072	0.243
11	0.2813	0.2806	0.2191	0.2190	0.246
12	0.2899	0.2927	0.2085	0.2089	0.243
13	0.2925	0.2913	0.2081	0.2081	0.243

Table 10.4.2. (*Continued*)

Chromosome	S_A/S	S_T/S	S_C/S	S_G/S	Harmonic mean
14	0.2939	0.2945	0.2057	0.2059	0.242
15	0.2899	0.2905	0.2098	0.2097	0.244
16	0.2950	0.2956	0.2046	0.2048	0.242
17	0.2862	0.2868	0.2136	0.2134	0.245
18	0.2929	0.2927	0.2069	0.2075	0.243
19	0.2875	0.2852	0.2139	0.2134	0.245
X	0.3038	0.3037	0.1962	0.1963	0.239
Y	0.3046	0.3065	0.1949	0.1940	0.238

10.5. The hyperbolic rules in all chromosomes of the plant *Arabidopsis thaliana*

One more model organism is the plant *Arabidopsis thaliana*. This small flowering plant has been used for over fifty years to study plant mutations and for classical genetic analysis. It became the first plant genome to be fully sequenced; it has a small genome of ~120 Mb. The section represents data about the analysis of single-stranded DNA sequences from the complete set of five chromosomes of this plant by the OS method. All initial data about the chromosomes were taken from GenBank (https://www.ncbi.nlm.nih.gov/genome/4, the column RefSeq).

The resulting data in Table 10.5.1 confirm that — for all the chromosomes — the model harmonic progressions $H_{A,1}(n) = S_A/n$, $H_{T,1}(n) = S_T/n$, $H_{C,1}(n) = S_C/n$, and $H_{G,1}(n) = S_N/n$ from the expression (10.2) practically coincide with the real sequences of total amounts of n-plets from the classes A_1-, T_1-, C_1-, and G_1-oligomers at $n = 1, 2, 3, \ldots, 20$. In all shown cases, the deviations of real sequences from model harmonic progressions are less than 0.6% as data in the tabular columns «Range %» indicates. This means that the formulated hyperbolic (harmonic) rules are fulfilled in the considered genome.

Table 10.5.2 shows data of normalized OS sequences for all chromosomes of *Arabidopsis thaliana*.

Table 10.5.1. The results of the analysis of all chromosomes of the plant *Arabidopsis thaliana* by the OS method. The left column shows symbols of chromosomes. S_A, S_T, S_C, and S_G refer to the quantities of monomers A, T, C, and G in appropriate chromosomes. The columns "Range %" show maximal deviations of members of real OS sequences of n-plets from corresponding members of the model hyperbolic sequences $H_{A,1}(n) = S_A/n$, $H_{T,1}(n) = S_T/n$, $H_{C,1}(n) = S_C/n$, and $H_{G,1}(n) = S_G/n$ at $n = 1, 2, 3, \ldots, 20$ (the model values are taken as 100%).

№	S_A	Range %	S_T	Range %	S_C	Range %	S_G	Range %
1	9709674	−0.275/0.103	9697113	−0.140/0.209	5435374	−0.130/0.303	5421151	−0.186/0.296
2	6315641	−0.035/0.198	6316348	−0.256/0.162	3542973	−0.406/0.252	3520766	−0.148/0.478
3	7484757	−0.121/0.101	7448059	−0.141/0.238	4258333	−0.283/0.200	4262704	−0.206/0.121
4	5940546	−0.155/0.239	5914038	−0.109/0.238	3371349	−0.222/0.333	3356091	−0.293/0.161
5	8621974	−0.123/0.184	8652238	−0.213/0.204	4832253	−0.515/0.073	4858759	−0.132/0.407

Table 10.5.2. Data of normalized OS sequences S_A/nS, S_T/nS, S_C/nS, and S_G/nS of all chromosomes of *Arabidopsis thaliana* are shown for comparison. Here $S = S_A + S_T + S_C + S_G$. Harmonic means of the values S_A/S, S_T/S, S_C/S, and S_G/S in each chromosome are also indicated.

Chromosome	S_A/S	S_T/S	S_C/S	S_G/S	Harmonic mean
1	0.3208	0.3204	0.1796	0.1791	0.230
2	0.3207	0.3207	0.1799	0.1788	0.230
3	0.3191	0.3176	0.1816	0.1817	0.231
4	0.3197	0.3183	0.1814	0.1806	0.231
5	0.3197	0.3209	0.1792	0.1802	0.230

10.6. Analysis of long genes by the oligomer sums method

Before proceeding to the analysis of prokaryotic genomes, it is useful to show the applicability of the OS method to the analysis of genes whose sequences are much shorter than DNA sequences in chromosomes. The application of the method unexpectedly reveals the phenomenon of regular rhythmic deviations in the sequences of real total sums of n-plets in the described genes from the corresponding model hyperbolic progressions.

Let us first consider the human *TTN* gene encoding the largest known protein Titin. Titin, also known as connectin, is important in the contraction of striated muscle tissues. Figs. 8.1–8.6 show some results of the analysis — by the OS method — of the nucleotide sequence of the *TTN* gene (numeric results will be represented below). Initial data on its nucleotide sequence are taken from the GenBank (https://www.ncbi.nlm.nih.gov/nuccore/X90568.1). This gene contains 26373 nucleotides A, 19569 nucleotides T, 17097 nucleotides C, and 18901 nucleotides G, that is $S_A = 26373$, $S_T = 19569$, $S_C = 17097$, and $S_G = 18901$ for the model hyperbolic progressions (2.2). It can be especially noted that, in this gene, the amounts of nucleotides A and T are significantly different (26373 and 19569), that is, Chargaff's second rule on their approximate equality in long sequences is not satisfied here since this nucleotide sequence is not enough long for Chargaff's rule.

Fig. 10.6.1 shows the sequences of the highly regular significant deviations of the real total amounts of n-plets, which start with the nucleotides T and G, from model hyperbolic progressions $S_T/n = 19569/n$ and $S_G/n = 18901/n$. One should note that all these significant deviations happen only at $n = 3, 6, 9, \ldots, 3m$, that is, only for cases of $3m$-plets (here $m = 1, 2, 3, \ldots$). Correspondingly these significant deviations are called «triplet-deviations».

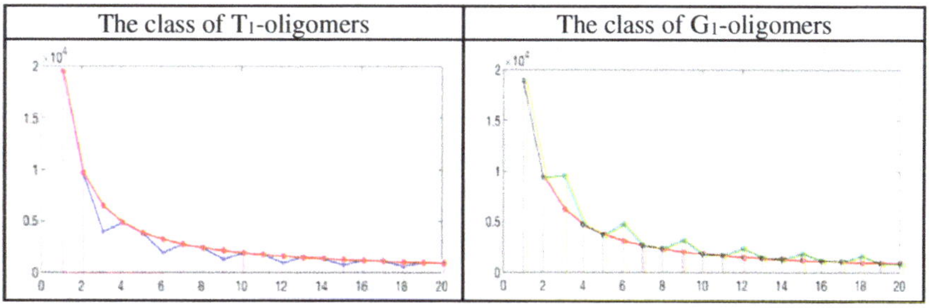

Fig. 10.6.1. Graphical representations of the analysis — by the OS method — of the human *TTN* gene. The OS sequences of its total amounts of n-plets, which start with the nucleotide T (left) and the nucleotide G (right), are shown. The red lines refer to model hyperbolic progressions S_T/n and S_G/n correspondingly, where $S_T = 19569$ and $S_G = 18901$ are quantities of nucleotides T and G in the gene; $n = 1, 2, 3, \ldots, 20$ as shown on the abscissa. The blue line (left) and the green line (right) with dots on them refer to the real OS sequences of the total amounts of such n-plets. The ordinate indicates the total amounts of n-plets.

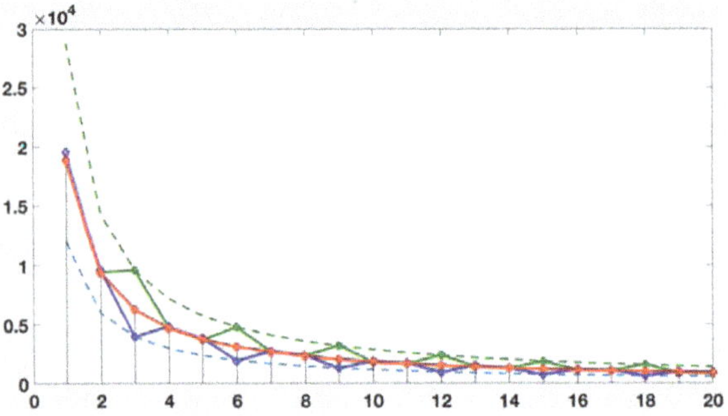

Fig. 10.6.2. Unity of two graphs from Fig. 10.6.1 for the *TTN* gene. The blue and green dot lines correspond to those additional hyperbolic progressions 11979/*n* and 28788/*n*, which model real total amounts of 3*m*-plets. Other parts of this united graph are the same as in Fig. 10.6.1.

Fig. 10.6.2 shows the unity of both graphs from Fig. 10.6.1 and demonstrates a few interesting features of the highly regular series of these triplet-deviations.

Firstly, one can see in Fig. 10.6.2 that, in classes of T_1-oligomers and G_1-oligomers, the triplet-deviations happen in opposite directions (or, figuratively speaking, in antiphase):

- in the class of T_1-oligomers, they decrease real values compared with model values of the hyperbolic progression 19569/*n*;
- in the class of G_1-oligomers, they increase real values in comparison with model values of the hyperbolic progression 18901/*n*.

Secondly, under triplet-deviations, real total amounts of 3*m*-plets from the classes of T_1- and G_1-oligomers belong correspondingly to other hyperbolic progressions 11979/*n* and 28788/*n*. These hyperbolic progressions are indicated by the blue and green dot lines in Fig. 10.6.2. Where did these numerators of model hyperbolas come from? Each of these numerators is associated with the total amount of triplets ($n = 3$) in an appropriate class of oligomers in this gene: the total amount of triplets starting with nucleotide T is equal to 3993, and that starting with G is equal to 9596. To calculate the first values of the model hyperbolas, each of these amounts of triplets must be tripled, giving the shown numerators 11979 and 28788.

Similar triplet-deviations exist in the OS representations not only of the *TTN* gene but also of other long genes, prokaryotic genomes, and viruses in different degrees as the author has discovered in the analysis of a limited set of nucleotide series by the OS method. In the genetic code system, triplets have an important meaning, which differ from other n-plets: they encode amino acids and punctuations of protein synthesis. One can believe that the phenomenon of triplet-deviations is related to this special meaning of triplets. For this reason, the deeper analysis of triplet-deviations in different species can be useful to study the secrets of the genetic system and biological evolution.

Fig. 10.6.2 demonstrated the highly regular rhythmic triplet-deviations for $n = 1, 2, 3, \ldots, 20$, but similar rhythmic triplet-deviations exist in a much wider range of values n.

Fig. 10.6.3 shows in graphical forms percentage values of the highly regular rhythmic deviations of the real total amounts of n-plets, which

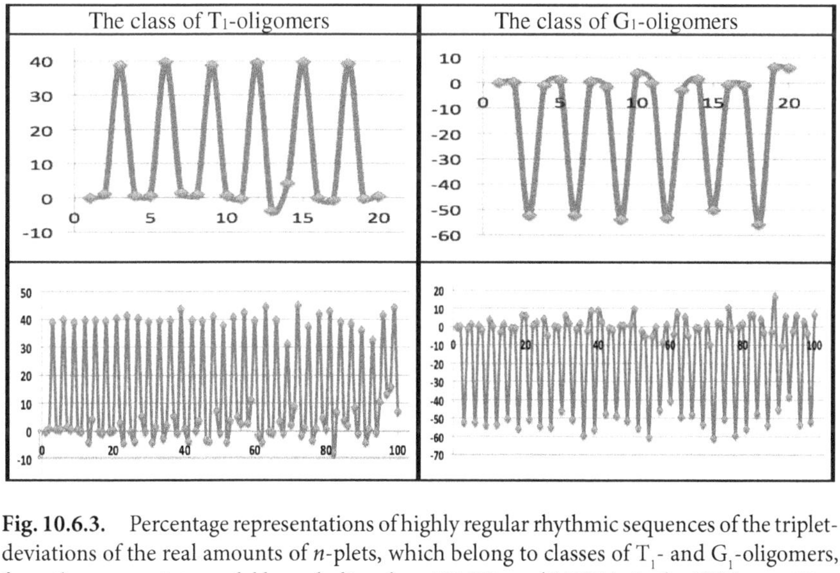

Fig. 10.6.3. Percentage representations of highly regular rhythmic sequences of the triplet-deviations of the real amounts of n-plets, which belong to classes of T_1- and G_1-oligomers, from the appropriate model hyperbolic values $19569/n$ and $18901/n$ in the *TTN* gene. Here $n = 1, 2, 3, \ldots, 20$ (upper row) and $n = 1, 2, 3, \ldots, 100$ (bottom row) as shown on the abscissa. The ordinate shows percentages of the deviations (the model values are taken as 100%).

start with the nucleotides T and G in the *TTN* gene, from the appropriate model values 19569/n and 18901/n. Two cases of the range of values n are represented: $n = 1, 2, 3, \ldots, 20$, and $n = 1, 2, 3, \ldots, 100$.

The nucleobases T and G are keto-nucleobases. Figs. 10.6.2 and 10.6.3 draw attention to the phenomenon of long-range correlations in the *TTN* gene between sequences of the triplet-deviations in classes of T_1-, and G_1-oligomers: the triplet-deviations in these sequences happen in opposite directions as abovementioned. Such binary oppositions, which meet in different long genes, prokaryotic genomes, and viruses regarding the classes of different N_1-oligomers (here N refers to A, T, C, or G), should be specially studied in the future since they bear important information and are associated with other binary-opposition features of molecular genetic systems.

The following Fig. 10.6.4 shows the OS sequences of the total amounts of n-plets, which start with two other nucleotides A and C in the

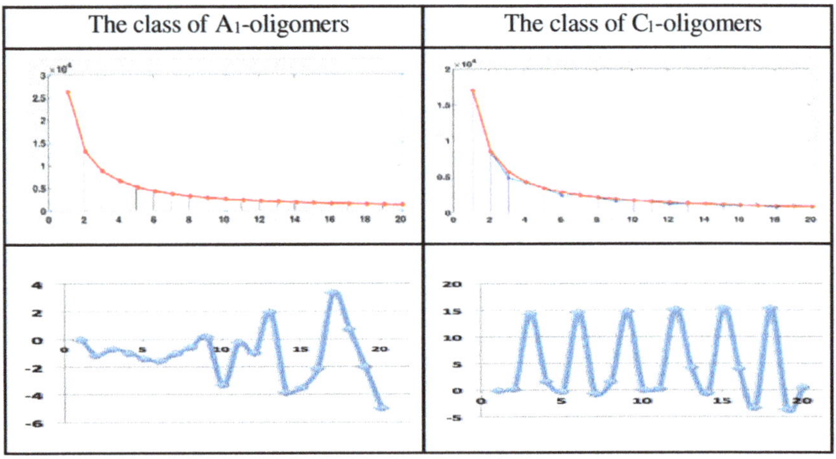

Fig. 10.6.4. Graphical representations of the results of the analysis — by the OS method — of the human *TTN* gene regarding the sequences of the total amounts of n-plets, which start with nucleotide A (left) and nucleotide C (right). Here $n = 1, 2, 3, \ldots, 20$ (on the abscissa). Upper row: the red lines refer to model hyperbolic progressions $S_A/n = 26373/n$ and $S_C/n = 17097/n$ correspondingly. The ordinate shows the total amounts of appropriate n-plets. The class of C_1-oligomers has regular sequences of the significant triplet-deviations at $3m$-plets shown by the blue line. Bottom row: percentage representations of the sequences of deviations for the real total amounts of n-plets of these classes from the appropriate model hyperbolic values 26373/n and 17097/n (the ordinate axes show these percentages). The model values are taken as 100%.

TTN gene. This gene contains 26373 nucleotides A and 17097 nucleotides C; correspondingly $S_A = 26373$ and $S_C = 17097$ for the model hyperbolic progressions (10.2).

One can see in Fig. 10.6.4 that the class of C_1-oligomers has regular sequences of the significant triplet-deviations at $3m$-plets shown by the blue line. The class of A_1-oligomers has no such regular sequences of significant deviations; besides, its deviations are essentially less than deviations in the class of C_1-oligomers. In the class of A_1-oligomers, the real and model values differ little from each other, and therefore, in Fig. 10.6.4, the red line of model values covers the line of real values.

Table 10.6.1 shows the numerical results of the analysis of the *TTN* gene by the OS method.

The coordinated deviations of all four OS sequences from their model harmonic progressions can be conveniently represented by the general sequence of harmonic mean values, which are calculated for their four corresponding members at each fixed n. Fig. 10.6.5 shows such a sequence for the human *TTN* gene. One can see the very regular rhythmic nature of this general sequence of harmonic mean values, expressively reflecting the phenomenon of agreed triplet-deviations under $3m$-plets in this gene.

Let us show briefly, for comparison, also the OS representations of two long human genes: *NEB* gene and *SYNE1* gene.

Figs. 10.6.6 and 10.6.7 show graphs with the results of the *NEB* gene by the OS method. Initial data on this gene were taken from https://www.ncbi.nlm.nih.gov/nuccore/X83957. This gene contains 7071 nucleotides A, 4478 nucleotides T, 4578 nucleotides C, and 4754 nucleotides G, that is $S_A = 7071$, $S_T = 4478$, $S_C = 4578$, and $S_G = 4754$ for the model hyperbolic progressions (10.2). It can be especially noted that, in this gene, the amounts of nucleotides A and T are significantly different (7071 and 4478), that is, Chargaff's second rule on their approximate equality in long sequences is not fulfilled here since this nucleotide sequence is not enough long.

Fig. 10.6.7 additionally draws attention to the phenomenon of long-range correlations in the *NEB* gene between sequences of the

Table 10.6.1. Real and model values to OS representations of the classes of A_1-, T_1-, C_1-, and G_1-oligomers in the human *TTN* gene are shown for $n = 1, 2, ..., 20$. The real total amounts of n-plets, which start with a certain nucleotide (A, T, C, or G), are indicated jointly with their model values $H_{A,1}(n) = 26373/n$, $H_{T,1}(n) = 19569/n$, $H_{C,1}(n) = 17097/n$, and $H_{G,1}(n) = 18901/n$ (in red). Δ% refers to deviations of real values from model values (the model values are taken as 100%).

n	1	2	3	4	5	6	7	8	9	10
A										
Real	26373	13334	8848	6656	5346	4463	3805	3315	2924	2724
Model	26373	13187	8791	6593	5275	4396	3768	3297	2930	2637
Δ%	0	-1.119	-0.648	-0.952	-1.354	-1.536	-0.993	-0.557	0.216	-3.287
T										
Real	19569	9677	3993	4857	3885	1964	2755	2426	1332	1943
Model	19569	9784.5	6523	4892	3914	3262	2796	2446	2174	1957
Δ%	0	1.099	38.786	0.721	0.736	39.782	1.451	0.823	38.740	0.710
C										
Real	17097	8522	4876	4199	3426	2431	2458	2101	1617	1707
Model	17097	8549	5699	4274	3419	2850	2442	2137	1900	1710
Δ%	0	0.310	14.441	1.761	-0.193	14.687	-0.638	1.690	14.880	0.158
G										
Real	18901	9437	9596	4773	3731	4798	2687	2400	3231	1820
Model	18901	9451	6300	4725	3780	3150	2700	2363	2100	1890
Δ%	0	0.143	-52.309	-1.011	1.302	-52.309	0.487	-1.582	-53.849	3.709

n	11	12	13	14	15	16	17	18	19	20
A										
Real	2403	2219	1989	1956	1819	1683	1499	1454	1415	1384
Model	2398	2198	2029	1884	1758	1648	1551	1465	1388	1319
Δ%	-0.228	-0.967	1.957	-3.833	-3.458	-2.104	3.375	0.762	-1.941	-4.956
T										
Real	1782	986	1563	1339	788	1224	1160	660	1032	974
Model	1779	1631	1505	1398	1305	1223	1151	1087	1030	978
Δ%	-0.169	39.537	-3.833	4.206	39.598	-0.077	-0.772	39.292	-0.199	0.455
C										
Real	1548	1207	1258	1227	963	1024	1038	803	932	849
Model	1554	1425	1315	1221	1140	1069	1006	950	900	855
Δ%	0.404	15.283	4.346	-0.474	15.511	4.170	-3.211	15.459	-3.574	0.684
G										
Real	1716	2416	1493	1330	1892	1190	1123	1635	933	890
Model	1718	1575	1454	1350	1260	1181	1112	1050	995	945
Δ%	0.132	-53.389	-2.688	1.487	-50.151	-0.735	-1.005	-55.706	6.211	5.825

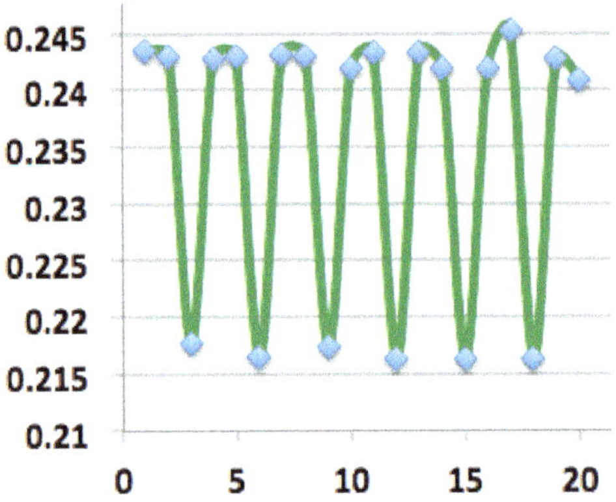

Fig. 10.6.5. The sequence of harmonic mean values of agreed deviations of all four OS sequences from their model harmonic progressions $H_{A,1}(n) = 26373/n$, $H_{T,1}(n) = 19569/n$, $H_{C,1}(n) = 17097/n$, and $H_{G,1}(n) = 18901/n$ in the human *TTN* gene. $n = 1, 2, \ldots, 20$ are plotted along the abscissa. The ordinate shows harmonic mean values.

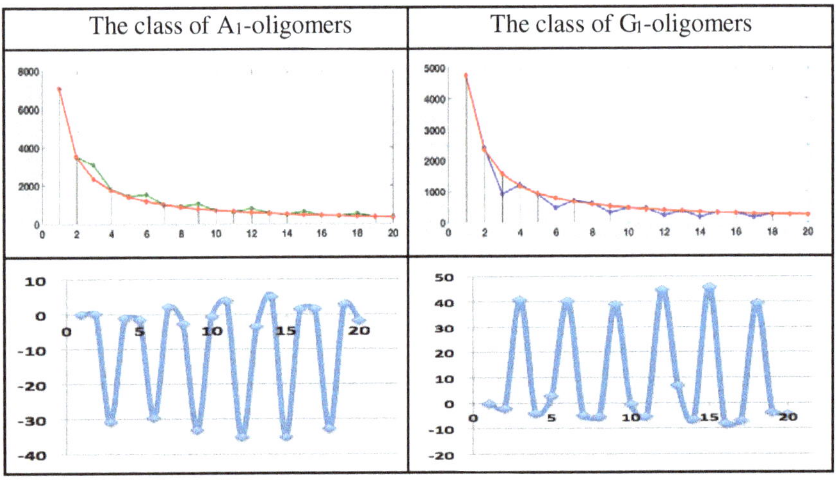

Fig. 10.6.6. Graphical representations of the analysis — by the OS method — of the human *NEB* gene: the sequences of the total amounts of n-plets, which start with the nucleotides A (left) and G (right), are shown. Here $n = 1, 2, 3, \ldots, 20$ (at the abscissa). Upper row: the red lines refer to model hyperbolic progressions $S_A/n = 7071/n$ (left) and $S_G/n = 4754/n$ (right). The ordinate shows the total amounts of appropriate n-plets. The highly regular sequences of the significant triplet-deviations at $3m$-plets are shown by the green line (left) and the blue line (right). Bottom row: percentage representations of the sequences of deviations for the real total amounts of n-plets of these classes from the appropriate model hyperbolic values $7071/n$ and $4754/n$ (the ordinate axes show these percentages). The model values are taken as 100%.

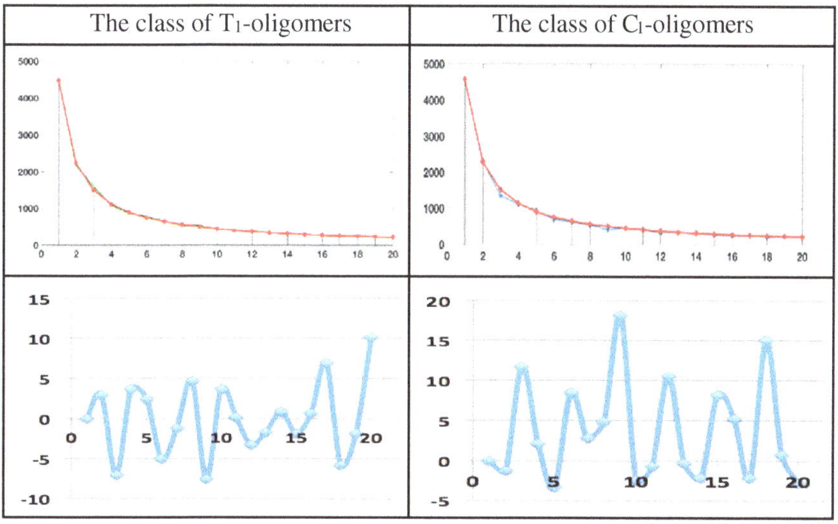

Fig. 10.6.7. Graphical representations of the results of the analysis — by the OS method — of the human *NEB* gene: the sequences of the total amounts of n-plets, which start with the nucleotides T (left) and C (right), are shown. Here $n = 1, 2, 3, \ldots, 20$ (on the abscissa). Upper row: the red lines refer to model hyperbolic progressions $S_T/n = 4478/n$ (left) and $S_C/n = 4578/n$ (right). The ordinate shows the total amounts of appropriate n-plets. The weakly regular sequences of the significant triplet-deviations at $3m$-plets are shown by the green line (left) and the blue line (right). Bottom row: percentage representations of the sequences of deviations for the real total amounts of n-plets of these classes from the appropriate model hyperbolic values $4478/n$ and $4578/n$ (the ordinate axes show these percentages). The model values are taken as 100%.

triplet-deviations in classes of A_1-, and G_1-oligomers: the triplet-deviations in these sequences happen in opposite directions.

By analogy with Fig. 10.6.6, Fig. 10.6.8 shows the sequence of harmonic mean values for agreed deviations of all four OS sequences from their model harmonic progressions for the human *NEB* gene. One can see the very regular rhythmic nature of this general sequence of harmonic mean values, reflecting the phenomenon of agreed triplet-deviations under $3m$-plets in this gene.

Figs. 10.6.9 and 10.6.10 show analysis of the human *SYNE1* gene by the OS method. Initial data on this gene were taken from https://www.ncbi.nlm.nih.gov/nuccore/NM_182961. This gene contains 8697 nucleotides A, 6032

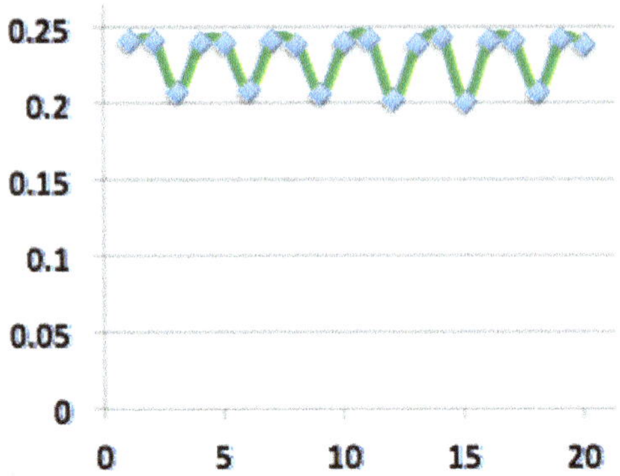

Fig. 10.6.8. The sequence of harmonic mean values of agreed deviations of all four OS sequences from their model harmonic progressions in the human *NEB* gene. $n = 1, 2, \ldots, 20$ are plotted along the abscissa. The ordinate shows harmonic mean values.

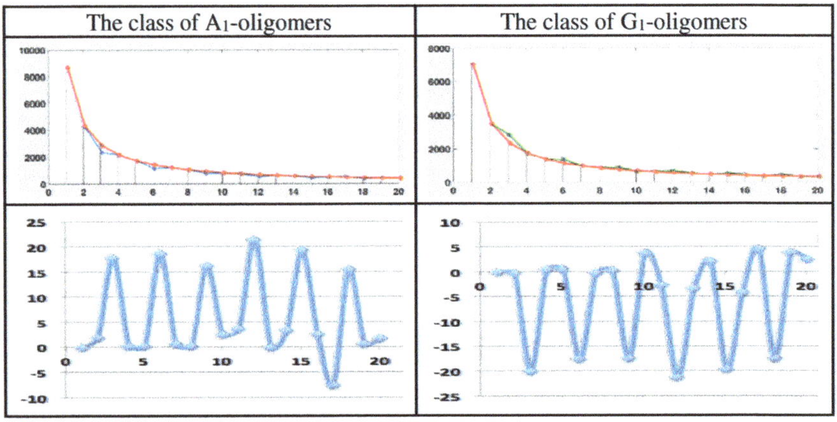

Fig. 10.6.9. Graphical representations of the analysis — by the OS method — of the human *SYNE1* gene: the sequences of the total amounts of n-plets, which start with the nucleotides A (left) and G (right), are shown. Here $n = 1, 2, 3, \ldots, 20$ (on the abscissa). Upper row: the red lines refer to model hyperbolic progressions $S_A/n = 8697/n$ (left) and $S_G/n = 7039/n$ (right). The ordinate shows the total amounts of appropriate n-plets. The highly regular sequences of the significant triplet-deviations at $3m$-plets are shown by the blue line (left) and the green line (right). Bottom row: percentage representations of the sequences of deviations for the real total amounts of n-plets of these classes from the appropriate model hyperbolic values $7071/n$ and $4754/n$ (the ordinate shows these percentages). The model values are taken as 100%.

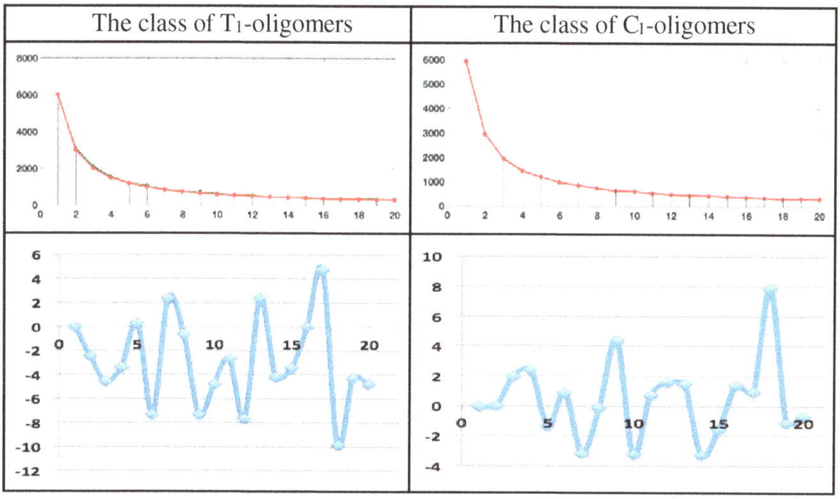

Fig. 10.6.10. Graphical representations of the analysis — by the OS method — of the human *SYNE1* gene: the sequences of the total amounts of n-plets, which start with the nucleotides T (at left) and C (at right), are shown. Here $n = 1, 2, 3, \ldots, 20$ (on the abscissa). Upper row: the red lines refer to model hyperbolic progressions $S_T/n = 6032/n$ (left) and $S_C/n = 5940/n$ (right). The ordinate axes show the total amounts of appropriate n-plets. The triplet-deviations in both of these classes are small in magnitude, and therefore, on these graphs, the model hyperbolic progressions (in red) practically hide under themselves the sequences of real total amounts of $3m$-plets. Bottom row: percentage representations of the sequences of deviations for the real total amounts of n-plets of these classes from the appropriate model hyperbolic values $6032/n$ and $5940/n$ (the ordinate shows these percentages). The model values are taken as 100%.

nucleotides T, 5940 nucleotides C, and 7039 nucleotides G, that is $S_A = 8697$, $S_T = 6032$, $S_C = 5940$, and $S_G = 7039$ for the model hyperbolic progressions (10.2). It can be especially noted that, in this gene, the amounts of nucleotides A and T are significantly different, as are the amounts of nucleotides C and G, and therefore Chargaff's second rule for long nucleotide sequences is not satisfied here since this nucleotide sequence is not long enough.

Fig. 10.6.10 additionally draws attention to the phenomenon of long-range correlations in the *SYNE1* gene between sequences of the triplet-deviations in classes of A_1-, and G_1-oligomers: the triplet-deviations in these sequences happen in opposite directions.

One should note that not all long genes have regular sequences of the pronounced triplet-deviations in their OS representations. The comparison analysis of the OS representations of different genes is a new research field.

One of the interesting topics for comparative analysis of genes by the OS method relates to the structure of histones, which is highly conservative in evolution. Histones are highly basic proteins found in eukaryotic cell nuclei that pack and order the DNA into structural units called nucleosomes. Histones are the chief protein components of chromatin, acting as spools around which DNA winds, and playing a role in gene regulation.

Figs. 10.6.11 and 10.6.12 show results of the analysis for one of the short genes of human histones by the OS method (this gene was randomly selected from multiple histone genes for analysis): H.sapiens H1.1 gene for histone H1, 1034 bp DNA (GenBank: X57130.1, https://www.ncbi.nlm.nih.gov/nuccore/X57130.1). The results confirm the implementation of the hyperbolic (harmonic) rule № 1 for this gene.

One can see in Fig. 10.6.12 the existence in this short histone gene some analog of those triplet-deviations related to $3m$-plets that were described above for long genes and shown in Figs. 10.6.1–10.6.10. In particular, the correlation exists in this short gene between two sequences of the triplet-deviations in the considered classes of C_1- and G_1-oligomers: the triplet-deviations in these sequences happen in opposite directions.

Fig. 10.6.12 shows the results of a similar analysis of the same histone gene regarding its classes of A_1- and T_1-oligomers.

One can see from Fig. 10.6.12 that rhythmic deviations in the classes of A_1- and T_1-oligomers are less regular and stable than in the classes of C_1- and G_1-oligomers in Fig. 10.6.11. To clarify the general picture of such properties of histone genes, systematic studies of a wide set of histone genes in their OS representations are required.

By analogy with Figs. 10.6.5 and 10.6.8, Fig. 10.6.13 shows the sequence of harmonic mean values of agreed deviations of all four OS sequences from their model harmonic progressions for the case of the human histone *H1.1* gene. One can see the regular rhythmic nature of this general sequence of harmonic mean values, reflecting the phenomenon of agreed triplet-deviations under $3m$-plets in this gene.

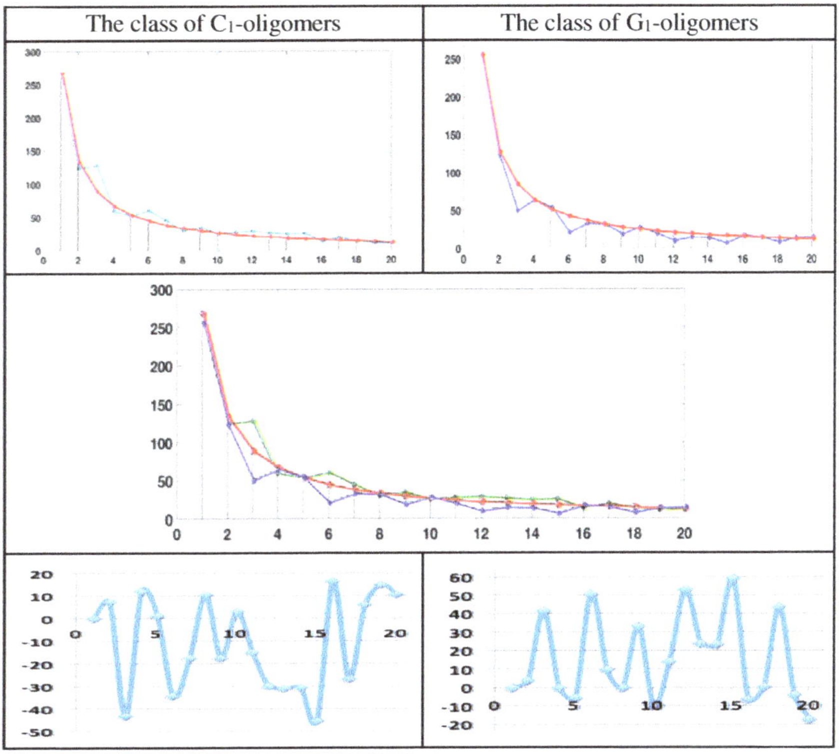

Fig. 10.6.11. Graphical representations of the analysis — by the OS method — of the human histone *H1.1* gene regarding its classes of C_1- and G_1-oligomers. The sequences of its total amounts of *n*-plets, which start with the nucleotides C (left) and G (right), jointly with model hyperbolic progressions are shown. Top row: the red lines refer to model hyperbolic progressions S_C/n and S_G/n correspondingly, where $S_C = 268$ and $S_G = 256$ are the quantities of nucleotides C and G in the gene; $n = 1, 2, 3, \ldots, 20$ as shown on the abscissa. The green line (left) and the blue line (right) with dots on them refer to the real OS sequences of the total amounts of such *n*-plets. The ordinate indicates the total amounts of *n*-plets. Middle row: the graph combining both graphs from the top row. Bottom row: graphs indicate the sequences of percentage deviations of the real total amounts of *n*-plets, which start with appropriate nucleotides C and G in the gene, from the model hyperbolic values S_C/n (left) and S_G/n (right) under $n = 1, 2, 3, \ldots, 20$ (on the abscissa). The ordinate axis indicates values of percentage deviations. The model values are taken as 100%.

Certain triplet-deviations between real and model values under $3m$-plets are also found in the OS representations of entire chromosomes of humans and other organisms, but in a much less pronounced form than shown in this section for individual genes.

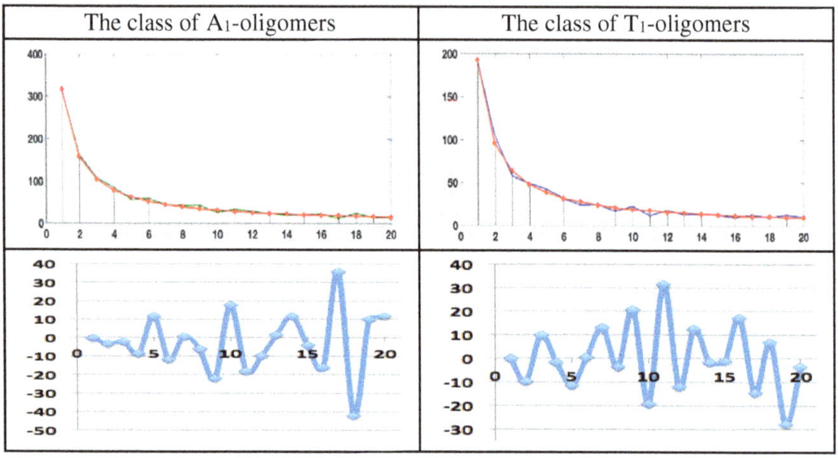

Fig. 10.6.12. Graphical representations of the results of the analysis — by the OS method — of the human histone *H1.1* gene regarding its classes of A_1- and T_1-oligomers. Top row: the red lines with red dots refer to model hyperbolic progressions S_A/n and S_T/n correspondingly, where $S_A = 317$ and $S_T = 193$ are the quantities of nucleotides A and T; $n = 1, 2, 3, \ldots, 20$ as shown on the abscissa. The green line (left) and the blue line (right) with dots on them refer to the real OS sequences of the total amounts of such n-plets. The ordinate indicates the total amounts of n-plets. Bottom row: graphs indicate the sequences of percent deviations of the real total amounts of n-plets, which start with appropriate nucleotides A and T in the gene, from the model hyperbolic values S_A/n and S_T/n under $n = 1, 2, 3, \ldots, 20$ (on the abscissa). The ordinate indicates values of percentage deviations. The model values are taken as 100%.

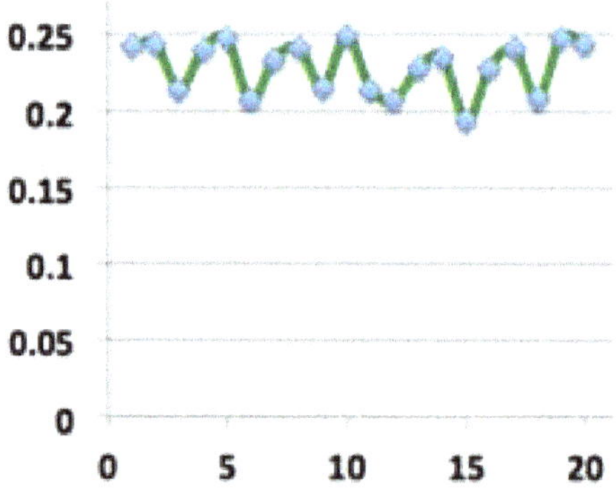

Fig. 10.6.13. The sequence of harmonic mean values of agreed deviations of all four OS sequences from their model harmonic progressions in the human histone *H1.1* gene. $n = 1, 2, \ldots, 20$ are plotted along the abscissa. The ordinate shows harmonic mean values.

10.7. The hyperbolic rules in genomes of different groups from Bacteria and Archaea

Let us turn now to prokaryotic genomes. The section represents the results of the analysis for nucleotide sequences of all 19 bacterial genomes of different groups both from Bacteria and Archaea, which are listed in the article on Chargaff's second rule [194, p. 2]: "*Nucleotide disparities for prokaryotic coding sequences were taken from bacterial genomes of different groups both from Bacteria and Archaea. All together 19 genomes were used: Aquifex aeolicus, Acidobacteria bacterium, Bradyrhizobium japonicum, Bacillus subtilis, Chlamydia trachomatis, Chromobacterium violaceum, Dehalococcoides ethenogenes, Escherichia coli, Flavobacterium psychrophilum, Gloeobacter violaceus, Helicobacter pillory, Methanosarcina acetivorans, Nanoarchaeum equitans, Syntrophus aciditrophicus, Streptomyces coelicolor, Sulfolobus solfataricus, Treponema denticola, Thermotoga maritima, and Thermus thermophiles.*"

Fig. 10.7.1 shows the results of the analysis of these prokaryotic genomes by the OS method. These results demonstrate that the hyperbolic rule No. 1 is fulfilled for all the listed genomes of prokaryotes: the model hyperbolic progressions $H_{A,1}(n) = S_A/n$, $H_{T,1}(n) = S_T/n$, $H_{C,1}(n) = S_C/n$, and $H_{G,1}(n) = S_G/n$ from the expression (10.2) practically coincide with the OS sequences of real total amounts of n-plets from the classes A_1-, T_1-, C_1-, and G_1-oligomers at $n = 1, 2, 3, \ldots, 20$. Because of this coincidence, the model hyperbolic progressions, which are represented by red lines in the graphs of Fig. 10.7.1, almost completely cover the sequences of real values (the blue lines in the lower graphs show in percentage slight alternating deviations of real values from model values).

The genomes are enumerated in Fig. 10.7.1 by the following numbers 1–19:

1) *Aquifex aeolicus* VF5, complete genome, 1551335 bp, accession AE000657, version AE000657.1, https://www.ncbi.nlm.nih.gov/nuccore/AE000657.1?report=genbank;

2) *Acidobacteria bacterium* KBS 146 M015DRAFT_scf7180000000004_quiver.1_C, whole genome shotgun sequence, 4996384 bp, accession JHVA01000001, https://www.ncbi.nlm.nih.gov/nuccore/JHVA01000001.1?report=genbank;

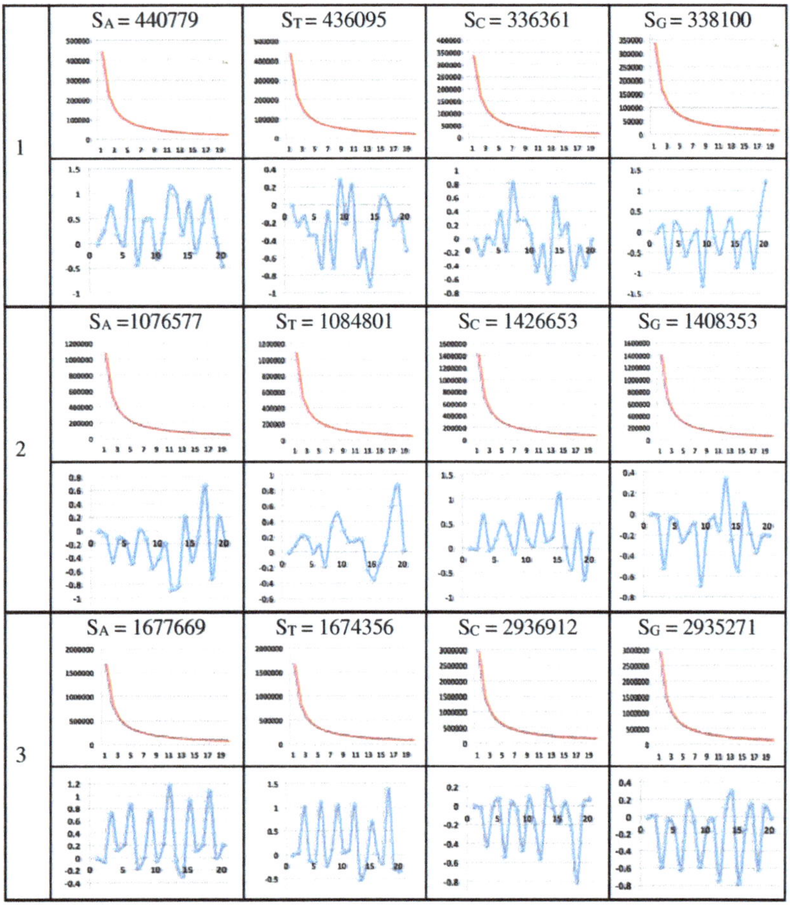

Fig. 10.7.1. Graphical representations of the results of the analysis — by the OS method — of 19 bacterial genomes of Bacteria and Archaea mentioned in [194, p. 2]. For each of the genomes two rows of resulting data are shown at $n = 1, 2, \ldots, 20$ plotted along the abscissa: the top rows demonstrate that model hyperbolic progressions S_A/n, S_T/n, S_C/n, S_G/n (red lines) almost completely cover the OS sequences of real values (the ordinate shows appropriate values); the bottom blue lines show in percentage slight alternating deviations of real values from model values. The left column indicates numbers denoting the genomes as explained in the text.

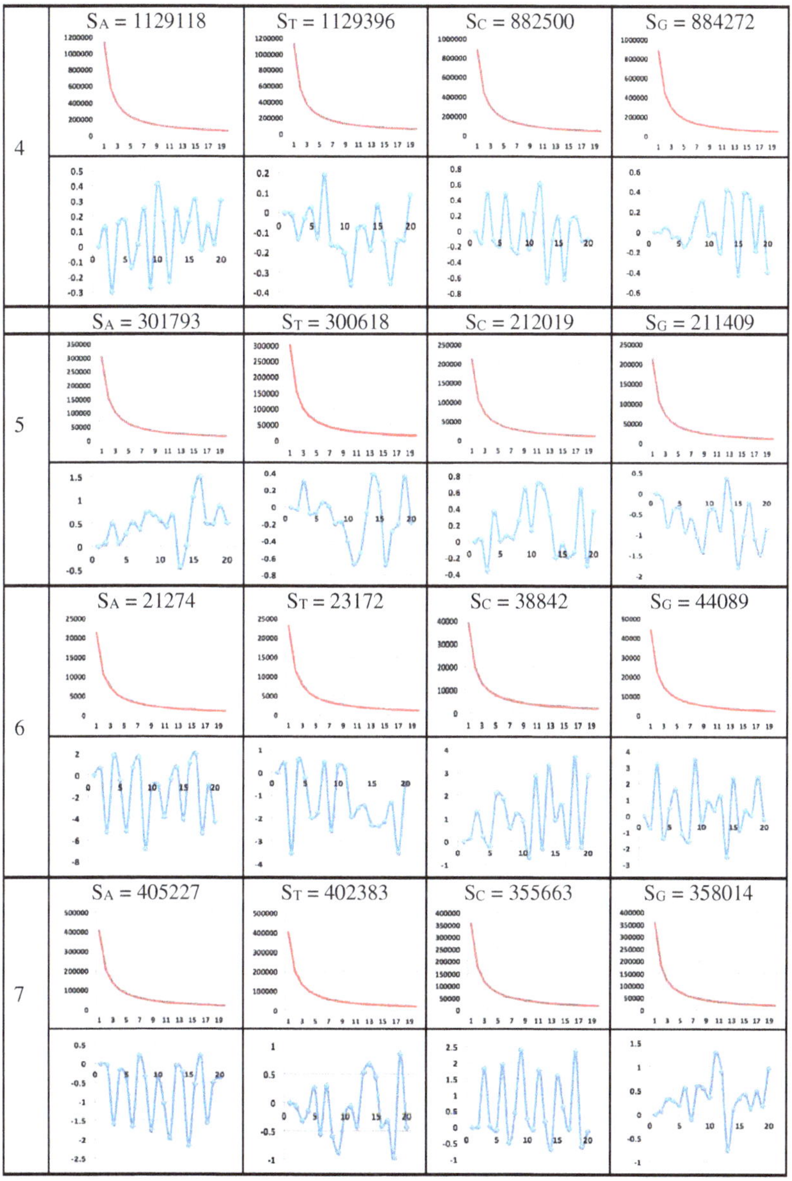

Fig. 10.7.1. *(Continued)*

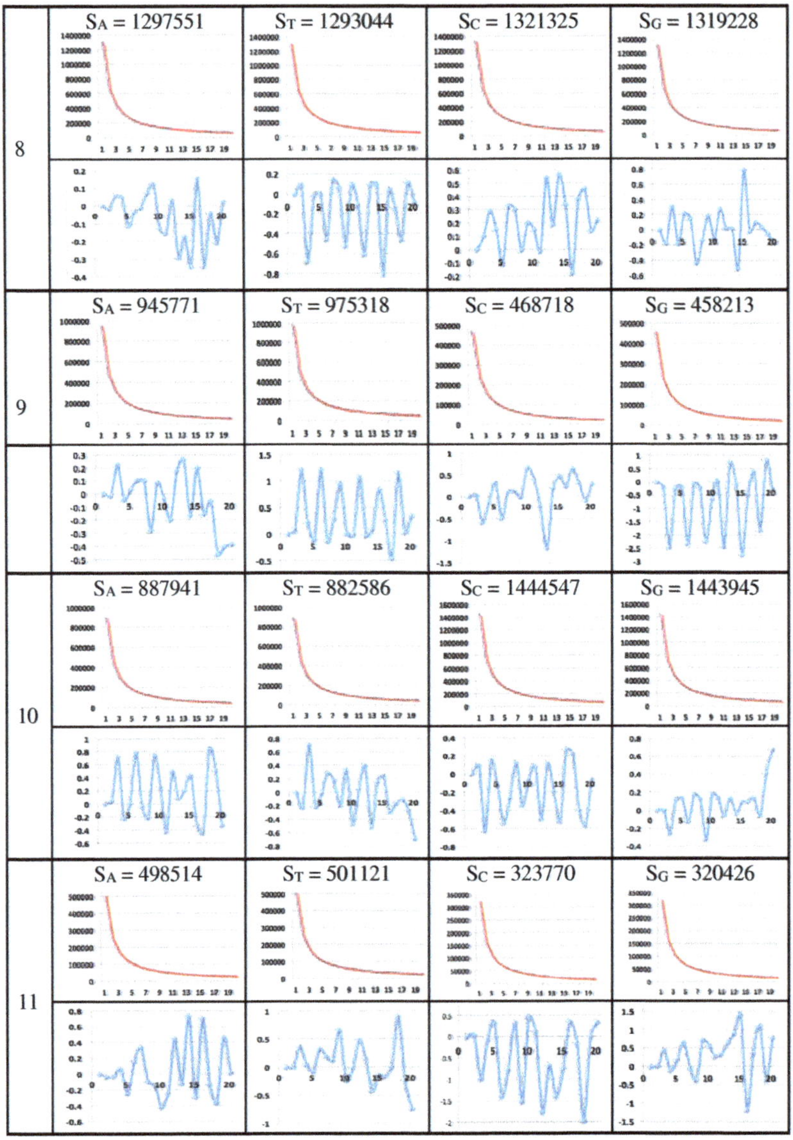

Fig. 10.7.1. (*Continued*)

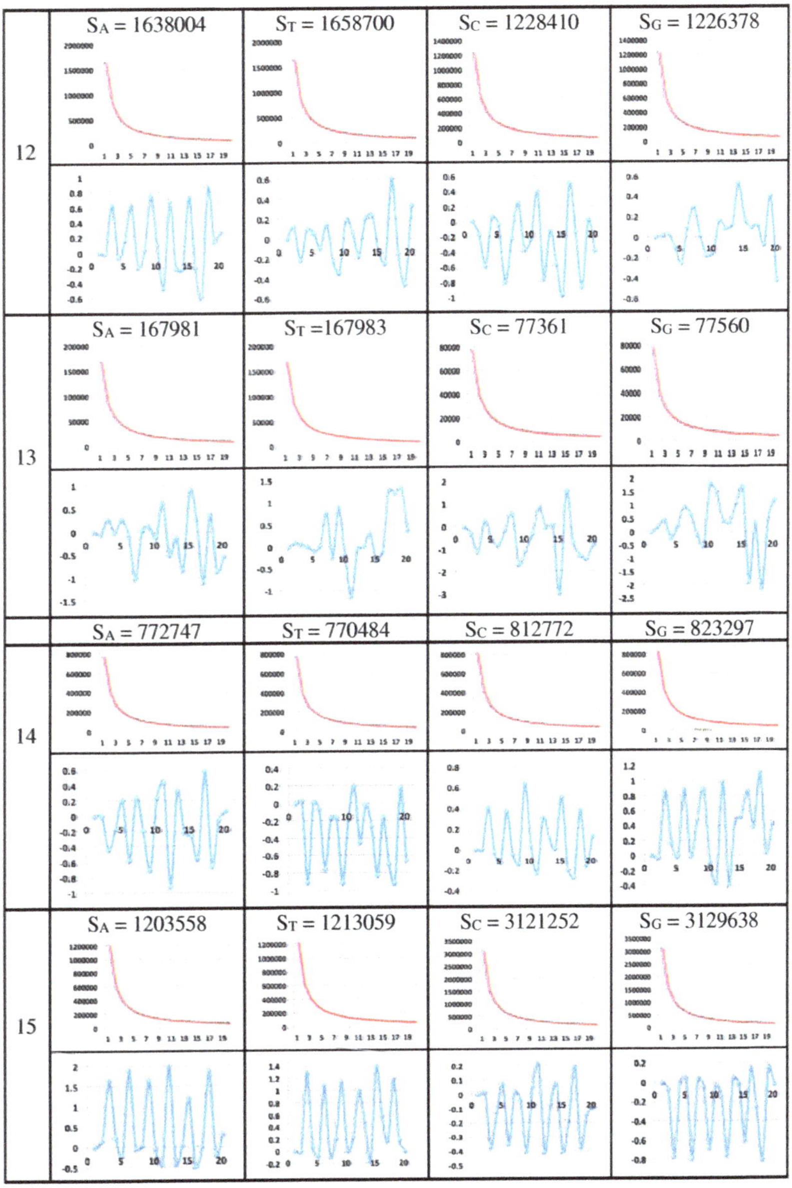

Fig. 10.7.1. (*Continued*)

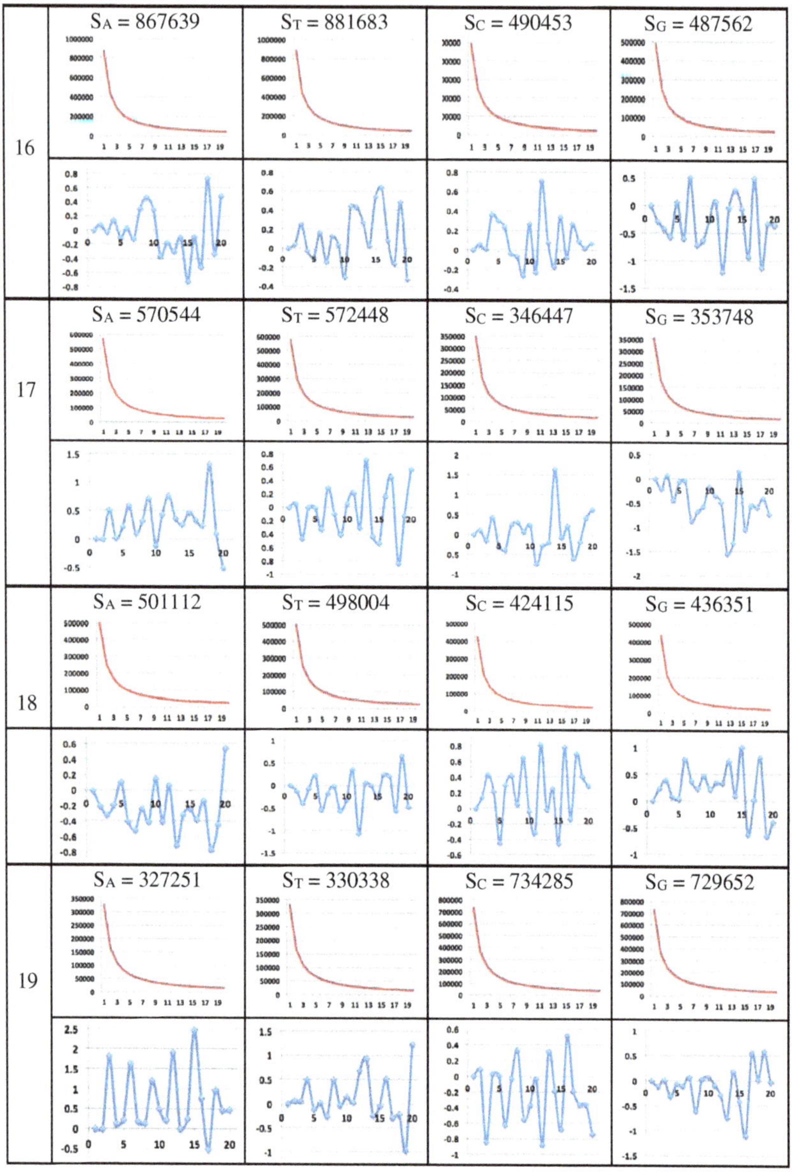

Fig. 10.7.1. (*Continued*)

3) *Bradyrhizobium japonicum* strain E109, complete genome, 9224208 bp, accession CP010313, https://www.ncbi.nlm.nih.gov/nuccore/CP010313.1?report=genbank;
4) *Bacillus subtilis* strain UD1022, complete genome, 4025326 bp, accession CP011534, https://www.ncbi.nlm.nih.gov/nuccore/CP011534.1?report=genbank;
5) *Chlamydia trachomatis* strain QH111L, complete genome, 1025839 bp, accession CP018052, https://www.ncbi.nlm.nih.gov/nuccore/CP018052.1?report=genbank;
6) *Chromobacterium violaceum* strain LK301, whole genome shotgun sequence, 127377 bp, accession LDUX01000001 version LDUX01000001.1, https://www.ncbi.nlm.nih.gov/nuccore/LDUX01000001.1?report=genbank;
7) *Dehalococcoides mccartyi* strain CG3, complete genome, NCBI Reference Sequence: NZ_CP013074.1, 1521287 bp, https://www.ncbi.nlm.nih.gov/nuccore/NZ_CP013074.1?report=genbank;
8) *Escherichia coli* CFT073, complete genome, GenBank: AE014075.1, 5231428 bp, https://www.ncbi.nlm.nih.gov/nuccore/AE014075.1?report=genbank;
9) *Flavobacterium psychrophilum* JIP02/86, complete genome, 2860382 bp, accession NC_009613, https://www.ncbi.nlm.nih.gov/nuccore/NC_009613.3;
10) *Gloeobacter violaceus* PCC 7421 DNA, complete genome, GenBank: BA000045.2, 4659019 bp, accession BA000045 AP006568-AP006583 version BA000045.2, https://www.ncbi.nlm.nih.gov/nuccore/BA000045.2?report=genbank;
11) *Helicobacter pylori*, NCBI Reference Sequence: NC_000921.1, complete genome, 1643831 bp, accession NC_000921 NZ_AE001440-NZ_AE001571 version NC_000921.1, https://www.ncbi.nlm.nih.gov/nuccore/NC_000921.1;
12) *Methanosarcina acetivorans* str. C2A, complete genome, 5751492 bp, accession AE010299 AE010656-AE011189 version AE010299.1, https://www.ncbi.nlm.nih.gov/nuccore/AE01029;

13) *Nanoarchaeum equitans* Kin4-M, complete genome, 490885 bp, accession AE017199 AACL01000000 AACL01000001 version AE017199.1, https://www.ncbi.nlm.nih.gov/nuccore/AE017199.1?report=genbank;
14) *Syntrophus aciditrophicus* SB, complete genome, 3179300 bp, accession CP000252, https://www.ncbi.nlm.nih.gov/nuccore/CP000252.1?report=genbank;
15) *Streptomyces coelicolor* A3(2) complete genome, 8667507 bp, accession AL645882, https://www.ncbi.nlm.nih.gov/nuccore/AL645882.2?report=genbank;
16) *Sulfolobus solfataricus* strain SULA, complete genome, 2727337 bp, accession CP011057, https://www.ncbi.nlm.nih.gov/nuccore/CP011057.1?report=genbank;
17) *Treponema denticola* SP33 supercont1.1, whole genome shotgun sequence, NCBI Reference Sequence: NZ_KB442453.1, 1850823 bp, accession NZ_KB442453 NZ_AGDZ01000000 version NZ_KB442453.1, https://www.ncbi.nlm.nih.gov/nuccore/NZ_KB442453.1?report=genbank;
18) *Thermotoga maritima* strain Tma200, complete genome, 1859582 bp, accession CP010967, https://www.ncbi.nlm.nih.gov/nuccore/CP010967.1?report=genbank;
19) *Thermus thermophilus* DNA, complete genome, strain: TMY, 2121526 bp, accession AP017920, https://www.ncbi.nlm.nih.gov/nuccore/AP017920.1?report=genbank

One can see from Fig. 10.7.1 that in some prokaryotic genomes (for example in №№ 3, 7, 9, and 15) the alternating small deviations of real values from model values are systematic and related to $3m$-plets; it seems to be analogous to the much stronger triplet-deviations described above for human genes in Figs. 10.6.1–10.6.7. Can a sign of the presence of such triplet-deviations in the genomes of some bacteria serve as a criterion for the selection of bacterial species for genetic engineering problems? It is one of many new questions arisen due to the discovery of the represented hyperbolic rules and the applications of the OS method.

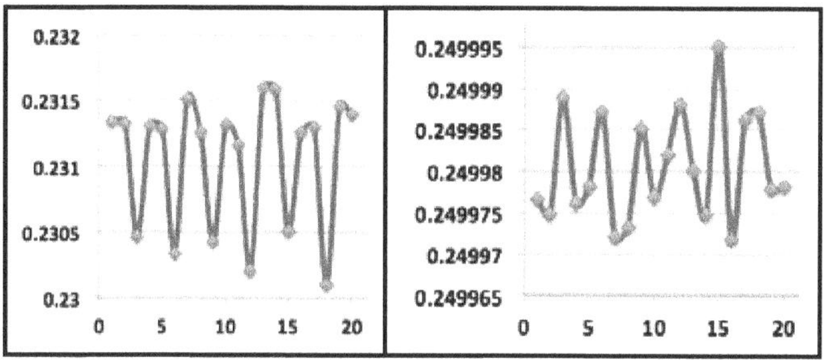

Fig. 10.7.2. The sequences of harmonic mean values of agreed deviations of all four OS sequences from their model harmonic progressions in the genomes of *Bradyrhizobium japonicum* strain E109 (left) and *Escherichia coli* CFT073 (right). $n = 1, 2, \ldots, 20$ are plotted along the abscissa. The ordinate shows harmonic mean values.

Fig. 10.7.2 shows examples of sequences of the harmonic mean values for two of these bacterial genomes. One can see triplet-deviations in these sequences at dots corresponding to $3m$-plets.

10.8. Analysis of genomes of microorganisms living in extreme environments

Of particular interest is the analysis of the genetic characteristics of microorganisms (extremophiles) living under extreme conditions of high and low temperatures, radiation, acidic and alkaline environments, drying, etc. Study of extremophiles is useful for many practical and theoretical problems. The https://en.wikipedia.org/wiki/Extremophile website contains a table of extremophiles. For the analysis of their genomes by the OS method, the author used 1–2 organisms from each category of the table. The initial data on the genomes were taken from the GenBank. Figs. 10.8.1–10.8.8 show the results of their analysis.

The resulting data in Figs. 10.8.1–10.8.8 shows the fulfillment of the hyperbolic rule №1 of oligomeric sums for all studied and presented extremophiles. The extremal living conditions of these microorganisms do not affect the subordination of their genomes to the described hyperbolic (harmonic) rules of algebraic invariance, which are true for the genomes of other prokaryotes and eukaryotes.

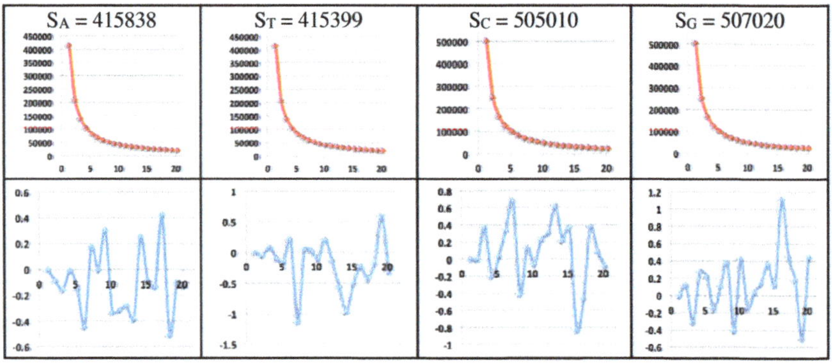

Fig. 10.8.1. The results of the analysis — by the OS method — of the extremophile *Pyrolobus fumarii 1A*, complete genome, 1843267 bp (this extremophile lives in submarine hydrothermal vents), https://www.ncbi.nlm.nih.gov/nuccore/NC_015931.1. All abscissa show the values $n = 1, 2, \ldots, 20$. The top row demonstrates that model hyperbolic progressions S_A/n, S_T/n, S_C/n, S_G/n (red lines) almost completely cover the OS sequences of real values (the ordinate axes show appropriate values). The bottom row shows in percentage slight alternating deviations of real values of the OS sequences from model values.

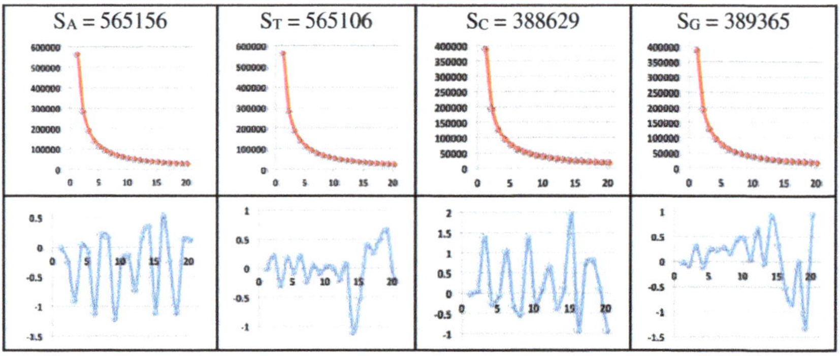

Fig. 10.8.2. The results of the analysis — by the OS method — of the extremophile *Pyrococcus furiosus DSM 3638*, complete genome, 1908256 (this extremophile lives in submarine hydrothermal vents), https://www.ncbi.nlm.nih.gov/nuccore/NC_003413.1. The explanation of these graphs is identical to the explanation for Fig. 10.8.1.

10. 9. Analysis of giant viruses by the oligomer sums method

This section represents examples of studying genomes of different viruses by the OS method. The focus is on giant viruses (Figs. 10.9.1–10.9.4).

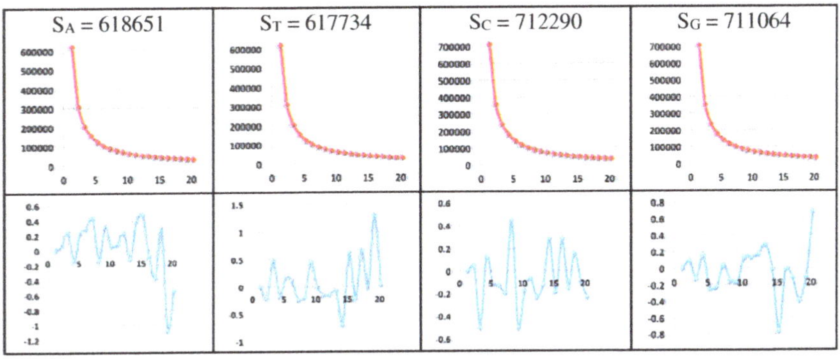

Fig. 10.8.3. The results of the analysis — by the OS method — of the extremophile *Synechococcus lividus PCC 6715* chromosome, complete genome, 2659739 bp (this extremophile lives in low temperature conditions), https://www.ncbi.nlm.nih.gov/nuccore/NZ_CP018092.1. The explanation of these graphs is identical to the explanation for Fig. 10.8.1.

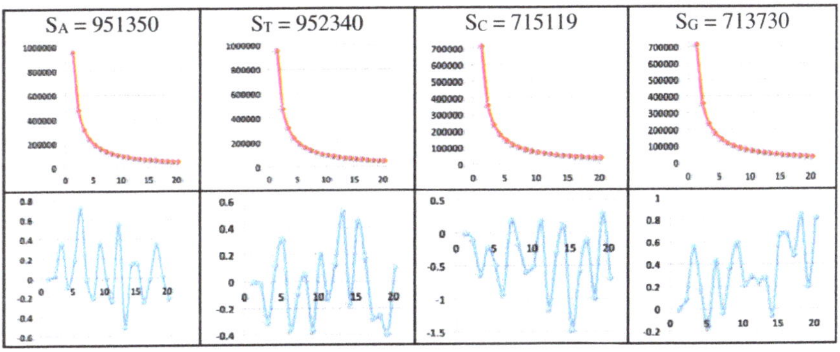

Fig. 10.8.4. The results of the analysis — by the OS method — of the extremophile *Psychrobacter alimentarius strain PAMC 27889* chromosome, complete genome, 3332539 bp (this extremophile lives in soda lakes), https://www.ncbi.nlm.nih.gov/nuccore/NZ_CP014945.1. The explanation of these graphs is identical to the explanation for Fig. 10.8.1.

The results, presented in this section, show the fulfillment of the hyperbolic (harmonic) rule No. 1 for the viruses considered and provide material for comparative analysis of different OS sequences.

10.10. Analysis of the COVID-19 virus by the oligomer sums method

Let us turn now to the analysis — by the OS method — of the COVID-19 virus, which led to a pandemic. The initial data on its nucleotide sequence

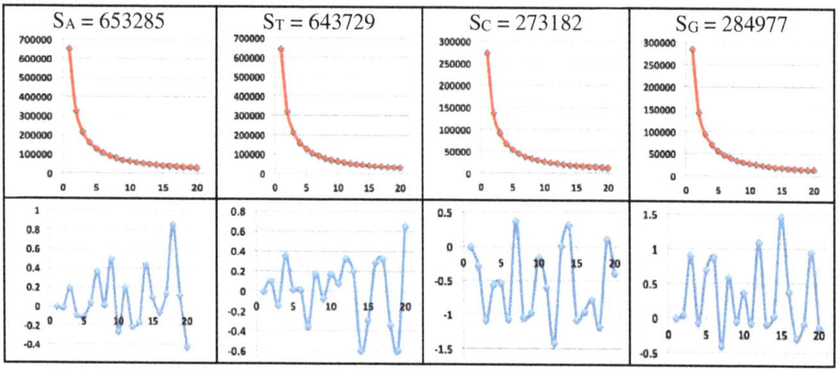

Fig. 10.8.5. The results of the analysis — by the OS method — of the extremophile *Clostridium paradoxum* JW-YL-7 = DSM 7308 strain JW-YL-7 ctg1, whole genome shotgun sequence, 1855173 bp (this extremophile lives in volcanic springs, acid mine drainage), https://www.ncbi.nlm.nih.gov/nuccore/LSFY01000001.1. The explanation of these graphs is identical to the explanation for Fig. 10.8.1.

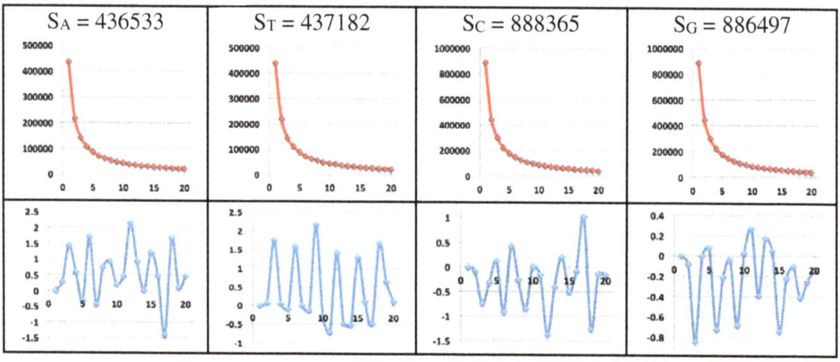

Fig. 10.8.6. The results of the analysis — by the OS method — of the extremophile *Deinococcus radiodurans R1* chromosome 1, complete sequence, 2648638 bp (this extremophile lives in conditions of cosmic rays, X-rays, and radioactive decay), https://www.ncbi.nlm.nih.gov/nuccore/NC_001263.1. The explanation of these graphs is identical to the explanation for Fig. 10.8.1.

was taken from the site https://www.ncbi.nlm.nih.gov/nuccore/MN908947.3, where the following is written about it: severe acute respiratory syndrome coronavirus 2 isolate Wuhan-Hu-1, complete genome, GenBank: MN908947.3, LOCUS MN908947, 29903 bp ss-RNA linear VRL 18-MAR-2020. Figs. 10.10.1–10.10.5 show results of such an analysis of the virus.

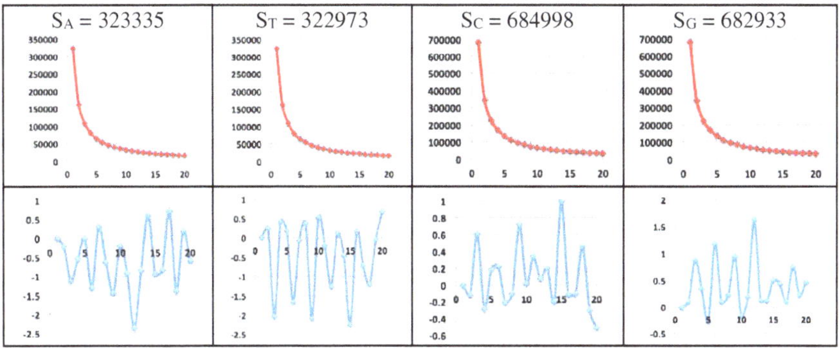

Fig. 10.8.7. The results of analysis — by the OS method — of the extremophile *Halobacterium sp. NRC-1*, complete genome, 2014239 bp (this extremophile lives in conditions of high salt concentration), https://www.ncbi.nlm.nih.gov/nuccore/NC_002607.1. The explanation of these graphs is identical to the explanation for Fig. 10.8.1.

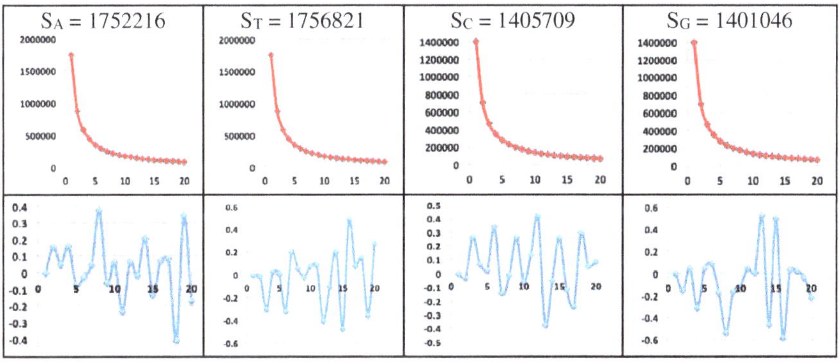

Fig. 10.8.8. The results of the analysis — by the OS method — of the extremophile *Chroococcidiopsis thermalis PCC 7203*, complete genome, 6315792 bp (this extremophile lives in conditions of desiccation), https://www.ncbi.nlm.nih.gov/nuccore/NC_019695.1. The explanation of these graphs is identical to the explanation for Fig. 10.8.1.

In particular, Figs. 10.10.1–10.10.4 show that this virus in its OS representations has under $n = 3, 6, 9, \ldots, 3m$ such deviations of real values from model values, which resemble the triplet-deviations in human genes described above in Figs. 10.6.1–10.6.10. Perhaps the harmfulness of this virus to humans is related to this similarity. It should also be noted that — in the classes of pyrimidines C_1- and T_1-oligomers (Figs. 10.10.2 and 10.10.3) — these deviations occur in opposite directions in a coordinated manner,

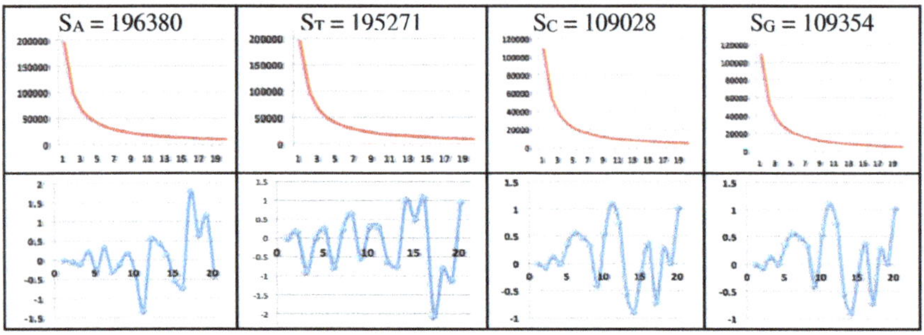

Fig. 10.9.1. The results of the analysis — by the OS method — of the giant virus *Pithovirus sibericum* isolate P1084-T, complete genome, 610033 bp, NCBI Reference Sequence: NC_023423.1 https://www.ncbi.nlm.nih.gov/nuccore/NC_023423.1. All abscissa show the values $n = 1, 2, \ldots, 20$. The top row demonstrates that model hyperbolic progressions S_A/n, S_T/n, S_C/n, S_G/n (red lines) almost completely cover the OS sequences of real values (the ordinate shows appropriate values). The bottom row shows in percentage slight alternating deviations of real values of the OS sequences from model values.

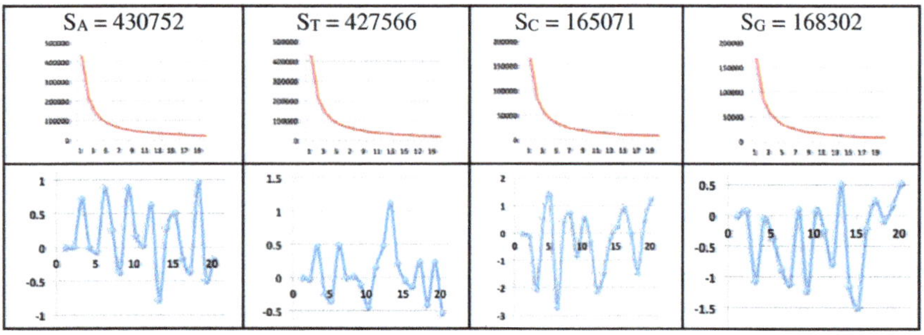

Fig. 10.9.2. The results of the analysis — by the OS method — of the giant virus *Acanthamoeba castellanii* mamavirus strain Hal-V, complete genome, 1191693 bp, GenBank: JF801956.1, https://www.ncbi.nlm.nih.gov/nuccore/JF801956.1. The explanation of these graphs is identical to the explanation for Fig. 10.9.1.

which indicates a particular consistency in the structure of the nucleotide sequence of this virus concerning pyrimidine classes.

By analogy with Figs. 10.6.2, 10.6.9, and 10.6.14, Fig. 10.10.5 shows the sequence of harmonic mean values of agreed deviations of all four OS sequences from their model harmonic progressions for the case of the coronavirus 2 isolate Wuhan-Hu-1. Two cases are shown: for $n = 1, 2, \ldots, 20$ and $n = 1, 2, \ldots, 100$. One can see the regular rhythmic nature of this

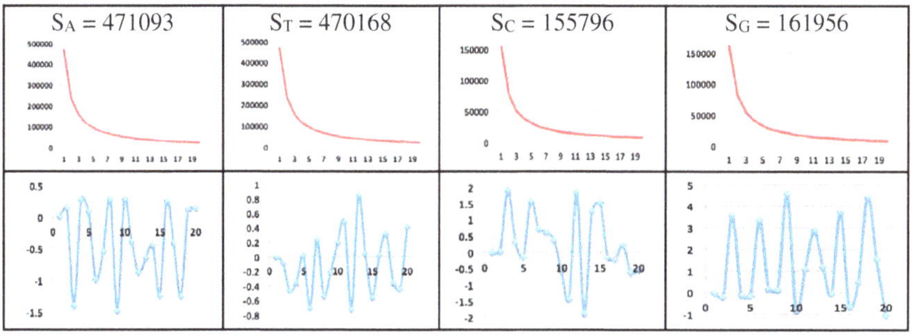

Fig. 10.9.3. The results of the analysis — by the OS method — the giant virus *Megavirus chiliensis*, complete genome, 1259197 bp, NCBI Reference Sequence: NC_016072.1, https://www.ncbi.nlm.nih.gov/nuccore/NC_016072.1. The explanation of these graphs is identical to the explanation for Fig. 10.9.1.

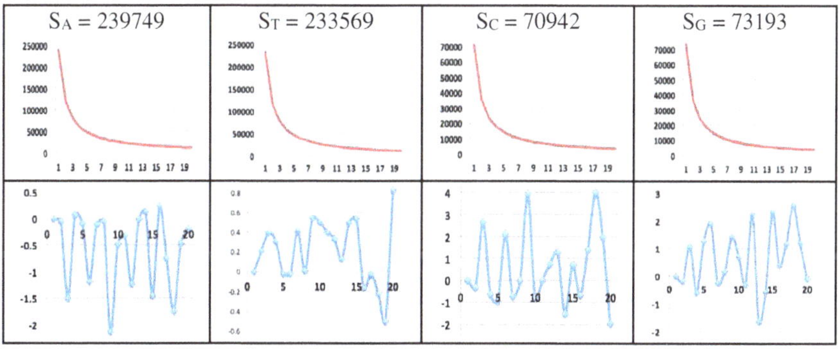

Fig. 10.9.4. The results of the analysis — by the OS method — the giant virus *Cafeteria roenbergensis* virus BV-PW1, complete genome, 617453 bp, NCBI Reference Sequence: NC_014637.1, https://www.ncbi.nlm.nih.gov/nuccore/NC_014637.1. The explanation of these graphs is identical to the explanation for Fig. 10.9.1.

general sequence of harmonic mean values, reflecting the phenomenon of agreed triplet-deviations under 3*m*-plets in this coronavirus.

10.11. DNA epi-chains and the hyperbolic rules for oligomer sums

This section presents the study of hyperbolic rules in epi-chains of long single-stranded DNA by the OS method. The notion "epi-chains" was described in Chapter 4, Fig. 4.3.1. The obtained results for the above-described hyperbolic

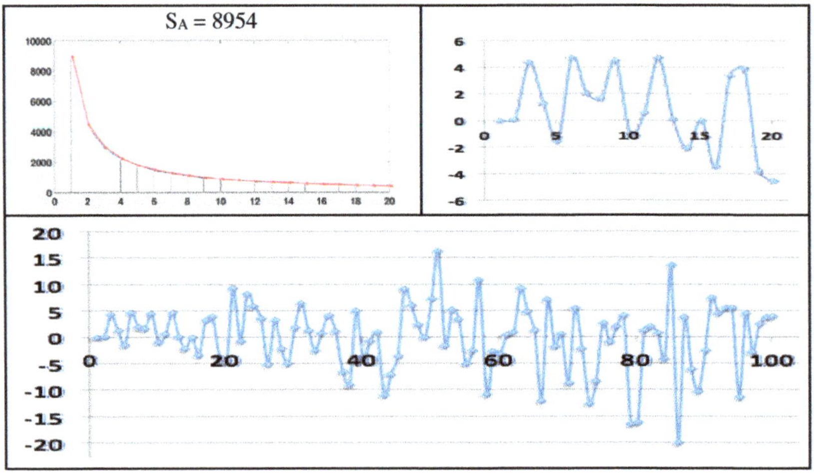

Fig. 10.10.1. The graphs for the OS sequences of n-plets from the class A_1-oligomers of the coronavirus 2 isolate Wuhan-Hu–1, complete genome, GenBank: MN908947.3, LOCUS MN908947, 29903 bp. In these graphs, the abscissa represents the values $n = 1, 2, 3, \ldots, 20$ (in top row) and $n = 1, 2, 3, \ldots, 100$ (in bottom row). Top left: the ordinate represents the set of phenomenological total amounts $\Sigma_{A,n,1}$ of n-plets beginning with the nucleotide A. Top right and bottom: deviations of real OS sequences $\Sigma_{A,n,1}$ from model hyperbolic progressions $S_A/n = 8954/n$ in percentages.

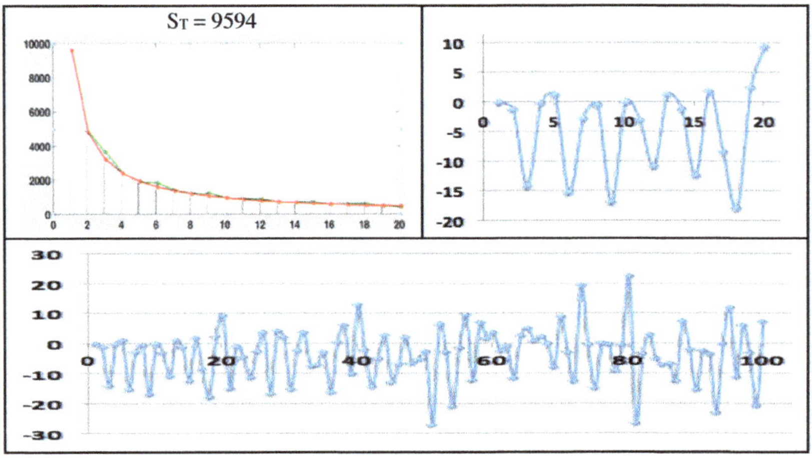

Fig. 10.10.2. The graphs for the OS sequences of n-plets from the class T_1-oligomers of the coronavirus 2 isolate Wuhan-Hu-1, complete genome, GenBank: MN908947.3, LOCUS MN908947, 29903 bp. In these graphs, the abscissa represents the values $n = 1, 2, 3, \ldots, 20$ (in top row) and $n = 1, 2, 3, \ldots, 100$ (in bottom row). Top left: the ordinate axis represents the set of phenomenological total amounts $\Sigma_{T,n,1}$ of n-plets beginning with the nucleotide T. Top right and bottom: deviations of the real OS sequence $\Sigma_{T,n,1}$ from the model hyperbolic progression $S_T/n = 9594/n$ in percentages.

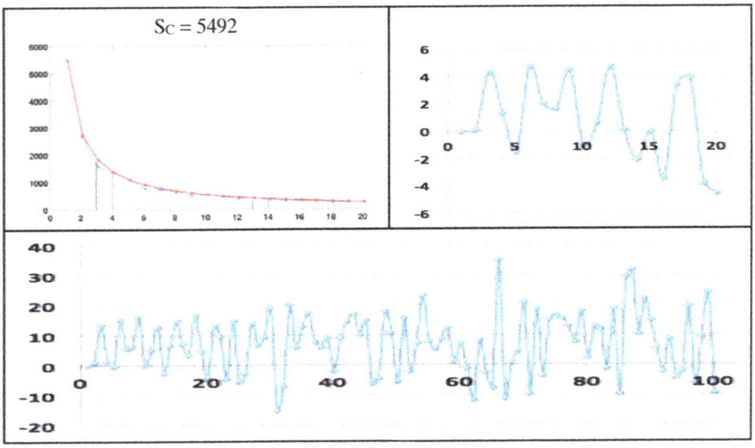

Fig. 10.10.3. The graphs for the OS sequences of n-plets from the class C_1-oligomers of the coronavirus 2 isolate Wuhan-Hu-1, complete genome, GenBank: MN908947.3, LOCUS MN908947, 29903 bp. In these graphs, the abscissa represents the values $n = 1, 2, 3, \ldots, 20$ (in top row) and $n = 1, 2, 3, \ldots, 100$ (in bottom row). Top left: the ordinate axis represents the set of phenomenological total amounts $\Sigma_{C,n,1}$ of n-plets beginning with the nucleotide C. Top right and bottom: deviations of the real OS sequences $\Sigma_{C,n,1}$ from the model hyperbolic progression $S_C/n = 5492/n$ in percentages.

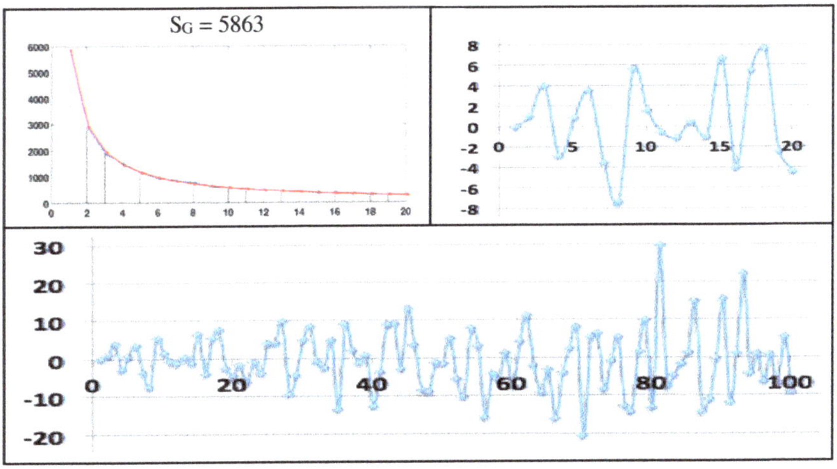

Fig. 10.10.4. The graphs for the OS sequences of n-plets from the class G_1-oligomers of the coronavirus 2 isolate Wuhan-Hu-1, complete genome, GenBank: MN908947.3, LOCUS MN908947, 29903 bp. In these graphs, the abscissa represents the values $n = 1, 2, 3, \ldots, 20$ (in top row) and $n = 1, 2, 3, \ldots, 100$ (in bottom row). Top left: the ordinate represents the set of phenomenological total amounts $\Sigma_{G,n,1}$ of n-plets beginning with the nucleotide G. Top right and bottom: deviations of the real OS sequence $\Sigma_{G,n,1}$ from the model hyperbolic progression $S_G/n = 5863/n$ in percentages.

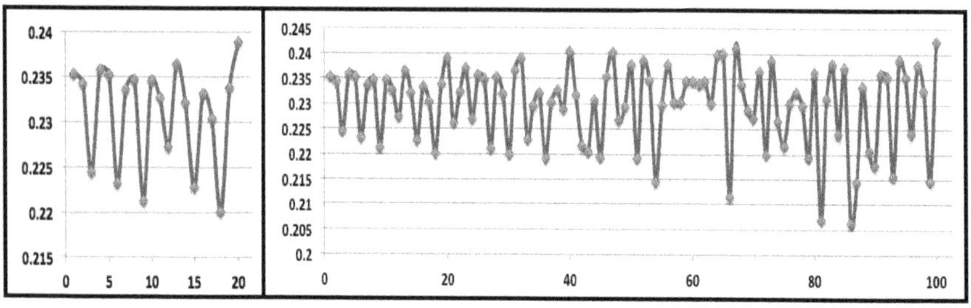

Fig. 10.10.5. The sequence of harmonic mean values of agreed deviations of all four OS sequences from their model harmonic progressions in the coronavirus 2 isolate Wuhan-Hu-1. The ordinate shows harmonic mean values. The left and right graphs show the cases of $n = 1, 2, \ldots, 20$ and $n = 1, 2, \ldots, 100$, which are plotted along the abscissa.

rules of OS for genomes are also fulfilled for these epi-chains; it gives new materials to the theme of fractal-like structures in genetics.

By definition, in a nucleotide sequence N_1 of any DNA strand with sequentially numbered nucleotides 1, 2, 3, 4, ... (Fig. 4.3.1), epi-chains of different orders k are those sub-sequences that contain only those nucleotides, whose numeration differ from each other by natural number $k = 1, 2, 3, \ldots$ For example, in any single-stranded DNA, epi-chains of the third order are those three nucleotide sub-sequences $N_{3/1}$, $N_{3/2}$, and $N_{3/3}$, each of which has sequence numbers that differ by n = 3: these epi-chains contain nucleotides with numerations 1, 4, 7, ... or 2, 5, 8, ... or 3, 6, 9, ..., respectively. Each DNA epi-chain of kth order (if $k = 2, 3, 4, \ldots$) contains k times fewer nucleotides than the DNA strand and has its own arrangements of nucleobases A, T, C, and G. But unexpectedly, despite these differences, OS sequences of the total amounts of those n-plets, which start with A, T, C, or G, are modeled by very similar hyperbolic progressions as in the complete DNA strand and its epi-chains [116, 155, 198, 232].

Figs. 10.11.1–10.11.5 explains these results in graphical forms by examples of the OS representations of epi-chains $N_{2/1}$, $N_{3/1}$, $N_{4/1}$, $N_{10/1}$, and $N_{50/1}$ in the human chromosome №1 (the OS representation of this complete chromosome was presented above in Figs. 10.1.1–10.1.2).

Figs. 10.11.1–10.11.5 show that in these epi-chains, which are sparse sub-sequences of the complete DNA sequence, the same hyperbolic rule № 1 is fulfilled, which was formulated above for complete DNA sequences in

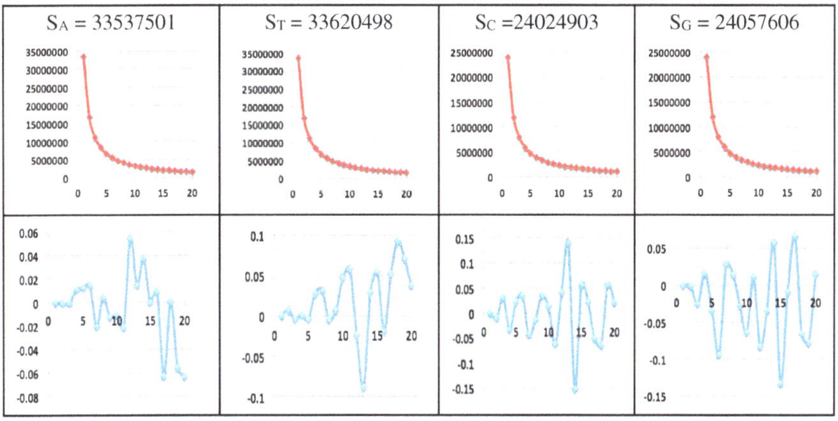

Fig. 10.11.1. The results of the analysis — by the OS method — of the nucleotide sequence of the epi-chain of the second order $N_{2/1}$, which consists of nucleotides with serial numerations 1-3-5-7-9-... in the DNA sequence of the human chromosome № 1. All abscissa show the values $n = 1, 2, \ldots, 20$. The top row demonstrates that the model hyperbolic progressions S_A/n, S_T/n, S_C/n, S_G/n (red lines) almost completely cover the OS sequences of real total amounts of those n-plets, which start with A, T, C, or G in this epi-chain correspondingly (the ordinate shows appropriate amounts). The bottom row shows in percentage slight alternating deviations of real values of the OS sequences from model values.

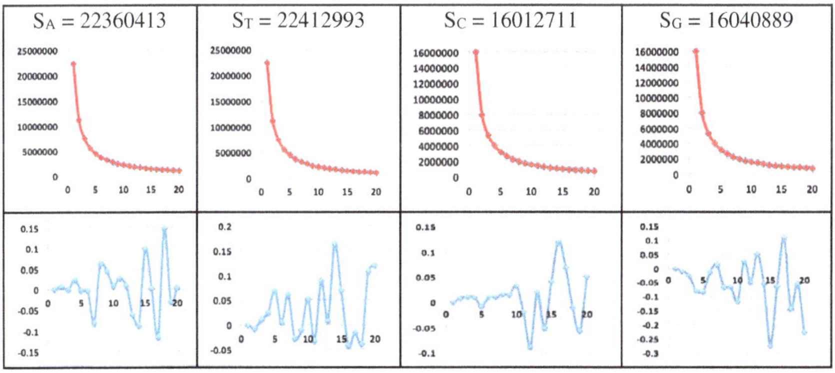

Fig. 10.11.2. The results of the analysis — by the OS method — of the nucleotide sequence of the epi-chain of the third order $N_{3/1}$, which consists of nucleotides with serial numerations 1-4-7-10-13-... in the DNA sequence of the human chromosome № 1. The top row demonstrates that the model hyperbolic progressions S_A/n, S_T/n, S_C/n, S_G/n (red lines) almost completely cover the OS sequences of real total amounts of those n-plets, which start with A, T, C, or G in this epi-chain correspondingly. The bottom row shows in percentage slight alternating deviations of real values of the OS sequences from model values. All denotations are the same as in Fig. 10.11.1.

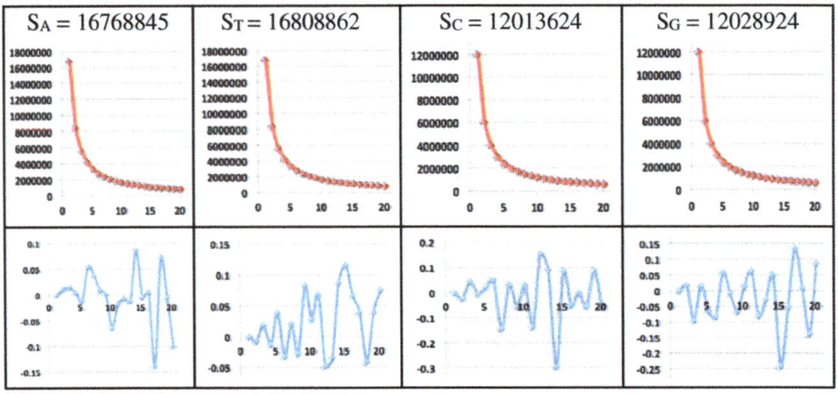

Fig. 10.11.3. The results of the analysis — by the OS method — of the nucleotide sequence of the epi-chain of the 4th order $N_{4/1}$, which consists of nucleotides with serial numerations 1-5-9-13-... in the DNA sequence of the human chromosome № 1. The top row demonstrates that the model hyperbolic progressions S_A/n, S_T/n, S_C/n, S_G/n (red lines) almost completely cover the OS sequences of real total amounts of those n-plets, which start with A, T, C, or G in this epi-chain correspondingly. The bottom row shows in percentage slight alternating deviations of real values of the OS sequences from model values. All denotations are the same as in Fig. 10.11.1.

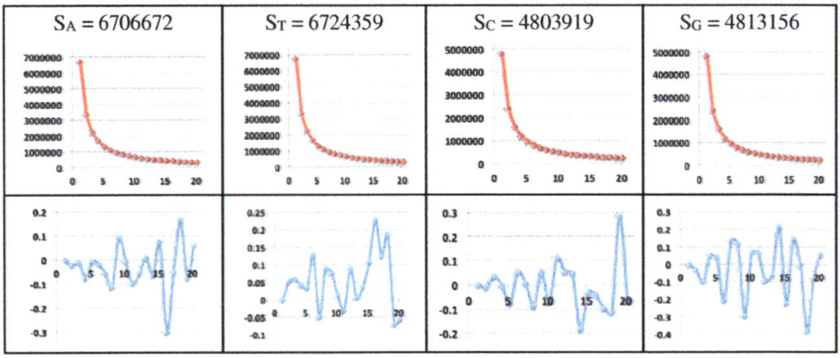

Fig. 10.11.4. The results of the analysis — by the OS method — of the nucleotide sequence of the epi-chain of the 10th order $N_{10/1}$, which consists of nucleotides with serial numerations 1-11-21-31-41-... in the DNA sequence of the human chromosome № 1. The top row demonstrates that the model hyperbolic progressions S_A/n, S_T/n, S_C/n, S_G/n (red lines) almost completely cover the OS sequences of real total amounts of those n-plets, which start with A, T, C, or G in this epi-chain correspondingly. The bottom row shows in percentage slight alternating deviations of real values of the OS sequences from model values. All denotations are the same as in Fig. 10.11.1.

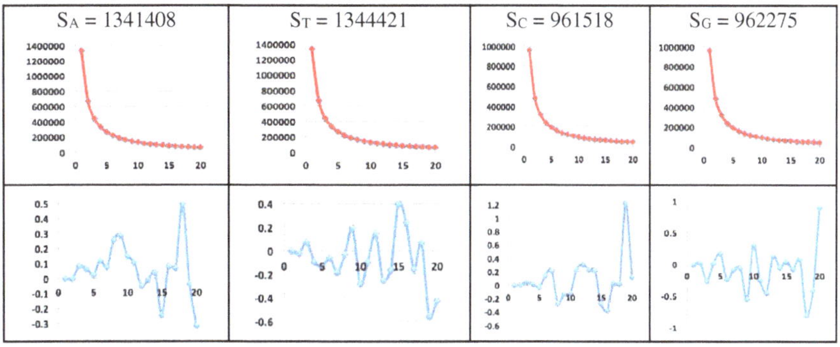

Fig. 10.11.5. The results of the analysis — by the OS method — of the nucleotide sequence of the epi-chain of the 50th order $N_{50/1}$, which consists of nucleotides with serial numerations 1-51-101-151-201-... in the DNA sequence of the human chromosome № 1. The top row demonstrates that the model hyperbolic progressions S_A/n, S_T/n, S_C/n, S_G/n (red lines) almost completely cover the OS sequences of real total amounts of those *n*-plets, which start with A, T, C, or G in this epi-chain correspondingly. The bottom row shows in percentage slight alternating deviations of real values of the OS sequences from model values. All denotations are the same as in Fig. 10.11.1.

eukaryotic and prokaryotic genomes. The rule is fulfilled in these epi-chains with the same high accuracy as in the complete DNA of the sequence.

Similar results were obtained by the author in the study of epi-chains in the single-stranded DNA of other analyzed genomes (see corresponding data in [238]). These results allow formulating the fourth hyperbolic (or harmonic) rule of eukaryotic and prokaryotic genomes, which is considered by the author as a candidate for the role of a universal genetic rule (it is necessary to further investigate the widest variety of genomes to verify the degree of its universality).

The fourth hyperbolic rule (about interrelations of oligomers in epi-chains of long DNA sequences):

- In any nuclear chromosome of eukaryotic genomes and in prokaryotic genomes, the hyperbolic rules №№ 1 and 2 are fulfilled not only for the complete nucleotide sequences but also for their epi-chains of the order *k* (where *k* = 2, 3, 4, ... is not too large compared to the full length of the nucleotide sequence).

The numeric data, graphically represented in Figs. 10.11.1–10.11.5, are contained in [198].

10.12. Additional data to the hyperbolic rules in genomic DNAs

Till now we considered OS sequences, whose members are total amounts of n-plets, which start with a certain «attributive» nucleotide, for example, with the nucleotide A. In this case, we calculate the total amounts of oligomers in the following sets: 4 duplets AT, AC, AG, AA; 16 triplets ATT, ATC, ATG, ACC, ...; and so on. But what results arise if one calculates, in the same genome, the total amounts in other sets of n-plets having the same attributive nucleotide A at their second positions, that is, the following sets: 4 duplets TA, CA, GA, AA; 16 triplets TAT, TAC, TAG, CAC, ...; and so on for $n = 2$, 3, 4, ...? And what results arise if one calculates, in the same genome, total amounts in the sets of n-plets, which have the same nucleotide at their third positions, that is, the following sets: 16 triplets TTA, TCA, TGA, CCA, ...; 64 tetraplets TTTA, TCTA, TGCA, ...; and so on for $n = 3, 4, 5, ...$? It turns out that in all such cases the resulting OS sequences will be practically identical to the hyperbolic-like OS sequence of the total amounts of n-plets with the same attributive nucleotide at their first position [116, 232].

Tables 10.12.1–10.12.2 show examples of the comparisons of different OS sequences calculated for DNA of the human chromosome №1 in three locations of attributive nucleotides in its n-plets: 1) at the first position in n-plets; 2) at the second position; 3) at the third position. These OS sequences correspond to the cases of n-textual representations of the chromosomal DNA as sequences of n-plets of equal lengths, where $n = 1, 2, ..., 10$.

One can see from the shown results that the differences Δ% of the corresponding members of these three OS sequences from each other are less than 0.1%, that is, these OS sequences are practically identical. These differences were calculated for each n by the formulas $\Delta\% = 100(1 - \text{Pos1}/\text{Pos2})\%$ and $\Delta\% = 100(1 - \text{Pos1}/\text{Pos3})\%$ where Pos1, Pos2, and Pos3 refer

Table 10.12.1. The comparison of the OS sequences of the total amounts of n-plets, which have the nucleotide N (A, T, C, or G) at their first position (the row "Pos. 1") and at their second position (the row "Pos. 2") in the human chromosome №1. Δ% shows the percentage of differences between the corresponding total amounts of considered n-plets in corresponding n-textual representations of the chromosomal DNA. The comparison begins with duplets, since there is no second position in monoplets. Cases of $n = 1, 2, \ldots, 10$ are presented.

n	1	2	3	4	5	6	7	8	9	10
A										
Pos. 1	67070277	33537501	22360413	16768845	13413532	11179286	9584038	8383461	7453552	6706672
Pos. 2	—	33532776	22353979	16767465	13413514	11174459	9578118	8383936	7452356	6704047
Δ%		0.014	0.029	0.008	0.000	0.043	0.062	-0.006	0.016	0.039
T										
Pos. 1	67244164	33620498	22412993	16808862	13445360	11207274	9606748	8405040	7470145	6724359
Pos. 2	—	33623666	22411166	16811071	13445910	11206100	9610249	8405351	7472348	6724456
Δ%		0.009	-0.008	0.013	0.004	-0.010	0.036	0.004	0.029	0.001
C										
Pos. 1	48055043	24024903	16012711	12013624	9612227	8005708	6865944	6008215	5336968	4803919
Pos. 2	—	24030140	16021444	12015843	9615911	8012553	6865662	6005986	5338638	4808410
Δ%		0.022	0.055	0.018	0.038	0.085	-0.004	-0.037	0.031	0.093
G										
Pos. 1	48111528	24057606	16040889	12028924	9625086	8021235	6869132	6013412	5348337	4813156
Pos. 2	—	24053922	16040412	12025875	9620866	8020389	6871831	6014853	5345656	4811187
Δ%		-0.015	-0.003	-0.025	-0.044	-0.011	0.039	0.024	-0.050	-0.041

Table 10.12.2. The comparison of the OS sequences of the total amounts of *n*-plets, which have the nucleotide N (A, T, C, or G) at their first position (the row "Pos. 1") and at their third position (the row "Pos. 3") in the human chromosome №1. Δ% shows the percentage of differences between the corresponding total amounts of considered *n*-plets in corresponding *n*-textual representations of the chromosomal DNA. The comparison begins with triplets since there is no third position in monoplets and duplets. Cases of $n = 1, 2, \ldots, 10$ are presented.

n	**1**	**2**	**3**	**4**	**5**	**6**	**7**	**8**	**9**	**10**
A										
Pos. 1	67070277	33537501	22360413	16768845	13413532	11179286	9584038	8383461	7453552	6706672
Pos. 3	—	—	22355885	16768656	13414900	11178695	9578685	8383657	7450656	6710255
Δ%	I		0.020	0.001	−0.010	0.005	0.056	−0.002	0.039	−0.053
T										
Pos. 1	67244164	33620498	22412993	16808862	13445360	11207274	9606748	8405040	7470145	6724359
Pos. 3	—	—	22420005	16811636	13448900	11208158	9604848	8406144	7472996	6723773
Δ%			−0.031	−0.017	−0.026	−0.008	0.020	−0.013	−0.038	0.009
C										
Pos. 1	48055043	24024903	16012711	12013624	9612227	8005708	6865944	6008215	5336968	4803919
Pos. 3	—	—	16020888	12011279	9611721	8010304	6867877	6005835	5342246	4803498
Δ%			−0.051	0.020	0.005	−0.057	−0.028	0.040	−0.099	0.009
G										
Pos. 1	48111528	24057606	16040889	12028924	9625086	8021235	6869132	6013412	5348337	4813156
Pos. 3	—	—	16030227	12028682	9620676	8016348	6874449	6014493	5343102	4810570
Δ%			0.066	0.002	0.046	0.061	−0.077	−0.018	0.098	0.054

to values indicated in the rows Pos. 1, Pos. 2, and Pos. 3. Here the results are presented only for n-plets with $n = 2, 3, 4, \ldots, 10$ but similar situations of practical coincidences for the corresponding members of the considered OS sequences are also true for larger n.

These phenomenological data about the oligomer cooperative organization in genomic DNAs give a significant extension to the hyperbolic rule №1 regarding the hyperbolic-like OS sequences of the total amounts of n-plets, which have the same attributive nucleotide at their kth position (not only in their first position). These results and the extended rules additionally open up the deep connections of genomic sequences with the harmonic progression (10.4) and discover new aspects of the algebraic harmony of living bodies.

Another impressive numeric data about n-textual representations of genomic DNAs concern quantitative interrelations in groupings of different n-plets, which start from the same duplet, or triplet, etc. [116]. In particular, the amount S_2 of any of 16 duplets NN is algebra-harmonically interrelated with the total amounts S_3, S_4, S_5, ... of oligomers in the following sets: 4 triplets, which start with this attributive duplet NN; 16 tetraplets, which start with this attributive duplet NN; 64 pentaplets, which start with this attributive duplet NN; and so on (here N refers to any nucleotide A, C, G, or T). This interrelation is again based on the harmonic progression (10.4). More precisely, according to the model prediction, the ratios of these total amounts S_2/S_3, S_2/S_4, S_2/S_5, ... should be correspondingly equal to the ratios of the second member 1/2 of the harmonic progression (10.4) to its subsequent members 1/3, 1/4, 1/5, ... that is to values 3/2, 4/2, 5/2, ... Table 10.12.3 presents the comparison of the amount S_2 of each of 16 duplets to the total amounts S_3, S_4, S_5 of n-plets ($n = 3, 4, 5$), which start with this n-plet, in the human chromosome №1.

The rows in the left part of Table 10.12.3 show very different numeric series of total amounts, which are individual in each of the rows. But the right part shows that in each row its amounts are interrelated identically based

Table 10.12.3. The comparison of total amounts $S_2 = \Sigma(NN)$ of each of 16 duplets NN to the total amounts S_3 of 4 triplets, S_4 of 16 tetraplets, and S_5 of 64 pentaplets, which start with such attributive duplet NN, is shown for *n*-textual representations of DNA of the human chromosome №1. The left part of the table indicates the values of the corresponding total amounts. The right part contains appropriate values of the ratios S_2/S_3, S_2/S_4, and S_2/S_5, which are equal to the same magnitudes 1.5, 2.0, and 2.5 for the cases of all 16 duplets. Here N refers to any of the nucleotides A, T, C, and G.

DUPLETS	TRIPLETS	TETRAPLETS	PENTAPLETS	S_2/S_3	S_2/S_4	S_2/S_5
$S_2 = \Sigma(AA)$	$S_3 = \Sigma(AAN)_4$	$S_4 = \Sigma(AANN)_{16}$	$S_5 = \Sigma(AANNN)_{64}$			
10952057	7300222	5476855	4381298	1.50	2.00	2.50
$S_2 = \Sigma(AT)$	$S_3 = \Sigma(ATN)_4$	$S_4 = \Sigma(ATNN)_{16}$	$S_5 = \Sigma(ATNNN)_{64}$			
8561194	5706906	4280647	3420561	1.50	2.00	2.50
$S_2 = \Sigma(AC)$	$S_3 = \Sigma(ACN)_4$	$S_4 = \Sigma(ACNN)_{16}$	$S_5 = \Sigma(ACNNN)_{64}$			
5799729	3868541	2899991	2322063	1.50	2.00	2.50
$S_2 = \Sigma(AG)$	$S_3 = \Sigma(AGN)_4$	$S_4 = \Sigma(AGNN)_{16}$	$S_5 = \Sigma(AGNNN)_{64}$			
8224510	5484720	4111320	3289579	1.50	2.00	2.50
$S_2 = \Sigma(TA)$	$8_3 = \Sigma(TAN)_4$	$S_4 = \Sigma(TANN)_{16}$	$S_5 = \Sigma(TANNN)_{64}$			
7274275	4849731	3636741	2909412	1.50	2.00	2.50
$S_2 = \Sigma(TT)$	$S_3 = \Sigma(TTN)_4$	$S_4 = \Sigma(TTNN)_{16}$	$S_5 = \Sigma(TTNNN)_{64}$			
11026157	7346507	5511908	4409900	1.50	2.00	2.50
$S_2 = \Sigma(TC)$	$S_3 = \Sigma(TCN)_4$	$S_4 = \Sigma(TCNN)_{16}$	$S_5 = \Sigma(TCNNN)_{64}$			
6923689	4617788	3461837	2768794	1.50	2.00	2.50
$S_2 = \Sigma(TG)$	$S_3 = (TGN)_4$	$S_4 = \Sigma(TGNN)_{16}$	$S_5 = \Sigma(TGNNN)_{64}$			
8396349	5598933	4198342	3357218	1.50	2.00	2.50
$S_2 = \Sigma(CA)$	$S_3 = \Sigma(CAN)_4$	$S_4 = \Sigma(CANN)_{16}$	$S_5 = \Sigma(CANNN)_{64}$			
8382478	5591208	4191829	3354600	1.50	2.00	2.50
$S_2 = \Sigma(CT)$	$S_3 = \Sigma(CTN)_4$	$S_4 = \Sigma(CTNN)_{16}$	$S_5 = \Sigma(CTNNN)_{64}$			
8221421	5477836	4111963	3289510	1.50	2.00	2.50
$S_2 = \Sigma(CC)$	$8_3 = \Sigma(CCN)_4$	$S_4 = \Sigma(CCNN)_{16}$	$S_5 = \Sigma(CCNNN)_{64}$			
6233384	4153642	3117570	2492824	1.50	2.00	2.50
$S_2 = \Sigma(CG)$	$S_3 = (CGN)_4$	$S_4 = (CGNN)_{16}$	$S_5 = \Sigma(CGNNN)_{64}$			
1187593	789995	592235	475262	1.50	2.01	2.50
$S_2 = \Sigma(GA)$	$S_3 = \Sigma(GAN)_4$	$S_4 = \Sigma(GANN)_{16}$	$S_5 = \Sigma(GANNN)_{64}$			
6923938	4612792	3462012	2768171	1.50	2.00	2.50

Table 10.12.3. (*Continued*)

DUPLETS	TRIPLETS	TETRAPLETS	PENTAPLETS	S_2/S_3	S_2/S_4	S_2/S_5
$S_2 = \Sigma(GT)$	$S_3 = \Sigma(GTN)_4$	$S_4 = \Sigma(GTNN)_{16}$	$S_5 = (GTNNN)_{64}$			
5814874	3879880	2906516	2325903	1.50	2.00	2.50
$S_2 = \Sigma(GC)$	$S_3 = \Sigma(GCN)_4$	$S_4 = \Sigma(GCNN)_{16}$	$S_5 = (GCNNN)_{64}$			
5073325	3381454	2536422	2032200	1.50	2.00	2.50
$S_2 = \Sigma(GG)$	$S_3 = \Sigma(GGN)_4$	$S_4 = \Sigma(GGNN)_{16}$	$S_5 = \Sigma(GGNNN)_{64}$	1.50	2.00	2.50
6245451	4166742	3123944	2498784			

on the numeric series of the ratios 1.5, 2.0, and 2.5, which serves here as a general invariant for the cases of all 16 duplets. But this sequence of ratios exists in the harmonic progression (10.4): 1, 1/2, 1/3, 1/4, 1/5, …, where the ratios of its second member 1/2 to its third, fourth and fifth members (that is, 1/3, 1/4, and 1/5) give this series 3/2, 4/2, and 5/2. Similar results are true for all other human chromosomes and for all those genomes, which were analyzed [198].

Here we note that the following numeric interconnections also exist in complete genomic sequences:

- The amount S_3 of any of 64 triplets NNN is algebra-harmonically interrelated with the total amounts S_4, S_5, S_6, … of oligomers in the following sets: 64 tetraplets, which start with the attributive triplets NNN; 256 pentaplets, which start with the attributive triplets NNN; 1024 six-plets, which start with the attributive triplets NNN; … The ratios of these total amounts S_3/S_4, S_3/S_5, S_3/S_6, … should be correspondingly equal to the ratios of the third member 1/3 of the harmonic progression (10.4) to its subsequent members 1/4, 1/5, 1/6, …, that is, to values 4/3, 5/3, 6/3, …
- The amount S_4 of any of 256 tetraplets NNNN is algebra-harmonically interrelated with the total amounts of S_5, S_6, S_7, … of oligomers in the following sets: 256 pentaplets, which start with the attributive tetraplets NNNN; 1024 six-plets, which start with the attributive tetraplets NNNN; 4906 seven-plets, which start with the attributive tetraplets NNNN, …

The ratios of these total amounts S_4/S_5, S_4/S_6, S_4/S_7, ... should be correspondingly equal to the ratios of the fourth member 1/4 of the harmonic progression (10.4) to its subsequent members 1/5, 1/6, 1/7, ..., that is to values 5/4, 6/4, 7/4, ...

- And so on (the length of attributive oligomers NN... N in the considered sets of *n*-plets should not be too large compared to the full length of the genomic sequence).

Similar phenomenological rules exist not only for the listed cases, when the considered attributive nucleotides, duplets, triplets, etc., occupy the first position in *n*-plets of the considered sets, but also for cases where these attributive nucleotides or oligomers occupy the second position, third position, etc. (see corresponding rules about collective probabilities in oligomer tetra-groups for cases of attributive oligomers in different positions of *n*-plets in the article [70]). The obtained phenomenological data enlarge significantly the list of hyperbolic rules in genomes and lead to new tools and opportunities to study genetic structures. They confirm that the eukaryotic and prokaryotic genomes represent a regular algebraic fractal-like net with important participation of the harmonic progression (10.4) in interconnections of its parts. This allows us to say about the algebraic harmony in living bodies.

One should emphasize the following important aspect of the OS representations of genomic sequences. Each nucleotide of a DNA sequence is a participant of those sets of its different *n*-plets (duplets, triplets, etc.), whose total amounts are members of OS sequences of this DNA; in other words, each DNA nucleotide makes its small contribution immediately to many members of the OS sequences. Figuratively speaking, each DNA nucleotide is "smeared" (or distributed) over many members of the DNA OS sequence (this "smearing" is also true for each DNA duplet, triplet, etc.). Correspondingly, OS sequences reflect a sort of an interrelation over all *n*-plets in DNA sequences. Or, in other words, the OS method represents any long nucleotide sequence as a multi-partite (or many-body) system having a cooperative state regarding many its interrelated oligomers of different lengths $n = 1, 2, 3, \ldots$ This has some analogies with the well-known problem of multi-partite entanglement in quantum informatics described, for example, in [376–379].

Quantum entanglement is the physical phenomenon that occurs when a pair or group of particles is generated, interact, or share spatial proximity in a way such that the quantum state of each particle of the pair or group cannot be described independently of the state of the others. In quantum informatics, entangled states play very important roles. The study and use of entangled states are one of the main problems of quantum computing: "*...entanglement is a key element in effects such as quantum teleportation, fast quantum algorithms, and quantum error-correction. It is, in short, a resource of great utility in quantum computation and quantum information. ... Entangled states play a crucial role in quantum computation and quantum information*" [15, p. xxiii and p. 96].

Quantum systems with many degrees of freedom are ubiquitous in nature, particularly in the context of condensed matter theory. "*It is hence not surprising that important classes of states, such as ground states of local Hamiltonians, are multi-partite entangled states. ... Recent years have seen an enormous increase in interest at the intersection of quantum information and condensed matter theory that stems from the insight that notions of entanglement are crucial in the understanding of quantum phases of matter... Another family of quantum many-body states that can be efficiently described is the classes of bosonic and fermionic Gaussian states. They both arise naturally in the context of quantum many-body models in condensed matter physics, but their bosonic variant is also highly useful in quantum optics when it comes to describing systems constituted of several quantum modes of light... Relatedly, multi-partite entangled states serve as resources to a number of important protocols in quantum information theory in which more than two parties come together. A prominent example of such a multi-party quantum protocol is quantum secret sharing, in which a message is distributed to several parties in such a way that no subset is able to read the message, but the entire collection of parties is. … Multi-partite entanglement does not only facilitate processing or transmission of information but also allow for applications in metrology*" [376, pp. 15, 18, 20, 23]. The entanglement refers to the non-local properties of quantum states that cannot be explained classically.

One can suppose that in eukaryotic and prokaryotic genomes we have some special case of multi-partite entangled states, which are realized not in groups of many particles, but in genomic systems of many oligomers. This can be termed as "genomic entanglement" or "genomic tetra-entanglement" since genomic sequences contain 4 kinds of nucleotides. It should be emphasized that we do not declare an existence of ordinary physical quantum entanglement in the genomes, but only that the mathematical apparatus of the theory of quantum informatics is suitable for modeling the considered genetic sequences.

10.13. Regarding the application of the oligomer sums method to long protein sequences

Till now we considered applications of the OS method to the analysis of long single-stranded DNA sequences. Such DNA sequences consist of 4 kinds of nucleotides, and corresponding 4 equivalency classes of A_1-, T_1-, C_1-, G_1-oligomers are analyzed. This section discusses opportunities to apply this method for the similar revealing of possible algebra-harmonic features in primary sequences of 20 amino acids in long proteins. Each long sequence of amino acids (for example, ArgSerThrGlyPheLysLeuSerMetAla…) can be represented either as a sequence of monomers (Arg-Ser-Thr-Gly-Phe-Lys-Leu-Ser-Met-Ala-…), duplets (ArgSer-ThrGly-PheLys-LeuSer-MetAla-…), triplets (ArgSerThr-GlyPheLys-LeuSerMet-…), and so on. When analyzing long DNA sequences of nucleotides, which consist of 4 kinds of nucleotides, we considered 4 equivalency classes of A_1-, T_1-, C_1-, G_1-oligomers. By analogy, in the case of sequences of 20 kinds of amino acids, we will analyze 20 equivalency classes, each of which is defined by the corresponding amino acid and combines all oligomers, which start with this amino acid.

For example, the amino acid Ala defines the equivalency class of Ala_1-oligomers, which includes all *n*-plets starting with this amino acid: the set of Ala_1-duplets contains all 20 duplets which start with Ala (AlaAla, AlaArg, AlaAsn, …, AlaCys); the set of Ala_1-triplets contains all 400 triplets which start with Ala (AlaAlaAla, AlaAlaArg, …, AlaCysCys); and so on.

The application of the OS method to the analysis of any long amino acid sequence and their 20 classes of the oligomer equivalency is as follows:

- For any of the 20 classes of the oligomer equivalency, the total amount Σ of its defining amino acid and total amounts of all those *n*-plets (n = 2, 3, 4, …) that have this acid in their first position (or in other fixed positions) are calculated;
- The sequence of these phenomenological amounts is compared with the model hyperbolic sequence Σ/n of this equivalency class, where $n = 1, 2, 3, \ldots$

Let us explain the proposed application of the OS method by an example of the analysis of the primary amino acid sequence of the protein titin, which is one of the longest proteins. Titin is important in the contraction of striated muscle fibers and is the third most abundant protein in the muscle (after myosin and actin). Below some results of the analysis of titin by the OS method are presented [198]. Fig. 10.13.1 shows 20 graphs demonstrating the OS sequences for each of the 20 amino acids combined in the single general amino acid sequence of titin. Each of these 20 graphs presents data for one of the kinds of amino acids and shows number Σ of this amino acid in titin and also two sequences: one of them (in blue) corresponds to the sequence of the real total amounts of those *n*-plets, which start with this amino acid, and the second sequence (in red) corresponds to the model hyperbolic sequence Σ/n (here $n = 1, 2, 3, \ldots, 10$).

One can see from Fig. 10.13.1 that, in titin, for each of the 20 amino acids its sequence (in blue) of phenomenological values for total amounts of those *n*-plets, which start with this amino acid, approximately coincides

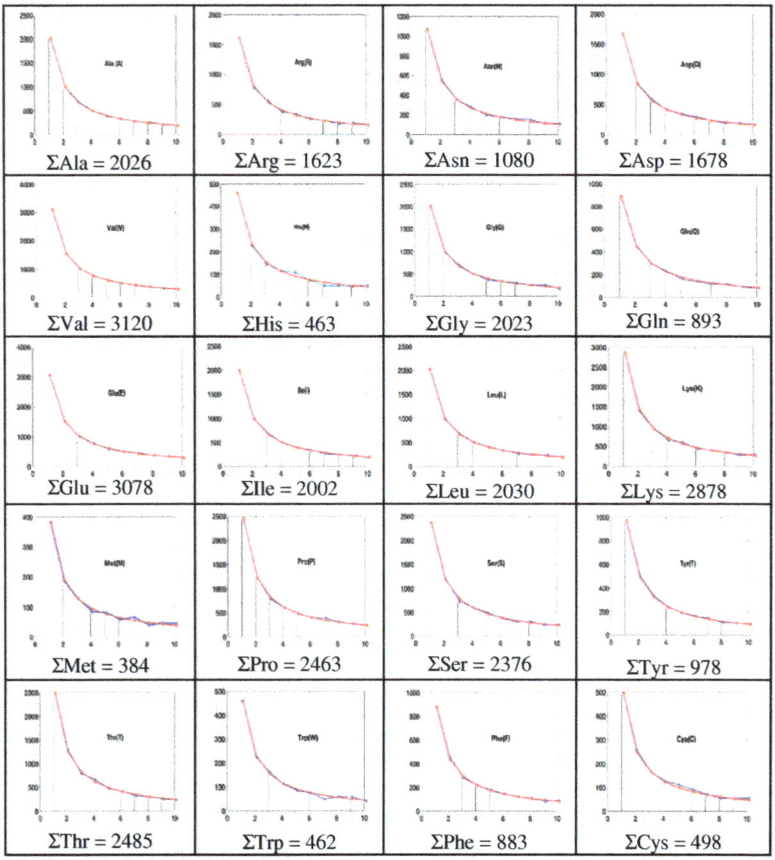

Fig. 10.13.1. Graphs of analysis results of the human protein Titin by the OS method for each of the 20 equivalency classes, which are defined by its 20 kinds of amino acids. Each graph shows a sequence (in blue) of real total amounts of n-plets, which start with this amino acid, and also a model hyperbolic sequence Σ/n (in red), where Σ refers to a number of this amino acid ($n = 1, 2, \ldots, 10$). The abscissa shows the values n; the ordinate shows total amounts of the corresponding n-plets which start with this amino acid. Initial data on this protein were taken from https://www.ncbi.nlm.nih.gov/protein/ACN81321.1.

with the corresponding model hyperbolic sequence Σ/n (in red) or slightly fluctuates around it. In the considered case of titin, the accuracy of the coincidence for the sequences of real and model values is lower than in the case of genomic DNAs described above. This seems to be due to the relatively short length of the titin amino acid sequence compared to the lengths of

Table 10.13.1. Examples of numerical data about OS sequences concerning two equivalency classes of Ala_1-oligomers and Arg_1-oligomers in the human protein titin. Graphic presentations of corresponding OS sequences are shown in Fig. 10.13.1 at the very top.

n	1	2	3	4	5	6	7	8	9	10
Ala										
Real	2026	1016	698	506	394	343	287	261	232	206
Model	2026	1013	675	506.5	405	338	289	253	225	203
Δ%	−0.3	−3.4	0.1	2.8	−1.6	0.8	−3.1	−3.1	−1.7	−0.3
Arg										
Real	1623	777	564	379	346	254	234	177	192	170
Model	1623	812	541	406	325	271	232	203	180	162
Δ%	0	4.3	−4.3	6.6	−6.6	6.1	−0.9	12.8	−6.5	−4.7

genomic nucleotide sequences. The graphs in the figure show that the largest deviations of the sequences of real values from sequences of model values occur in cases of amino acids whose number is minimal: the number of amino acids His is 463, Met — 384, Trp — 462, Cys — 498. Moreover, the deviations of the real values of OS from model values are relatively small for small values $n = 2, 3$, but with an increase in the length of oligomers at $n = 4, 5, \ldots, 10$, these deviations can increase (the number of corresponding n-plets decreases with increasing n).

Table 10.13.1 gives examples of real and model numeric values for the classes Ala_1- and Arg_1-oligomers from the first graphs in Fig. 10.13.1.

The study of the amino acid sequences of long proteins by this OS method should be continued to allow comparative analysis of various proteins.

10.14. Some concluding remarks

The article [239] contains numerical results of analysis by the oligomer sums method of a wide list of genomic DNAs of eukaryotes and prokaryotes.

As is known, mutations and the pressure of natural selection influence the genomic sequences of nucleotides. For these reasons, one can assume that because of many million years of biological evolution, genomic sequences, due to various influences, receive a completely random structure as a whole. This chapter provides additional evidence that, despite mutations, the pressure of natural selection, and other evolutionary factors, the nucleotide sequences of the eukaryotic and prokaryotic genomes have universal algebraic invariants. One can believe that the algebraic unity of living organisms is found (this should be tested further on a greater number of genomes). New mathematical tools and approaches for an in-depth study of this world and its evolution appear. In particular, the OS method can be used for the analysis of amino acid primary sequences in long proteins.

The discovery of the algebraic genomic invariants gives new knowledge about the unity of the world of all living organisms and about the features of biological evolution. This concerns additionally the problem of the origin of life, since the following natural question arises: where and how did these genomic algebraic invariants come from, which are expressed in the described hyperbolic (harmonic) rules and related to the quantum-information model if they exist even in the genomes of archaea and bacteria? The obtained results are interesting also for discussions concerning various well-known theories of biological evolution: Darwinism, nomogenesis, orthogenesis, etc. Some of these results are briefly described in the letter [232].

Researchers of the genetic system study the Nature system of storage, processing, and transmission of information, which has no direct analogies in modern science and technology, but which is studied based on analogies with their achievements. The disclosure of informational patents of living nature can make an important contribution to scientific and technological progress.

It should be noted that the genomic hyperbolic rules are cardinally different from well-known hyperbolic Zipf's law. Zipf's law was originally formulated in terms of quantitative linguistics, stating that given some corpus of natural language utterances, the frequency of any word is inversely proportional to its rank in the frequency table. In linguistics and other fields, Zipf's law speaks on the frequency of encounter for separate words or objects. In contrast, the hyperbolic rules of the genomes focus on OS sequences of the total amounts of *n*-plets and the genomic tetra-entanglement, that is, on the relative number not of separate oligomers, but the whole sums of sets of different *n*-plets distributed in the genomic sequence, where each separate nucleotide is a part of many oligomer sets existing simultaneously (each nucleotide is a distributed participant of many members of the appropriate genomic OS sequence at once and makes a contribution to each of them). From the quantum-information model, OS sequences serve as quantum-information characteristics of genomic sequences.

The presented study is a continuation of researches on symmetries in biological objects. This study further illustrates the effectiveness of symmetry analysis in natural systems. No wonder the theory of symmetries is one of the foundations of modern mathematical natural science. The presented results reveal the existence of a new number of symmetries in eukaryotic and prokaryotic genomes. The described data give a wide list of hyperbolic rules in genomes and lead to new tools and opportunities to study genetic structures. They confirm that the eukaryotic and prokaryotic genomes represent a regular algebraic fractal-like net with important participation of the harmonic progression (10.4) in interconnections of its parts. This additionally allows us to say about the algebraic harmony in living bodies.

Appendix 10.I. Numerical data on epi-chains of DNA of the human chromosome № 1

This appendix shows numerical data about epi-chains represented graphically above in Figs. 10.11.1–10.11.5 for *n*-textual representations of the DNA under n = 1, 2, …, 20.

Table 10.A.1. The results of the analysis — by the OS method — of the nucleotide sequence of the epi-chain of the second order $N_{2/1}$, which consists of nucleotides with serial numerations 1-3-5-7-9-… in the DNA sequence of the human chromosome №1. The table demonstrates that the model hyperbolic progressions S_A/n, S_T/n, S_C/n, S_G/n (in red) almost completely coincide with the OS sequences of real total amounts of those *n*-plets, which start with A, T, C, or G in this epi-chain correspondingly. Differences between the corresponding values in these numerical sequences are expressed by small percentage values Δ%. $n = 1, 2, \ldots, 20$.

n	1	2	3	4	5	6	7	8	9	10
A										
Real	33537501	16768845	11179286	8383461	6706672	5588773	4792078	4192017	3726860	3354107
Model	33537501	16768751	11179167	8384375	6707500	5589584	4791072	4192188	3726389	3353750
Δ%	0	-0.001	-0.001	0.011	0.012	0.015	-0.021	0.004	-0.013	-0.011
T										
Real	33620498	16808862	11207274	8405040	6724359	5601854	4801395	4202773	3735327	3360459
Model	33620498	16810249	11206833	8405125	6724100	5603416	4802928	4202562	3735611	3362050
Δ%	0	0.008	-0.004	0.001	-0.004	0.028	0.032	-0.005	0.008	0.047
C										
Real	24024903	12013624	8005708	6008215	4803919	4002753	3433636	3003511	2668499	2402186
Model	24024903	12012451.5	8008301	6006226	4804981	4004151	3432129	3003113	2669434	2402490
Δ%	0	-0.010	0.032	-0.033	0.022	0.035	-0.044	-0.013	0.035	0.013
G										
Real	24057606	12028924	8021235	6013412	4813156	4013372	3435824	3006763	2673815	2407301
Model	24057606	12028803	8019202	6014402	4811521	4009601	3436801	3007201	2673067	2405761
Δ%	0	-0.001	-0.025	0.016	-0.034	-0.094	0.028	0.015	-0.028	-0.064

n	11	12	13	14	15	16	17	18	19	20
A										
Real	3049510	2793265	2579432	2394635	2235831	2095893	1974049	1863181	1766123	1677938
Model	3048864	2794792	2579808	2395536	2235833	2096094	1972794	1863195	1765132	1676875
Δ%	−0.021	0.055	0.015	0.038	0.000	0.010	−0.064	0.001	−0.056	−0.063
T										
Real	3054627	2802390	2588494	2400749	2240133	2101692	1976639	1866079	1768246	1680401
Model	3056409	2801708	2586192	2401464	2241367	2101281	1977676	1867805	1769500	1681025
Δ%	0.058	−0.024	−0.089	0.030	0.055	−0.020	0.052	0.092	0.071	0.037
C										
Real	2185450	2001255	1845496	1718676	1600752	1501210	1413949	1335576	1263741	1200991
Model	2184082	2002075	1848069	1716065	1601660	1501556	1413230	1334717	1264469	1201245
Δ%	−0.063	0.041	0.139	−0.152	0.057	0.023	−0.051	−0.064	0.058	0.021
G										
Real	2186817	2006465	1851228	1717404	1605990	1503735	1414212	1337414	1267181	1202700
Model	2187055	2004801	1850585	1718400	1603840	1503600	1415153	1336534	1266190	1202880
Δ%	0.011	−0.083	−0.035	0.058	−0.134	−0.009	0.067	−0.066	−0.078	0.015

Table 10.A.2. The results of the analysis — by the OS method — of the nucleotide sequence of the epi-chain of the third order $N_{3/1}$, which consists of nucleotides with serial numerations 1-4-7-10-13-... in the DNA sequence of the human chromosome № 1. The table demonstrates that the model hyperbolic progressions S_A/n, S_T/n, S_C/n, S_G/n (in red) almost completely coincide with the OS sequences of real total amounts of those n-plets, which start with A, T, C, or G in this epi-chain correspondingly. Differences between the corresponding values in these numerical sequences are expressed by small percentage values Δ%. $n = 1, 2, \ldots, 20$.

n	1	2	3	4	5	6	7	8	9	10
A										
Real	A Real	22360413	11179286	7453552	5588773	4472245	3726860	3196917	2793265	2483348
Model	Model	22360413	11180207	7453471	5590103.25	4472083	3726736	3194345	2795052	2484490
Δ%	Δ%	0	0.008	-0.001	0.024	-0.004	-0.003	-0.081	0.064	0.046
T										
Real	22412993	11207274	7470145	5601854	4479492	3735327	3199876	2802390	2490602	2240133
Model	22412993	11206496.5	7470998	5603248	4482599	3735499	3201856	2801624	2490333	2241299
Δ%	0	-0.007	0.011	0.025	0.069	0.005	0.062	-0.027	-0.011	0.052
C										
Real	16012711	8005708	5336968	4002753	3202830	2668499	2287279	2001255	1778911	1600752
Model	16012711	8006355.5	5337570	4003178	3202542	2668785	2287530	2001589	1779190	1601271
Δ%	0	0.008	0.011	0.011	-0.009	0.011	0.011	0.017	0.016	0.032
G										
Real	16040889	8021235	5348337	4013372	3210839	2673815	2291215	2006465	1783466	1605990
Model	16040889	8020445	5346963	4010222	3208178	2673482	2291556	2005111	1782321	1604089
Δ%	0	-0.010	-0.026	-0.079	-0.083	-0.012	0.015	-0.068	-0.064	-0.119

n	**11**	**12**	**13**	**14**	**15**	**16**	**17**	**18**	**19**	**20**
A										
Real	2032220	1863181	1721074	1598554	1489212	1397489	1316829	1240400	1177210	1117975
Model	2032765	1863368	1720032	1597172	1490694	1397526	1315318	1242245	1176864	1118021
Δ%	0.027	0.010	–0.061	–0.087	0.099	0.003	–0.115	0.149	–0.029	0.004
T										
Real	2038200	1866079	1723940	1598312	1493164	1401402	1318617	1245654	1178340	1119290
Model	2037545	1867749	1724076	1600928	1494200	1400812	1318411	1245166	1179631	1120650
Δ%	–0.032	0.089	0.008	0.163	0.069	–0.042	–0.016	–0.039	0.109	0.121
C										
Real	1455982	1335576	1231496	1144356	1067083	999584	941265	889682	843256	800223
Model	1455701	1334393	1231747	1143765	1067514	1000794	941924	889595	842774	800636
Δ%	–0.019	–0.089	0.020	–0.052	0.040	0.121	0.070	–0.010	–0.057	0.052
G										
Real	1457871	1337414	1233259	1146423	1072344	1003210	942530	892426	844720	803863
Model	1458263	1336741	1233915	1145778	1069393	1002556	943582	891161	844257	802044
Δ%	0.027	–0.050	0.053	–0.056	–0.276	–0.065	0.111	–0.142	–0.055	–0.227

Table 10.A.3. The results of the analysis — by the OS method — the nucleotide sequence of the epi-chain of the 4th order $N_{4/1}$, which consists of nucleotides with serial numerations 1-5-9-13-… in the DNA sequence of the human chromosome № 1. The table demonstrates that the model hyperbolic progressions S_A/n, S_T/n, S_C/n, S_G/n (in red) almost completely coincide with the OS sequences of real total amounts of those *n*-plets, which start with A, T, C, or G in this epi-chain correspondingly. Differences between the corresponding values in these numerical sequences are expressed by small percentage values Δ%. $n = 1, 2, \ldots, 20$.

n	**1**	**2**	**3**	**4**	**5**	**6**	**7**	**8**	**9**	**10**
A										
Real	16768845	8383461	5588773	4192017	3354107	2793265	2394635	2095893	1863181	1677938
Model	16768845	8384423	5589615	4192211.25	3353769	2794808	2395549	2096106	1863205	1676885
Δ%	0	0.011	0.015	0.005	-0.010	0.055	0.038	0.010	0.001	-0.063
T										
Real	16808862	8405040	5601854	4202773	3360459	2802390	2400749	2101692	1866079	1680401
Model	16808862	8404431	5602954	4202216	3361772	2801477	2401266	2101108	1867651	1680886
Δ%	0	-0.007	0.020	-0.013	0.039	-0.033	0.022	-0.028	0.084	0.029
C										
Real	12013624	6008215	4002753	3003511	2402186	2001255	1718676	1501210	1335576	1200991
Model	12013624	6006812	4004541	3003406	2402725	2002271	1716232	1501703	1334847	1201362
Δ%	0	-0.023	0.045	-0.003	0.022	0.051	-0.142	0.033	-0.055	0.031
G										
Real	12028924	6013412	4013372	3006763	2407301	2006465	1717404	1503735	1337414	1202700
Model	12028924	6014462	4009641	3007231	2405785	2004821	1718418	1503616	1336547	1202892
Δ%	0	0.017	-0.093	0.016	-0.063	-0.082	0.059	-0.008	-0.065	0.016

n	11	12	13	14	15	16	17	18	19	20
A										
Real	1524710	1397489	1290062	1196717	1117975	1047993	987755	930924	882614	839279
Model	1524440	1397404	1289911	1197775	1117923	1048053	986403	931603	882571	838442
Δ%	−0.018	−0.006	−0.012	0.088	−0.005	0.006	−0.137	0.073	−0.005	−0.100
T										
Real	1527023	1401402	1293440	1199582	1119290	1049849	988367	934203	884323	839809
Model	1528078	1400739	1292989	1200633	1120591	1050554	988757	933826	884677	840443
Δ%	0.069	−0.047	−0.035	0.088	0.116	0.067	0.039	−0.040	0.040	0.075
C										
Real	1093622	999584	923273	860649	800223	751218	706684	667789	631744	601012
Model	1092148	1001135	924125	858116	800908	750852	706684	667424	632296	600681
Δ%	−0.135	0.155	0.092	−0.295	0.086	−0.049	0.000	−0.055	0.087	−0.055
G										
Real	12028924	6013412	4013372	3006763	2407301	2006465	1717404	1503735	1337414	1202700
Model	12028924	6014462	4009641	3007231	2405785	2004821	1718418	1503616	1336547	1202892
Δ%	0	0.017	−0.093	0.016	−0.063	−0.082	0.059	−0.008	−0.065	0.016

Table 10.A.4. The results of the analysis — by the OS method — of the nucleotide sequence of the epi-chain of the 10th order $N_{10/1}$, which consists of nucleotides with serial numerations 1-11-21-31-41-… in the DNA sequence of the human chromosome № 1. The table demonstrates that the model hyperbolic progressions S_A/n, S_T/n, S_C/n, S_G/n (in red) almost completely coincide with the OS sequences of real total amounts of those n-plets, which start with A, T, C, or G in this epi-chain correspondingly. Differences between the corresponding values in these numerical sequences are expressed by small percentage values Δ%. $n = 1, 2, …, 20$.

n	1	2	3	4	5	6	7	8	9	10
A										
Real	6706672	3354107	2235831	1677938	1341408	1117975	958626	839279	744475	670703
Model	6706672	3353336	2235557	1676668	1341334	1117779	958096	838334	745186	670667
Δ%	0	–0.023	–0.012	–0.076	–0.005	–0.018	–0.055	–0.113	0.095	–0.005
T										
Real	6724359	3360459	2240133	1680401	1344421	1119290	961102	839809	746575	672348
Model	6724359	3362179.5	2241453	1681090	1344872	1120727	960623	840545	747151	672436
Δ%	0	0.051	0.059	0.041	0.034	0.128	–0.050	0.088	0.077	0.013
C										
Real	4803919	2402186	1600752	1200991	961518	800223	686222	601012	533486	480738
Model	4803919	2401959.5	1601306	1200980	960784	800653	686274	600490	533769	480392
Δ%	0	–0.009	0.035	–0.001	–0.076	0.054	0.008	–0.087	0.053	–0.072
G										
Real	4813156	2407301	1605990	1202700	962275	803863	686639	600918	536368	481023
Model	4813156	2406578	1604385	1203289	962631	802193	687594	601645	534795	481316
Δ%	0	–0.030	–0.100	0.049	0.037	–0.208	0.139	0.121	–0.294	0.061

n	11	12	13	14	15	16	17	18	19	20
A										
Real	610306	559209	515854	479353	446769	420435	394716	371969	353254	335131
Model	609697	558889	515898	479048	447111	419167	394510	372593	352983	335334
Δ%	−0.100	−0.057	0.008	−0.064	0.077	−0.303	−0.052	0.167	−0.077	0.060
T										
Real	611496	559871	517229	480135	447813	419315	395062	372883	354165	336406
Model	611305	560363	517258	480311	448291	420272	395551	373576	353914	336218
Δ%	−0.031	0.088	0.006	0.037	0.107	0.228	0.124	0.185	−0.071	−0.056
C										
Real	436216	400115	369357	343754	320358	300365	282859	267188	252122	240344
Model	436720	400327	369532	343137	320261	300245	282583	266884	252838	240196
Δ%	0.115	0.053	0.047	−0.180	−0.030	−0.040	−0.098	−0.114	0.283	−0.062
G										
Real	437262	401484	370485	343053	321595	300385	283136	268414	253519	240527
Model	437560	401096	370243	343797	320877	300822	283127	267398	253324	240658
Δ%	0.068	−0.097	−0.065	0.216	−0.224	0.145	−0.003	−0.380	−0.077	0.054

Table 10.A.5. The results of the analysis — by the OS method — of the nucleotide sequence of the epi-chain of the 50th order $N_{50/1}$, which consists of nucleotides with serial numerations 1-51-101-151-201-… in the DNA sequence of the human chromosome № 1. The table demonstrates that the model hyperbolic progressions S_A/n, S_T/n, S_C/n, S_G/n (in red) almost completely coincide with the OS sequences of real total amounts of those *n*-plets, which start with A, or T, or C, or G in this epi-chain correspondingly. Differences between the corresponding values in these numerical sequences are expressed by small percentage values Δ%. $n = 1, 2, \ldots, 20$.

n	**1**	**2**	**3**	**4**	**5**	**6**	**7**	**8**	**9**	**10**
A										
Real	1341408	670703	446769	335131	268213	223299	191485	167237	148619	133939
Model	1341408	670704	447136	335352	268282	223568	191630	167676	149045	134141
Δ%	0	0.000	0.082	0.066	0.026	0.120	0.076	0.262	0.286	0.150
T										
Real	1344421	672348	447813	336406	269243	224202	192407	168101	149090	134818
Model	1344421	672210.5	448140	336105	268884	224070	192060	168053	149380	134442
Δ%	0	−0.020	0.073	−0.089	−0.133	−0.059	−0.181	−0.029	0.194	−0.280
C										
Real	961518	480738	320358	240344	192359	160018	137048	120522	106967	96272
Model	961518	480759	320506	240380	192304	160253	137360	120190	106835	96152
Δ%	0	0.004	0.046	0.015	−0.029	0.147	0.227	−0.276	−0.123	−0.125
G										
Real	962275	481023	321595	240527	192109	160749	137576	120343	107506	95930
Model	962275	481138	320758	240569	192455	160379	137468	120284	106919	96228
Δ%	0	0.024	−0.261	0.017	0.180	−0.231	−0.079	−0.049	−0.549	0.309

n	11	12	13	14	15	16	17	18	19	20
A										
Real	121816	111840	103193	95777	89643	83769	78850	74151	70625	67280
Model	121946	111784	103185	95815	89427	83838	78906	74523	70600	67070
Δ%	0.107	−0.050	−0.008	0.040	−0.241	0.082	0.071	0.499	−0.035	−0.313
T										
Real	122336	111872	103678	96184	89269	83822	79208	74638	71151	67505
Model	122220	112035	103417	96030	89628	84026	79084	74690	70759	67221
Δ%	−0.095	0.146	−0.252	−0.160	0.401	0.243	−0.157	0.070	−0.554	−0.422
C										
Real	87210	79875	73792	68526	64277	60322	56542	53407	49983	48018
Model	87411	80127	73963	68680	64101	60095	56560	53418	50606	48076
Δ%	0.230	0.314	0.231	0.224	−0.274	−0.378	0.032	0.020	1.231	0.120
G										
Real	87691	80548	73923	68766	64118	60188	56555	53892	50857	47678
Model	87480	80190	74021	68734	64152	60142	56604	53460	50646	48114
Δ%	−0.242	−0.447	0.133	−0.047	0.052	−0.076	0.087	−0.809	−0.417	0.906

Table 10.A.6. The normalized values $S_N/(S_A + S_T + S_C + S_G)$ of amounts S_A, S_T, S_C, and S_G of each nucleotide A, T, C, and G are practically identical in all considered epi-chains of different orders 1, 2, 3, 10, and 50 in the human chromosome № 1, that is, they are independent of the epi-chain orders. Here N refers to any nucleotide.

Epi-ch.	$S_A/(S_A+S_T+S_c+S_G)$	$S_T/(S_A+S_T+S_c+S_G)$	$S_c/(S_A+S_T+S_c+S_c)$	$S_c/(S_A+S_T+S_c+S_c)$
$N_{1/1}$	0.2910	0.2918	0.2085	0.2087
$N_{2/1}$	0.2910	0.2917	0.2085	0.2088
$N_{3/1}$	0.2910	0.2917	0.2084	0.2088
$N_{4/1}$	0.2910	0.2917	0.2085	0.2088
$N_{10/1}$	0.2910	0.2918	0.2084	0.2088
$N_{50/1}$	0.2910	0.291	0.2086	0.2088

Table 10.A.6 shows that normalized values of amounts S_A, S_T, S_C, and S_G of each nucleotide A, T, C, and G are practically identical in all considered epi-chains of the human chromosome №1, that is, they are independent of the epi-chain order.

Chapter 11

Fractal-like Features of Genomic DNAs

In inherited biological structures, many fractal-like patterns are known, the ensembles of which gradually arise and become more complex during the ontogenetic development of organisms from a unicellular to a multicellular state. According to one of the definitions of fractals, a fractal is a structure consisting of parts that are in some sense similar to the whole [380]. In developmental biology, the formation of inherited fractal-like patterns during ontogenesis can be considered as one of the options for implementing the ancient principle "like gives rise to like" (see Chapter 1 on the previously unknown various relationships of this principle with the genetic coding system, revealed by matrix analysis of molecular genetic families of binary oppositions).

There is an extensive set of publications about fractals in biology and medicine by many authors. In particular, many authors published their ideas and materials about relations of genomes with fractal structures in different aspects [189, 326, 381–386]. For example, the work [387] shows an existence of fractal globules in the three-dimensional architecture of whole genomes, where spatial chromosome territories exist and where maximally dense packing is provided on the basis of a special fractal packing, which provides the ability to easily fold and unfold any genomic locus. One should note that, by contrast to this article, in this book we present not spatial packing of whole genomes in the form of fractal globules but quite another thing: we present the fractal organization of long DNA texts, in particular, in connection with epi-chains of genomic DNAs and fractal genetic nets or trees, which are related to tetra-group symmetries of these texts [242, 344]. We believe that the discovery of fractal-like properties of DNAs texts related to their tetra-group symmetries can lead to new ideas in theoretical and application areas, including problems of artificial intelligence and in-depth study of genetic phenomena for medical and biotechnological tasks. The data available in scientific publications show the existence of fractal-like structures at various levels and branches of the hereditary organization of living bodies. Especially, oncological processes are associated with fractals: fractal patterns arise in a cell surface when a healthy human cell turns cancerous [384]. The article "Fractals and cancer" [382] shows a possible connection of cancer structures with fractal-like patterns generated by the tensor product of matrices, which is used in quantum mechanics and algebraic biology.

Recall that in the book we have repeatedly met with data on fractal-like structures in the organization of the genetic system:

— In Chapter 3, about the tensor product having the property of inheritance of mosaic structure of the original matrix under its tensor exponentiation, it was emphasized that the tensor product of matrices is connected with fractals [18, Chapter 10];
— In Chapter 4, about epi-chains of nucleotides in DNAs, the fractal-like character of genomic DNAs was noted (the articles [242, 344] additionally consider the so-called "fractal genetic nets" connected with epi-chains of genomic DNAs);
— In Chapter 6, about the doctrine of evolution based on bio-antenna arrays, the theme of fractal antennas in modern engineering was mentioned, which gives new ideas for understanding biological phenomena;
— In Chapter 7, it was noted that each genomic DNA can be considered as a fractal-like system of many interrelated quantum information mandalas, presenting genomic DNAs in the graphical form; in addition, each of the individual stochastic mandalas of separate cellules at different tiers in such quantum information mandalas has a fractal-like structure as well;

— In Chapter 11, it was noted that eukaryotic and prokaryotic genomes represent a regular algebraic fractal-like net with important participation of the harmonic progression in interconnections of its parts.

This chapter provides additional material on the topic of heritable fractal-like structures in living organisms.

11.1. Tensor product of matrices and fractal-like patterns

Let us return to Chapter 3 where it was noted that tensor exponentiation of mosaic matrices can produce fractal-like mosaic matrices of enlarged orders [18, Chapter 10]. This property of the tensor exponentiation leads to a possibility of formally constructing so-called tensor-matrix memory, which can be used for modeling purposes in matrix genetics and algebraic biology. It is convenient to demonstrate this by using the Hadamard (2 × 2)-matrix and its tensor exponentiations.

As mentioned in Chapter 2, one type of binary opposition in the DNA alphabet is the opposition of the nucleotide T to the other three nucleotides A, C, G:

1) During the transition from DNA to RNA, only this nucleotide T is replaced by another nucleotide (uracil U);
2) Only this nucleotide T does not have a functionally important amino group NH_2.

Representing this binary opposition as A = C = G = +1, T = −1, we obtain the representation of the symbolic genetic matrix [C, A; T, G] (from Chapter 1, Fig. 1.2.1) as a Hadamard matrix H_2 in Fig. 11.1.1.

Hadamard matrices H raised to tensor powers (n) always generate new Hadamard matrices $H^{(n)}$, examples of which are shown in Fig. 11.1.2. These examples show fractal repetitions of the original black and white mosaic.

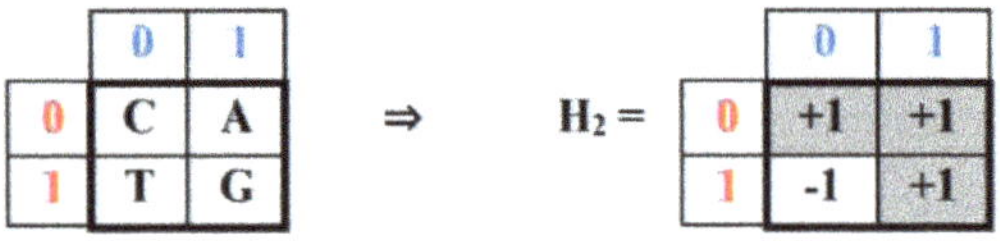

Fig. 11.1.1. Transformation of the genetic matrix [C, A; T, G] into the Hadamard matrix $H_2 = [1, 1; -1, 1]$ based on the molecular binary opposition described in the text.

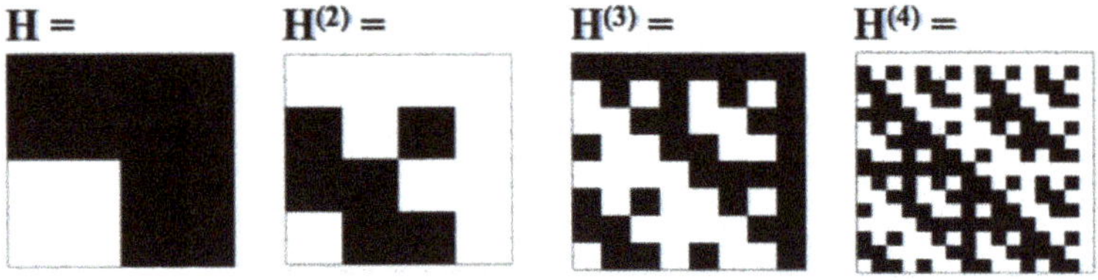

Fig. 11.1.2. Graphical images of Hadamard matrices $H^{(n)}$. Black cells contain entries +1, and white cells contain entries −1.

The rows of Hadamard matrices form complete orthogonal systems of Walsh functions. Thousands of publications have been devoted to their effective applications in error-correcting coding of information. The alphabetic systems of DNA and RNA, with their binary-opposition features, are related to many families of Hadamard matrices. This is interesting due to the important role of Hadamard matrices in quantum mechanics (unitary Hadamard operators), quantum computers (Hadamard gates), spectral analysis, multi-channel spectrometers, and error-correcting information coding, where a whole class of error-correcting codes is based on them, etc.

The fractal properties of tensor families of Hadamard matrices can be represented by graphs where entries −1 are replaced by entries 0 in the matrices. Then Hadamard matrices take the form of adjacency matrices known in graph theory [388]. The tensor family of the corresponding adjacency matrices has the form $[1, 1; 0, 1]^{(n)}$.

In the graphs of these adjacency matrices, the cooperative link groups of the graph of the previous matrix can be "inherited" (or transmitted) into the graphs of matrices of higher tensor degrees as Fig. 11.1.3 illustrates. This inheritance of cooperative link groups in graphs of a tensor family of the adjacency matrices is conditionally called "tensor-matrix memory".

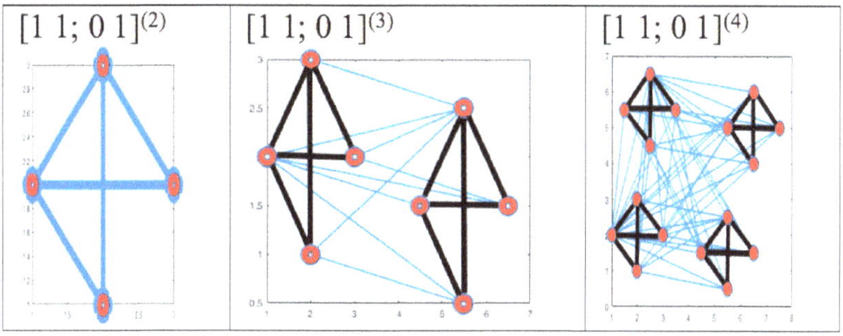

Fig. 11.1.3. The graphs of the adjacency matrices of the tensor family $[1, 1; 0, 1]^{(n)}$, which show "inheritance" of the cooperative link groups along the series of the matrices of the tensor family.

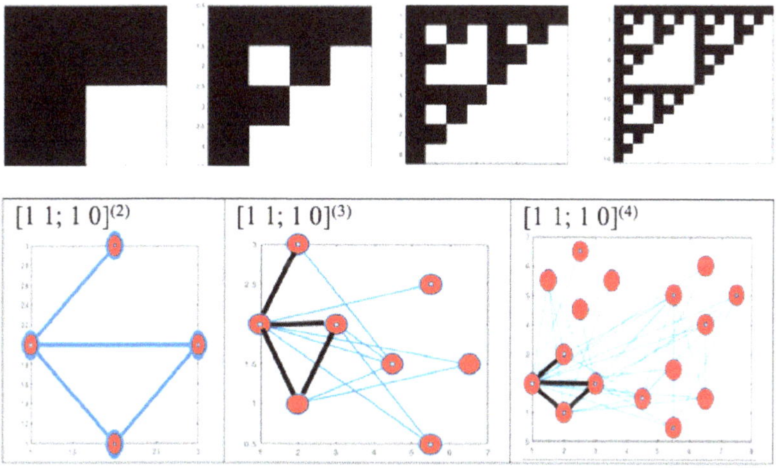

Fig. 11.1.4. The tensor family of Hadamard matrices $[1, 1; 1, -1]^{(n)}$ under $n = 1, 2, 3, 4$ (top) and corresponding tensor family of adjacency matrices $[1, 1; 1, 0]^{(n)}$ with corresponding family of graphs (bottom).

In the genetic system, much is associated with cyclic permutations. Taking a different matrix of the DNA alphabet with cyclically shifted positions of nucleotides [A, G; C, T], we obtain another tensor family of Hadamard matrices $[1, 1; 1, -1]^{(n)}$ and another family of corresponding adjacency matrices $[1, 1; 1, 0]^{(n)}$ with a different tensor family of graphs and a different tensor-matrix memory as Fig. 11.1.4 illustrates.

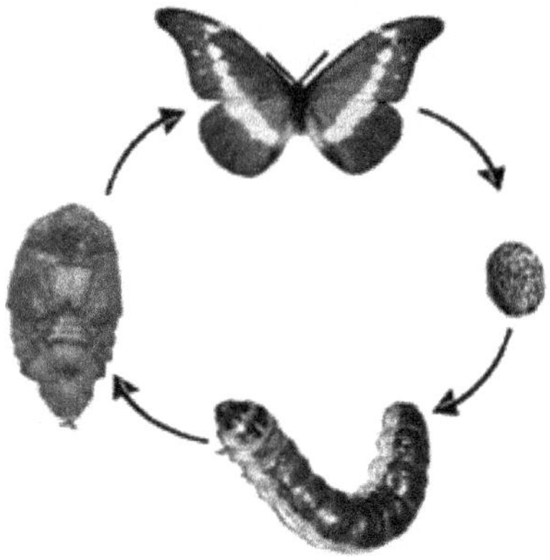

Fig. 11.1.5. The cycle of the metamorphosis of butterflies: butterfly-egg-caterpillar-chrysalis-butterfly.

The life of organisms is accompanied by cyclical changes. For example, the metamorphosis of butterflies includes a cycle: butterfly-egg-caterpillar-chrysalis-... (Fig. 11.1.5). The cells of these different bio-bodies contain the same DNA, but are cyclically read in different ways with some permutations. At the same time, the caterpillar, becoming a pupa, does not feed, and the pupa is rebuilt into a butterfly, in fact, due to rearrangements of the molecular components already present in it under the dictates of DNA information and epigenesis factors.

To the topic of cooperative groups of connections and networks of cyclic tensor-matrix memory, one can note the studies that are being conducted in cognitive neuroscience on networks of cooperative groups of neurons. Perhaps, in the future, it will make sense to compare networks of cooperative groups of connections in genetics and neuroscience, since the nervous system appeared relatively recently in biological evolution. Billions of organisms perfectly existed and still exist, without having nerve fibers, and still successfully solving the problems of finding food, escaping from predators, etc. The nervous system did not appear in evolution from scratch but, apparently, as a superstructure over the corresponding networks of molecular genetic structures with their families of binary oppositions. At the same time, the nervous system, like other types of excitable tissues, is also organized according to the principle of binary oppositions, presented in the physiological law "all-or-none": a nerve cell or a muscle fiber gives only the answer "yes" or "no" under the action of different stimuli by analogy with the Boolean variables. If a stimulus is above a certain threshold, a nerve or muscle fiber will fire with full response. Essentially, there will either be a full response or there will be no response [389, 390].

11.2. Regarding genomic fractals and vegetative propagation

In Chapter 7, it was noted that each genomic DNA is a fractal-like system, which can be graphically illustrated in the comfortable form of many interrelated quantum information mandalas, where each of the cellules at different mandala tiers can be represented by its individual stochastic fractal-like mandala. Thus, these genomic mandalas are fractal-like in nature, similar to the long-known fractal-like structures of morphogenesis: fern, spruce with its multi-tiered needle structures, cabbage *Brassica oleracea*, etc. (Fig. 11.2.1). Each next part of such a structure is potentially capable of producing again a similar fractal-like structure in the developing multi-parametric body whose number of freedom degrees and dimensionality of the configuration space

Fig. 11.2.1. Examples of biological fractals. Left: a step-by-step development of a dichotomic fractal-like structure, where each new part of the structure with dichotomous branching generates a new dichotomous branching at the next step in the development of the system. Middle: a fractal model of fern (from https://en.wikipedia.org/wiki/Barnsley_fern, permission granted under the Creative Commons Attribution-Share Alike 4.0 International license). Right: cabbage *Brassica oleracea* (from https://en.wikipedia.org/wiki/Romanesco_broccoli, permission granted under the Creative Commons Attribution-Share Alike 4.0 International license).

are increased correspondingly. In our opinion, this fundamental possibility of separate parts of a growing biological body to generate new parts having analogical structures is one of the most important secrets of living nature, which relates to the ancient principle "like begets like" (see Chapter 1 on the connection of the genetic coding system with this principle).

The features of the genomic mandalas testify — in comfortable graphical forms — to the fractal-like organization of genomic DNAs. Mandelbrot introduced the following definition of fractals: "*A fractal is a rough or fragmented geometric shape that can be split into parts, each of which is (at least approximately) a reduced-size copy of the whole*" [380]. In accordance with this definition, literary sources usually describe a creation of fractals by recursion algorithms inside a two-dimensional space or other spaces of fixed dimensions. Well-known examples of such fractals are two-dimensional Sierpiński carpet (https://en.wikipedia.org/wiki/Sierpi%C5%84ski_carpet) and three-dimensional Menger sponge (https://en.wikipedia.org/wiki/Menger_sponge). If one considers biological bodies (for example, trees or seashells) as ready structures in three-dimensional physical space, their configurations can be frequently interpreted as fractal-like ones in accordance with Mandelbrot's definition. But if you consider a step-by-step ontogenetic development of multi-parametric living bodies, whose number of parameters and degrees of freedom become more and more and whose configuration spaces are correspondingly grown in dimensionalities, such definition is not applicable since you have another situation: a dimensionality of a space is not fixed but is tensor growing. A theory of such biological fractals, whose similar fragments are distributed in different subspaces of a tensor-growing configuration space, should be constructed in the future.

It seems that for the studying of stochastic organization of genomic DNAs, stochastic fractals of Schramm-Löwner evolution can be useful [391]. They are connected with conformally invariant fractal curves that arise in critical two-dimensional models of statistical mechanics, such as the Ising model and percolation. It can be recalled that the topic of conformal bio-symmetries is meaningful and has several unexpected manifestations in biological phenomena. Conformal bio-symmetries are associated with morphogenesis, the Poincaré disk model of hyperbolic geometry, split-quaternions by Cockle, and features of algebraic holography in the family of structured DNA alphabets [12, 13, 32, 71].

11.3. Some concluding remarks

The presented results testify that already at the deepest level of biological organization — at the level of genomic DNAs associated with quantum informatics — fractal principles work, which totally permeate the overlying inherited biostructures and which are also connected to stochastic mechanisms of Gestalt biology. Fragments of the fractal-like organization of genomic DNAs are also presented in Chapters 3, 4, 6, 7, and 10.

These results were obtained by applications of the matrix-tensor methods to analysis of molecular-genetic structures and inherited physiological phenomena. We believe that these algebraic methods should be applied further to give also other interesting results helping to understand interrelations of fractal-like phenomena on different levels of biological organization. The expected results of this direction of research can be of significant applied value, for example, in connection with the phenomenon of vegetative reproduction, which is related to many problems of increasing the productivity of crops and which is one of the manifestations of the principle "like begets like".

Chapter 12

Genetic Biomechanics and United-Hypercomplex Systems

All biomechanical systems of the body are genetically inherited, and therefore must be structurally consistent with the genetic coding system to be transmitted in the form encoded to the next generations. Because of this, their structural features can bear the imprints of genetic coding structures. This chapter is devoted to the problem of studying these structural relationships between biomechanical and genetic systems. This area of research is conditionally called "genetic biomechanics". There are many published books on biomechanics in the world, but none of them deals with these relationships between biomechanics and genetic structures. Accordingly, the materials of this chapter are devoted to genetic biomechanics as a new scientific direction and interesting results already obtained in it. The main attention is paid to the algebraic modeling of the kinematics of multi-parametric or multi-block biomechanical bodies.

Living bodies are multi-parametric or multi-block. For example, the human body consists of many kinematic blocks — arms, legs, fingers, head, torso. A person can move these parts of the body in concert or relatively independently of each other. In the multi-dimensional vector configuration space of the parameters of such a multi-block body, the named movements of the blocks correspond to changes in the coordinates of the multi-dimensional vector describing the set of studied parameters of this multi-block structure, that is, changes in the subspaces of this vector space of parameters. To model the biomechanical kinematics of multi-block bodies in vector parameter spaces, it is important to find matrix operators that would make it possible to describe in the language of vector analysis such coordinated or relatively independent movements of a set of blocks of a single body in configuration spaces having non-limited dimensionalities. At the same time, it is desirable that these matrix operators form such closed equivalence classes, within which they would be related to each other by algebraic operations that are common for working with hypercomplex numbers (primarily, complex and hyperbolic). Given the increase in the dimension of the configuration space of body parameters in the course of its ontogenetic development, the desired types of matrix operators should have an additional property: when they are tensorially expanded, they must retain the ability to similarly model changes inside the tensorially expanded configuration space of parameters.

It turns out that the system of structured DNA alphabets in their matrix form of presentation is associated with matrix operators of this type; in other words, the genetic structures themselves suggest the way to the construction of such operators and their tensor extensions [339, 340]. These operators are related to the concept of multi-dimensional united-hypercomplex numeric systems, the description of the properties and applications of which is presented in this chapter.

12.1. The connection of the genetic coding system with united-hypercomplex systems

The main task of the mathematical natural sciences is the creation of mathematical models of natural systems. Development of models and formalized theories depends highly on those mathematical notions and instruments on which they are based.

At the beginning of the nineteenth century the following opinion existed: the world possesses a single real geometry (Euclidean geometry) and a single arithmetic. But this opinion was neglected after the discovery of non-Euclidean geometries and quaternions by Hamilton. Science understood that different natural systems could possess their own individual geometries and arithmetic (see this theme in the book [392]). The example of Hamilton, who was unlucky during 10 years of attempts to solve the task of describing transformations of three-dimensional space by means of three-dimensional algebras, is a very demonstrative one. This example says that, if a scientist does not guess correctly what types of numeric systems are adequate for the natural system investigated by him, he can waste many years without any result. One can add that the effectiveness of analytical models for geometrical and physical-geometrical properties of separate natural systems (including laws of conservation, theories of oscillations and waves, theories of potentials and fields, etc.) depends on the adequacy of the numerical systems employed.

The idea about special mathematical peculiarities of living matter existed long ago. We also believe that living bodies live in their own bio-information space, which has specific algebraic and geometric properties. In our opinion many difficulties of modern science in understanding genetic and other biological systems are determined by approaches to them from the non-adequate algebraic standpoints, which were developed formerly for other systems.

But how one can find appropriate algebraic approaches in biology if various species of organisms differ from each other so significantly in their morphogenetic and many other features? The discovery of the genetic code, basic features of which are general for all biological organisms, has allowed hope that such mathematical problems can be solved by means of investigation of genetic code structures. It seems an important task to investigate from different standpoints what systems of multi-dimensional numbers are connected or can relate to ensembles of parameters of the genetic code and with relevant bio-information spaces.

Various kinds of complex and hypercomplex numbers — complex numbers, double (or hyperbolic) numbers, quaternions and others — are used in different branches of modern science. They have played the role of the magic tool for development of theories and calculations in problems of heat, light, sounds, fluctuations, elasticity, gravitation, magnetism, electricity, current of liquids, quantum-mechanical phenomena, special theory of relativity, nuclear physics, etc.

$$Z = a_0 \times 1 + a_1 \times i, \quad \text{or } Z = a_0 \times \begin{vmatrix} 1 & 0 \\ 0 & 1 \end{vmatrix} + a_1 \times \begin{vmatrix} 0 & 1 \\ -1 & 0 \end{vmatrix} = \begin{vmatrix} a_0 & a_1 \\ -a_1 & a_0 \end{vmatrix}$$

×	1	i
1	1	i
i	i	−1

Fig. 12.1.1. Representations of complex numbers Z in its linear form (left) and matrix forms (right) where the matrix [1, 0; 0, 1] = E is the identity matrix and the matrix [0, 1; −1, 0] represents the imaginary unit "*i*". Bottom: the multiplication table of basic elements of complex numbers.

Each kind of hypercomplex number has its own multiplication table of basic elements. As is known, complex and hypercomplex numbers can be written not only in linear forms but also in matrix forms of their representation (in cases of associative algebras). Hypercomplex numbers are a generalization of complex numbers. The latter have one real unit "1" and one imaginary unit "*i*" with the property $i^2 = -1$. Complex numbers are written in a linear form and also in a matrix form (Fig. 12.1.1).

As is known, hypercomplex numbers have a single real unit and additional quantity of imaginary units, all of which in the total are called basic elements of hypercomplex numbers. The expression (12.1) shows their linear form in the general case:

$$a_0\mathrm{E} + a_1\mathrm{e}_1 + a_2\mathrm{e}_2 + \cdots + a_n\mathrm{e}_n \qquad (12.1)$$

where E denotes the real unit (or the identity matrix E in the case of the matrix representation of hypercomplex numbers), $\mathrm{e}_1, \mathrm{e}_2, \ldots, \mathrm{e}_n$ denote imaginary units, and $a_0, a_1, \ldots, a_n$ denote real coefficients. The set of these basic elements $(\mathrm{E}, \mathrm{e}_1, \mathrm{e}_2, \ldots, \mathrm{e}_n)$ should be closed under multiplication. This means that multiplication of any two basic elements with each other always gives again one of these elements or their linear superposition [58]. The results of the multiplication of the basic elements are represented in the corresponding multiplication table, which is individual for each type of hypercomplex number. If a system (12.1) does not possess the real unit E, which is mutual for all terms in the expression (12.1), it is not a system of hypercomplex numbers.

Recently it was discovered in [339] that the structured alphabets of DNA and RNA are related to special $(2^n \times 2^n)$-matrix operators. Various decompositions of these matrix operators show their connection with special 2^n-dimensional numeric systems having the following features:

1) they are not hypercomplex numbers in the whole because of absence of the identity matrix E in their decompositions (that is, the general real unit is absent among basic elements of such systems);
2) they contain two or more blocks, each of whose is represented by a sparse matrix, which itself represents complex numbers or hypercomplex numbers. Each block contains an individual set of basic elements, which play the role of its own real and imaginary units but only inside this separate block; because of this reason we call them local-real units and local-imaginary units.

In other words, two or more different systems of complex or hypercomplex numbers exist in this case not separately but are combined into a single structure by means of a summation of sparse matrices. Such multi-dimensional numeric systems are called "multi-block united-hypercomplex systems" (or, more briefly, "united-hypercomplex numbers" or simply "U-numbers") [339]. Such systems of multi-block U-numbers are useful mathematical instruments, which possess interesting properties to model biological phenomena and multi-parametrical systems in general. One can think that U-numbers play an important role in bio-information phenomena since they relate to the phenomenology of DNA and RNA alphabets. Some examples of application of united-hypercomplex numeric systems to model biological phenomena are given below. From the modeling standpoint, the hypothesis arises that inherited bio-informational aspects of organization of living matter are ruled by systems of 2^n-blocks U-hypercomplex numbers. The described results evoke associations with the famous Pythagoras's slogan "numbers rule the world".

12.2. United-complex numbers and the genetic Rademacher matrix of 16 duplets

Now let us turn to the genetic Rademacher matrix $[C, A; T, G]^{(2)}$ of 16 duplets (Fig. 1.4.3 in Chapter 1) and show that it is connected with such an algebraic numeric system, which does not possess the real unit E but possesses the local-imaginary units.

This matrix W_2 can be decomposed into a sum of 4 sparse matrices e_0, e_1, e_2, e_3, for example, in the way shown in Fig. 12.2.1. It is interesting that this variant of the decomposition consists of two sets of matrices, each of which is closed under multiplication unexpectedly: the first set consists of matrices e_0 and e_1; the second set e_2 and e_3. Fig. 12.2.1 shows multiplication tables for both sets, in each of which multiplication between its elements always gives an element from the same set. These multiplication tables coincide with the multiplication table of systems of complex numbers in Fig. 12.1.1. Thus, this Rademacher genetic matrix is the union of two complementary matrix representations of complex numbers, just as DNA is the union of two complementary molecular strands.

It means that each of the expressions $Z_L = a_0e_0 + a_1e_1$ and $Z_R = a_2e_2 + a_3e_3$ represents its own system of complex numbers in the unusual form of the sparse (4×4)-matrix with their 2 independent parameters a_0, a_1 and a_2, a_3

$$\left|\begin{array}{cccc} 1, & -1, & 1, & -1 \\ 1, & 1, & -1, & -1 \\ 1, & -1, & 1, & -1 \\ -1, & -1, & 1, & 1 \end{array}\right| = \left|\begin{array}{cccc} 1, & 0, & 0, & 0 \\ 0, & 1, & 0, & 0 \\ 1, & 0, & 0, & 0 \\ 0, & -1, & 0, & 0 \end{array}\right| + \left|\begin{array}{cccc} 0, & -1, & 0, & 0 \\ 1, & 0, & 0, & 0 \\ 0, & -1, & 0, & 0 \\ -1, & 0, & 0, & 0 \end{array}\right| + \left|\begin{array}{cccc} 0, & 0, & 1, & 0 \\ 0, & 0, & 0, & -1 \\ 0, & 0, & 1, & 0 \\ 0, & 0, & 0, & 1 \end{array}\right| + \left|\begin{array}{cccc} 0, & 0, & 0, & -1 \\ 0, & 0, & -1, & 0 \\ 0, & 0, & 0, & -1 \\ 0, & 0, & 1, & 0 \end{array}\right| = e_0+e_1+e_2+e_3.$$

*	e_0	e_1
e_0	e_0	e_1
e_1	e_1	$-e_0$

*	e_2	e_3
e_2	e_2	e_3
e_3	e_3	$-e_2$

Fig. 12.2.1. The decomposition of the genetic Rademacher matrix (from Fig. 1.4.3) into 4 sparse matrices e_0, e_1, e_2, e_3 (at the top). Multiplication tables of the two matrices e_0 and e_1, and of e_2 and e_3, are shown at the bottom. Both these multiplication tables coincide with the multiplication table of basic elements of complex numbers in Fig. 12.1.1.

$$Z_L = \begin{vmatrix} a_0 & -a_1 & 0 & 0 \\ a_1 & a_0 & 0 & 0 \\ a_0 & -a_1 & 0 & 0 \\ -a_1 & -a_0 & 0 & 0 \end{vmatrix}; \quad Z_L^{-1} = ((a_0^2 + a_1^2)^{-1})*(a_0e_0 - a_1e_1) = ((a_0^2 + a_1^2)^{-1})* \begin{vmatrix} a_0, & a_1, & 0, & 0 \\ -a_1, & a_0, & 0, & 0 \\ a_0, & a_1, & 0, & 0 \\ a_1, & -a_0, & 0, & 0 \end{vmatrix}$$

$$Z_R = \begin{vmatrix} 0 & 0 & a_2 & -a_3 \\ 0 & 0 & -a_3 & -a_2 \\ 0 & 0 & a_2 & -a_3 \\ 0 & 0 & a_3 & a_2 \end{vmatrix}; \quad Z_R^{-1} = ((a_2^2 + a_3^2)^{-1})*(a_2e_2 - a_3e_3) = ((a_2^2 + a_3^2)^{-1})* \begin{vmatrix} 0, & 0, & a_2, & a_3 \\ 0, & 0, & a_3, & -a_2 \\ 0, & 0, & a_2, & a_3 \\ 0, & 0, & -a_3, & a_2 \end{vmatrix}$$

Fig. 12.2.2. Left: two systems of complex numbers $Z_L = a_0e_0 + a_1e_1$ and $Z_R = a_2e_2 + a_3e_3$ in their form of (4 × 4)-matrices. Right: inverse matrices Z_L^{-1} and Z_R^{-1} in relation to matrices e_0 and e_2, which play the role of identity matrices in the systems Z_L and Z_R correspondingly.

correspondingly (Fig. 12.2.2). It can be formulated also in the following way: each of the systems Z_L and Z_R is isomorphic to the classical system of complex numbers. Inside the system Z_L, the matrix e_0+e_1 represents the complex number with unit coordinates ($a_0 = a_1 = 1$). Inside the system Z_R, the matrix e_2+e_3 represents the complex number with unit coordinates ($a_2 = a_3 = 1$).

The classical identity matrix E = [1 0 0 0; 0 1 0 0; 0 0 1 0; 0 0 0 1] is absent in the set of matrices Z_L and Z_R, where each matrix has zero determinant. Consequently, the usual notion of the inverse matrix Z_L^{-1} or Z_R^{-1} in relation to E (as $Z_LZ_L^{-1} = E$ or $Z_RZ_R^{-1} = E$) cannot be defined in accordance with the famous theorem about inverse matrices for matrices having zero determinant in the case of the complete set of matrices [17, Chapter 6, § 4]. In our case we analyze not the complete set of (4 × 4)-matrices but very limited special sets of matrices Z_L and Z_R. The set Z_L has the matrix e_0 (Fig. 12.2.1), which possesses all properties of the identity matrix for any matrix Z_L since $e_0Z_L = Z_Le_0 = Z_L$ and $e_0^2 = e_0$. In the frame of the set of matrices Z_L, where locally the matrix e_0 plays the role of the real unit, one can define — for any non-zero matrix Z_L — its inverse matrix Z_L^{-1} in relation to the matrix e_0 based on equations: $Z_LZ_L^{-1} = Z_L^{-1}Z_L = e_0$ (Fig. 12.2.2, right).

By analogy, the set of matrices Z_R has the matrix e_2 (Fig. 12.2.1), which possesses all properties of the identity matrix for any matrix Z_R since $e_2Z_R = Z_Re_2 = Z_R$ and $e_2^2 = e_2$. In the frame of the set of matrices Z_R, where locally the

$Z_LZ_R = [0, 0, a_0a_2+a_1a_3, a_1a_2-a_0a_3$	$Z_RZ_L = [a_0a_2+a_1a_3, a_0a_3-a_1a_2, 0, 0$
$0, 0, a_1a_2-a_0a_3, -a_0a_2-a_1a_3$	$a_1a_2-a_0a_3, a_0a_2+a_1a_3, 0, 0$
$0, 0, a_0a_2+a_1a_3, a_1a_2-a_0a_3$	$a_0a_2+a_1a_3, a_0a_3-a_1a_2, 0, 0$
$0, 0, a_0a_3-a_1a_2, a_0a_2+a_1a_3]$	$a_0a_3-a_1a_2, -a_0a_2-a_1a_3, 0, 0]$

Fig. 12.2.3. Multiplication of two matrices, where one matrix is taken from the system Z_L and the second matrix from Z_R (Fig. 12.2.2), is non-commutative.

matrix e_2 plays the role of the real unit, one can define — for any non-zero matrix Z_R — its inverse matrix Z_R^{-1} in relation to the matrix e_2 on the basis of equations: $Z_RZ_R^{-1} = Z_R^{-1}Z_R = e_2$ (Fig. 12.2.2, right).

Multiplication of two members from the same set Z_L (or Z_R) is commutative as it is true for complex numbers. But multiplication of two members from different sets (one member from Z_L and one from Z_R) is not commutative and gives a new matrix from one of these sets (Fig. 12.2.3).

Multiplication of the complex number $Z_L = a_0e_0 + a_1e_1$ with its conjugate complex number $a_0e_0 - a_1e_1$ gives the norm of this complex number $(a_0^2 + a_1^2)$ e_0. Multiplication of the complex number $Z_R = a_2e_2 + a_3e_3$ with its conjugate complex number $a_2e_2 - a_3e_3$ gives the norm of this complex number: $(a_2^2 + a_3^2)e_2$.

One should emphasize that the total set of sparse matrices e_0, e_1, e_2, e_3 does not contain the identity matrix E = [1,0,0,0; 0,1,0,0; 0,0,1,0; 0,0,0,1] and therefore the total sum $Z_2 = a_0e_0 + a_1e_1 + a_2e_2 + a_3e_3$ does not represent a system of hypercomplex numbers. It contains two blocks of complex numbers $(a_0e_0 + a_1e_1)$ and $(a_2e_2 + a_3e_3)$. By this reason it is one of many types of 2^n-blocks united-complex and united-hypercomplex numbers, which — we believe — reflect the specificity of inherited structures of living organisms and which — we propose — should be used to model biological phenomena in the field of algebraic biology.

The set of matrices e_0, e_1, c_2, e_3 is closed under multiplication and defines the multiplication table of U-complex numbers Z_2 in Fig. 12.2.4. In this system Z_2, the matrices e_0 and e_1 are conditionally called as local-real unit and the local-imaginary unit correspondingly in the block $Z_L = a_0e_0 + a_1e_1$; the matrices e_2 and e_3 are called the local-real unit and the local-imaginary unit in the block $Z_R = a_2e_2 + a_3e_3$.

$Z_2 = Z_L + Z_R = [a_0, -a_1, a_2, -a_3$
$a_1, a_0, -a_3, -a_2$
$a_0, -a_1, a_2, -a_3$
$-a_1, -a_0, a_3, a_2]$;

*	e_0	e_1	e_2	e_3
e_0	e_0	e_1	e_2	e_3
e_1	e_1	$-e_0$	$-e_3$	e_2
e_2	e_0	e_1	e_2	e_3
e_3	$-e_1$	e_0	e_3	$-e_2$

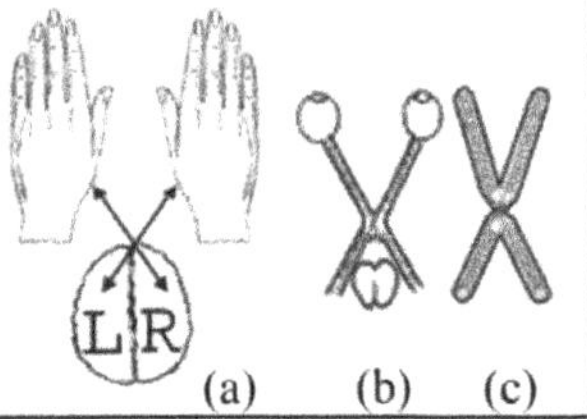

Fig. 12.2.4. Left: the matrix representation of united-complex numbers $Z_2 = Z_L + Z_R$. Middle: the multiplication table of 4 sparse matrices e_0, e_1, e_2, e_3 from Fig. 12.2.1. Right: the cross-shaped schemes of some morpho-functional structures in human organism: (a) the cross-shaped connections of brain hemispheres with the left and the right halves of a human body; (b) the cross-shaped structure of optic nerves from eyes in brain; (c) the cross-shaped structure of a chromosome.

This multiplication table has a cross-shaped structure with an element of a fractal plexus in relative locations of sets of the basic units:

- the structure of both (2×2)-quadrants along the main diagonal coincides with the multiplication table of basic elements of complex numbers; in opposition to this, the structure of both (2×2)-quadrants along the second diagonal coincides with the known matrix representation of complex numbers [a, b; –b, a] (Fig. 12.1.1);
- all cells of the main diagonal of the table contain only local-real units e_0 and e_2; all cells of the second diagonal contain only local-imaginary units e_1 and e_3;
- each of the four (2×2)-sub-quadrants of the table also has a cross-shaped structure since its main diagonal contains only local-real units e_0 or e_2 and its second diagonal contains only local-imaginary units e_1 and e_3.

Speaking in passing, one can suppose that the cross-like form of the multiplication table of the 2-block united-complex numbers (Fig. 12.2.4) is associated with the well-known phenomena of inherited cross-shaped structures of physiological objects (see examples in Fig. 12.2.4, right):

- the cross-shaped connection between the hemispheres of the human brain and the halves of the human body managed by them: the left hemisphere serves the right half of the body and the right hemisphere serves the left half of the body [393–396];

- The system of optic cranial nerves from two eyes possesses the cross-shaped structures as well: the optic nerves transfer information from the right half of the field of vision into the left hemisphere of the brain, and information from the left half of the field of vision into the right hemisphere. The same is true for the hearing system [397, Chapter 9]. One can suppose that these inherited physiological phenomena are connected with genetic cross-shaped structures, which participate in providing noise immunity properties of genetic systems;
- cross-shaped chromosomes, etc.

Ordinary united-hypercomplex systems can be written in the following linear form (12.2) in the case of using ordinary multiplication in definition of their separate blocks (below we will also meet tensor-U-hypercomplex numbers, where tensor multiplication participates in definition of separate blocks of united-hypercomplex numbers):

$$(a_0 \mathrm{E}_1 + a_1 \mathrm{e}_1 + \cdots + a_k \mathrm{e}_k) + (b_0 \mathrm{E}_2 + b_1 \mathrm{q}_1 + \cdots + \mathrm{b}_m \mathrm{q}_m) \\ + \cdots + (d_0 \mathrm{E}_\mathrm{n} + d_1 \mathrm{j}_1 + \cdots + d_p \mathrm{j}_p) \quad (12.2)$$

where each expression in brackets represents its own complex or hypercomplex system; $\mathrm{E}_1, \mathrm{E}_2, \ldots, \mathrm{E}_\mathrm{n}$ are local-real units; $\mathrm{e}_1, \ldots, \mathrm{e}_\mathrm{k}, \mathrm{q}_1, \ldots, \mathrm{q}_m, \mathrm{j}_1, \ldots, \mathrm{j}_p$ are local-imaginary units; $a_0, \ldots, a_k, b_0, \ldots, b_m, d_0, \ldots, d_p$ — real coefficients. It is obvious that U-hypercomplex systems (12.2) are a generalization of hypercomplex systems (12.1). The known Frobenius theorem for hypercomplex systems [58] does not correspond to U-hypercomplex systems in the whole.

The system of U-numbers $\mathrm{Z}_2 = \mathrm{Z}_\mathrm{L} + \mathrm{Z}_\mathrm{R}$ has operations of addition, subtraction and non-commutative multiplication (Fig. 12.2.5): applications of any of these operations for 2-blocks U-complex numbers Z_2 gives a new 2-blocks U-complex number of the same type. But this system has no operation of division.

$$\begin{bmatrix} a_0, -a_1, & a_2, -a_3 \\ a_1, a_0, & -a_3, -a_2 \\ a_0, -a_1, & a_2, -a_3 \\ -a_1, -a_0, & a_3, a_2 \end{bmatrix} * \begin{bmatrix} b_0, -b_1, & b_2, -b_3 \\ b_1, b_0, & -b_3, -b_2 \\ b_0, -b_1, & b_2, -b_3 \\ -b_1, -b_0, & b_3, b_2 \end{bmatrix} = \begin{bmatrix} c_0, -c_1, & c_2, -c_3 \\ c_1, c_0, & -c_3, -c_2 \\ c_0, -c_1, & c_2, -c_3 \\ -c_1, -c_0, & c_3, c_2 \end{bmatrix}$$

where $c_0 = a_0b_0 - a_1b_1 + a_2b_0 + a_3b_1$,
$c_1 = a_0b_1 + a_1b_0 + a_2b_1 - a_3b_0$,
$c_2 = a_0b_2 + a_1b_3 + a_2b_2 - a_3b_3$,
$c_3 = a_0b_3 - a_1b_2 + a_2b_3 + a_3b_2$

Fig. 12.2.5. Multiplication of two matrices with the described structure of 4-parametrical U-complex numbers $Z_2 = Z_L + Z_R$ (from Fig. 12.2.3) gives a new matrix of the same location of its 4 parameters c_0, c_1, c_2 and c_3.

$[x_0, x_1, x_2, x_3] \times Z_L = [a_0x_0+a_0x_2+a_1x_1-a_1x_3, a_0x_1-a_1x_0-a_0x_3-a_1x_2, 0, 0]$
$[x_0, x_1, x_2, x_3] \times Z_R = [0, 0, a_2x_0+a_2x_2-a_3x_1+a_3x_3, a_2x_3-a_3x_0-a_2x_1-a_3x_2]$

Fig. 12.2.6. Multiplication of matrix operators Z_L and Z_R (Fig. 12.2.3) with a vector $X = [x_0, x_1, x_2, x_3]$ from the right side leads to a selective manage of subspaces.

Results of actions of the matrix operators Z_L (or Z_R) on a four-dimensional vector $X = [x_0, x_1, x_2, x_3]$ depend on the side of the action: results of multiplication from right side $X \times Z_L$ and from left side $Z_L \times X$ differ and they have important properties concerning two-dimensional subspaces of configurational four-dimensional spaces (Figs. 12.2.6 and 12.2.7).

In the case of multiplication from the right side (Fig. 12.2.6), the possibility is revealed for a selective manage of vectors of two-dimensional subspaces of configurational four-dimensional spaces. The action $[x_0, x_1, x_2, x_3] \times Z_L$ gives the result that non-zero vectors exist only in the subspace (x_0, x_1), where they can be managed arbitrary by means of parameters a_0 and a_1 of the operator Z_L. If the parameters a_0 and a_1 are functions of time, then the vector $[x_0, x_1, x_2, x_3] \times Z_L$ moves along different trajectories in the plane (x_0, x_1) of the four-dimensional space $[x_0, x_1, x_2, x_3]$. In other words, the corresponding 2-parametrical subsystem of the 4-parametrical system is running under the selective manage of the operator Z_L. It leads to valuable capabilities to model multi-parametrical biological systems and bio-informational spaces, where different subsystems and their configurational subspaces are managed independently. In particular, it gives a new approach for understanding Mendel's law of independent inheritance of traits. The

$Z_L \times [x_0, x_1, x_2, x_3]^T = [a_0x_0-a_1x_1; a_0x_1+a_1x_0; a_0x_0-a_1x_1; -(a_0x_1+a_1x_0)]$
$Z_R \times [x_0, x_1, x_2, x_3]^T = [a_2x_2-a_3x_3; -(a_2x_3+a_3x_2); a_2x_2-a_3x_3; a_2x_3+a_3x_2]$

Fig. 12.2.7. Multiplication of the matrix operators Z_L and Z_R (Fig. 12.2.3) with a vector X = $[x_0, x_1, x_2, x_3]$ from the left side leads to effects of twinning.

work shows that such block-hypercomplex operators can be generalized into 2^n-block-hypercomplex operators by means of their tensor multiplication with appropriate square matrices.

A similar situation of the selective manage is true for results of the action $[x_0, x_1, x_2, x_3] \times Z_R$ (Fig. 12.2.6): in this case non-zero vectors exist only in the subspace (x_2, x_3), where they can be managed arbitrary by means of parameters a_2 and a_3 of the operator Z_R. If the parameters a_2 and a_3 are functions of time, then the vector $[x_0, x_1, x_2, x_3] \times Z_R$ moves along different trajectories in the plane (x_2, x_3) of the four-dimensional space $[x_0, x_1, x_2, x_3]$.

In the case of multiplication from the left side, the interesting effects of twinning appear (Fig. 12.2.7).

Fig. 12.2.7 shows that multiplication from the left side $Z_L \times [x_0, x_1, x_2, x_3]^T$ generates a new vector $[y_0, y_1, y_2, y_3]$ whose components y_0 and y_2 are identical to each other $(a_0x_0 - a_1x_1)$ and whose components y_1 and y_3 are identical to each other up to sign: $y_1 = a_0x_1 + a_1x_0$ and $y_3 = -(a_0x_1 + a_1x_0)$ correspondingly. Such mathematical peculiarities of the multiplication from the left side $Z_L \times [x_0, x_1, x_2, x_3]^T$ can be called effects of twinning. All components of the resulting vector $[y_0, y_1, y_2, y_3]$ in this case are independent of components x_2 and x_3 of the initial vector.

Analogically, multiplication from the left side $Z_R \times [x_0, x_1, x_2, x_3]^T$ generates a new vector $[y_0, y_1, y_2, y_3]$ whose components y_0 and y_2 are identical to each other $(a_2x_2 - a_3x_3)$ and whose components y_1 and y_3 are identical to each other up to sign: $-(a_2x_3 + a_3x_2)$ and $a_2x_3 + a_3x_2$ correspondingly. All components of the resulting vector $[y_0, y_1, y_2, y_3]$ are independent of components x_0 and x_1 of the initial vector.

Results of actions of the matrix operators $Z_2 = Z_L + Z_R$ on a four-dimensional vector $X = [x_0, x_1, x_2, x_3]$ depend on the side of the action

$[x_0, x_1, x_2, x_3] \times Z_2 = [a_0x_0+a_0x_2+a_1x_1-a_1x_3,\ a_0x_1-a_1x_0-a_0x_3-a_1x_2,\ a_2x_0+a_2x_2-a_3x_1+a_3x_3,\ a_2x_3-a_3x_0-a_2x_1-a_3x_2]$
$Z_2 \times [x_0, x_1, x_2, x_3]^T = [a_0x_0-a_1x_1+a_2x_2-a_3x_3,\ a_0x_1+a_1x_0-a_2x_3-a_3x_2,\ a_0x_0-a_1x_1+a_2x_2-a_3x_3;\ a_2x_3-a_1x_0-a_0x_1+a_3x_2]$

Fig. 12.2.8. Multiplication of the matrix operators $Z_2 = Z_L + Z_R$ with a vector $X = [x_0, x_1, x_2, x_3]$ from the right and left sides.

$$\begin{bmatrix} 1 & 0 & 0 & -1 \\ 0 & 1 & -1 & 0 \\ 0 & 0 & 0 & 0 \\ 0 & 0 & 0 & 0 \end{bmatrix} + \begin{bmatrix} 0 & -1 & 1 & 0 \\ 1 & 0 & 0 & -1 \\ 0 & 0 & 0 & 0 \\ 0 & 0 & 0 & 0 \end{bmatrix} + \begin{bmatrix} 0 & 0 & 0 & 0 \\ 0 & 0 & 0 & 0 \\ 0 & -1 & 1 & 0 \\ -1 & 0 & 0 & 1 \end{bmatrix} + \begin{bmatrix} 0 & 0 & 0 & 0 \\ 0 & 0 & 0 & 0 \\ 1 & 0 & 0 & -1 \\ 0 & -1 & 1 & 0 \end{bmatrix} = q_0+q_1+q_2+q_3.$$

*	q_0	q_1
q_0	q_0	q_1
q_1	q_1	$-q_0$

*	q_2	q_3
q_2	q_2	q_3
q_3	q_3	$-q_2$

$$Q = a_0q_0+a_1q_1+a_2q_2+a_3q_3 = \begin{vmatrix} a_0, & -a_1, & a_1, & -a_0 \\ a_1, & a_0, & -a_0, & -a_1 \\ a_3, & -a_2, & a_2, & -a_3 \\ -a_2, & -a_3, & a_3, & a_2 \end{vmatrix}$$

Fig. 12.2.9. Top left: the decomposition of the genetic Rademacher matrix (Fig. 1.4.3 in Chapter 1) into 4 sparse matrices q_0, q_1, q_2, q_3. Top right: multiplication tables for the pair q_0 and q_1 and for the pair q_2 and q_3. Bottom: the matrix representation of the system of 2-blocks U-complex numbers $Q = a_0q_0 + a_1q_1 + a_2q_2 + a_3q_3$.

(Fig. 12.2.8) and they join together results of actions of separate operators Z_L and Z_R (see Fig. 12.2.5). The above-described property of the selective manage of two-dimensional subspaces of the four-dimensional space is conserved in the case of multiplication by Z_2 from the right side. The described effect of twinning is also conserved in the case of multiplication by Z_2 from the left side.

One should note the existence of another decomposition of the genetic Rademacher matrix $[C, A; T, G]^{(2)}$ of 16 duplets (Chapter 1, Fig. 1.4.3) into a sum of another set of 4 sparse matrices q_0, q_1, q_2, q_3 (Fig. 12.2.9). Each of its two subsets (q_0 and q_1) and (q_2 and q_3) is closed in relation to multiplication and corresponds again to the multiplication table of complex numbers (Fig. 12.2.9, right).

Correspondingly expressions $a_0q_0 + a_1q_1 = Q_L$ and $a_2q_2 + a_3q_3 = Q_R$ represent two new systems of complex numbers by analogy with the described case of the systems Z_L and Z_R. The total matrix $Q = Q_L + Q_R$ represents a new variant of 2-blocks U-complex numeric systems (Fig. 12.2.9). One can check that those algebraic properties of both systems of 2-blocks U-complex numbers (Z_L+Z_R and Q_L+Q_R) are similar.

One can note that the set of 8 duplets related to non-zero entries of the sparse matrices e_0 and e_1 (that is, CC, CG, TC, TG and TT, TA, CA, CT) and the set of 8 duplets related to non-zero entries of the sparse matrices e_2 and e_3 (that is, GG, GC, AG, AC and AA, AT, GT, GA) are complementary to each other in DNA double helices. A similar relation holds for interconnections between the set of sparse matrices q_0 and g_1 and the set of sparse matrices q_2 and q_3. It means that the noted decompositions of the Rademacher matrix (Figs. 12.2.1 and 12.2.9), which reveal the connection of the genetic system with united-complex numbers, are not accidental but correspond to the phenomenon of the complementarity of nucleobases in double-stranded DNAs. And vice versa, it means also that this phenomenon of the nucleobases complementary in DNAs — jointly with peculiarities of genetic code degeneracy — can be considered as existing due to its connection with algebraic united-complex numbers.

Let us turn now to another type of two-dimensional numbers, which are known under the name hyperbolic numbers (their synonyms are split-complex numbers, Lorentz numbers, double numbers, perplex numbers). Fig. 12.2.10 shows the usual (2 × 2)-matrix representation of hyperbolic numbers $D = a \times 1 + b \times j$, the imaginary unit of which «j» satisfies the condition $j^2 = +1$.

The genetic Rademacher matrix of 16 duplets (Chapter 1, Fig. 1.4.3) can be also decomposed into another set of 4 sparse matrices s_0, s_1, s_2, s_3,

$$D = a^* \begin{vmatrix} 1, 0 \\ 0, 1 \end{vmatrix} + b^* \begin{vmatrix} 0, 1 \\ 1, 0 \end{vmatrix}$$

*	1	j
1	1	j
j	j	1

Fig. 12.2.10. The matrix representation of hyperbolic numbers. The first matrix presents the real unit and the second matrix presents the imaginary unit "j" of hyperbolic numbers. The multiplication table of the algebra of hyperbolic numbers is shown on the right.

which represents the 2-block united-hyperbolic number S with unit values of its coordinates (Fig. 12.2.11).

The set of matrices s_0 and s_1 is closed under multiplication and defines the multiplication table of hyperbolic numbers (Fig. 12.2.11). The same is true for the set of matrices s_2 and s_3. Correspondingly, the expression $S = a_0 s_0 + a_2 s_1 + a_2 s_2 + a_3 s_3$ represents the 4-parametrical system of 2-block united-hyperbolic numbers (Fig. 12.2.11). Here s_0 and s_2 are local-real units and s_1 and s_3 are local-imaginary units (by analogy with the described case of local-real and local-imaginary units in 2-block U-complex numbers (Fig. 12.2.1)).

The Rademacher matrix of 16 duplets (Chapter 1, Fig. 1.4.3) can be decomposed into one more set of 4 sparse matrices p_0, p_1, p_2, p_3, which leads to another system of 2-block U-hyperbolic numbers (Fig. 12.2.12).

$$\begin{vmatrix} 1 & -1 & 1 & -1 \\ 1 & 1 & -1 & -1 \\ 1 & -1 & 1 & -1 \\ -1 & -1 & 1 & 1 \end{vmatrix} = \begin{vmatrix} 1 & -1 & 0 & 0 \\ 0 & 0 & 0 & 0 \\ 0 & 0 & 1 & -1 \\ 0 & 0 & 0 & 0 \end{vmatrix} + \begin{vmatrix} 0 & 0 & 1 & -1 \\ 0 & 0 & 0 & 0 \\ 1 & -1 & 0 & 0 \\ 0 & 0 & 0 & 0 \end{vmatrix} + \begin{vmatrix} 0 & 0 & 0 & 0 \\ 1 & 1 & 0 & 0 \\ 0 & 0 & 0 & 0 \\ 0 & 0 & 1 & 1 \end{vmatrix} + \begin{vmatrix} 0 & 0 & 0 & 0 \\ 0 & 0 & -1 & -1 \\ 0 & 0 & 0 & 0 \\ -1 & -1 & 0 & 0 \end{vmatrix} = s_0+s_1+s_2+s_3$$

*	s_0	s_1
s_0	s_0	s_1
s_1	s_1	s_0

*	s_2	s_3
s_2	s_2	s_3
s_3	s_3	s_2

;

$$S = a_0s_0+a_2s_1+a_2s_2+a_3s_3 = \begin{vmatrix} a_0, & -a_0, & a_1, & -a_1 \\ a_2, & a_2, & -a_3, & -a_3 \\ a_1, & -a_1, & a_0, & -a_0 \\ -a_3, & -a_3, & a_2, & a_2 \end{vmatrix}$$

Fig. 12.2.11. Top: one of the possible decompositions of the Rademacher matrix of 16 duplets into 4 sparse matrices s_0, s_1, s_2, s_3. Bottom left: multiplication tables of the set of matrices s_0 and s_1 and the set of matrices s_2 and s_3 coincide with the multiplication table of hyperbolic numbers in Fig. 12.2.10. Bottom right: the matrix representation of 2-blocks U-hyperbolic numbers S.

$$\begin{vmatrix} 1 & -1 & 1 & -1 \\ 1 & 1 & -1 & -1 \\ 1 & -1 & 1 & -1 \\ -1 & -1 & 1 & 1 \end{vmatrix} = \begin{vmatrix} 1 & 0 & 0 & 0 \\ 0 & 0 & 0 & -1 \\ 1 & 0 & 0 & 0 \\ 0 & 0 & 0 & 1 \end{vmatrix} + \begin{vmatrix} 0 & 0 & 0 & -1 \\ 1 & 0 & 0 & 0 \\ 0 & 0 & 0 & -1 \\ -1 & 0 & 0 & 0 \end{vmatrix} + \begin{vmatrix} 0 & -1 & 0 & 0 \\ 0 & 0 & -1 & 0 \\ 0 & -1 & 0 & 0 \\ 0 & 0 & 1 & 0 \end{vmatrix} + \begin{vmatrix} 0 & 0 & 1 & 0 \\ 0 & 1 & 0 & 0 \\ 0 & 0 & 1 & 0 \\ 0 & -1 & 0 & 0 \end{vmatrix} = p_0+p_1+p_2+p_3$$

*	p_0	p_1
p_0	p_0	p_1
p_1	p_1	p_0

*	p_2	p_3
p_2	p_2	p_3
p_3	p_3	p_2

;

$$P = a_0p_0+a_2p_1+a_2p_2+a_3p_3 = \begin{vmatrix} a_0, & -a_2, & a_3, & -a_1 \\ a_1, & a_3, & -a_2, & -a_0 \\ a_0, & -a_2, & a_3, & -a_1 \\ -a_1, & -a_3, & a_2, & a_0 \end{vmatrix}$$

Fig. 12.2.12. Top: decomposition of the Rademacher matrix of 16 duplets into 4 sparse matrices p_0, p_1, p_2, p_3. Bottom left: multiplication tables of the set of matrices p_0 and p_1 and the set of matrices p_2 and p_3 coincide with the multiplication table of hyperbolic numbers in Fig. 12.2.10. Bottom right: the matrix expression of 2-block U-hyperbolic numbers P.

The set of matrices p_0 and p_1 is closed under multiplication and defines the multiplication table of hyperbolic numbers (Fig. 12.2.12). The same is true for the set of matrices p_2 and p_3. Correspondingly the expression $P = a_0p_0 + a_2p_1 + a_2p_2 + a_3p_3$ represents the 4-parametrical system of 2-block united-hyperbolic numbers (Fig. 12.2.12). Here p_0 and p_2 are local-real units and p_1 and p_3 are local-imaginary units.

12.3. United-hypercomplex numbers and the genetic matrix of 64 triplets

Let us consider now the Rademacher matrix of 64 triplets (Chapter 1, Fig. 1.4.3) and its decomposition into a sum of 8 sparse matrices r_0, r_1, r_2, r_3, r_4, r_5, r_6, r_7 (Fig. 12.3.1).

Here each of the following 4 pairs of matrices — (r_0, r_1), (r_2, r_3), (r_4, r_5), (r_6, r_7) — is closed under multiplication and defines the multiplication table of complex numbers (Fig. 12.1.1). Correspondingly, the following expression represents the 8-parametrical system of 4-blocks U-complex numbers: $a_0r_0+ a_1r_1+a_2r_2+a_3r_3+a_4r_4+a_5r_5+a_6r_6+a_7r_7$. Algebraic, symmetrical and permutation properties of this system are similar in many aspects to the above-described

1,1,-1,-1,1,1,-1,-1		1,0, 0,0,0,0,0,0		0,0,-1,0,0,0,0,0		0,1,0, 0,0,0,0,0	
1,1,-1,-1,1,1,-1,-1		1,0, 0,0,0,0,0,0		0,0,-1,0,0,0,0,0		0,1,0, 0,0,0,0,0	
1,1,1,1,-1,-1,-1,-1		0,0, 1,0,0,0,0,0		1,0, 0,0,0,0,0,0		0,0,0, 1,0,0,0,0	
1,1,1,1,-1,-1,-1,-1	=	0,0, 1,0,0,0,0,0	+	1,0, 0,0,0,0,0,0	+	0,0,0, 1,0,0,0,0	+
1,1,-1,-1,1,1,-1,-1		1,0, 0,0,0,0,0,0		0,0,-1,0,0,0,0,0		0,1,0, 0,0,0,0,0	
1,1,-1,-1,1,1,-1,-1		1,0, 0,0,0,0,0,0		0,0,-1,0,0,0,0,0		0,1,0, 0,0,0,0,0	
-1,-1,-1,-1,1,1,1,1		0,0,-1,0,0,0,0,0		-1,0, 0,0,0,0,0,0		0,0,0,-1,0,0,0,0	
-1,-1,-1,-1,1,1,1,1		0,0,-1,0,0,0,0,0		-1,0, 0,0,0,0,0,0		0,0,0,-1,0,0,0,0	

0, 0,0,-1,0,0,0,0		0,0,0,0,1,0, 0,0		0,0,0,0, 0,0,-1,0		0,0,0,0,0,1,0, 0		0,0,0,0,0, 0,0,-1
0, 0,0,-1,0,0,0,0		0,0,0,0,1,0, 0,0		0,0,0,0, 0,0,-1,0		0,0,0,0,0,1,0, 0		0,0,0,0,0, 0,0,-1
0, 1,0, 0,0,0,0,0		0,0,0,0,0,0,-1,0		0,0,0,0,-1,0, 0,0		0,0,0,0,0,0,0,-1		0,0,0,0,0,-1,0, 0
0, 1,0, 0,0,0,0,0	+	0,0,0,0,0,0,-1,0	+	0,0,0,0,-1,0, 0,0	+	0,0,0,0,0,0,0,-1	+	0,0,0,0,0,-1,0, 0
0, 0,0,-1,0,0,0,0		0,0,0,0,1,0, 0,0		0,0,0,0, 0,0,-1,0		0,0,0,0,0,1,0, 0		0,0,0,0,0, 0,0,-1
0, 0,0,-1,0,0,0,0		0,0,0,0,1,0, 0,0		0,0,0,0, 0,0,-1,0		0,0,0,0,0,1,0, 0		0,0,0,0,0, 0,0,-1
0,-1,0, 0,0,0,0,0		0,0,0,0,0,0, 1,0		0,0,0,0, 1,0, 0,0		0,0,0,0,0,0,0, 1		0,0,0,0,0, 1,0, 0
0,-1,0, 0,0,0,0,0		0,0,0,0,0,0, 1,0		0,0,0,0, 1,0, 0,0		0,0,0,0,0,0,0, 1		0,0,0,0,0, 1,0, 0

Fig. 12.3.1. The decomposition of the Rademacher matrix of 64 triplets into the sum of 8 sparse matrices $r_0+r_1+r_2+r_3+r_4+r_5+r_6+r_7$.

*	r_0	r_1	r_2	r_3	r_4	r_5	r_6	r_7
r_0	r_0	r_1	r_2	r_3	r_4	r_5	r_6	r_7
r_1	r_1	$-r_0$	r_3	$-r_2$	$-r_5$	r_4	$-r_7$	r_6
r_2	r_0	r_1	r_2	r_3	r_4	r_5	r_6	r_7
r_3	r_1	$-r_0$	r_3	$-r_2$	$-r_5$	r_4	$-r_7$	r_6
r_4	r_0	r_1	r_2	r_3	r_4	r_5	r_6	r_7
r_5	$-r_1$	r_0	$-r_3$	r_2	r_5	$-r_4$	r_7	$-r_6$
r_6	r_0	r_1	r_2	r_3	r_4	r_5	r_6	r_7
r_7	$-r_1$	r_0	$-r_3$	r_2	r_5	$-r_4$	r_7	$-r_6$

Fig. 12.3.2. The cross-shaped multiplication table of matrices $r_0, r_1, \ldots, r_7$ from Fig. 12.3.1.

properties of systems of 2-blocks U-complex numbers. The total set of matrices $r_0, r_1, \ldots, r_7$ is closed under multiplication and corresponds to the multiplication table in Fig. 12.3.2.

This multiplication table also has a cross-shaped structure with elements of a fractal plexus in relative locations of sets of basic units:

- two (4×4)-quadrants along its main diagonal contain all (2×2)-sub-quadrants, the structure of which coincides with the multiplication table of basic elements of complex numbers; opposite to this, two (4×4)-quadrants along its second diagonal contain all (2×2)-sub-quadrants, the structure of which coincides with the cross-shaped matrix representation of complex numbers themselves $[a, b; -b, a]$ (Fig. 12.1.1);
- the main diagonal of the (8×8)-table contains only cells with local-real units r_0, r_2, r_4, r_6 having even indices; opposite to this, its second diagonal contains only cells with local-imaginary units r_1, r_3, r_5, r_7 having odd indices;

- inside each of the four (4 × 4)-quadrants its main diagonal contains only cells with local-real units; opposite to this, its second diagonal contains only cells with local-imaginary units;
- inside each of the sixteen (2 × 2)-sub-quadrants its main diagonal also contains only cells with one of the local-real units; opposite to this, its second diagonal contains only cells with one of the local-imaginary units.

One can note that cross-shaped structures of the multiplication table in Fig. 12.3.2, which is connected with the Rademacher matrix of 64 triplets, are also associated with known phenomena of inherited cross-shaped schemes of morpho-functional systems (see examples in Fig. 12.2.4).

This system of 4-blocks U-complex numbers (Fig. 12.3.1) gives the special separation of the whole set of 64 triplets into 8 subsets, each of whose members relates to one of these matrices r_0, r_1, …, r_7. Fig. 12.3.3 shows the symbolic writing of these 8 subsets, each member of which contains 8 triplets.

Fig. 12.3.3 additionally shows remarkable symmetries in the natural distribution of 20 amino acids and stop codons among the 8 local units r_i of this system of 4-blocks U-complex numbers: each of the local-real units r_0, r_2, r_4, r_6 is associated with 3 amino acids encoded by a triplet having a

Triplets in the sparse matrices	**Corresponding amino acids**
$r_0 \leftrightarrow$ {CCC, CGC, CCT, CGT, TCC, TGC, TCT, TGT}	**Pro**, **Arg**, **Pro**, **Arg**, **Ser**, Cys, **Ser**, Cys
$r_1 \leftrightarrow$ {CAC, CTC, CAT, CTT, TAC, TTC, TAT, TTT}	His, **Leu**, His, **Leu**, Tyr, Phe, Tyr, Phe
$r_2 \leftrightarrow$ {CCG, CGG, CCA, CGA, TCG, TGG, TCA, TGA}	**Pro**, **Arg**, **Pro**, **Arg**, **Ser**, Trp, **Ser**, Trp
$r_3 \leftrightarrow$ {CAG, CTG, CAA, CTA, TAG, TTG, TAA, TTA}	Gln, **Leu**, Gln, **Leu**, Stop, Leu, Stop, Leu
$r_4 \leftrightarrow$ {GCC, GGC, GCT, GGT, ACC, AGC, ACT, AGT}	**Ala**, **Gly**, **Ala**, **Gly**, **Thr**, Ser, **Thr**, Ser
$r_5 \leftrightarrow$ {GAC, GTC, GAT, GTT, AAC, ATC, AAT, ATT}	Asp, **Val**, Asp, **Val**, Asn, Ile, Asn, Ile
$r_6 \leftrightarrow$ {GCG, GGG, GCA, GGA, ACG, AGG, ACA, AGA}	**Ala**, **Gly**, **Ala**, **Gly**, **Thr**, Stop, **Thr**, Stop
$r_7 \leftrightarrow$ {GAG, GTG, GAA, GTA, AAG, ATG, AAA, ATA}	Glu, **Val**, Glu, **Val**, Lys, Met, Lys, Met

Fig. 12.3.3. Left: the representation of the conformity of numeric matrices r_0, r_1, …, r_7 (Fig. 12.3.1) and genetic triplets in the symbolic matrix $[C, A; T, G]^{(3)}$ (Table 1.2.1 and Fig. 1.3.1 of Chapter 1). Right: sets of amino acids and stop signals (from Table 1.2.1 and Fig. 1.3.1 of Chapter 1) for each of 8 local units r_i of the system of 4-blocks U-complex numbers (Fig. 12.3.1). Each amino acid encoded by a triplet with a strong root is denoted by bold and underlined.

strong root and 1 amino acid encoded by a triplet having a weak root; it is opposite to the case of local-imaginary units r_1, r_3, r_5, r_7, each of which is associated with 3 amino acids encoded by a triplet having a weak root and 1 amino acid encoded by a triplet having a strong root.

12.4. Diagonal united-hypercomplex numbers and their tensor families

Do such systems of U-complex numbers, which possess operations of commutative multiplication and division, exist? This section gives a positive answer on the question and shows systems with these operations and additionally with divisors of zero. These systems were revealed as a result of the matrix-algebraical study of genetic Rademacher matrices of *n*-plets.

In the Rademacher matrix W_2 of 16 duplets (Fig. 1.4.3 of Chapter 1) the mosaic of two quadrants along the main diagonal is mirror-antisymmetric to that along the second diagonal in relation to the middle vertical line. The operation of the mirror-antisymmetric transformation is given by the matrix $J = [0, 0, 0, -1; 0, 0, -1, 0; 0, -1, 0, 0; -1, 0, 0, 0]$, whose square is equal to the identical matrix E, that is, $J^2 = E$. Fig. 12.4.1 illustrates that the matrix W_2 is the sum of 2 sparse matrices W_{20} and W_{21} of diagonal types, non-zero parts of which coincide correspondingly with both quadrants along the main diagonal and with both quadrants along the second diagonal.

The mirror-antisymmetric transformation J transforms the matrix W_{21} into the matrix W_{20}. Correspondingly $W_2 = W_{20} + W_{21} = W_{20} + J \times W_{20} = W_{20} \times (E + J)$, where E+J is the hyperbolic number with unit coordinates since the set of matrices E and J is closed under multiplication and its multiplication

$$W_2 = W_{20} + W_{21} = \begin{bmatrix} 1 & -1 & 1 & -1 \\ 1 & 1 & -1 & -1 \\ 1 & -1 & 1 & -1 \\ -1 & -1 & 1 & 1 \end{bmatrix} = \begin{bmatrix} 1 & -1 & 0 & 0 \\ 1 & 1 & 0 & 0 \\ 0 & 0 & 1 & -1 \\ 0 & 0 & 1 & 1 \end{bmatrix} + \begin{bmatrix} 0 & 0 & 1 & -1 \\ 0 & 0 & -1 & -1 \\ 1 & -1 & 0 & 0 \\ -1 & -1 & 0 & 0 \end{bmatrix}$$

Fig. 12.4.1. Decomposition of the Rademacher matrix W_2 of 16 duplets into the sum of two matrices W_{20} and W_{21}.

$$\begin{bmatrix}1&-1&0&0\\1&1&0&0\\0&0&1&-1\\0&0&1&1\end{bmatrix} = \begin{matrix}1&0&0&0\\0&1&0&0\\0&0&0&0\\0&0&0&0\end{matrix} + \begin{matrix}0&-1&0&0\\1&0&0&0\\0&0&0&0\\0&0&0&0\end{matrix} + \begin{matrix}0&0&0&0\\0&0&0&0\\0&0&1&0\\0&0&0&1\end{matrix} + \begin{matrix}0&0&0&0\\0&0&0&0\\0&0&0&-1\\0&0&1&0\end{matrix} = d_0+d_1+d_2+d_3$$

*	d_0	d_1
d_0	d_0	d_1
d_1	d_1	$-d_0$

*	d_2	d_3
d_2	d_2	d_3
d_3	d_3	$-d_2$

Fig. 12.4.2. Top: the decomposition of the matrix W_{20} from Fig. 12.4.1: $W_{20} = d_0 + d_1 + d_2 + d_3$. Bottom: multiplication tables for the pair d_0 and d_1 and for the pair d_2 and d_3 coincide with the multiplication table of complex numbers.

$V_L = a_0d_0+a_1d_1 =$ [a_0, $-a_1$, 0, 0 a_1, a_0, 0, 0 0, 0, 0, 0 0, 0, 0, 0]	$V_R = a_2d_2+a_3d_3 =$ [0, 0, 0, 0 0, 0, 0, 0 0, 0, a_2, $-a_3$ 0, 0, a_3, a_2]	$V = V_L+V_R = a_0d_0+a_1d_1+a_2d_2+a_3d_3 =$ [a_0, $-a_1$, 0, 0 a_1, a_0, 0, 0 0, 0, a_2, $-a_3$ 0, 0, a_3, a_2]
$V_L^{-1} = (a_0^2 + a_1^2)^{-1} * (a_0d_0-a_1d_1)$; $V_LV_L^{-1}=V_L^{-1}V_L=d_0$, where d_0 – the local-identity matrix for V_L.	$V_R^{-1} = (a_2^2 + a_3^2)^{-1} * (a_2d_2-a_3d_3)$; $V_RV_R^{-1}=V_R^{-1}V_R=d_2$, where d_2 – the local-identity matrix for V_R.	$V^{-1} = (a_0^2+a_1^2)^{-1}*(a_0d_0-a_1d_1) + (a_2^2+a_3^2)^{-1}*(a_2d_2-a_3d_3) = V_L^{-1} + V_R^{-1}$. $VV^{-1} = V^{-1}V = E$, where E = [1,0,0,0; 0,1,0,0; 0,0,1,0; 0,0,0,1] – the identity matrix.

Fig. 12.4.3. Upper row: the system of diagonal 2-blocks U-complex numbers V, which unites systems of complex numbers V_L and V_R. Lower row: expressions for inverse matrices V_L^{-1}, V_R^{-1} and V^{-1}.

table coincides with the multiplication table of hyperbolic numbers. The matrix W_{20} is 2-blocks U-complex number of diagonal type with identity coordinates from the standpoint of its decomposition into 4 sparse matrices d_0, d_1, d_2, d_3 (Fig. 12.4.2).

The pair of matrices d_0 and d_1 forms a set which is closed under multiplication and which defines the multiplication table of complex numbers (Fig. 12.4.2, bottom). The same is true for the pair of matrices d_2 and d_3. These basis elements d_0, d_1, d_2, d_3 define the following view of two systems of complex numbers V_L and V_R represented by (4 × 4)-matrices: $V_L = a_0d_0 + a_1d_1$ and $V_R = a_2d_2 + a_3d_3$. Fig. 12.4.3 shows the corresponding general view of the system of 2-blocks diagonal U-complex numbers

$V = a_0 d_0 + a_1 d_1 + a_2 d_2 + a_3 d_3$ having 4 parameters (a_0, a_1, a_2, a_3). One should mention that the whole set of matrices d_0, d_1, d_2, d_3 is not closed under multiplication and does not represent any hypercomplex number.

The classical identity matrix E = [1 0 0 0; 0 1 0 0; 0 0 1 0; 0 0 0 1] is absent in the set of matrices V_L and V_R, where — beside this — each matrix has cannot determinant (Fig. 12.4.3). Consequently, the notion of the inverse matrix V_L^{-1} or V_R^{-1} can't be defined in relation to the identity matrix E. But we analyze not the complete set of (4 × 4)-matrices but limited special sets of matrices V_L and V_R. The system V_L has the matrix d_0 (Fig. 12.4.2), which possesses all properties of its identity matrix for any matrix V_L since $d_0 V_L = V_L d_0 = V_L$ and $d_0^2 = d_0$. In the frame of the set of matrices V_L, where locally the matrix d_0 plays the role of the identity matrix (the local-identity matrix), one can define — for any non-zero matrix V_L — the inverse matrix V_L^{-1} in relation to the matrix d_0 on the basis of equations: $V_L V_L^{-1} = V_L^{-1} V_L = d_0$. Such inverse matrix is defined by the following expression: $V_L^{-1} = (a_0^2 + a_1^2)^{-1} \times (a_0 d_0 - a_1 d_1)$ (Fig. 12.4.3).

By analogy, the set of matrices V_R has the matrix d_2 (Fig. 12.4.2), which possesses all properties of the identity matrix for any matrix V_R since $d_2 V_R = V_R d_2 = V_R$ and $d_2^2 = d_2$. In the frame of the set of matrices V_R, where the matrix d_2 is the local-identity matrix, one can define — for any non-zero matrix V_R — the inverse matrix V_R^{-1} in relation to the matrix d_2 on the basis of the equation: $V_R V_R^{-1} = V_R^{-1} V_R = d_2$.

Any of the non-zero 2-blocks diagonal U-complex numbers $V = V_L + V_R = a_0 d_0 + a_1 d_1 + a_2 d_2 + a_3 d_3$ has its inverse number V^{-1} of the same type in relation to the ordinary identity matrix E = [1, 0, 0, 0; 0, 1, 0, 0; 0, 0, 1, 0; 0, 0, 0, 1]: $VV^{-1} = V^{-1}V = E$, where $V^{-1} = (a_0^2 + a_1^2)^{-1} \times (a_0 d_0 - a_1 d_1) + (a_2^2 + a_3^2)^{-1} \times (a_2 d_2 - a_3 d_3) = V_L^{-1} + V_R^{-1}$. In other words, in this system the inverse U-number V^{-1} is equal to the sum of inverse numbers V_L^{-1} and V_R^{-1}.

The system of 2-blocks diagonal U-complex numbers V has operations of addition, subtraction, commutative multiplication and division (division is defined by multiplication with the inverse diagonal U-numbers V^{-1}). This system has divisors of zero; it means that the system has non-zero matrices, the product of which gives zero. Indeed, the product of non-zero matrices V_L and V_R is equal to a zero matrix: $V_LV_R = V_RV_L = 0$. (If hyperbolic numbers are used in both diagonal blocks of the matrix V in Fig. 12.4.2 instead of complex numbers, the system of diagonal 2-blocks U-split-hyperbolic numbers arises, which also possesses operations of addition, subtraction, commutative multiplication and division and which has divisors of zero).

Extensions of diagonal 2-block U-complex numbers into diagonal 2^n-blocks U-complex numbers can be easily done by means of tensor multiplication $[1\ 0;\ 0\ 1]^{(n)} \otimes [1, -1;\ 1, 1]$ by similarly using decomposition (Fig. 12.4.2). The same is true for extensions of diagonal U-hyperbolic numbers, etc. Norms and the scalar product of vectors can be defined in multi-dimensional spaces, which correspond to such systems of diagonal U-complex numbers [340]. Additional materials about properties of U-hypercomplex numbers one can see in [340].

12.5. Applications of united-hypercomplex systems in biomechanics

One of many problems studied traditionally in biomechanics is the problem of locomotion based on innate algorithms. As is known, newborn turtles and crocodiles, when hatched from eggs, crawl with quite coordinated movements to water without any training from anybody. Centipedes crawl by means of coordinated movements of tens and hundreds of their legs based on inherited algorithms of control. Celled organisms, which have no nervous systems and muscles, move by means of perfectly coordinated motions of cilia on their surfaces (the genetically inherited "dances of cilia"). In this locomotion, using certain kinds of kinematics of multi-block bodies, a huge number of muscle fibers, nerve cells, contractile proteins, enzymes, and so forth are acting in concert, by analogy with the coordinated work of a plurality of parts of computers. Different kinds of normal and pathological kinds of multi-block kinematics of human and animal locomotion can be modeled using matrix operators presenting multi-block united-hypercomplex numeric systems.

Let's give an example of such modeling of the behavior of an 8-parameter system, the state vector of which in the eight-dimensional configuration space has the form $\overline{V} = [a_0(t), a_1(t), a_2(t), a_3(t), a_4(t), a_5(t), a_6(t), a_7(t)]$. Let 4 pairs of parameters, for example, a_0 and a_4, a_1 and a_5, a_2 and a_6, a_3 and a_7, be changed independently of each other. To describe the behavior of this system, we need a corresponding eight-dimensional matrix operator, whose action on the vector $\overline{V}$ can model this system. Fig. 12.5.1 shows such a matrix operator M(*t*), which can be decomposed into the sum of 8 sparse matrices a_0m_0, a_1m_1, a_2m_2, a_3m_3, a_4m_4, a_5m_5, a_6m_6, a_7m_7. The matrix M(*t*) represents 4-blocks U-complex number, where each of the 4 pairs of matrices m_0 and m_4, m_1 and m_5, m_2 and m_6, m_3 and m_7 contains two matrices, the set of which is closed relative to multiplication and defines the multiplication table of the algebra of complex numbers. In these 2-parametric pairs, matrices m_0, m_1, m_2, and m_3 represent local-real unit, and matrices m_4, m_5, m_6, and m_7 represent local-imaginary unit of complex numbers.

In an 8-parametric system whose behavior is described by such operator M(t), the time behavior of each of the four named 2-parameter subsystems is visually depicted by the movement of a point in one of the four coordinate planes (x_0, x_4), (x_1, x_5), (x_2, x_6), (x_3, x_7) of the configuration space of the system and can be defined completely independently of the other three subsystems. To set the trajectory and the rate of movement of each point along one or

M(t) =

$a_0(t)$	$-a_1(t)$	$a_2(t)$	$-a_3(t)$	$-a_4(t)$	$a_5(t)$	$a_6(t)$	$-a_7(t)$
$a_0(t)$	$a_1(t)$	$a_2(t)$	$a_3(t)$	$-a_4(t)$	$-a_5(t)$	$a_6(t)$	$a_7(t)$
$-a_0(t)$	$a_1(t)$	$a_2(t)$	$-a_3(t)$	$a_4(t)$	$-a_5(t)$	$a_6(t)$	$-a_7(t)$
$-a_0(t)$	$-a_1(t)$	$a_2(t)$	$a_3(t)$	$a_4(t)$	$a_5(t)$	$a_6(t)$	$a_7(t)$
$a_4(t)$	$-a_5(t)$	$-a_6(t)$	$a_7(t)$	$a_0(t)$	$-a_1(t)$	$a_2(t)$	$-a_3(t)$
$a_4(t)$	$a_5(t)$	$-a_6(t)$	$-a_7(t)$	$a_0(t)$	$a_1(t)$	$a_2(t)$	$a_3(t)$
$-a_4(t)$	$a_5(t)$	$-a_6(t)$	$a_7(t)$	$-a_0(t)$	$a_1(t)$	$a_2(t)$	$-a_3(t)$
$-a_4(t)$	$-a_5(t)$	$-a_6(t)$	$-a_7(t)$	$-a_0(t)$	$-a_1(t)$	$a_2(t)$	$a_3(t)$

Fig. 12.5.1. Matrix operator M(t) represents a 4-blocks U-complex number.

another parametrically determined individual trajectory, it is only necessary to set the functions $a_k(t)$ in the operator M(t) (here $k = 0, 1, \ldots, 7$).

For example, independent movements of points in the planes (x_0, x_4), (x_1, x_5), (x_2, x_6), and (x_3, x_7) of eight-dimensional space $[x_0, x_1, \ldots, x_7]$ occur according to known types of parametric curves — "cardiode", "clover leaf", "rose" and " logarithmic spiral" — if the functions $a_k(t)$ have the following form: $a_0(t) = \cos(t) \times (1 + \cos(t))$; $a_4(t) = \sin(t) \times (1 + \cos(t))$; $a_1(t) = \cos(t) + \cos(t) \times \cos(3 \times t) + \cos(t) \times \sin^2(3 \times t)$; $a_5(t) = \sin(t) + \sin(t) \times \cos(3 \times t) + \sin(t) \times \sin^2(3 \times t)$; $a_2(t) = 2 \times \cos(t) \times \sin(5/3 \times t)$; $a_6(t) = \sin(t) \times \sin(5/3 \times t)$; $a_3(t) = \cos(t) \times (13/12)^t$; $a_7(t) = \sin(t) \times (13/12)^t$ (see Fig. 12.5.2).

Obviously, along with the presented description of a multi-parameter system with independent behavior of subsystems, multi-block U-hypercomplex numbers can be applied to modeling the behavior of systems with partially dependent or fully consistent behavior of parameter subsystems. To do this, it is necessary to establish, based on additional data, the relationship between the functions $a_k(t)$ related to different subspaces.

In biomechanics, there are many problems associated with the question of the interrelationship between individual parts of the body. For example,

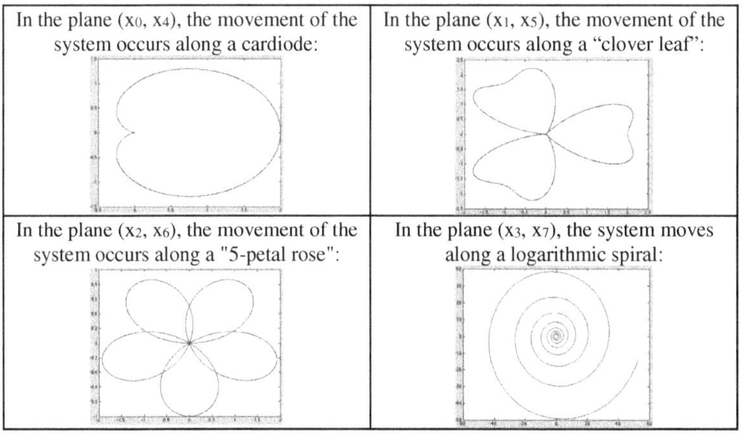

Fig. 12.5.2. An example of describing the time behavior of an 8-parameter system using the matrix operator M(t) from Fig. 12.5.1. In each indicated plane of the eight-dimensional configuration space of a given system, changes in its parameters occur regardless of changes in the parameters of the other planes.

these include the problem of atrial fibrillation, which occurs when the physiologically normal interrelation among the parts of the heart is lost. The vital activity of the organism is based on the coordinated work of many systems: respiratory, cardiac, nervous, etc. There are diagnostic tools for monitoring the operation of individual subsystems, for example, electrocardiography, pulsography, electroencephalography, etc. Of course, the operation of each of the subsystems of a multi-block or multi-parameter system can be modeled algebra-geometrically separately using, for example, the well-known apparatus of complex and hypercomplex numbers that have a relatively small dimension. But in biomechanics, physiology and medicine there are many tasks of studying the emergent properties of a system of many parts that make up a single whole and can be in varying degrees of consistency with each other; at the same time, many parts of such a system are often forced to change collectively under the influence of external factors, for example, when an organism enters an extreme situation. For this kind of problems, the apparatus of multi-block U-numbers and their matrix representations is preferable, allowing within the framework of one matrix the state of the entire system, together with its parts. At the same time, emergent changes in the entire system and its subsystems under the influence of external factors can be modeled by multiplying the time-variable U-number, which describes the initial state of the system, by another U-number, the result of which the emerging state of the system will be described by a new time-variable U-number. Such algebra-biological studies can reveal the existence of diseases and pathologies associated precisely with the emergent properties of the organism as a system containing many functional subsystems.

U-number systems, which have arisen in research on genetic biomechanics and matrix genetics, are also useful in group robotics. This direction of modern technology is associated with the tasks of the collective behavior of many robots and the collective behavior of ants, bees and other organisms. Particular attention is paid to flocks of cyborg insects. For example, cyborg cockroaches are being created [398], the route of which can be set remotely by radio control. Flocks of these cyborgs are designed to search for people inside collapsed buildings, detect radioactive and chemical contamination in cities, etc. (there are numerous videos and articles on the Internet about these cyborg cockroaches).

12.6. Some concluding remarks

Based on the matrix-algebraic study of genetic code systems we reveal and propose a new mathematical tool in the form of multi-block U-hypercomplex numbers. This tool seems to be useful for mathematical modeling of different structures. The vastness of its applications will be determined in large part by those researchers who want to apply it for different tasks. The introduced systems of multi-block U-numbers define new — for mathematical natural sciences — types of multi-dimensional spaces (U-numerical spaces), some of which are vector spaces with special interrelations of their vectors.

After the discovery of non-Euclidean geometries and Hamilton quaternions, it is known that different natural systems can possess their own geometry and algebra [392]. But what kinds of algebra are appropriate for living organisms in their bio-informational aspects? We propose that inherited bio-informational aspects of organization of living matter is governed by systems of 2^n-block U-hypercomplex numbers. It seems that many difficulties of modern bioinformatics and mathematical biology are connected with utilizing — for biological structures — inadequate algebras, which were developed for other natural systems. Hamilton had similar difficulties in his attempts to describe transformations in a three-dimensional space by means of three-dimensional numbers while this description needs 4-parametrical quaternions.

The proposed systems of 2^n-blocks U-numbers (U-hypercomplex numbers) generalize systems of complex and hypercomplex numbers, which become a particular case of appropriate systems of multi-block U-numbers. Hypercomplex numbers, beginning from quaternions of Hamilton, were created as an extension of complex numbers by means of including new independent imaginary units in addition to a single imaginary unit of complex numbers. In our studies of genetic matrices, we paid attention on the alternative way of extensions of complex numbers: such extension is provided by means of including new independent real units (one or more real units, which are called local-real units) together with their particular complect of imaginary units (local-imaginary units). As far as we know, the proposed systems of U-numbers have not yet been encountered in mathematical natural science, and therefore they are the basis of a new class of mathematical models, primarily in biology. In mathematical natural science and computer science, single-block hypercomplex numbers are traditionally used; the presented genetic-algebraic research draw attention to multi-block systems of hypercomplex numbers and corresponding matrix operators that allow working with multi-dimensional spaces, different subspaces of which are governed by different systems of hypercomplex numbers.

The modern mathematical natural sciences and technologies are based on using multi-dimensional numbers, which are called in a general case as hypercomplex numbers. The results presented in our book have led to their generalization in a matrix form of united-hypercomplex numbers. It is natural to think that such generalization of the numerical basis can lead to useful extensions in different sciences and technologies giving rise to appropriate extended theories and new practical applications in them. As is known, matrices have been introduced in theoretical physics due to the creation of matrix mechanics by Heisenberg. From that time matrices became one of the most important tools in mathematical natural sciences and informatics. U-numbers in their matrix representations further strengthen the value of matrices in science.

In our view, the mathematics of U-numbers can be used for further developing of algebraic biology, theoretical physics and informatics in accordance with the famous statement by P. Dirac, who taught that the creation of a physical theory must begin with the mathematical theory: "*If this theory is really beautiful, then it necessarily will appear as a fine model of important physical phenomena. It is necessary to search for these phenomena to develop applications of the beautiful mathematical theory and to interpret them as predictions of new laws of physics*" [399]. According to Dirac, all new physics, including relativistic and quantum, are developing in this way.

Materials of this book reinforce this point of view that living matter in its informational fundamentals is organized on algebraic bases. By analogy with the known fact that molecular foundations of molecular genetics turned up unexpectedly very simple, perhaps algebraic foundations of living matter are also relatively simple. Materials of our book give additional data to the theme and works about parallelisms in functioning of living organisms and computers and about special features of reading of genetic information [312, 346, 400–403, etc.]. These materials are useful, in particular, for developing genetic biomechanics, personalized genetics, artificial intelligence, etc.

Multi-block structure, which is genetically inherited, is one of the most characteristic features of a plurality of living bodies on different levels and branches of biological evolution. One can believe that systems of multi-block U-hypercomplex numbers represent a relevant mathematical tool for modeling such inherited structures and that structural organization of living matter can be considered as a territory controlled by U-numbers. But non-living substances also consist of various blocks of different levels up to atoms and elementary particles (one of the most important statements in the history of science has been made by Democritus approximately in the fourth century BC: "*All bodies are composed of discrete units — atoms*"). It is not excluded that the proposed systems of multi-block U-numbers will be also useful in theoretical physics to study block structures in non-living matter.

The notion of "number" is the main notion of mathematics and mathematical natural sciences. "*Complexity of a civilization is reflected in complexity of numbers used by this civilization*" [404]. "*Number is one of the most fundamental concepts not only in mathematics, but also in all natural sciences. Perhaps, it is the more primary concept than such global categories as time, space, substance or a field*" [405].

Pythagoras studied figurate numbers and formulated the idea "*all things in the world are numbers*" or "*number rules the world*". B. Russell noted that he did not know of any other man who has been as influential as Pythagoras was in the sphere of thought [406]. From this standpoint, there is no more fundamental scientific idea in the world than this idea about a basic meaning of numbers. As Heisenberg noted, modern physics is moving along the same path which the Pythagoreans walked [407]. Our genetic-mathematical research has led to new systems of multi-dimensional numbers and given new materials to the great idea by Pythagoras since these materials allow putting forward the hypothesis that genetic-informational organization of living matter are governed by multi-dimensional U-numbers (conditionally and briefly speaking — in the manner of Pythagoras's slogan — "U-numbers rule living matter"). Research in this direction should be sustained.

Chapter 13

Musical Harmony in Parameters of Molecular Genetics

Music is a game with acoustic resonances to which people are remarkably predisposed. For tens of thousands of years, people create musical instruments, adjusting them to specific systems of resonances. Over the centuries, people have learned to combine individual instruments and singers into orchestras and choirs as coordinated oscillating systems with an increased number of degrees of freedom. An organism can be seen as a musical synthesizer with multiple settings of inherited resonant modes. Leibniz declared that music is the arithmetic of the soul. Considering that music is represented by systems of resonances, one can reformulate this declaration: systems of resonances are the arithmetic of the soul, which computes without being aware of it. This chapter describes connections of numerical parameters of the molecular-genetic systems with ratios of the musical harmony of Pythagorean musical tunings and so-called Fibonacci-stage musical tunings. These connections were revealed due to the study of binary oppositions in the DNA parameters. The described results additionally testify that molecular-genetic systems are organized on ensembles of resonances and that binary sets of resonances are the basis of binary genetic computers. They give materials for a new slogan: a living body is a musical instrument (a synthesizer with an abundance of rearrangements of resonant modes). These results are used in the Moscow State Tchaikovsky Conservatory to create so-called "genetic music". In particular, these results can be used for tasks of music therapy, which is very popular around the world.

13.1. Inherited feeling of musical harmony and Pythagorean quint scales

From ancient times, understanding the phenomenon of music and building musical structures were associated with mathematics. The creator of the first computer, Leibniz, wrote: "*Music is a secret arithmetical exercise and the person who indulges in it does not realize that he is manipulating numbers*" and "*music is the pleasure the human mind experiences from counting without being aware that it is counting*" (http://thinkexist.com/quotes/g._wilhelm_leibniz/).

The scientific studies of physiological mechanisms of musical perception took place long ago. Science has been dealing with the physiological mechanisms of music perception for a long time [408]. Four-month-old infants turn toward a source of pleasant sounds (consonances) and away from a source of unpleasant sounds (dissonances). There is no specialized center of music in the human brain; a sense of love for music can be considered dispersed throughout the body, similar to the dispersion of genetic DNA molecules throughout all of its cells. More than 30 thousand years ago, long before the advent of arithmetic, our ancestors already played stone flutes and bone harps. For example, the bone flute found in France is at least 32 thousand years old. The enjoyment of music is usually explained by the fact that it gives rise to emotions and feelings. Aristotle tried to understand how rhythms and melodies, being only sounds, resemble states of mind.

For Europeans the idea of musical harmony is basically connected with the name Pythagoras. The Pythagorean musical 7-stage scale (a heptatonic scale) contains seven notes named do (C), re (D), mi (E), fa (F), sol (G), la (A), si (B). These seven notes are interrelated among themselves by their frequencies not in an accidental manner, but form a regular uniform ensemble. Indeed, it is well known that the seven notes from appropriate octaves form a geometric progression on the basis of the quint ratio 3:2 between frequencies of the adjacent members of this sequence (Fig. 13.1.1). The quint 3:2, which is the ratio between frequencies of the third and second

fa (F)	do (C)	sol (G)	re (D^1)	la (A^1)	mi (E^2)	si (B^2)
87	130	196	293	440	660	990
$(3/2)^{-3}$	**$(3/2)^{-2}$**	**$(3/2)^{-1}$**	**$(3/2)^{0}$**	**$(3/2)^{1}$**	**$(3/2)^{2}$**	**$(3/2)^{3}$**

Fig. 13.1.1. The quint sequence of the 7 notes of the Pythagorean musical scale is presented. The first row shows the notes. The second row shows their frequencies. The third row shows the ratios between the frequencies of these notes to the frequency 293 Hz of the note re (D^1). The designation of notes is given on the Helmholtz system. Values of frequencies are approximated to integers.

harmonics of an oscillated string, plays the role of the factor of this geometric progression.

Pythagoras has created the mathematical foundations of ancient Greek music, borrowing to a certain degree some ancient knowledge on musical harmony. His theory used the discovery that the frequency of a vibrating string is inversely proportional to its length and that musical consonances can be represented by the ratios of small integer numbers, first of all the octave ratio 2:1 and the quint ratio 3:2. These ideas became fundamental concepts of all music theory from antiquity to modern times. In ancient China the ratio 3/2, traditionally termed as the quint ratio (or the pure perfect fifth), was used as the fundament of quint music scales. After the ancient Chinese, Pythagoreans also considered numbers 2 and 3 as the female and male numbers (or Yin and Yang numbers), which can give birth to new musical tones in their interconnection. Ancient Greeks attached an extraordinary significance to search for the quint 3:2 in natural systems because of their thoughts about musical harmony in the organization of the world. For example, Archimedes considered as the best result of his life a detection of the quint 3/2 between volumes and surfaces of a cylinder and a sphere entered in it [409].

Such thoughts about the key significance of musical harmony in the organization of the world has existed since ancient times. For example, one can quote here a classical work of the Chinese literature *Spring and Autumn* by Lu Bu We about the fundamental role of music and numbers 3 and 2 as numbers of Heaven and Earth: "*The origins of music lie far back in the past. Music arises from Measure and is rooted in the great Oneness. ... Music is founded on the harmony between Heaven and Earth*" [410]. Pythagoreans thought about musical intervals in the planetary system and all around. Kepler wrote the famous book *Harmonices Mundi*, etc. Modern atomic physics found the harmonic ratios in spectral series by T. Lyman in the atom of hydrogen, which has been named "music of atomic spheres" by Einstein and A. Sommerfeld [409]. The importance of Pythagorean ideas about the role of musical harmony was emphasized also by Feynman [411, Chapter 50]. Many prominent crystallographers noted the connection between the structure of crystals and musical harmony [13].

The idea about a possible participation of musical harmony in the organization of biological organisms is not new for modern biophysics. For example, the famous Russian biophysicist S. Shnoll wrote: "*From possible consequences of interaction of macromolecules of enzymes, which are carrying out conformational (cyclic) fluctuations, we shall consider pulsations of pressure — sound waves. The range of numbers of turns of the majority of enzymes corresponds to acoustic sound frequencies. We shall consider ... a fantastic picture of "musical interactions" among biochemical systems, cells, bodies, and a possible physiological role of these interactions. ...It leads to pleasant thoughts about nature of hearing, about an origin of musical perception and about many other things, which belong to the area of biochemical aesthetics already*" [136]. This term "biochemical aesthetics" proposed by Shnoll reflects many materials of this chapter.

For most Europeans from antiquity, quint scales in music relate to this Pythagorean mathematical theory of musical harmony and divisions of vibrating strings in the quint ratio 3:2. In general, the Pythagorean scale is any scale which can be constructed from only quint ratios 3:2 and octaves 2:1 [412, p. 163]. One of the known Pythagorean scales is a pentatonic scale, which is a five-stage music scale, all the sounds of which can be arranged in quint ratios. An example is the following set of 5 notes with their sound frequencies from Fig. 13.1.1: do(C)-sol(G)-re(D^1)-la(A^1)-mi(E^2) or, respectively, 130-196-293-440-660 Hz. Other examples of Pythagorean scales are tetratonic and tritonic scales, which are correspondingly 4-stage and 3-stage music scales, all the sounds of which can be arranged by the quint ratio, for instance, 130-196-293-440 Hz for the tetratonic scale and 130-196-293 Hz for the tritonic scale.

The historical fact is that these Pythagorean musical scales based on the quint ratio were used by different civilizations around the world long before Pythagoras, i.e., without knowledge of any mathematical laws [413–417]. For example, the pentatonic scale is the foundation of traditional music of the Chinese, Vietnamese, Mongols, Turkic peoples (Bashkirs, Tatars, Chuvashes, etc.), the Inca Empire and the peoples of the South Andes in general. Pentatonics is also found in European musical folklore and in the oldest layers of the Russian folk song (especially in the so-called calendar ritual songs). Tetratonic music was noted as common in Polynesia and Melanesia. Tetratonic scales were known, for example, among the Plains Indians, the Arapaho, Blackfoot, Crow, Omaha, Kiowa, Pawnee, Sioux, some Plateau tribes, the Creek Indians, and in the Great Basin region among the Washo, Ute, Paiute, and Shoshone. In the Southwest, the Navajo people also largely used the pentatonic and tetratonic, occasionally tritonic, scales. Tetratonic as well as tritonic scales were commonly used by the tribal people of India, such as the Juang and Bhuyan of Orissa state [418]. Tetratonic scales are generally associated with prehistoric music [419].

These and other data indicate that our affinity for music and musical creativity is biological in nature and the sense of musical harmony is based on innate mechanisms. But what is there in living organisms that determines the special attraction of musical scales on the basis of the quint ratio 3/2 for representatives of various civilizations and epochs? A possible answer lies in the structural features of DNA molecules that are carriers of genetic information in humans and other living organisms. The fact is that the parametric structure of DNA molecules is connected in many ways with the quint ratio 3/2 and with numbers 3 and 2 at various levels of their parametric

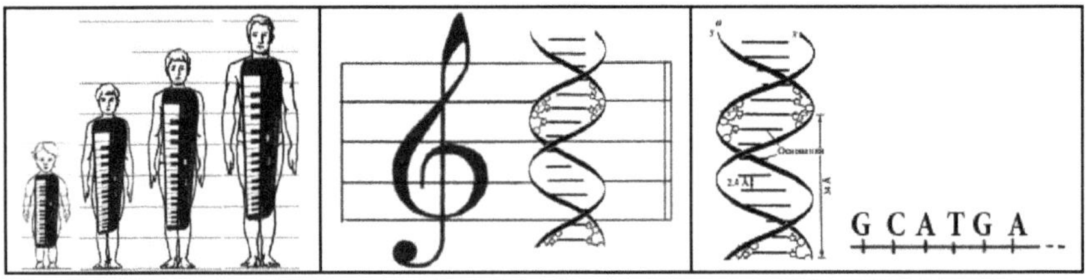

Fig. 13.1.2. Symbolic illustrations of the connection of the genetic system and inherited physiology with musical harmony.

organization [12, 13]. Let us say now about this relation between the musical harmony and structures of genetic molecules. Fig. 13.1.2 illustrates the connection of musical harmony with the genetic system, inherited feelings of musical harmony, and structural features of DNA molecules.

13.2. Quints of hydrogen bonds in DNA

In DNA double helices, nucleobases A-T and C-G form complementary pairs with 2 and 3 hydrogen bonds in them, respectively. From the standpoint of its sequence of two and three hydrogen bonds, each DNA molecule is a long chain of numbers 2 and 3 of a type 32232332 .... (Fig. 13.1.1, right).

The genetic code encodes sequences of 20 amino acids in proteins by means of 64 triplets (three-letter words) that represent all possible combinations of these four letters (ATC, TTA, ...). Since A = T = 2, C = G = 3, each triplet has a numerical representation for the product of number of hydrogen bonds of its constituent letters. For example, the triplet ACT is represented by the number $2 \times 3 \times 2 = 12$. Each of the 64 triplets is represented by one of such numbers of hydrogen bonds $2^3 = 8$, $2^2 \times 3 = 12$, $2 \times 3^2 = 18$, $3^3 = 27$, the pairwise relations between which are equal to the quint 3/2 in varying integer degrees (by analogy with music tetratonic scales), for example, $27/8 = (3/2)^3$, $18/8 = (3/2)^2$, etc.

Considering pairs of adjacent triplets, a DNA molecule appears as a quint sequence of 7 kinds of numbers of hydrogen bonds with the following numeric representation: $2^6 = 64$, $2^5 \times 3 = 96$, $2^4 \times 3^2 = 144$, $2^3 \times 3^3 = 216$, $2^2 \times 3^4 = 324$, $2 \times 3^5 = 486$, $3^6 = 729$. Pairwise ratios in this series of numbers are equal to the quint 3/2 in the same powers as in the Pythagoras 7-stage scale in Fig. 13.1.1. If, for example, the frequency of 87 Hz of the note "F" is compared with the first number 64 of this series, then all other numbers of this series will correspond precisely to the other frequencies of the Pythagoras scale. Then any sequence of triplets (e.g., insulin gene GGC-ATC-GTT-GAA-CAG-TGT-...) can be associated uniquely with a sequence of notes of Pythagoras 7-stage scale (figuratively speaking, we have "music of genes in the Pythagoras scale").

Accordingly, each DNA molecule as a chain of hydrogen bonds is characterized by its own sequences of the quint 3/2 in different integer degrees. By analogy with quint musical scales, depending on the chosen lengths of DNA nucleobase fragments, we have — on the basis of considered hydrogen bonds — various systems for transmitting information signals with quint-power relations between signals.

The quint ratios are realized in DNA not only for the hydrogen bonds of complementary nucleobases, but also for several other parameters, such as sums of atoms in the rings of purines and pyrimidines (numbers 9 and 6 with their ratio 3/2), or sums of protons in the rings of complementary nitrogenous bases (numbers 60 and 40 with their ratio 3/2), and others. Chains of these parameters in DNA form their own sequences of quint ratios, which are similar to sequences of note frequencies in quint scales of music. In other words, Nature created DNA as a plexus of various sequences of quint ratios ("a quint polyphony of DNA"). The harmony of the parametric organization of the genetic system is akin to the musical harmony of the Pythagorean scales. These different kinds of such genetic music at different parameters and levels of genetic system permits one to reproduce a musical polyphonic party for each gene and for other parts of the genetic system. These musical-like sequences were created by nature itself. Each gene and protein have their own genetic music composition (or, briefly, "genomusic"). The natural music of genes can be reproduced in acoustical diapason not only for aesthetic pleasure but, perhaps, also for medical therapy, theoretical needs, etc. This natural genomusic and its compositions can be connected to deep physiological archetypes, which were introduced into science by the creator of analytic psychology C. Jung.

Let us return to the tensor family of symbolic matrices [C, A; T, G]$^{(n)}$ in Chapter 1, Fig. 1.2.1, which were constructed on the basis of binary-oppositional attributes of DNA nucleobases.

Replacing the symbols of complementary nucleobases with the numerical values of their hydrogen bonds transforms the symbolic matrix of 4 nucleobases [C, A; T, G] into the numerical matrix Q shown in Fig. 13.2.1. Also its tensor powers $Q^{(2)}$ and $Q^{(3)}$ are shown, where each n-plet is represented by the product of the hydrogen bonds of its nucleobases. For example, the duplet CA, where C = 3 and A = 2, is represented by 3 × 2 = 6.

Each genetic matrix of the family [3, 2; 2, 3]$^{(n)}$ demonstrates the quint (or the perfect fifth) principle of its structure because they have the quint ratio 3:2 at different levels: between numerical sums in top and bottom quadrants, sub-quadrants, sub-sub-quadrants, etc., including quint ratios between neighbor numbers in them. For example, [3 2; 2 3]$^{(3)}$ contains 4 numbers — 27, 18, 12, 8 — with the quint ratio between them: 27/18 = 18/12 = 12/8 = 3/2.

$$Q = \begin{vmatrix} 3 & 2 \\ 2 & 3 \end{vmatrix}; \quad Q^{(2)} = \begin{vmatrix} 9 & 6 & 6 & 4 \\ 6 & 9 & 4 & 6 \\ 6 & 4 & 9 & 6 \\ 4 & 6 & 6 & 9 \end{vmatrix}; \quad Q^{(3)} = \begin{vmatrix} 27 & 18 & 18 & 12 & 18 & 12 & 12 & 8 \\ 18 & 27 & 12 & 18 & 12 & 18 & 8 & 12 \\ 18 & 12 & 27 & 18 & 12 & 8 & 18 & 12 \\ 12 & 18 & 18 & 27 & 8 & 12 & 12 & 18 \\ 18 & 12 & 12 & 8 & 27 & 18 & 18 & 12 \\ 12 & 18 & 8 & 12 & 18 & 27 & 12 & 18 \\ 12 & 8 & 18 & 12 & 18 & 12 & 27 & 18 \\ 8 & 12 & 12 & 18 & 12 & 18 & 18 & 27 \end{vmatrix}$$

Fig. 13.2.1. The beginning of the family [3, 2; 2, 3]$^{(n)}$ of quint matrices of DNA hydrogen bonds, which are based on numbers of DNA hydrogen bonds (C = G = 3, A = T = 2).

Each quint matrix $[3\ 2;\ 2\ 3]^{(n)}$ contains $(n+1)$ kinds of numbers from a geometric progression, the factor of which is equal to the quint 3/2:

$$
\begin{array}{ll}
[3\ 2;\ 2\ 3]^{(1)} & ®\ 3, 2 \\
[3\ 2;\ 2\ 3]^{(2)} & ®\ 9, 6, 4 \\
[3\ 2;\ 2\ 3]^{(3)} & ®\ 27, 18, 12, 8 \\
\cdots\cdots\cdots\cdots\cdots & \\
[3\ 2;\ 2\ 3]^{(6)} & ®\ 729, 486, 324, 216, 144, 96, 64 \\
\cdots\cdots\cdots\cdots\cdots &
\end{array}
\qquad (13.1)
$$

Let us write out these kinds of numbers in columns for each genetic matrix $[3\ 2;\ 2\ 3]^{(n)}$ to arrive at the "genetic" triangle, which is shown on the left part of the expression (13.2):

$$
\begin{array}{cccccc}
3 & 9 & 27 & 81 & 243 & \ldots \\
2 & 6 & 18 & 54 & 162 & \ldots \\
 & 4 & 12 & 36 & 108 & \ldots \\
 & & 8 & 24 & 72 & \ldots \\
 & & & 16 & 48 & \ldots \\
 & & & & 32 & \ldots
\end{array}
\qquad
\begin{array}{cccccc}
1 & 3 & 9 & 27 & & \\
 & 2 & & & & \\
 & & 4 & & & \\
 & & & 8 & & \\
 & & & & 16 & \\
 & & & & & \ldots
\end{array}
\qquad (13.2)
$$

On the right side of (13.2) the historically famous Plato's triangle is demonstrated. This triangle was utilized by ancient Greeks to create the Pythagorean musical scale based on its main proportions. One can see the analogy between the "genetic" triangle and Plato's triangle.

Moreover, as J. Kappraff (USA) has informed one of the authors of this book in his private letter, this genetic triangle, which was obtained from the matrices of the genetic code, was known many centuries ago: it is identical to the famous triangle which was published 2,000 years ago by Nichomachus of Gerasa in his famous book *Introduction into Arithmetic.* Nichomachus belonged to Pythagorean society, and this triangle was famous for centuries as the basis of the Pythagorean theory of musical harmony and aesthetics. In accordance with this triangle, the Parthenon [373] and other great architectural objects were created because architecture was interpreted as non-movement music, and music was interpreted as dynamic architecture. Nichomachus of Gerasa was one of the greats in the theory of musical harmony and aesthetics. The Cambridge library has the ancient picture where Nichomachus is shown together with other great persons in this field: Pythagoras, Plato and Boeticus (http://www.jcsparks.com/painted/boethius.html). One can find more details about the triangle by Nichomachus of Gerasa in [374, 375]. This unexpected connection of times makes additionally probable the adequacy of the matrix research of genetic systems and the assumed connection of genetic systems with the Pythagorean musical scale, reflected unconsciously in Nichomachus' triangle.

The set of numbers in each genetic matrix $[3, 2; 2, 3]^{(n)}$ reproduces fragments of the geometric progressions with the quint factor 3/2. By this reason, sequences of such kinds of numbers (13.1) can be compared to quint sequences of musical notes in quint musical tunings (tritonic, tetratonic, pentatonic, ...) mentioned above and including the 7-stage Pythagorean musical tuning (Fig. 13.1.1). A similar comparison was already made in this section for analogical series of numbers generated by the products of numbers of hydrogen bonds in *n*-plets. The matrix [3, 2; 2, 3] is the matrix representation of hyperbolic number 3+2j, where j is the imaginary unit of hyperbolic numbers satisfying the condition $j^2 = +1$ [58]. The tensor family of matrices $[3, 2; 2, 3]^{(n)}$ represents the family of corresponding 2^n-dimensional extensions of this hyperbolic number, which were proposed to use in algebraic-vector analysis of musical works [80].

13.3. Quint matrices, the golden ratio, and golden matrices

In biology, a genetic system provides the self-reproduction of biological organisms in their generations. In mathematics, the "golden ratio" (or "golden section", "divine proportion") and its properties were a mathematical symbol of self-reproduction from the Renaissance, and they were studied by Leonardo da Vinci, Kepler and many other prominent thinkers (see details in [13, 420]).

Is there any connection between these two systems? Yes, we demonstrate such unexpected connection. The golden section is the value $\varphi = (1+5^{0.5})/2 = 1.618\ldots$ (sometimes the inverse of this value is called the golden section in literature). If the simplest quint genetic matrix [3 2; 2 3] is raised to the power 1/2 in the ordinary sense (that is, if we take the square root), the result is the matrix $[\varphi, \varphi^{-1}; \varphi^{-1}, \varphi] = [3, 2; 2, 3]^{1/2}$, the matrix elements of which are equal to the golden section and its inverse value. This new matrix represents the hyperbolic number $\varphi + j \times \varphi^{-1}$. And if any other quint matrix $[3, 2; 2, 3]^{(n)}$ is raised to the power ½ in the ordinary sense, the result is the matrix $[\varphi, \varphi^{-1}; \varphi^{-1}, \varphi]^{(n)} = ([3, 2; 2, 3]^{(n)})^{1/2}$, the matrix elements of which are equal to the golden section in various integer powers with elements of symmetry among these powers (Fig. 13.3.1).

A ratio between adjacent numbers in numerical sequences inside each of such matrices $[\varphi, \varphi^{-1}; \varphi^{-1}, \varphi]^{(n)}$ (for example, $\ldots\varphi^3, \varphi^1, \varphi^{-1}, \varphi^{-3}\ldots$) is equal to φ^2 (or φ^{-2}) always. The same ratio φ^2 exists in regular 5-stars (Fig. 13.3.2) as a ratio between sides of the adjacent stars entered in each other. Below we will use the name "pentagram musical scales" for new musical scales connected with the golden matrices. In view of this, let us

$$[3\ 2;\ 2\ 3]^{1/2} = \begin{vmatrix} \varphi & \varphi^{-1} \\ \varphi^{-1} & \varphi \end{vmatrix} ; \quad ([3\ 2;\ 2\ 3]^{(2)})^{1/2} = \begin{vmatrix} \varphi^{2} & \varphi^{0} & \varphi^{0} & \varphi^{-2} \\ \varphi^{0} & \varphi^{2} & \varphi^{-2} & \varphi^{0} \\ \varphi^{0} & \varphi^{-2} & \varphi^{2} & \varphi^{0} \\ \varphi^{-2} & \varphi^{0} & \varphi^{0} & \varphi^{2} \end{vmatrix}$$

$([3\ 2;\ 2\ 3]^{(3)})^{1/2} =$

φ^3	φ^1	φ^1	φ^{-1}	φ^1	φ^{-1}	φ^{-1}	φ^{-3}
φ^1	φ^3	φ^{-1}	φ^1	φ^{-1}	φ^1	φ^{-3}	φ^{-1}
φ^1	φ^{-1}	φ^3	φ^1	φ^{-1}	φ^{-3}	φ^1	φ^{-1}
φ^{-1}	φ^1	φ^1	φ^3	φ^{-3}	φ^{-1}	φ^{-1}	φ^1
φ^1	φ^{-1}	φ^{-1}	φ^{-3}	φ^3	φ^1	φ^1	φ^{-1}
φ^{-1}	φ^1	φ^{-3}	φ^{-1}	φ^1	φ^3	φ^{-1}	φ^1
φ^{-1}	φ^{-3}	φ^1	φ^{-1}	φ^1	φ^{-1}	φ^3	φ^1
φ^{-3}	φ^{-1}	φ^{-1}	φ^1	φ^{-1}	φ^1	φ^1	φ^3

Fig. 13.3.1. The beginning of the tensor (Kronecker) family of the golden matrices $[\varphi\ \varphi^{-1}; \varphi^{-1}\ \varphi]^{(n)} = ([3\ 2; 2\ 3]^{(n)})^{1/2}$, where $\varphi = (1+5^{0.5})/2 = 1.618\ldots$ is the golden section.

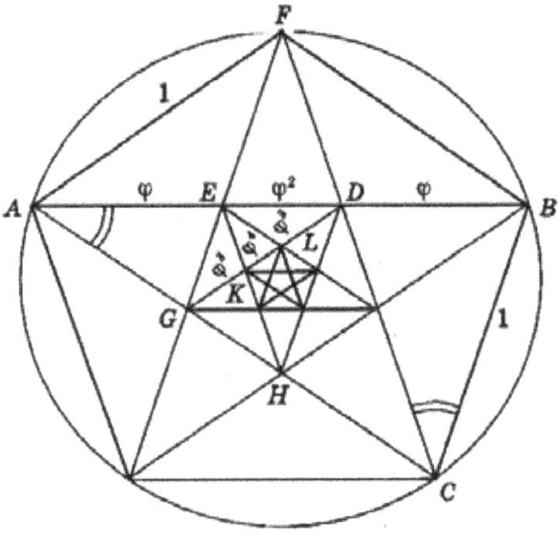

Fig. 13.3.2. Sizes of pentagrams, which are entered in each other, differ by scale factor φ^2 (from [13]).

recall that the pentagram and its metaphysical associations were explored by the Pythagoreans who considered it an emblem of perfection and health. Pythagoreans swore by it and used the pentagram as a distinctive sign of belonging to their community.

The value φ^2 (or φ^{-2}) is also well known in the inherited phyllotaxis laws: the ideal angle in phyllotaxis laws, which is termed as "the Fibonacci angle", is equal to φ^{-2} [142, § 2.2.1]. In Chapter 5, it was presented that the value $\varphi^2/2$, which is named the golden wurf, relates to invariants of proportions of three-block parts for kinematic schemes of the human body at different ages.

The golden section is presented in fivefold-symmetrical objects of biological bodies (flowers, etc.), but are forbidden in classical crystallography. It exists as well in many figures of modern generalized crystallography: quasicrystals by D. Shechtman, R. Penrose's mosaics, dodecahedra of ensembles of water molecules, icosahedral figures of viruses, biological phyllotaxis laws, etc. [420]. The article [421] shows that about 1-nm-wide ice chains that nucleate on metal surface Cu(110) are built from a face-sharing arrangement of water pentagons. The pentagon structure is favored over others because it maximizes water-metal bonding while maintaining a strong hydrogen-bonding network. It reveals an unanticipated structural adaptability of water-ice films.

Unexpected connections were discovered between the golden section and the micro-world of quantum mechanics, which includes genetic molecules. The article [144] describes that the chain of atoms in certain circumstances acts like a nanoscale guitar string. *Science Daily* gives a special title in its information about this discovery: "Golden ratio discovered in quantum world: hidden symmetry observed for the first time in solid state matter". The principal author of this paper R. Coldea noted: "*Here the tension comes from the interaction between spins causing them to magnetically resonate. For these interactions we found a series (scale) of resonant notes: the first two notes show a perfect relationship with each other. Their frequencies (pitch) are in the ratio of 1.618..., which is the golden ratio famous from art and architecture. ... It reflects a beautiful property of the quantum system — a hidden symmetry. Actually quite a special one called E8 by mathematicians, and this is its first observation in a material*" (http://www.sciencedaily.com/releases/2010/01/100107143909.htm).

The theme of the golden ratio in genetic matrices seems to be important because many physiological systems and processes are connected with it. It is known that proportions of the golden ratio characterize many physiological processes: cardiovascular processes, respiratory processes, electric activities of brain, locomotion activity, etc. The golden ratio has been described and investigated for a long time in phenomena of aesthetic perception as well. Considering these facts, the golden ratio should be considered as the candidate for the role of one of the basic elements in an inherited interlinking of physiological subsystems, which provides unity of an organism. The matrix relation between the golden section φ and significant parameters of genetic codes testifies in favor of a molecular-genetic clue providing such physiological phenomena.

13.4. Golden wurf and Fibonacci-stage scales for applications in music creativity

Many theorists of music paid attention to the connection of the structure of many musical compositions of prominent composers with the golden ratio

$\varphi = (1+5^{0.5})/2 = 1.618\ldots$ The results of matrix genetics, which were described above, reveal a new direction of thoughts about a relation between the golden ratio and music because structures of the genetic code are connected with the golden section.

By analogy with a quint matrix $P_{MULT}^{(n)}$, which contains a sequence of (n+1)-kinds of numbers from a geometric progression with the quint factor 3/2, a corresponding golden genetic matrix $\Phi^{(n)}$ (Fig. 13.3.1) contains a sequence of (n+1)-kinds of numbers from a geometric progression, the factor of which is equal to $\varphi^2 = 2.618\ldots$:

$$\Phi^{(1)} \Rightarrow \varphi^1, \varphi^{-1}$$
$$\Phi^{(2)} \Rightarrow \varphi^2, \varphi^0, \varphi^{-2}$$
$$\Phi^{(3)} \Rightarrow \varphi^3, \varphi^1, \varphi^{-1}, \varphi^{-3} \quad (13.3)$$

The previous paragraph demonstrated that the tensor family of the quint matrices is connected with the Pythagorean musical scale. Now we turn to the tensor family of the golden matrices (Fig. 13.3.1) and to the geometric progressions with the factor φ^2. Is it possible to apply the described Pythagorean algorithm to such geometric progressions with factor φ^2 to arrive at a new musical (or mathematical) scale, where only two interval factors exist (as its tone and semitone intervals) by analogy with the Pythagorean musical scale? Investigation of this question seems to be important because such a new scale or scales can be essential for a theory of musical harmony and for attempts to create musical compositions with increased physiological activity.

After research of this question the positive result was obtained: yes, it is possible every time, when you take one of the Fibonacci numbers 2, 3, 5, 8, 13 as the first member of such a geometric progression (the situation becomes more difficult for the greater Fibonacci numbers 21, 34, …). Mathematical scales, which are formed in these cases, possess such quantities of their tone and semitone intervals, which are equal to Fibonacci numbers as well.

Such interrelated Fibonacci-stage scales, each of which has interval factors of two kinds only, were named "the scales of the golden wurf" (or "pentagram scales") [12, 13]. Let us consider the example of the 8-stage scale of the golden wurf. We should construct a new mathematical scale of frequencies, which fills up the octave {1, 2} by means of the Pythagorean algorithm with the irrational factor φ^2 of a geometric progression instead of the quint ratio 3/2. As a result, we should arrive at such a scale which possesses two kinds of interval factors (tone and semitone intervals) only by analogy with the Pythagorean musical scale. One can note that the factor $\varphi^2 = 2.618\ldots$ exceeds the considered interval of the octave {1, 2}. Therefore, it is comfortable to use from the very beginning the twice smaller factor $\varphi^2/2 = p = 1.309\ldots$, the value of which belongs to this octave interval. It is easy to check that the final sequence (13.4) of the wurf scale does not depend on whether we use the factor φ^2 or $\varphi^2/2$, which are equivalent to each other in the given problem. This factor $p = \varphi^2/2$ has been known in the field of investigations of biological symmetries and invariants for a long time under the name of the golden wurf [32, 143]. In Chapter 5 we already discussed the golden wurf in connection with morphogenetic phenomena and bio-symmetries.

Now let us construct the 8-stage scale of the golden wurf by means of the analogue of the described Pythagorean algorithm, using the factor $p = \varphi^2/2$ in the initial geometric progression (instead of the quint factor 3/2). All three steps of the Pythagorean algorithm are reproduced:

1. Taking the first eight members of such a geometric progression with the factor $p = \varphi^2/2$, which begins from the inverse value of this factor: p^{-1}, p^0, p^1, p^2, p^3, p^4, p^5, p^6;
2. Returning into the octave interval {1, 2} for those members of this sequence, values of which overstep the limits of this interval; this returning is made for these values by means of their multiplication or division with the number 2. As a result of this operation, the new sequence is obtained (this sequence can be named "the geometric

progression with return to the octave"): $2 \times p^{-1}$, p^0, p^1, p^2, $p^3/2$, $p^4/2$, $p^5/2$, $p^6/4$;

3. The permutation of these seven members in accordance with their increasing values from 1 up to 2 (the number 2 is included in this sequence as the end of the octave):

$$1,\ p^3/2,\ p^6/4,\ p^1,\ p^4/2,\ 2 \times p^{-1},\ p^2,\ p^5/2,\ 2 \qquad (13.4)$$

This final sequence (13.4) satisfies the initial condition concerning the existence of two kinds of interval factors only. Indeed, it is easy to check that all ratios of adjacent members of this sequence are equal to two values only, which play the role of the interval factors. For this sequence (13.4) the tone interval is $T = p^3/2 = 1.1215\ldots$ and the semitone interval is $S = 4 \times p^{-5} = 1.0407\ldots$ The sequence of these interval factors is T-T-S-T-S-T-T-S. This sequence fills all of the octave in accuracy: $(p^3/2)^5 \times (4 \times p^{-5})^3 = 2$. The quantities of various interval factors are equal to Fibonacci numbers here. The 3 semitone intervals, 5 tone intervals and 8 interval factors all exist here. It is interesting that if we take a non-Fibonacci number (for example, 4, 6 or 9) as the first member of the initial geometric progression on the first step of the Pythagorean algorithm, final sequences arise which have more than two kinds of interval factors.

Let us compare the classical 7-stage Pythagorean musical scale with the obtained 8-stage scale of the golden wurf. Fig. 13.4.1 shows the minimal difference between the sequences (musical scales) of the tone and semitone intervals inside the octave for both scales. The initial and final parts of both sequences coincide completely, and only one additional semitone interval

T	T	S	T		T	T	S
T	T	S	T	S	T	T	S

Fig. 13.4.1. Sequences of interval factors in the 7-stage Pythagorean scale of C major (upper row) and in the 8-scale of the golden wurf. Tone intervals are marked by T, semitone intervals by S.

arises in the middle part of the octave. This additional semitone interval exists because the factor "p" is less than the quint factor.

Using the sequence (13.4) of the intervals, one can construct the sequence of tones (musical notes), which is named the "wurf-scale of C major" by analogy with the Pythagorean scale of C major (Fig. 13.4.2). A choice of frequencies for these tones of the first octave is made in such a way that this scale contains the frequency 440 Hz, which corresponds to note "la" in the Pythagorean scale and in equal temperament scale, and which is used traditionally for tuning in musical instruments. Fig. 13.4.2 compares the Pythagorean 7-step scale C major and 8-stage scale of the golden wurf for the first octave. Considering a minimal difference between both scales, the majority of the notes of the wurf scale are named by analogy with the appropriate notes of the Pythagorean scale but with the letter "m" in the end (for instance, "rem" instead "re"). The additional fifth note is named "pim".

This scale of the golden wurf, which was constructed in connection with parameters of the genetic code, possesses many analogies with the Pythagorean musical scale by their internal symmetries and proportions. Its main difference from the Pythagorean scale is connected with irrational values of its interval factors. This wurf scale could not be constructed by Pythagoreans who did not know irrational numbers. Irrational factors are used also in the modern equal-temperament scale. According to some data, ancient Chinese knew about the equal-temperament scale, but neglected it

260.7 Do_1	293.3 Re	330.0 Mi	347.6 Fa		391.1 Sol	440 La	495.0 Si	521.5 Do_2
256.8 Dom_1	288.0 Rem	323.0 Mim	336.1 Fam	376.8 Pim	392.3 Solm	440 Lam	493.5 Sim	513.6 Dom_2

Fig. 13.4.2. The upper row demonstrates the frequencies of the tones in the 7-stage Pythagorean scale of C major in the first octave. The bottom row demonstrates the frequencies of the tones in 8-stage scale of the golden wurf of C major in the similar octave. Numbers mean frequencies in Hz. The names of the notes are given [12, 13].

preferring the Pythagorean scale, in which they saw cosmic and biological importance.

The history of attempts at creation of new musical scales includes names of many prominent scientists: Kepler, Descartes, Leibniz, Euler, etc. But these authors had no possibility to use the data about the genetic code in their attempts. The above-mentioned Fibonacci-stage scales were initially constructed as pure mathematical series, but then specialists of the Moscow State Tchaikovsky Conservatory found that this mathematical series are interesting from the point of view of their musical harmony. In our opinion, the data about the genetic code allow one to create new musical scales with physiological potentials. Perhaps it is not accidental that the helical structure of the human ear cochlea shows the golden wurf proportion in its three turns as Fig. 5.4.4 in Chapter 5 shows.

The helical structure of the human ear cochlea, having three turns of different lengths, shows — concerning the cross-ratio of lengths of its three turns — practically the value of the golden wurf (Fig. 13.4.3).

The constructed Fibonacci-stage scales (with 5, 8, and 13 stages) and their applications are studied now in the «Center for Interdisciplinary Studies of Musical Creativity» of the Moscow Conservatory as a basis of the new direction in musical culture. Pieces of "genetic music" written in these Fibonacci-stage scales by professional composers of the Conservatory are performed at concerts in Russia and other countries. For example, a video of

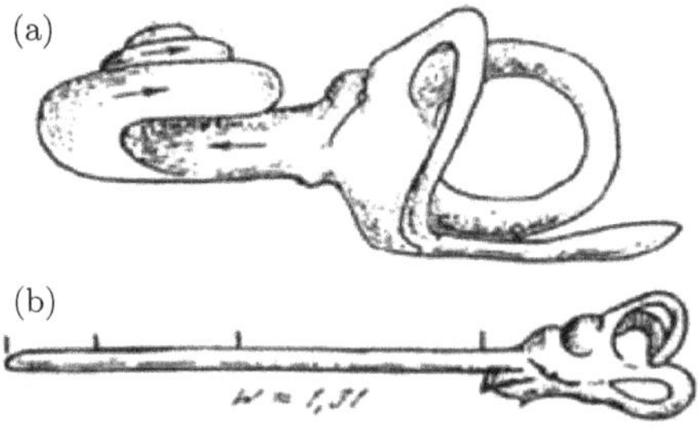

Fig. 13.4.3. The helical structure of the human ear cochlea (from [181]) with the golden wurf proportion concerning the lengths of its three turns: (a) — cochlea coiled into a helix; (b) — cochlea uncoiled into a straight line.

a genetic music concert from the Moscow Conservatory in Vienna, Austria in 2015 one can see on YouTube [http://youtu.be/gagKLDuO9z8]. One can find more details on the theme of genetic musical scales and genetic music in [12, 13, 137].

13.5. Some concluding remarks

The facts described in this chapter about relations of the genetic systems with musical harmony are essential for the problem of genetic bases of aesthetics and inborn feeling of harmony. According to the words of Feynman about the feeling of musical harmony, "*we may question whether we are any better off than Pythagoras in understanding why [stressed] only certain sounds are pleasant to our ear. The general theory of aesthetics is probably no further advanced now than in the time of Pythagoras*" [411, Chapter 50].

The ancient question about the meaning and role of music in the nature of man and the universe excites many modern scientists. For example, A. Bandyopadhyay, analyzing the question "Where does music exist?", argues that universe is an informational fractal and that consciousness is music [422].

A cultural direction of "genetic art" (or, briefly, "genoart") can be developed additionally due to these data of matrix genetics. Genoart has many patterns, which are revealed by matrix genetics, and can be used to create new works of art, design, architectural and musical compositions. For example, the quint genomatrices can be presented in the form of color mosaics if matrix numbers are replaced by colors. It is possible to see regular complications of color mosaics along the family of the genomatrices with an increase of their tensor powers. The discovery of the connection of the genetic code with the golden section shows the molecular-genetic base of many known facts about aesthetic meanings of the golden section. The described facts give new materials for the question about architectural canons, where the golden ratio has been used for a long time; for example, the famous modulor by Le Corbusier ([423]; http://en.wikipedia.org/wiki/

Le_Corbusier) is based on the golden ratio. The pentagram Fibonacci-stage scales can be additionally utilized for architectural proportions (in the role of "pentagram modulor").

Interesting additional data about the special role of the quint ratio 3/2 in parameters of the set of amino acids and in frequencies of electromagnetic influences on biological bodies one can see in the articles [21, 209, 210].

Chapter 14 Geno-Logical Aspects of Genetic Coding

Chapter 2 presented the theme of the algebraic bit-reversible holography connected with binary oppositions in structured alphabets of DNA. Now we continue analysis of holographic-like properties of biological organisms in another way by using the known notion of logic holography based on Walsh functions and binary oppositions, which are connected with the structures of the genetic coding system as it was shown first of all in Chapters 1 and 2.

14.1. The genetic system and logic holography based on Walsh functions

Physical holography, which possesses the highest properties of noise immunity, is based on a record of standing waves from two coherent physical waves of the object beam and of the reference beam. But physical waves can be modeled digitally. Correspondingly noise immunity and other properties of optical and acoustical holography can be modeled digitally, in particular, by using Walsh functions and logic operations concerning dyadic groups of binary numbers. This can be made on the basis of discrete electrical or other signals without any application of physical waves. The pioneer work on «holography by Walsh waves» is [9]. The work was devoted to Walsh waves (or Walsh functions), which propagate through electronic circuits — composed of logical and analog elements — by the analogy with the optical Fourier transform holography. In this digital Walsh holography, objects, whose digital holograms should be made, are represented in the form of 2^n-dimensional vectors. Each component of these vectors corresponds to one of 2^n input channels of appropriate electric circuits; the same is true for 2^n output channels, which are related to components of resulting vectors. Examples of electrical circuits for logic holography are shown in the works [9, 96]. Due to application of Walsh transformations, information about such vectors is written in each component of the appropriate hologram, which is also a 2^n-dimensional vector, to provide non-local characters of storing information.

One should specially note that Walsh functions are closely related to dyadic groups since Walsh functions are characters of the dyadic groups [424]. Therefore, the Fourier analysis on dyadic groups is defined in terms of Walsh functions. In the same way, the discrete Walsh functions are characters of the finite dyadic groups, on which the switching functions are defined. Therefore, the Fourier analysis for switching functions considered as a subset of complex valued functions is formulated in terms of Walsh functions [425].

This digital Walsh holography under the title «logical holography» was also considered later in [96, 426, 427]. All these and other works about logical Walsh holography considered possibilities of its applications in engineering technologies of digital signals processing without any supposition of its application in biology, in particular, in genetics. In our book deep connections of the genetic code system with Walsh functions, Hadamard matrices, dyadic groups and logical modulo-2 addition are described in accordance with [12, 36, 49, 50]. Due to presented materials, one can believe that principles of logical holography and appropriate logic operations are one of the bases for the genetic system.

This leads to a new class of mathematical models of genetic structures and phenomena on the basis of logical holography and appropriate logical operations. Correspondingly the theory of «genetic logical holography» is developed, where mathematics of logical holography and logical operations is used for modeling genetic phenomena. The mathematical basis of this modeling approach is lattice functions, logical operations with them, dyadic spaces, dyadic groups of binary numbers, logic modulo-2 addition, dyadic convolution, dyadic derivatives as well as isomorphic and homomorphic transformations of Walsh in close relation with peculiarities of molecular-genetic systems. In our opinion, a realization of digital-like mechanisms of the logical holography in biological organisms is created by Nature on the basis of their binary computers on resonances [117, 118, 133]. The new kind of mathematics in modeling genetic phenomena gives possibilities of new heuristic associations and new understanding of the natural phenomena. Let us show a few examples of mathematical models and approaches in this new field of research.

14.1.1. *The model of zipper-like reproduction of DNA molecules*

This example № 1 concerns the proposed model of the phenomenon of zipper-like reproduction of DNA molecules. A prerequisite for the preservation of existing genetic information in a series of successive generations of cells and organisms is replication of genetic material in DNA. Replication of DNA strands takes place before each division of normal DNA-containing structures in eukaryotes (nuclei, mitochondria and plastids), and also before each division of bacterial cells and before each reproduction of DNA viruses. Then doubled DNA in the process of segregation is distributed equally between the two daughter cell nuclei or bacterial cells. The process of replication of genetic information and its subsequent dichotomous segregation equally between child objects can be repeated as many times as you like. In this scheme nature provides a very high degree of noise immunity of the transmitted information, which can be compared with the highest noise immunity of information in holographic recordings.

The replication of DNA resembles a zipper mechanism with its teeth, which may cling to each other or unzip from each other step by step. The work [428] considers four molecular bases of DNA as individual teeth of the zipper and studies the mechanism of the DNA zipper by means of the whole isolation of these bases so that there is no interaction with anything else. In this work the authors were able to see, in particular, the frequencies of vibrations of hydrogen bonds in molecules. The DNA zipper has specificity: two pairs of complementary teeth exist in it (complementary bases C-G and A-T). Moreover the pair C-G differs from the pair A-T in two aspects: 1) the pair C-G is characterized by an odd number of hydrogen bonds (3 bonds) and the pair A-T by an even number of hydrogen bonds (2 bonds); 2) the families of triplets, whose first letter is C or G, are opposite to the families of triplets, whose first letter is A or T/U, on the ratio of amounts of triplets with strong and weak roots (3:1 or 1:3 in Fig. 1.4.2 of Chapter 1).

The following model of a vector zipper with its two pairs of complementary teeth can be proposed from logical holography. Let us take the constant Walsh function in the form of a four-dimensional vector V, which is the sum of 4 spare vectors $\bar{C}, \bar{A}, \bar{G}, \bar{T}$ (14.1):

$$\bar{V} = [1,1,1,1] = [1,0,0,0] + [0,1,0,0] + [0,0,1,0] + [0,0,0,1] = 2\bar{C} + 2\bar{A} + 2\bar{G} + 2\bar{T}, \tag{14.1}$$

where $\bar{C} = [0.5,0,0,0]$, $\bar{G} = [0,0,0.5,0]$, $\bar{A} = [0,0.5,0,0]$, $\bar{T} = [0,0,0,0.5]$.

The algorithm of logical holography operates with object vectors and reference vectors by analogy with classical holography, which is based on interference of object waves with reference waves. We will describe this algorithm following the work [96]. The algorithm consists of two parts: 1) generation of logic hologram of a vector signal; 2) restoration of the vector signal from logic hologram.

The generation of logic hologram consists of the following three steps, which are illustrated by us for the case of the vector $\bar{V}$ (14.1).

In the first step, an object vector and a reference vector are taken (both of them have identical 2^n-dimensionality but they differ from each other). Then their sum is considered for further transformations. For our example №1, we will generate holograms for each of the four object vectors $\bar{C}, \bar{A}, \bar{G}, \bar{T}$ from (14.1). For the case of object vectors $\bar{C}$ and $\bar{G}$, the role of the reference vector will play the vector $\bar{P}_1 = [0, 0, 0, 1]$; for the case of object vectors $\bar{A}$ and $\bar{T}$, the role of the reference vector will play the vector $\bar{P}_2 = [1, 0, 0, 0]$ (the reference vectors are more «intensive» than the object vectors by analogy with the classical holography [96]). In general, the non-zero component of the reference vector can be equal to a real number other than 1 but for simplicity here we follow numerical examples in the work, where this component is equal to 1 [96]. In our example № 1 corresponding sums of the object vectors and the reference vectors are equal to:

$$\bar{C}+\bar{P}_1 = [0.5, 0, 0, 1],\ \bar{G}+\bar{P}_1 = [0, 0, 0.5, 1],$$
$$\bar{A}+\bar{P}_2 = [1, 0.5, 0, 0],\ \bar{T}+\bar{P}_2 = [1, 0, 0, 0.5] \tag{14.2}$$

In the second step, the sum of the object vector and the reference vector is multiplied by one of the Hadamard matrices of the order 2^n to receive the Walsh-Hadamard spectrum in relation to this Hadamard matrix. For our example, taking into account the difference between complementary pairs C-G and A-T, we will use the Hadamard (4×4)-matrix $H_{CG} = [1, -1; 1, 1]^{(2)}$ for the vectors $\bar{C}+\bar{P}_1$ and $\bar{G}+\bar{P}_1$; oppositely we will use Hadamard (4×4)-matrix $H_{AT} = [1, 1; 1, -1]^{(2)}$ for the vectors $\bar{A}+\bar{P}_2$ and $\bar{T}+\bar{P}_2$ (14.1). In our example, the Walsh-Hadamard spectra W_C, W_G, W_A and W_C are equal to:

$$W_C = (\bar{C}+\bar{P}_1) \times H_{CG} = [1.5, 0.5, 0.5, 1.5]$$
$$W_G = (\bar{G}+\bar{P}_1) \times H_{CG} = [1.5, 0.5, 1.5, 0.5]$$
$$W_A = (\bar{A}+\bar{P}_2) \times H_{AT} = [1.5, 0.5, 1.5, 0.5]$$
$$W_T = (\bar{T}+\bar{P}_2) \times H_{AT} = [1.5, 0.5, 0.5, 1.5] \tag{14.3}$$

In the third step, a logical hologram is constructed for each of these spectra W_C, W_G, W_A and W_T (14.3) by means of exponentiation in square of each of its components:

$$Hol_C = [2.25, 0.25, 0.25, 2.25]$$
$$Hol_G = [2.25, 0.25, 2.25, 0.25]$$
$$Hol_A = [2.25, 0.25, 2.25, 0.25]$$
$$Hol_T = [2.25, 0.25, 0.25, 2.25] \tag{14.4}$$

The algorithm of restoration of vector signals from logic holograms also consists of three steps. In the first step, the Walsh-Hadamard spectrum of each reference vector is obtained in relation to the Hadamard matrix, which was used at the stage of generating the logical hologram. In our example, these spectra are equal to:

$$\overline{P}_1 \times H_{CG} = [1, 1, 1, 1];$$
$$\overline{P}_2 \times H_{AT} = [1, 1, 1, 1] \qquad (14.5)$$

In the second step, such spectrum should be component-wise multiplied with the appropriate logic hologram. In our example, the results of these actions are the following:

$$\mathrm{Hol}_{C}. \times [1, 1, 1, 1] = [2.25, 0.25, 0.25, 2.25]$$
$$\mathrm{Hol}_{G}. \times [1, 1, 1, 1] = [2.25, 0.25, 2.25, 0.25]$$
$$\mathrm{Hol}_{A}. \times [1, 1, 1, 1] = [2.25, 0.25, 2.25, 0.25]$$
$$\mathrm{Hol}_{T}. \times [1, 1, 1, 1] = [2.25, 0.25, 0.25, 2.25] \qquad (14.6)$$

In the third step, the result of such multiplication should be multiplied by the transposed version of the same Hadamard matrix, and the result must be divided by the numerical order of the matrix. In the result the object vector is restored together with the reference vector but with a new ratio 1/1.25 = 4/5 between their non-zero components by analogy with classical holography [96]. In our example, the result of these actions by using the same Hadamard matrices are the following:

$$R_C = (\mathrm{Hol}_{C}. \times [1, 1, 1, 1] \times (H_{CG}^{T})/4 = [1, 0, 0, 1.25] = [1, 0, 0, 0] + [0, 0, 0, 1.25] = 2 \times \overline{C} + 1.25 \times \overline{P}_1$$

$$R_G = (\mathrm{Hol}_{G}. \times [1, 1, 1, 1] \times (H_{CG}^{T})/4 = [0, 0, 1, 1.25] = [0, 0, 1, 0] + [0, 0, 0, 1.25] = 2 \times \overline{G} + 1.25 \times \overline{P}_1$$

$$R_A = (\mathrm{Hol}_{A}. \times [1, 1, 1, 1] \times (H_{AT}^{T})/4 = [1.25, 1, 0, 0] = [0, 1, 0, 0] + [1.25, 0, 0, 0] = 2 \times \overline{A} + 1.25 \times \overline{P}_2$$

$$R_T = (Hol_T. \times [1, 1, 1, 1] \times (H_{AT}^T)/4 = [1.25, 0, 0, 1] = [0, 0, 0, 1] + [1.25, 0, 0, 1] = 2 \times \overline{T} + 1.25 \times \overline{P}_2 \quad (14.7)$$

Expressions (14.7) show that in the result the object vectors $\overline{C}, \overline{A}, \overline{G}, \overline{T}$ and the reference vectors $\overline{P}_1$ and $\overline{P}_2$ are restored, though with their additional weight coefficients 2 and 1.25, which are not essential for vector processing in electric circuits of logical holography. In classical holography the reference beam also exists at the output. In (14.7) the sum of restored object vectors is equal to the initial Walsh function in [26]: $2 \times \overline{C} + 2 \times \overline{A} + 2 \times \overline{G} + 2 \times \overline{T} =$ [1, 1, 1, 1]. The theory of logical holography, including its noise immunity, is described in the work [96].

For us the following property of replication in logical Walsh holography, which was not noted in previous works of other authors, is interesting especially. In classical holography each separate half of the hologram allows restoring the object beam and reference beam with a certain deterioration of their quality. Let us study now what properties are revealed when, for instance, only the left half of the logical hologram is used for restoring the vector signal?

The left half (or right half) of each logical hologram from (14.4) can be denoted by the symbol beginning with the letter L (or R). Using such symbols, the left and right halves of the holograms (14.4) are represented in the following forms:

$$\begin{aligned} LHol_C &= [2.25, 0.25, 0, 0] & RHol_C &= [0, 0, 0.25, 2.25] \\ LHol_G &= [2.25, 0.25, 0, 0] & RHol_G &= [0, 0, 2.25, 0.25] \\ LHol_A &= [2.25, 0.25, 0, 0] & RHol_A &= [0, 0, 2.25, 0.25] \\ LHol_T &= [2.25, 0.25, 0, 0] & RHol_T &= [0, 0, 0.25, 2.25] \end{aligned} \quad (14.8)$$

The following expressions (14.9a-d, left) show vector signals $SLHol_C$, $SLHol_G$, $SLHol_A$ and $SLHol_T$, which are restored from $LHol_C$, $LHol_G$, $LHol_A$ and $LHol_T$ (14.8) by means of the algorithm of restoration described above. Expressions (14.9a-d, right) show vector signals $SRHol_C$, $SRHol_G$, $SRHol_A$ and

$SRHol_T$, which are restored from $RHol_C$, $RHol_G$, $RHol_A$ and $RHol_T$ (14.8) by means of the same algorithm of restoration. In these expressions D means the matrix of dyadic shifts [0, 0, 1, 0; 0, 0, 0, 1; 1, 0, 0, 0; 0, 1, 0, 0], and E means the identity matrix [1, 0, 0, 0; 0, 1, 0, 0; 0, 0, 1, 0; 0, 0, 0, 1]. The ratio 0.5/0.625 of non-zero components of the restored object vectors to non-zero components of the restored reference vectors is equal again to 4/5.

$$SLHol_C = [0.5, 0.625, 0.5, 0.625] = [0.5, 0, 0.5, 0] + [0, 0.625, 0, 0.625] = \bar{C} + \bar{G} + 0.625 \times (D + E) \times \bar{P}_1$$
$$SRHol_C = [0.5, -0.625, -0.5, 0.625] = [0.5, 0, 0, 0] + [0, 0, 0, 0.625] + [0, -0.625, -0.5, 0] = \bar{C} + 0.625 \times \bar{P}_1 + [0, -0.625, -0.5, 0] \quad (14.9a)$$

$$SLHol_G = [0.5, 0.625, 0.5, 0.625] = [0.5, 0, 0.5, 0] + [0, 0.625, 0, 0.625] = \bar{C} + \bar{G} + 0.625 \times (D + E) \times \bar{P}_1$$
$$SRHol_G = [-0.5, -0.625, 0.5, 0.625] = [0, 0, 0.5, 0] + [0, 0, 0, 0.625] + [-0.5, -0.625, 0, 0] = \bar{G} + 0.625 \times \bar{P}_1 + [-0.5, -0.625, 0, 0] \quad (14.9b)$$

$$SLHol_A = [0.625, 0.5, 0.625, 0.5] = [0, 0.5, 0, 0.5] + [0.625, 0, 0.625, 0] = \bar{A} + \bar{T} + 0.625 \times (D + E) \times \bar{P}_2$$
$$SRHol_A = [0.625, 0.5, -0.625, -0.5] = [0, 0.5, 0, 0] + [0.625, 0, 0, 0] + [0, 0, -0.625, -0.5] = \bar{A} + 0.625 \times \bar{P}_2 + [0, 0, -0.625, -0.5] \quad (14.9c)$$

$$SLHol_T = [0.625, 0.5, 0.625, 0.5] = [0, 0.5, 0, 0.5] + [0.625, 0, 0.625, 0] = \bar{A} + \bar{T} + 0.625 \times (D + E) \times \bar{P}_2$$
$$SRHol_T = [0.625, -0.5, -0.625, 0.5] = [0, 0, 0, 0.5] + [0.625, 0, 0, 0] + [0, -0.5, -0.625, 0] = \bar{T} + 0.625 \times \bar{P}_2 + [0, -0.5, -0.625, 0] \quad (14.9d)$$

From the left column (14.9a-b) one can see the following. From the left half of the logical hologram Hol_C (14.6) for the object vector $\bar{C}$, this algorithm of logical holography restores the pair of the object vectors $\bar{C} + \bar{G}$ (14.9a).

Also, from the left half of the logical hologram Hol_G (14.6) for the object vector $\bar{G}$ this algorithm restores the same pair of the object vectors $\bar{C} + \bar{G}$ (14.9b). By this reason the pair of the object vectors $\bar{C}$ and $\bar{G}$ is interpreted as an analogue of the complementary pair of cytosine C and guanine G in this modeling approach.

From the left half of the logical hologram Hol_A (14.6) for the object vector $\bar{A}$, this algorithm restores the pair of the object vectors $\bar{A} + \bar{T}$ (14.9c). Also from the left half of the logical hologram Hol_T (14.6) for the object vector $\bar{T}$ this algorithm restores the same pair of the object vectors $\bar{A} + \bar{T}$ (14.9d). By this reason the pair of vectors $\bar{A}$ and $\bar{T}$ is interpreted as an analogue of complementary pair of adenine A and thymine T in this modeling approach.

These results allow proposing the following model of the phenomenon of zipper-like reproduction of DNA molecules (Fig. 14.1.1). The model supposes that biological mechanisms embody the described algorithm of logical holography in this phenomenon.

In the case of a separate DNA chain (Fig. 14.1.1a), each nitrogenous base in the form of appropriate vectors ($\bar{C}, \bar{G}, \bar{A}$, or $\bar{T}$) is read as an appropriate

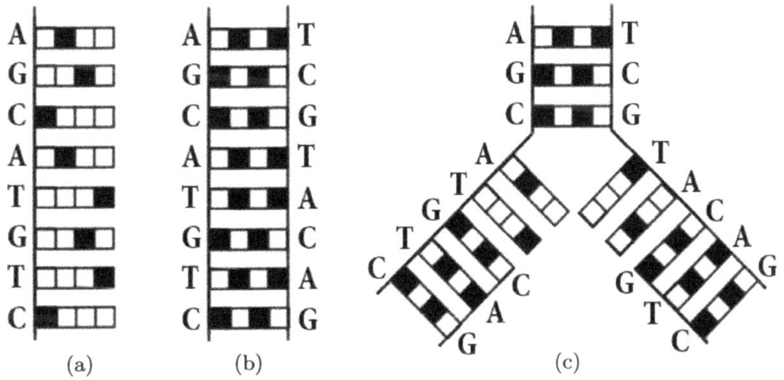

Fig. 14.1.1. The illustration to the model of the zipper-like reproduction of DNA. Nitrogenous bases of DNA are denoted as informational vectors (14.1): $\bar{C} = [0.5, 0, 0, 0]$, $\bar{G} = [0, 0, 0.5, 0]$, $\bar{A} = [0, 0.5, 0, 0]$, $\bar{T} = [0, 0, 0, 0.5]$. These vectors are shown schematically as structures with 4 cells in each, where each black cell means 0.5 and each white cell means 0. (a) A separate DNA chain; (b) two complementary chains of DNA; (c) unzipping two chains of DNA.

logic hologram (14.6), from the left half of which (14.8, left) the described algorithm restores the complementary pair $\bar{C}+\bar{G}$ (19.4b) or $\bar{A}+\bar{T}$ (14.9c). Such stepwise reading of each of the next nitrogenous bases on the DNA chain provides the zipper-like construction of double chains of DNA (14.1.1b).

The second part of this model concerns unzipping of two chains of DNA (Fig. 14.1.1c). In this case each of the nitrogenous bases on a DNA chain is read again as an appropriate logic hologram (14.6), from the right half of which (14.8, right) the described algorithm restores only one of the vectors $\bar{C}$, $\bar{G}, \bar{A}$ or $\bar{T}$ (14.9) without its complementary pair, but with additional vectorial "trash" like [0, −0.625, −0.5, 0] in (14.9a), which does not correspond to any of the considered vectors $\bar{C}$, $\bar{G}, \bar{A}, \bar{T}, \bar{P}_1$ or $\bar{P}_2$. In this model, such result of the restoration means unzipping the complementary pair $\bar{C}+\bar{G}$ or $\bar{A}+\bar{T}$.

14.1.2. *The model of different repetitions in DNA sequences*

This example of applications of algebraic tools in the theory of logic holography in genetics is devoted to phenomena of different repetitions in nucleotide sequences of DNA and RNA where a great number of complementary palindromes and simple repetitions exists [51, 52]. For instance, families of repetitive sequences occupy about one-third of the human genome. The importance of the problem of repeats in genetic sequences is reflected in the fact that 6,000 articles were published on this subject in the 20 years before 1991 [53]. One can propose a new approach for modeling and comparative analyses of the families of repetitive sequences. The proposed approach concerns, first, reproduction of fragments of nucleotide sequences but also some other biological phenomena where similar kinds of repetitions (simple repetitions or with dyadic shifts) are realized.

One should be reminded of the difference in notions of an ordinary and a complementary palindrome. By definition, an ordinary palindrome is a string that reads the same from beginning and from the end. By contrast, a complementary palindrome in molecular genetics is a fragment of a chain of DNA or RNA, which becomes an ordinary palindrome, if each symbol in one half of the fragment is replaced by its complementary symbol ($A \leftrightarrow T$, $C \leftrightarrow G$) [51]. For instance, AGCTCGCGAGCT is a complementary palindrome. In molecular genetics the term «mirror repetition» is sometimes used for ordinary palindromes, which also exist in genetic strings.

One of many applications of classical holography is devoted to the reproduction of numerous copies of flat templates required to produce printed circuits, etc. [429]. In this case a special holographic operation is used, which has its analogue in mathematics under the title "convolution of two functions". For example, if such convolution of a flat image of a triangle with images of three points is made, the result is an image of three triangles [429].

In logical holography the analogue of this operation is the dyadic convolution of two 2^n-dimensional vectors $\overline{X}$ and $\overline{Y}$, which is known in the theory of discrete signal processing and which is calculated by means of the following expression, where $\oplus$ means logical modulo-2 addition [26]:

$$Z(m) = n^{-1} \times \sum_{h=0}^{n-1} X(h)Y(m \oplus h), \quad m = 0,\ 1,\ \ldots,\ n^{-1} \qquad (14.10)$$

Our model uses the dyadic convolution (14.10) of two 2^n-dimensional vectors $\overline{X}$ and $\overline{Y}$, where some components of $\overline{X}$ represent nitrogenous bases A, C, G, and T. Taking into account the opposite character of the bases, which form complementary pairs (A and T, C and G) on two opposite strands of DNA, we assume $T = -A$, $G = -C$. Let us consider dyadic convolution between, for instance, the spare vector $\overline{X} = [A, G, C, A, 0, 0, 0, 0]$ and variants of the reproducing vector $\overline{Y}$ (Table 14.1.1). We call the symbolic part A, G, C, A in $\overline{X}$ as the "template" (or "word", or "kernel"), which can be reproduced in variants by means of its dyadic convolution with variants of the reproducing vector $\overline{Y}$. In general this genetic template may be very different in its length, composition and order of its nitrogenous bases. The reproducing vector $\overline{Y}$ is an analogue of the image of points in the mentioned example from classical

Table 14.1.1. The dyadic convolution of two vectors $\overline{X}$ and $\overline{Y}$.

№	$\overline{X}$	$\overline{Y}$	Result of dyadic convolution
1	[A,G,C,A,0,0,0,0]	[8,0,0,0,0,0,0,−8]	[A,G,C,A,−A,−C,−G,−A] = [A,G,C,A,T,G,C,T]
2	[A,G,C,A,0,0,0,0]	[8,0,0,0,8,0,0,0]	[A,G,C,A,A,G,C,A]
3	[A,G,C,A,0,0,0,0]	[8,0,0,0,0,0,0,8]	[A,G,C,A,A,C,G,A]
4	[A,G,C,A,0,0,0,0]	[8,0,0,0,0,0,8,0]	[A,G,C,A,C,A,A,G]
5	[A,G,C,A,0,0,0,0]	[8,0,0,0,0,8,0,0]	[A,G,C,A,G,A,A,C]
6	[A,G,C,A,0,0,0,0]	[8,0,0,0,−8,0,0,0]	[A,G,C,A,−A,−G,−C,−A] = [A,G,C,A,T,C,G,T]
7	[A,G,C,A,0,0,0,0]	[8,0,0,0,0,−8,0,0]	[A,G,C,A,−G,−A,−A,−C] = [A,G,C,A,C,T,T,G]
8	[A,G,C,A,0,0,0,0]	[8,0,0,0,0,0,−8,0]	[A,G,C,A,−C,−A,−A,−G] = [A,G,C,A,G,T,T,C]

holography about reproduction of images of a single triangle into its many copies. The results of dyadic convolutions of eight-dimensional vectors $\overline{X}$ and $\overline{Y}$ are shown in Table 14.1.1.

These results (Table 14.1.1) show that the operation of dyadic convolution of the same vector $\overline{X}$, which comprises the template AGCT, with different reproducing vectors $\overline{Y}$ generates different types of regularly organized vectors, which correspond to the well-studied types of repetitions in genetic sequences, as well as to their more complex dyadic-shift variants, which previously did not attract attention of researchers as we can judge:

- the row №1 shows an example of obtaining a complementary palindrome;
- the row №2 shows an example of obtaining a simple repetition of the template;
- the row №3 shows an example of an ordinary palindrome (or a mirror repetition);
- the rows №4,5 show examples of non-trivial repetitions, where the second half of vectors repeats the template of the first half but with dyadic-shift permutations of components;
- the row №6 shows an example of a simple complementary repetition, where the second half of the vector contains the complementary repetition of the first half;

- the rows №7,8 show examples of non-trivial complementary repetitions, where the second half of the vectors contain the complementary repetitions of the first half but with dyadic-shift permutations of components.

Recalling the meaning of «dyadic-shift permutations of vector components», Fig. 14.1.2 illustrates such permutations of components in the four-dimensional vector [a_1, a_2, a_3, a_4] by means of the appropriate dyadic-shift matrix (in Chapter 1, Fig. 1.5.1 shows the dyadic-shift matrix for the case of eight-dimensional vectors). Each row of this matrix represents a vector with a dyadic shift of its components in comparison with vectors in other rows (the set of components is identical in all the vectors).

Complementary palindromes, simple repetitions, mirror repetitions, examples of which are shown in rows №№1–3 in Table 14.1.1, are known in nucleotide sequences since they were studied previously in works of different authors [51]. But more complex kinds of repetitions of templates, which are generated by the dyadic convolution and examples of which are shown in rows №№ 4–8, were not studied in nucleotide sequences as we can judge. The theory of genetic logical holography predicts that the last kinds of repetitions also exist in nucleotide sequences since all these known and unknown kinds of repetitions of templates are only different variants of the same mechanism of logical holography and dyadic convolutions. This prediction should be checked in further research.

It is obvious that similar applications of the method of dyadic convolution for reproduction of templates and for generating various repetitions can be used not only for eight-dimensional vectors but also for any 2^n-dimensional vectors for modeling and analysis of genetic sequences. This method of logical

a_1	a_2	a_3	a_4
a_2	a_1	a_4	a_3
a_3	a_4	a_1	a_2
a_4	a_3	a_2	a_1

Fig. 14.1.2. The dyadic-shift matrix for the case of four-dimensional vectors.

holography allows modeling long sequences, in which an initial template is repeated many times in different parts of the sequences. For instance, let us consider the 16-dimensional vector [A,G,C,A,0,0,0,0,0,0,0,0,0,0,0,0] with the template A, G, C, A and the reproducing vector [16,0,0,0,0,0,0,–16,0,0,0,0,16,0,0,0]. The result of their dyadic convolution is the following vector with nitrogenous bases (we assume again that A = –T, C = –G): [A, G, C, A, –A, –C, –G, –A, 0, 0, 0, 0, A, G, C, A] = [A, G, C, A, T, G, C, T, 0, 0, 0, 0, A, G, C, A]. Here in the result we get the new 16-dimensional vector, the first half of which is a complementary palindrome, and the last part is the simple repetition of the template A, G, C, and A. In our opinion, the proposed method of the dyadic convolution can be used in modeling and analysis of phenomena of jumping genes, exons and introns, etc., that is, different phenomena related with transpositions and structural interrelations of fragments of DNA and RNA. Also, we propose to use the method of dyadic convolution for comparison analysis of different genetic sequences, which contain repetitions and which are realized at different levels and branches of biological evolution.

Holographic methods in engineering allow quickly detecting individual elements in a huge image. The theory of genetic logic holography allows assuming that one of the secrets of noise immunity of genetic informatics is based on similar possibilities of genetic holography. Each DNA molecule can be only one element of the whole logic hologram of a certain kind in organisms (inside an organism many different logic holograms can exist for its different organs). In this case, the genetic logic holography — by analogy with classic holography — also allows giving integral estimates of the entire molecule of DNA and promptly compares different molecules of DNA from different parts of an organism. If one DNA molecule mutates, its genetic information — as a result of a comparison of this DNA with other DNA molecules by means of holographic methods — will be found to be incorrect for further usage in the organism.

The above examples, of course, do not exhaust the possibilities of applying the algebraic methods of logical holography to the study of a genetic system. The application of these methods in genetics makes it possible to confirm many assumptions about the implementation of holographic properties in inherited physiological macrostructures, which were previously made by many authors, for example [59, 74, 429].

14.2. The geno-logic code in DNA and epigenetics

In the middle of the twentieth century a great discovery was made about the genetic code as a code correspondence between the set of 64 triplets and the set of 20 amino acids and stop signals. From that time the genetic code is understood as a rule, in accordance with which informational fragments of DNA (or RNA) in the form of triplets define corresponding amino acids and stop codons, that is, biochemical elements and punctuation signs of protein synthesis. This classical biochemical understanding of the genetic code can be named «the first genetic code».

Different authors supposed that other kinds of genetic coding can also exist. For example, a supposition about the histone code is well known (https://en.wikipedia.org/wiki/Histone_code#cite_note-Jenuwein-1). The histone code is a hypothesis that the transcription of genetic information encoded in DNA is in part regulated by chemical modifications to histone proteins, primarily on their unstructured ends. Together with similar modifications such as DNA methylation it is part of the epigenetic code (see, for example, [99]). No mathematical approaches to model such additional kinds of genetic code were proposed.

This section presents the original idea that DNA (and RNA) are not only carriers of the genetic code of amino acids but also they are participants of the second type of genetic encoding that is termed «geno-logic code» [36, 146]. This geno-logic code of DNA and of epigenetic mechanisms is based on systems of Boolean functions, their Walsh-Hadamard spectra and logical operations. DNA is an important part of this integrated coding system, peculiarities of which are reflected in structures of DNA alphabets and in features of the degeneration of the genetic code of amino acids. The proposed approach is related to mathematics that allows creating a wide class of mathematical models in genetics.

The system of the geno-logic code exists in parallel with the genetic code of amino acids and participates in its functioning, we believe. As is known, for the creation of a computer, a usage of material substances for its hardware is not sufficient but logical operations should be also included. These logical operations can successfully work in different hardware from very different materials. The same situation is true for living bodies, where genetic systems should provide genetic information not only about material substances (proteins) but also about logic of interrelated operations in biological processes. One can think that the first genetic code defines material aspects of biological bodies and the geno-logic code defines logic rules and functions of their operating work. In the proposed new modeling approach about the geno-logic code, separate molecular-genetic elements (monoplets, duplets, triplets, etc.) are represented as Boolean functions or systems of these functions.

Existence of the geno-logic code allows you to interpret the work of genetic systems using analogies with computers, which also operate on the basis of Boolean functions and logic operations. As is known, Boolean functions in computers can be realized by means of triggers (the states "yes" or "no"), which can regulate flows of electric current, optical signals, liquids, etc.; flows of different substances can be used for constructing computers to provide systems of Boolean functions playing a key role in them. Achievements of computer technologies and mathematics are used in the proposed direction of bioinformatics research. An important role in bioinformatics research belongs to so-called spectral logic on Walsh-Hadamard spectra from the theory of information processing and digital communication [425, 430, 431]. As is known, spectral methods of development of logical schemes give ways to create devices of manage and communication, which possess effective resources of self-monitoring, self-correction, adaptability and ability of structural adjustment under changes of external conditions (biological organisms possess similar properties).

Spectral methods are easily realized by computers for logical systems of any degree of complexity. Walsh functions and spectra are used in the creation of multi-channel systems of communication, where different signals are simultaneously transmitted via each communication channel (multiplexing).

In computers, logical systems of Boolean functions provide a huge variety of computer functions. By analogy, the geno-logic code of DNA and RNA provides an inheritance of a huge variety of biological functions. In our opinion, it is the code of providing biological variety in the course of biological evolution in which systems of genetic Boolean functions are evolving. We believe that knowledge about the geno-logic code will be useful for understanding those genetic phenomena, which cannot be explained from knowledge of the code correspondence among triplets and amino acids. In our opinion, the epigenetic code is related to the geno-logic code. Genetics can be additionally developed as a science about genetic systems of logic functions. Results of this development can be used not only for deeper understanding of living matter but also for progress in the fields of artificial intellect and artificial life, where mathematical logic plays a key role. In this case those computer systems and theoretical models should be developed, which are based on genetic systems of logic functions, which constitute a relative small part of the infinite number of possible systems of logic functions. As is known, artificial intellect, which possesses an ability of reproducing features of biological intellect, cannot be constructed without usage of mathematical logic [432], and so the idea of the geno-logic code is very natural.

The idea of the geno-logic code was generated due to studying structures of the genetic alphabets and some other inherited biological phenomena taking into account the following facts:

1) The basic alphabet of DNA (and RNA) is closely related with binary numbers since it is represented by the set of 4 specific polyatomic molecular constructions, which bears the symmetric system of pairs of binary-oppositional attributes and forms three binary sub-alphabets

(see Chapter 1). This peculiarity of the alphabet is associated with thoughts about biological computers on the basis of binary-oppositional resonances of genetic molecules [118];

2) The set of alphabets of genetic n-plets is connected with the dyadic groups of $2n$-bit binary numbers ($n = 1, 2, 3, \ldots$). Really each of such dyadic groups contains 2^{2n} members by analogy with a corresponding alphabet of genetic n-plets: the alphabet of monoplets (the nitrogenous bases of DNA or RNA) contains $2^2 = 4$ members; the alphabet of duplets contains $2^4 = 16$ members; the alphabet of triplets contains $2^6 = 64$ members; …;
3) Dyadic groups of binary numbers are deeply related with Walsh functions from the standpoint of mathematical theory of characters on dyadic groups: Walsh functions are the characters of dyadic groups; groups of Walsh functions are isomorphic to appropriate dyadic groups [27, 424, 425]. But phenomenological structures of the alphabets of duplets and triplets in their matrix representations demonstrate their relations with the set of Walsh functions, which includes Rademacher functions as its particular cases;
4) Binary numbers are the basis of Boolean algebra, on which technical computers with their trigger devices work. Hypotheses about analogies between functioning of living organisms and technical computers have existed long ago (see for example [312, 397, 433, 434]). The results, presented in our book, of studying molecular-genetic systems lead to the evidences that genetic systems work on the basis of Boolean algebra; it is important since the level of the genetic code is deeper than secondary levels of inherited nervous systems or separate kinds of proteins such as tubulin, trigger properties of which were used in the theory of consciousness in publications [119, 303, 433, 435]. According to the concept of systemic-resonant genetics [119], binary-oppositional kinds of molecular resonances can be the natural basis of biocomputers working with genetic systems of Boolean functions.

The presented idea of the geno-logic coding is confirmed, in particular, by the well-known physiological law "all-or-none" for excitable tissues: by analogy with triggers in computers, a nerve cell or muscle fiber give only their answers "yes" or "no" under action of different stimulus. If a stimulus is above a certain threshold, a nerve or muscle fiber will fire with full response. Essentially, there will either be a full response or no response at all [389, 390]. A separate muscle, which contains many muscle fibers, can reduce its length to different degrees due to the combined work of the plurality of its muscle fibers. The nervous system can also account for stimulus of a different force due to the combined action of its many nerve fibers (and due to the ability to change the frequency of the generation of nerve impulses at their fixed amplitude).

In view of these data, the human body (and other organisms) can be interpreted as an immense network of trigger devices, which provide systems of logic functions and logic operations on them. In light of this it is not so surprising that the genetic system, that provides transmission of corresponding logic networks along the chain of generations, is also built on the principles of Boolean algebra of logic, including the above-described logic holography based on Walsh functions.

One can note additionally that from the standpoint of the usual genetic code (the table of correspondence between 64 triplets and 20 amino acids) it is unclear why the alphabets of n-plets of DNA and RNA are related with dyadic groups of binary numbers and Walsh functions. But from the standpoint of the geno-logic code, this relationship plays a key role since dyadic groups are closely related not only with Boolean algebra of logic but also with Walsh functions, which are algebraic characters of dyadic groups. By this reason the dyadic-group structure of the set of DNA alphabets connects genetics with Boolean algebra of logic by the natural way.

The genetic code has a strong degeneracy: 64 triplets encode 20 amino acids and a stop signal. Correspondingly a few triplets encode each of the amino acids. By this reason any fragment of a sequence of amino acids in proteins has many different variants of its encoding by different triplets. For example, let us consider the short sequence of 3 amino acids: Ser-Pro-Leu. As is known, the amino acid Ser can be encoded by 6 triplets (TCC, TCT, TCA, TCG, AGC, AGT), the amino acid Pro by 4 triplets (CCC, CCT,

CCA, CCG) and the amino acid Leu by 6 triplets (CTC, CTT, CTA, CTG, TTA, TTG). Due to this, the short sequence Ser-Pro-Leu can be encoded by means of 144 (= $6 \times 4 \times 6$) different variants of a sequence of 3 triplets: TCC-CCC-CTC, TCT-CCT-CTG, AGT-CCT-TTG, etc. If more and more long sequences of amino acids are considered (some proteins have sequences with many thousands of amino acids in them), the number of variants of encoding increases rapidly to astronomical quantities.

Why does living matter require such a tremendously excessive number of encoding options, which can greatly complicate the work of reliable coding system? This is a difficult question without a satisfactory explanation in modern science.

Considering that in the geno-logic code no degeneracy exists (see below), we think that a good answer to this question is the following. DNA (RNA) molecules are carriers of at least two different genetic codes: in one of them triplets encode amino acids, and in the second code the same triplets (or n-plets in general case) encode genetic systems of Boolean functions. In this case, existence of different variants of sequences of triplets, which encode the same amino acid sequence, has a sense since any two different sequences of triplets, which encode the same amino acid sequence, simultaneously encode different systems of Boolean functions. One can suppose that the amino acid sequence, encoded by the first set of triplets, is designed to work in cooperation with a special genetic system of Boolean functions (which is encoded by the first set of triplets); by contrast, the same amino acid sequence, encoded by the second set of triplets, is designed to work in cooperation with another genetic system of Boolean functions (which is encoded by the second set of triplets). This implies that identical proteins are intended for use in different genetic systems of Boolean functions if they are encoded by different sequences of triplets. From this point of view the fate of the protein inside an organism depends on its membership of a particular genetic system of Boolean functions. It leads to the conclusion that genetic systems of Boolean functions dominate over the set of encoded proteins.

Such a standpoint gives new abilities to understand some genetic phenomena. For example, the existence of introns, which do not encode proteins (unlike exons in nucleotide sequences), is explained by the fact that introns encode a realization of genetic systems of Boolean functions without adding amino acids into proteins. The known problem of jumping genes [436] can also be related to the existence of two kinds of genetic code in DNA: jumping of DNA fragments need changing of encoded systems of Boolean functions without changing the amino acid sequence in the encoded protein.

For developing theory of the geno-logic code one needs to study possible variants of natural relations of molecular-genetic alphabets with systems of Boolean functions and their methods of analysis in mathematics, including spectral logic with Walsh functions, Kronecker (or tensor) product and Fibonacci transforms [425, 431, 437]. These known mathematical methods pay special attention to Walsh-Hadamard spectra and dyadic autocorrelation characteristics of systems of Boolean functions. Different classes of Boolean functions exist: linear, self-dual, anti-self-dual, threshold, etc. Knowledge of a class of Boolean functions is important for the synthesis of devices that implement complex logic functions or their systems since for each class specific synthesis methods are known. Dyadic autocorrelation characteristics, which coincide with dyadic convolutions [26], are traditionally used to get this knowledge. Below we will indicate Walsh-Hadamard spectra and autocorrelation characteristics for each case of Boolean functions or their systems in the representation of genetic elements as elements of algebra of logic.

One can recall that for a 2^n-dimensional dimensional vector $\overline{V}$ its Walsh-Hadamard spectrum $\overline{S}$ (in relation to a certain Hadamard matrix H of the same order 2^n) is calculated by means of the expression (14.11):

$$\overline{S} = \overline{V} \times H \tag{14.11}$$

The dyadic autocorrelation characteristic (the dyadic convolution) Z(*m*) of a real-valued function X(*m*), where $m = 0, 1, 2, \ldots, N-1$, is calculated [26] by means of the expression (14.12):

$$Z(m) = \left(\frac{1}{N}\right) \times \sum_{h=0}^{N-1} X(h)X(m \oplus h) \qquad (14.12)$$

where $\oplus$ means the logical operation of modulo-for addition for binary numbers.

Let us describe the first of possible variants of the representation of elements of molecular-genetic alphabets in forms of Boolean functions and their systems for further studying and classifications. In Fig. 1.2.1 of Chapter 1, DNA nucleotides A, C, G, and T were binary enumerated: cytosine C is denoted by number 00 (it bears the attributes "amino" and "pyrimidine"); adenine A by 01 ("amino" and "purine"); thymine T (or uracil U in RNA) by 10 ("keto" and "pyrimidine"); guanine G by 11 ("keto" and "purine").

In the described variant of algebraic-logical representation of molecular elements of the genetic system we will follow the methods of spectral logic, which have been developed for engineering purposes [425, 431]. According to these general methods, in our case the 2-bit inputs X_0X_1, the following Boolean functions $f_C(X_0X_1)$, $f_A(X_0X_1)$, $f_T(X_0X_1)$, and $f_G(X_0X_1)$ can be defined which represent, respectively, cytosine, adenine, thymine and guanine in the form of four-dimensional Boolean vectors (Table 14.2.1).

We now refer to the alphabet 16 of duplets, each of which can be represented as an ordered system of the two individual Boolean functions

Table 14.2.1. Boolean functions $f_C(X_0X_1)$, $f_A(X_0X_1)$, $f_T(X_0X_1)$ and $f_G(X_0X_1)$, which represent cytosine C, adenine A, thymine T, and guanine G.

X_0X_1	$f_C(X_0X_1)$	$f_A(X_0X_1)$	$f_T(X_0X_1)$	$f_G(X_0X_1)$
00	1	0	0	0
01	0	1	0	0
10	0	0	1	0
11	0	0	0	1

from Table 14.2.1, which correspond to nitrogenous bases in the first and second positions of the duplet. Systems of Boolean functions are analyzed and synthesized by so-called step functions Φ, which are associated with trigger states of electrical networks [425, § 1.2]. Step functions Φ are obtained from the lattice functions f(z), representing the studied system of Boolean functions, by means of the following expression:

$$\Phi(z) = f(k), \qquad \text{where } z \in [k, k+1), k = 0, 1, \ldots, n-1, \qquad (14.13)$$

where n means dimensionality of Boolean vectors.

Let us explain this with an example of the duplet CG, for which we construct a lattice function f_{CG} (Table 14.2.2). Each component of the function f_{CG} is equal to the decimal expression of a binary number, composed of individual components of Boolean functions $f_C(X_0X_1)$ and $f_G(X_0X_1)$ from the left side in the same row of the table. The step function Φ_{CG}, which corresponds to the lattice function f_{CG}, is shown on the right side of Table 14.2.2.

The table in Table 14.2.3 shows the data about lattice functions, which represent 16 duplets of DNA, their spectra Walsh-Hadamard (with respect to the Hadamard matrix $[1, 1; 1, -1]^{(2)}$) and dyadic autocorrelation characteristics of the corresponding step functions. Dyadic autocorrelations coincide with dyadic convolutions and are calculated according to (14.12).

By analogy, for triplets and n-plets ($n = 4, 5, 6, \ldots$) one can also construct their representation in the form of systems of Boolean functions on the

Table 14.2.2. The example of the representation of the duplet CG by means of the lattice function $f_{CG} = [2, 0, 0, 1]$ and the corresponding step function Φ (on the right).

$X = X_0X_1$	$f_C(X_0X_1)$	$f_G(X_0X_1)$	f_{CG}
00	1	0	2
01	0	0	0
10	0	0	0
11	0	1	1

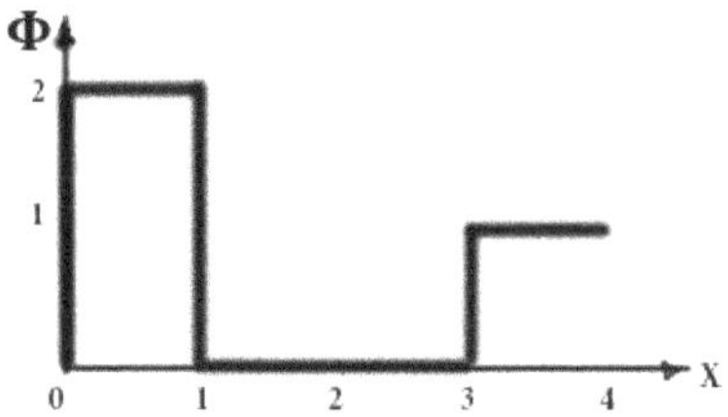

Table 14.2.3. Lattice functions of 16 duplets, their Walsh-Hadamard spectra and dyadic autocorrelations.

Duplet	Lattice function f	Walsh-Hadamard spectrum	Dyadic autocorrelation
CC	[3, 0, 0, 0]	[3, 3, 3, 3]	[2.25, 0, 0, 0]
CA	[2, 1, 0, 0]	[3, 1, 3, 1]	[1.25, 1, 0, 0]
AC	[1, 2, 0, 0]	[3, −1, 3, −1]	[1.25, 1, 0, 0]
AA	[0, 3, 0, 0]	[3, −3, 3, −3]	[2.25, 0, 0, 0]
CT	[2, 0, 1, 0]	[3, 3, 1, 1]	[1.25, 0, 1, 0]
CG	[2, 0, 0, 1]	[3, 1, 1, 3]	[1.25, 0, 0, 1]
AT	[0, 2, 1, 0]	[3, −1, 1, −3]	[1.25, 0, 0, 1]
AG	[0, 2, 0, 1]	[3, −3, 1, −1]	[1.25, 0, 1, 0]
TC	[1, 0, 2, 0]	[3, 3, −1, −1]	[1.25, 0, 1, 0]
TA	[0, 1, 2, 0]	[3, 1, −1, −3]	[1.25, 0, 0, 1]
GC	[1, 0, 0, 2]	[3, −1, −1, 3]	[1.25, 0, 0, 1]
GA	[0, 1, 0, 2]	[3, −3, −1, 1]	[1.25, 0, 1, 0]
TT	[0, 0, 3, 0]	[3, 3, −3, −]	[2.25, 0, 0, 0]
TG	[0, 0, 2, 1]	[3, 1, −3, −1]	[1.25, 1, 0, 0]
GT	[0, 0, 1, 2]	[3, −1, −3, 1]	[1.25, 1, 0, 0]
GG	[0, 0, 0, 3]	[−, −3, −3, 3]	[2.25, 0, 0, 0]

basis of individual Boolean functions of nitrogenous bases. Let us explain this with an example of the triplet CAG, for which we construct a lattice function f_{CAG} (Table 14.2.4). Each component of the function f_{CAG} is equal to the decimal expression of a binary number, composed of components of Boolean functions $f_C(X_0X_1)$, $f_A(X_0X_1)$, and $f_G(X_0X_1)$ from the left side in the same row of the table. The step function Φ_{CAG}, which corresponds to the lattice function f_{CAG}, is shown on the right side of Table 14.2.4.

The table in Table 14.2.5 shows data about lattice functions, which represent 64 triplets of DNA, their Walsh-Hadamard spectra (relative Hadamard matrix $[1, 1; 1, -1]^{(2)}$) and dyadic autocorrelations of corresponding step functions.

The language of systems of Boolean functions can be also used to study the phenomenon of the symmetrical division of the set of 16 duplets into two equal subsets with 8 «strong» duplets and 8 «weak» duplets

Table 14.2.4. The example of the representation of the triplet CAG by means of the lattice function f_{CAG} = [4, 2, 0, 1] and the corresponding step function Φ (on the right).

$X = X_0X_1$	$f_C(X)$	$f_A(X)$	$f_G(X)$	f_{CAG}
00	1	0	0	4
01	0	1	0	2
10	0	0	0	0
11	0	0	1	1

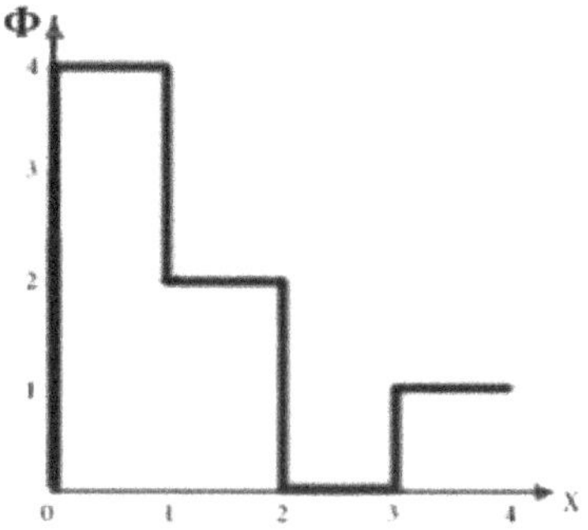

Table 14.2.5. Lattice functions of 64 triplets, their Walsh-Hadamard spectra and dyadic autocorrelations.

Triplet	Its lattice function f	Walsh-Hadamard spectrum	Dyadic autocorrelation
CCC	[7, 0, 0, 0]	[7, 7, 7, 7]	[12.25, 0, 0, 0]
CCA	[6, 1, 0, 0]	[7, 5, 7, 5]	[9.25, 3, 0, 0]
CAC	[5, 2, 0, 0]	[7, 3, 7, 3]	[7.25, 5, 0, 0]
CAA	[4, 3, 0, 0]	[7, 1, 7, 1]	[6.25, 6, 0, 0]
ACC	[3, 4, 0, 0]	[7, −1, 7, −1]	[6.25, 6, 0, 0]
ACA	[2, 5, 0, 0]	[7, −3, 7, −3]	[7.25, 5, 0, 0]
AAC	[1, 6, 0, 0]	[7, −5, 7, −5]	[9.25, 3, 0, 0]
AAA	[0, 7, 0, 0]	[7, −7, 7, −7]	[12.25, 0, 0, 0]
CCT	[6, 0, 1, 0]	[7, 7, 5, 5]	[9.25, 0, 3, 0]
CCG	[6, 0, 0, 1]	[7, 5, 5, 7]	[9.25, 0, 0, 3]
CAT	[4, 2, 1, 0]	[7, 3, 5, 1]	[5.25, 4, 2, 1]
CAG	[4, 2, 0, 1]	[7, 1, 5, 3]	[5.25, 4, 1, 2]
ACT	[2, 4, 1, 0]	[7, −1, 5, −3]	[5.25, 4, 1, 2]
ACG	[2, 4, 0, 1]	[7, −3, 5, −1]	[5.25, 4, 2, 1]
AAT	[0, 6, 1, 0]	[7, −5, 5, −7]	[9.25, 0, 0, 3]
AAG	[0, 6, 0, 1]	[7, −7, 5, −5]	[9.25, 0, 3, 0]
CTC	[5, 0, 2, 0]	[7, 7, 3, 3]	[7.25, 0, 5, 0]
CTA	[4, 1, 2, 0]	[7, 5, 3, 1]	[5.25, 2, 4, 1]
CGC	[5, 0, 0, 2]	[7, 3, 3, 7]	[7.25, 0, 0, 5]
CGA	[4, 1, 0, 2]	[7, 1, 3, 5]	[5.25, 2, 1, 4]

(Continued)

Table 14.2.5. (*Continued*)

Triplet	Its lattice function f	Walsh-Hadamard spectrum	Dyadic autocorrelation
ATC	[1, 4, 2, 0]	[7, −1, 3, −5]	[5.25, 2, 1, 4]
ATA	[0, 5, 2, 0]	[7, −3, 3, −7]	[7.25, 0, 0, 5]
AGC	[1, 4, 0, 2]	[7, −5, 3, −1]	[5.25, 2, 4, 1]
AGA	[0, 5, 0, 2]	[7, −7, 3, −3]	[7.25, 0, 5, 0]
CTT	[4, 0, 3, 0]	[7, 7, 1, 1]	[6.25, 0, 6, 0]
CTG	[4, 0, 2, 1]	[7, 5, 1, 3]	[5.25, 1, 4, 2]
CGT	[4, 0, 1, 2]	[7, 3, 1, 5]	[5.25, 1, 2, 4]
CGG	[4, 0, 0, 3]	[7, 1, 1, 7]	[6.25, 0, 0, 6]
ATT	[0, 4, 3, 0]	[7, −1, 1, −7]	[6.25, 0, 0, 6]
ATG	[0, 4, 2, 1]	[7, −3, 1, −5]	[5.25, 1, 2, 4]
AGT	[0, 4, 1, 2]	[7, −5, 1, −3]	[5.25, 1, 4, 2]
AGG	[0, 4, 0, 3]	[7, −7, 1, −1]	[6.25, 0, 6, 0]
TCC	[3, 0, 4, 0]	[7, 7, −1, −1]	[6.25, 0, 6, 0]
TCA	[1, 1, 4, 0]	[7, 5, −1, −3]	[5.25, 1, 4, 2]
TAC	[1, 2, 4, 0]	[7, 3, −1, −5]	[5.25, 1, 2, 4]
TAA	[0, 3, 4, 0]	[7, 1, −1, −7]	[6.25, 0, 0, 6]
GCC	[3, 0, 0, 4]	[7, −1, −1, 7]	[6.25, 0, 0, 6]
GCA	[2, 1, 0, 4]	[7, −3, −1, 5]	[5.25, 1, 2, 4]
GAC	[1, 2, 0, 4]	[7, −5, −1, 3]	[5.25, 1, 4, 2]
GAA	[0, 3, 0, 4]	[7, −7, −1, 1]	[6.25, 0, 6, 0]
TCT	[2, 0, 5, 0]	[7, 7, −3, −3]	[7.25, 0, 5, 0]
TCG	[2, 0, 4, 1]	[7, 5, −3, −1]	[5.25, 2, 4, 1]
TAT	[0, 2, 5, 0]	[7, 3, −3, −7]	[7.25, 0, 0, 5]
TAG	[0, 2, 4, 1]	[7, 1, −3, −5]	[5.25, 2, 1, 4]
GCT	[2, 0, 1, 4]	[7, −1, −3, 5]	[5.25, 2, 1, 4]
GCG	[2, 0, 0, 5]	[7, −3, −3, 7]	[7.25, 0, 0, 5]
GAT	[0, 2, 1, 4]	[7, −5, −3, 1]	[5.25, 2, 4, 1]
GAG	[0, 2, 0, 5]	[7, −7, −3, 3]	[7.25, 0, 5, 0]
TTC	[1, 0, 6, 0]	[7, 7, −5, −5]	[9.25, 0, 3, 0]
TTA	[0, 1, 6, 0]	[7, 5, −5, −7]	[9.25, 0, 0, 3]
TGC	[1, 0, 4, 2]	[7, 3, −5, −1]	[5.25, 4, 2, 1]
TGA	[0, 1, 4, 2]	[7, 1, −5, −3]	[5.25, 4, 1, 2]
GTC	[1, 0, 2, 4]	[7, −1, −5, 3]	[5.25, 4, 1, 2]

Table 14.2.5. (*Continued*)

Triplet	Its lattice function f	Walsh-Hadamard spectrum	Dyadic autocorrelation
GTA	[0, 1, 2, 4]	[7, −3, −5, 1]	[5.25, 4, 2, 1]
GGC	[1, 0, 0, 6]	[7, −5, −5, 7]	[9.25, 0, 0, 3]
GGA	[0, 1, 0, 6]	[7, −7, −5, 5]	[9.25, 0, 3, 0]
TTT	[0, 0, 7, 0]	[7, 7, −7, −7]	[12.25, 0, 0, 0]
TTG	[0, 0, 6, 1]	[7, 5, −7, −5]	[9.25, 3, 0, 0]
TGT	[0, 0, 5, 2]	[7, 3, −7, −3]	[7.25, 5, 0, 0]
TGG	[0, 0, 4, 3]	[7, 1, −7, −1]	[6.25, 6, 0, 0]
GTT	[0, 0, 3, 4]	[7, −1, −7, 1]	[6.25, 6, 0, 0]
GTG	[0, 0, 2, 5]	[7, −3, −7, 3]	[7.25, 5, 0, 0]
GGT	[0, 0, 1, 6]	[7, −5, −7, 5]	[9.25, 3, 0, 0]
GGG	[0, 0, 0, 7]	[7, −7, −7, 7]	[12.25, 0, 0, 0]

Table 14.2.6. Genetic matrices for nitrogenous bases and duplets of DNA, which reproduce the matrices from Fig. 1.2.1 in Chapter 1 with their cell contents slightly modified. The dyadic-shift numberings of all monoplets and duplets are indicated in brackets.

⊕	0	1
0	C(0)	A(1)
1	T(1)	G(0)

⊕	00	01	10	11
00	CC (00)	CA (01)	AC (10)	AA (11)
01	CT (01)	CG (00)	AT (11)	AG (10)
10	TC (10)	TA (11)	GC (00)	GA (01)
11	TT (11)	TG (10)	GT (11)	GG (00)

(see Chapter 1). This division is related to a degeneration of the genetic code and encoding of amino acids by triplets. Let us represent separate families of 16 duplets using the genetic matrices in Table 14.2.6, where 4 monoplets and 16 duplets have their dyadic-shift numberings, which were introduced due to the binary-oppositional attributes of nitrogenous bases of DNA. Each n-plet has its dyadic-shift numbering produced by the modulo-2 addition of binary numberings of its rows and columns. For example, in the matrix of 16 duplets (Table 14.2.6), the duplet TA belongs to the row enumerated by 10 and the column enumerated by 01; correspondingly, its dyadic-shift numbering is 11, since $10 \oplus 01 = 11$.

Each member of the complementary pair C and G has its dyadic-shift numeration 0 inside the dyadic group of one-digit binary numbers X_0: 0, 1 (Table 14.2.6, left). Each member of the second complementary pair A and T has its dyadic-shift numeration 1 inside the same dyadic group. Correspondingly any member of the complementary pair C and G can be represented by the same Boolean function f_{CG} (two-dimensional Boolean vector) and any member of the second complementary pair A and T can be represented by the same Boolean function f_{AT} (Table 14.2.7).

The set of 16 duplets is divided into 4 subsets, each of which contains 4 duplets with identical two-digit binary numbers (Table 14.2.6, right):

— the subset with binary numeration 00 includes 4 strong duplets CC, CG, GC, GG; each of them can be represented by the lattice function f_{00} of the system of two identical Boolean functions f_{CG} and f_{CG} (Tables 14.2.8 and 14.2.9);

— the subset with binary numeration 01 includes 2 weak duplets CA, GA and 2 strong duplets CT, GA; each of them can be represented by the

Table 14.2.7. Boolean functions f_{CG} and f_{AT}, which represent correspondingly members of the complementary pairs C-G and A-T.

X_0	f_{CG}	f_{AT}
0	1	0
1	0	1

Table 14.2.8. The example of the representation of the subset of duplets CC, CG, GC, GG by means of the lattice function $f_{00} = [3, 0]$.

X_0	f_{CG}	f_{CG}	f_{00}
0	1	1	3
1	0	0	0

Table 14.2.9. Lattice functions of 4 subsets of duplets, their Walsh-Hadamard spectra (relative to the Hadamard matrix [1, 1; 1, −1]) and dyadic autocorrelations.

Subsets of duplets	Lattice function	Walsh-Hadamard spectrum	Dyadic autocorrelation
CC, CG, GC, GG	f_{00} = [3, 0]	[3, 3]	[4.5, 0]
CA, CT, GA, GT	f_{01} = [2, 1]	[3, 1]	[2.5, 2]
AC, TC, AG, TG	f_{10} = [1, 2]	[3, −1]	[2.5, 2]
AA, AT, TA, TT	f_{11} = [0, 3]	[3, −3]	[4.5, 0]

lattice function f_{01} of the system of two different Boolean functions f_{CG} and f_{AT} (Table 14.2.9);

— the subset with binary numeration 10 includes 2 strong duplets AC, TC and 2 weak duplets AG, TG; each of them can be represented by the lattice function f_{10} of two different Boolean functions f_{AT} and f_{CG} (Table 14.2.9);

— the subset with binary numeration 11 includes 4 weak duplets AA, AT, TA, TT; each of them can be represented by the lattice function f_{11} of the system of two identical Boolean functions f_{AT} and f_{AT} (Table 14.2.9).

Table 14.2.8 shows an example of a construction of the lattice function f_{00} of the system of two identical Boolean functions f_{CG} and f_{CG}.

In this spectral-logic approach, genetic lattice functions are connected with genetic information, which is duplicated under mitotic divisions of cells. Does an appropriate mathematical operation exist in the spectral logic for corresponding duplication of genetic lattice functions? Yes, such operation exists and it is based on a simple repetition of Walsh-Hadamard transformations since the square of symmetric Hadamard matrices satisfies the following expression: $([1, 1; 1, -1]^{(n)})^2 = 2^n \times E$, where E is identity matrix, (n) the tensor power, $n = 1, 2, 3, \ldots$ For example, $[x_0, x_1] \times [1, 1; 1, -1]^2 = 2 \times [x_0, x_1]$, that is, this operation duplicates the initial vector $[x_0, x_1]$.

The geno-logic approach gives new abilities to create mathematical models of inherited cyclic processes. In genetics, the principle of selectivity in reading information from DNA, which allows a switching of protein synthesis of one kind into other kinds at different stages of development by means of different variants of information reading from the same DNA, is well known. By analogy a similar principle of selectivity in reading information from genetic lattice functions can be used in a simulation of cyclic processes, which is related with genetic lattice functions. One should recall here that spectral logic analyses and creates digital devices with Boolean signals (0 or 1) in sets of input channels X_i and output channels Y_k. The system of Boolean functions of output channels Y_k defines an appropriate lattice function and its step function, a Walsh-Hadamard spectrum of which is studied in the spectral logic. For example, such lattice function, which can characterize a current stage of an inherited structure, can be represented by the vector [4, 2, 0, 1] from Table 14.2.5. Its Walsh-Hadamard spectrum in relation to the Hadamard matrix $[1, 1; 1, -1]^{(2)}$ is represented by the vector $[4, 2, 0, 1] \times [1, 1; 1, -1]^{(2)} = [7, 1, 5, 3]$. But components of this lattice function can be read in changed order of priorities, generated, for instance, by means of simple cyclic permutations: [2, 0, 1, 4], [0, 1, 4, 2], [1, 4, 2, 0], [4, 2, 0, 1], each of which has its own Walsh-Hadamard spectrum. Such cyclic sequences of genetic lattice functions can be used in a simulation of inherited cyclic processes. Separate cyclic sequences of genetic lattice functions can be combined in different ways into a whole network to model many mutually coordinated cyclic processes. In particular, here one can wait for new knowledge on the biological meaning of those linear codes, which are related to Boolean functions. Additional details on applications of spectral logic to modeling genetic structures are in [36].

Of course, the described representations should be considered as initial materials for further research of genetic systems and biological organisms from the standpoint of Boolean algebra of logic and spectral logic for deeper understanding of living matter and development of "algebra-logical biology". One of the many secrets of living matter is the extremely effective work of enzymes inside living bodies. As is known, biological enzymes can accelerate a chemical reaction in 10^{10}–10^{14} times [10]. We believe that such efficiency of enzymes in biological bodies is related to cooperative algebra-logical mechanisms of the geno-logic coding. Here we additionally mention that there are only 4 kinds of active centers for a great number of enzymes. The set of these 4 kinds can be compared with 4 members of the dyadic group of 2-bit binary numbers 00, 01, 10, 11 by analogy with the case of 4 nitrogenous bases of DNA and RNA, which form complementary pairs, and by analogy with the case of 4 kinds of histones, which also form the complementary pairs H3-H4 and H2A-H2B.

From the described point of view, life is a complex system of logical functions and logical operations, which possesses properties of self-organization, self-development, noise immunity, communications, etc. If the genetic system is closely related with Boolean algebra of logic, then it is natural to think that our sensory systems (visual, auditory, tactile, etc.) get and process information from the world by means of Boolean variables and Boolean functions. George Boole created his mathematics of logic to describe the laws of thought: his book was titled *An Investigation of the Laws of Thoughts*, 1854. The reasoned statement about the existence of the geno-logic code shows that our genetically encoded body is created on the basis of the same laws of logic, on which our thoughts are constructed (the unity of the laws of thought and body). It gives new material for a discussion about an old problem: what is primary — thoughts or matter?

If our thought is built on the same laws on which our body is built, then all intellectual activities of man, including a creation of theoretical sciences, is subordinated to the logical laws of genetics. This intellectual activity is reminiscent of some animals that create instinctively their building structures (such as bees with their honeycombs and spiders with their webs). In this light, human intellectual work also resembles the innate ability of animals to motor movements with congenital coordination of many muscle units, etc. This new direction of biological research and comparative analysis includes, in particular, problems of musicology and so-called "genetic music", which is discussed in Chapter 13.

Here one can return to the mentioned fact that our body is a huge network of triggers of different types, different biological levels, different functionalities and material incarnations, including trigger subnets of genetic molecules, tubulin proteins, muscle fibers and neurons. The network of trigger devices of genetic molecules works to provide genetic information and genetic realization of all other trigger subnets of the body; the trigger subnet of tubulin proteins, which is very important for functional systems of the body tissue; the trigger subnet of muscle fibers, which provides the work of muscles; the trigger subnet of neurons, which is essential for information processing in the nervous system. But all these trigger subnets of an entire organism, which possess their relative autonomy, functional specificity, and different material embodiment, are able to work in concert on the principles of Boolean algebra of logic by analogy with computers.

The trigger network of genetic molecules participates in the genetic realization of all other trigger networks of the body, and — in this sense — it is the most fundamental, which defines algorithms and relationships of all the trigger networks of a whole organism. From this perspective, biological evolution can be represented as a process of self-organization and self-development of systems of biological trigger networks.

With respect to effects of physiotherapy, including methods of ancient oriental medicine (acupuncture, etc.), one can think that medical intervention in the functioning of the organism acts in many aspects through its impact on the network of logical functions of the body, causing its restructuring. This is similar to the situation with a computer, when one click on the desired key of its keyboard is able to change all its work due to activation of another way in its logical networks. We believe that gerontological phenomena of aging and death are also closely interfaced with the system of trigger networks of genetic logical functions.

The desire to reproduce in the technical device the operating principles of natural neural networks has given rise to an important technology of "neural computers". But trigger neural networks are just one of the types of genetically inherited trigger networks of the body (many organisms live perfectly without the nervous system). It seems that a similar desire to reproduce — in technical devices — operating principles of genetically inheritable sets of trigger networks is capable of generating a wider direction of "computers on genetically programmable systems of plenty of trigger networks" or briefly "geno-computers". Geno-computers differ from neurocomputers primarily in the fact that it is a union of many trigger networks endowed with different functions, different levels of autonomy, as well as the different characteristics of functional, spatial and temporal linkages with many other trigger networks of a unified device. By contrast to neurocomputers, which are created as imitators of properties of the nervous system, the creation of geno-computers is designed to imitate properties of biological bodies and their evolution, understood as a genetic process of self-organization and self-development of biological sets of trigger networks, including the emergence of new trigger networks (analogs of networks of neurons, muscle fiber networks, etc.).

14.3. The geno-logic code and Fibonacci numbers

Let us return to Fibonacci numbers, the Fibonacci matrix Q, and phyllotaxis laws of morphogenesis, which were considered in Chapter 3.

In mathematics the matrix Q = [1, 1; 1, 0] is called «the Fibonacci matrix» since its integer power "n" gives matrices with Fibonacci entries only (Fig. 14.3.1). The Fibonacci matrix Q sometimes is called the "growth matrix" since it was used many times in biomathematics for simulation of phyllotaxis phenomena [142]. From the standpoint of the geno-logic approach, it is natural to think that Fibonacci numbers and the Fibonacci matrix Q are related with the geno-logic code and correspondingly they

F_n	0	1	1	2	3	5	8	13	21	34	55	89	144	...
n	0	1	2	3	4	5	6	7	8	9	10	11	12	...

$$Q = \begin{vmatrix} 0, 1 \\ 1, 1 \end{vmatrix}; \qquad Q^n = \begin{vmatrix} F_{n-1}, F_n \\ F_n, \ F_{n+1} \end{vmatrix}$$

Fig. 14.3.1. The series of Fibonacci numbers $F_{n+2} = F_n + F_{n+1}$ and the Fibonacci matrix Q.

$[0, 1] \times Q = [1, 0]$	$[0, 1] \times Q^2 = [1, 1]$	$[0, 1] \times Q^3 = [2, 1]$	$[0, 1] \times Q^4 = [3, 2]$	...
$[1, 0] \times Q = [1, 1]$	$[1, 0] \times Q^2 = [2, 1]$	$[1, 0] \times Q^3 = [3, 2]$	$[1, 0] \times Q^4 = [5, 3]$	...
$[1, 1] \times Q = [2, 1]$	$[1, 1] \times Q^2 = [3, 2]$	$[1, 1] \times Q^3 = [5, 3]$	$[1, 1] \times Q^4 = [8, 5]$	...

Fig. 14.3.2. Multiplications of Boolean vectors with the Fibonacci matrix Q^n.

W_p	$V_0 \times H = [1, 1]$	$V_1 \times H = [2, 0]$	$V_2 \times H = [3, 1]$	$V_3 \times H = [5, 1]$	$V_4 \times H = [8, 2]$	...
p	0	1	2	3	4	...

Fig. 14.3.3. The additive series of Fibonacci vectors W_p, which are Walsh-Hadamard spectra of Fibonacci vectors V_k from Fig. 14.3.2.

have interesting mathematical interrelations with formalisms of spectral logic of Boolean functions described in [425, 431]. Below shows that it is true and such interrelations really exist. Let us demonstrate some examples. Here the following terms are used: vectors, all entries of which are Fibonacci numbers, are called "Fibonacci vectors"; vectors, all entries of which are Boolean variables 0 and 1, are called "Boolean vectors".

Rows of the Fibonacci matrix Q are Boolean vectors [1, 1] and [1, 0]. Multiplication of the $[2 \times 2]$-matrix Q^n with any Boolean two-dimensional non-zero vector [0, 1], [1, 0] or [1, 1] gives a Fibonacci vector (Fig. 14.3.2).

One can see from Fig. 14.3.3 that an additive series of Fibonacci vectors $V_{k+1} = V_k + V_{k-1}$ arises by analogy with the series of Fibonacci numbers (Fig. 14.3.1). In each vector V_k the sum of squares of its components is equal to a Fibonacci number: $F_n^{\,2} + F_{n+1}^{\,2} = F_{2n+1}$.

And what can one say about Walsh-Hadamard spectra of members of this series of Fibonacci vectors V_k in relation to the Hadamard matrix H = [1 1; 1 −1], that is, in other words, what type of vector arises from $V_k \times [1, 1; 1, -1]$? The answer is that Walsh-Hadamard spectra of Fibonacci vectors V_k are new Fibonacci vectors W_p, which form a new additive series $W_{p+1} = W_p + W_{p-1}$ (Fig. 14.3.3). In each vector W_p the sum of squares of its components is equal to a Fibonacci number with factor 2.

And what can one say about dyadic derivatives of J. Gibbs [438], which are used in spectral logic and which are expressed by the following formula:

$$f^{[1]}(x) = \frac{1}{2}\sum_{r=0}^{n-1}(f(x \oplus 2^r) \quad f(x))2^r \tag{14.14}$$

where $f^{[1]}(x)$ is the dyadic derivative of the first order and $f(x)$ is an analyzed lattice function (2^n-dimensional vector). Dyadic derivatives are closely related to Walsh functions [438].

Dyadic derivatives (14.14) of Fibonacci vectors V_k and W_p give new Fibonacci vectors $D(V_k)$ and $D(W_p)$, which form new additive series of Fibonacci vectors (in the case of $D(V_k)$, Fibonacci vectors have the weight factor 1/2) as Fig. 14.3.4 shows.

The wide theme of interrelations between 2^n-dimensional Fibonacci vectors and their dyadic derivatives needs to be published separately. One should mention that the Fibonacci (2×2)-matrix Q can be generalized into Fibonacci matrices of higher orders, for example, by means of Kronecker multiplications of Q with the matrix [1, 1; 1, 1] [332].

And how are Fibonacci vectors related to the spectral logic of systems of Boolean functions, which is described in [425, 431]? Spectral logic considers digital devices with input channels X_0, X_1, ..., X_n and output channels Y_0, Y_1, ..., Y_m. Signals in each of these channels can be only 0 or 1 (Boolean variables). Fig. 14.3.5 (left) shows a simple example of such a digital device and two examples of a typical tabular representation of systems of Boolean functions Y_0, Y_1 and Y_2. For each of 8 possible combinations for values of input Boolean variables X_0, X_1, X_2, each of output Boolean functions Y_0, Y_1 and Y_2 takes one of the possible values 0 or 1. Such correspondence is written

V_k	[1, 0]	[1, 1]	[2, 1]	[3, 2]	[5, 3]	[8, 5]	[13, 8]	...
$D(V_k)$	[1, −1]/2	[0, 0]/2	[1, −1]/2	[1, −1]/2	[2, −2]/2	[3, −3]/2	[5, −5]/2	...
W_p	[1, 1]	[2, 0]	[3, 1]	[5, 1]	[8, 2]	[13, 3]	[21, 5]	...
$D(W_p)$	[0, 0]	[1, −1]	[1, −1]	[2, −2]	[3, −3]	[5, −5]	[8, −8]	...

Fig. 14.3.4. Dyadic derivatives $D(V_k)$ and $D(W_p)$ of vectors V_k and W_p from Figs. 14.3.2 and 14.3.3.

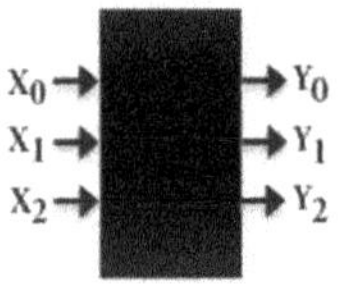

X_0	X_1	X_2	Y_0	Y_1	Y_2	S_1
0	0	0	0	1	1	**3**
0	0	1	0	1	0	**2**
0	1	0	0	0	0	**0**
0	1	1	0	0	0	**0**
1	0	0	0	0	0	**0**
1	0	1	0	0	0	**0**
1	1	0	0	0	0	**0**
1	1	1	0	0	0	**0**

X_0	X_1	X_2	Y_0	Y_1	Y_2	S_2
0	0	0	1	0	1	**5**
0	0	1	0	1	1	**3**
0	1	0	0	0	0	**0**
0	1	1	0	0	0	**0**
1	0	0	0	0	0	**0**
1	0	1	0	0	0	**0**
1	1	0	0	0	0	**0**
1	1	1	0	0	0	**0**

Fig. 14.3.5. Examples of representations of digital devices in spectral logic.

in rows of the table. This system of Boolean functions is characterized by one eight-dimensional vector S, which is called a step function and each component of which is equal to decimal notation of 3-bit binary number $Y_0Y_1Y_2$ in the same row. Walsh-Hadamard spectra of such step functions of systems of Boolean functions (switching functions) are studied in the spectral logic as key characteristics for analysis and synthesis of digital devices.

In Fig. 14.3.5, middle, the first table shows a case of digital devices with the system of only two non-zero Boolean functions Y_1 and Y_2, step function of which is the Fibonacci vector $S_1 = [3, 2, 0, 0, 0, 0, 0, 0]$. Its Walsh-Hadamard spectrum $S_1 \times [1, 1; 1, -1]^{(3)}$ is equal to the Fibonacci vector [5, 1, 5, 1, 5, 1, 5, 1]. The second table (Fig. 14.3.5, right) shows a case of digital devices with the system of three non-zero Boolean functions Y_0, Y_1 and Y_2, the step function of which $S_2 = [5, 3, 0, 0, 0, 0, 0, 0]$ is the Fibonacci vector with increased values of its non-zero components. Its Walsh-Hadamard spectrum $S_2 \times [1, 1; 1, -1]^{(3)}$ is equal to another Fibonacci vector [8, 2, 8, 2, 8, 2, 8, 2].

The Fibonacci step function S_2 is obtained from S_1 by means of the transformation $S_2 = S_1 \times Q_8$, where $Q_8 = ||q_{ij}||$ (here $i, j = 1, 2, \ldots, 8$) is the Fibonacci (8×8)-matrix with the following entries: $q_{11} = 1, q_{12} = 1, q_{21} = 1$, all other entries being equal to zero. It is obvious that for a case of a Fibonacci vector with higher values, for example, of the vector with Fibonacci numbers 55 and 34, we need to appeal to the system with more than three Boolean functions Y_0, Y_1 and Y_2 since the decimal value of the maximal 3-bit binary number 111 is equal to 7. Decimal number 55 is equal to 6-bit binary number

110111. Correspondingly we need a system with 6 Boolean functions Y_0, Y_1, Y_2, Y_3, Y_4 and Y_5 to represent number 55 as a component of a step function in tables of the spectral logic.

From the standpoint of spectral logic, one can model Fibonacci phyllotaxis phenomena on the basis of Fibonacci step functions (or their spectra) of genetic systems of Boolean functions. Iterative actions of Fibonacci matrices on Fibonacci vectors lead to increasing non-zero values in new Fibonacci vectors. To represent these increasing values in step functions of the spectral logic, systems with more and more quantities of Boolean functions should be organized. Such algorithmic growth of systems of Boolean functions can be interpreted as a part of ontogenesis of phyllotaxis structures in living bodies. We believe that the manifestation of Fibonacci numbers on very different levels and branches of biological evolution is related to geno-logic coding. Fibonacci matrices, which are sometimes called matrices of growth, act as breeding matrices for systems of Boolean functions. Fibonacci systems of Boolean functions define some archetypes of physiologic patterns, including phyllotaxis.

In this approach, morphogenetic and other onto- and phylogenetic phenomena are interpreted as consequences of an algorithmic growth of genetic systems of Boolean functions. In engineering technologies, the spectral logic of systems of Boolean functions has wonderful achievements in analysis and synthesis of digital devices, which have elements of artificial intellect for detection and correction of errors; adaptation to the external environment; training in a course of work; interrelation with other digital devices, etc. All these possibilities can now be transferred into the field of study and mathematical modeling of genetically inherited biological phenomena including bio-rhythmic processes and the ability of newborn animals to coordinate movements on the basis of inherited motion algorithms without training (genetic biomechanics). This transfer became possible due to the identification of the structural kinship of molecular-genetic systems with Walsh functions and mathematical formalisms of the spectral logic of systems of Boolean functions. It is this kinship is the foundation of the proposed doctrine about geno-logic coding in living matter, which allows developing spectral-logic genetics by analogy with the theory of digital devices [16, 133].

14.4. Some concluding remarks

This chapter shows that many aspects of genetic structures can be successfully modeled within the framework of the stated concept of geno-logical coding. This concept uses the language of spectral logic and Walsh functions, which have extensive applications in technical systems. These two different areas — biological and technical — of applications of the algebraic apparatus of spectral logic and Walsh functions are able to mutually enrich each other for further development of research in this direction. There are many definitions of life proposed by various authors including Schrodinger, Bohr and I. Prigogine. One of them emphasizes the important role of the genetic system: "*Life is a partnership between genes and mathematics*" [6]. From the standpoint of the theory of genetic logic holography, the following new definition of life seems to be interesting: life is a complex system of genetic Boolean functions and their spectral logic with Walsh-Hadamard transformations and genetic logical holograms. Briefly speaking, life is a system of genetic logic holograms.

Considering the presented results, one can see that genetic sequences are algebra-logic holograms, and in order to read the objective image hidden in them, it is necessary to subject them to a restoring transformation.

Conclusions

Biological bodies have a very complex and variable heterogeneous composition, which makes their mathematical modeling difficult. But at the same time, they have a regular property to pass their traits to their descendants, as it was first expressed in mathematical forms by Mendel back in 1865 and 1866.

Traditionally, mathematics is used in biological research as an auxiliary tool for processing experimental data in contrast to mathematics, which ensured the progress of modern physics, computer science, and other sciences about inanimate objects. But does such mathematics exist, which can generate similar progress in the field of knowledge of living matter by revealing its secrets as a mathematically organized entity? This book presents results confirming that matrix-tensor analysis together with matrix algebra and dyadic groups of binary numbers can serve as a basis of the desirable mathematical language for the genetic coding system and inherited physiological structures.

The materials of this book are aimed at including biology in the field of developed mathematical natural science. This goal is realized through the search and use of a mathematical language adequate for modeling genetic structures. It may be recalled here that mathematical formalisms have creative power. As Heisenberg noted, "*the Pythagoreans seem to have been the first to realize the creative force inherent in mathematical formulations*" [407]. Different areas of natural science may require different mathematical languages. This shows the experience of the formation of quantum mechanics, which required a different mathematical language for the phenomena of the microworld from classical mechanics.

It should be added that since the philosophical works of Martin Heidegger, there exists an idea that language is smarter than us. A rich language has an extensive history of development and effective application to reality. Using this or that language, we indirectly use all experience of its formation and applications to reality. As soon as the language is smarter than us, then a good language can guide us, suggesting new solutions and directions of search. The authors of this book encountered this feature of the well-developed and widely used scientific language of matrix-tensor analysis, which is one of the foundations of modern mathematical natural science: after receiving the first confirmations of its adequacy for modeling a genetic coding system [12, 13], the authors began to exploit its rich and interconnected algebraic capabilities in the new scientific field, that is, in matrix genetics, guided by inner features of this algebraic language and obtaining new and valuable biological results. In other words, we translate genetic structures into this algebraic language while receiving completely new meanings and universal regularities to which this language has brought us.

Applying this approach to genetic structures, for example, the following results are obtained, presented in this book:

- an algebraic realization of the ancient principle "like generates like" in the emergent algebraic properties of a multicomponent system of binary oppositions in genetic coding was discovered; this allows us to consider this principle as one of the key principles in genetics and all biology;
- the universal rules of the stochastic-deterministic organization of DNA nucleotide sequences in the genomes of higher and lower organisms were disclosed; an international project to study new biological symmetries presented by Petoukhov's rules was proposed by the international institute Symmetrion based in Hungary [439];
- the connection of the genetic coding system with algebraic holography, known in digital informatics and built on matrices and dyadic groups of binary numbers, was revealed;
- structural connections of the genetic coding system with the formalisms of the theory of resonances of oscillatory systems with many degrees of freedom are revealed; the concept of multi-resonant genetics was

formulated, correlating with the views of Pauling and other scientists on the important role of resonant interactions in molecular ensembles;
— the energy-information doctrine of evolution on bio-antenna arrays is formulated and substantiated, which allows a fresh look at the genetic code and the role of coordinated energy waves in biological self-organization; this doctrine connects the basis of inherited biological phenomena with amazing properties of antenna arrays used in modern techniques;
— proposed methods for algebraic modeling of biological dualism "stochastics-determinism" based on tensor-unitary transformations;
— new connections of genetic structures with the formalisms of quantum mechanics and quantum informatics are described;
— data on harmonic progression and fractal-like patterns in the information sequences of genomic DNA are presented;
— proposed united-hypercomplex numerical systems for the development of genetic biomechanics;
— the implementation of the relations of musical harmony and the golden section in the parametric organization of DNA information sequences is shown;
— links of the genetic coding system with spectral logic and logical holography on Walsh functions are presented.

The genetic system and inherited physiological structures and functions are closely interrelated. Therefore, the found algebraic language and patterns of the genetic coding system can be considered as the basis for synthesis of algebraic models of various spheres of life, including art with its long-standing attention to the laws of aesthetics of proportions, the aesthetic meaning of the golden section, statements about the connection of all types of arts with music, etc. This theme of the commonality of arts and science is beautifully presented in the famous book of the Nobel laureate G. Hesse *The Glass Bead Game*. This book is devoted to the problem of metatext for the synthesis of all branches of art into one universal art, in which, in particular, musical processes would be expressed by physical and mathematical formulas. This correlates with the statement of many thinkers of different eras about the connection between music and mathematics. It is characteristic that in living nature information at different levels of organization is transmitted in the form of chain-like texts: chains of DNA nucleotides, protein chains of amino acids, sequences of impulses in neurons, literary texts, musical texts, etc. Genomic DNA texts with their gestalt rules described in our book can be considered as metatexts in the analysis of biological texts of various types. Note that the phenomena of Gestalt psychology, which is an integral part of Gestalt biology, work for a variety of arts. In this regard, the totality of the genomic gestalt rules described in our book can provide new approaches to the analysis of the relationship between texts of different arts for the development of structural foundations of universal art in the future, about which Hesse wrote.

In conclusion, let us recall the well-known words of I. Kant that in every natural science theory there is as much truth as there is mathematics in it. The authors of this book have been guided by these words in the algebraic study of genetic phenomena.

Acknowledgments

The authors would like to express our sincere thanks to many of our colleagues who had worked with us in exploring the topics relevant to this book. Their names can be found in the references. Only those literatures closely related to our work are listed in the references. However, due to the large extent of subjects and latest advances of the studies, the references cited in our book are far from complete. Special acknowledgement also to the greatest library that is the World Wide Web, a constant reference source of materials. The authors apologize for any omission.

We want to thank the Mechanical Engineering Research Institute of the Russian Academy of Sciences, Moscow, Russia, and Halmos College of Arts and Sciences of Nova Southeastern University, Fort Lauderdale, USA for their support. The authors would like to acknowledge the help of our colleagues throughout this entire process from the Russian and Hungarian Academies of Sciences, International Symmetry Association, Department of Mathematics of Halmos College of Arts and Sciences of Nova Southeastern University, Moscow State Conservatory and its Center for Interdisciplinary Studies of Musical Creativity, and from the international interdisciplinary seminar "Algebraic Biology and Systems Theory", Moscow.

Special thanks also go to World Scientific, whose contributions throughout the entire process from initial proposal to final publication have been invaluable; especially to Mr. Shaun Tan Yi Jie and his team from World Scientific, who continuously provided us with prompt guidance and support throughout the book editing process.

Finally, we would like to give our special thanks to our families for their love and patience that enabled us to complete this work.

References

[1] Teilhard de Chardin, P. (1999). *The Human Phenomenon.* Sussex Academic.

[2] Capra, F. (2000). *The Tao of Physics: An Exploration of the Parallels Between Modern Physics and Eastern Mysticism.* Shambhala Publications, Inc.

[3] Poincaré, H. (2011). *Science and Hypothesis.* Walter Scott Publishing Co.

[4] May, R. M. (2004). Uses and abuses of mathematics in biology. *Science*, **303**, 790–793.

[5] Wigner, E. (1960). The unreasonable effectiveness of mathematics in the natural sciences. *Commun. Pure Appl. Math.*, **13**(1), 1–14.

[6] Stewart, I. (1999). *Life's Other Secret: The New Mathematics of the Living World.* Penguin.

[7] Fickett, J. W. and Burks, C. (1989). Development of a Database for Nucleotide Sequences. In *Mathematical Methods for DNA Sequences* (Ed. Waterman, M. S.), pp. 1–34. CRC Press.

[8] Chernavskiy, D. S. (2000). The problem of origin of life and thought from the viewpoint of the modern physics (in Russian). *Uspehi Phizicheskih Nauk*, **170**(2), 157–183.

[9] Morita, Y. and Sakurai, Y. (1973). Holography by Walsh waves. *Proceedings of the IV Symposium «Application of Walsh functions».* Catholic University of America, Washington, D. C., pp. 122–126.

[10] Varfolomeev, S. D. (2005). *Chemical Enzymology* (in Russian). Akademia.

[11] Fimmel, E., Danielli, A. and Strüngmann, L. (2013). On dichotomic classes and bijections of the genetic code. *J. Theor. Biol.*, **336**, 221–230.

[12] Petoukhov, S. V. (2008). *Matrix Genetics, Algebras of the Genetic Code, Noise Immunity* (in Russian). Moscow Publishing House.

[13] Petoukhov, S. V. and He, M. (2010). *Symmetrical Analysis Techniques for Genetic Systems and Bioinformatics: Advanced Patterns and Applications.* IGI Global.

[14] Stambuk, N. (1999). Circular coding properties of gene and protein sequences. *Croat. Chem. Acta*, **72**(4), 999–1008.

[15] Nielsen, M. A. and Chuang, I. L. (2010). *Quantum Computation and Quantum Information.* Cambridge University Press.

[16] Knuth, D. (1997). *The Art of Computer Programming. Volume 2: Seminumerical Algorithms*, 3rd edition. Addison-Wesley.

[17] Bellman, R. (1960). *Introduction in Matrix Analysis.* McGraw Hill.

[18] Gazale, M. J. (1999). *Gnomon: From Pharaohs to Fractals.* Princeton University Press.

[19] Hopcroft, J. E. and Ullman, J. D. (1979). *Introduction to Automata Theory, Languages and Computation.* Addison-Wesley.

[20] Frank-Kamenetskii, M. D. (1988). *The Most Important Molecule* (in Russian). Nauka.

[21] Boulay, J. Y. (2022). Numbering of the twenty proteinogenic amino-acids 3/2 ratios inside the genetic code. doi: 10.13140/RG.2.2.13001.83046.

[22] Rumer, Yu. B. (1968). Codon systematization in the genetic code. *Doklady Akademii Nauk SSSR*, **183**(1), 225–226.

[23] Fimmel, E. and Strüngmann, L. (2016). Yury Borisovich Rumer and his 'biological papers' on the genetic code. *Philos. Trans. R. Soc. A*, **374**(2063), 20150228.

[24] Alexits, G. (1961). *Convergence Problems of Orthogonal Series.* Pergamon.

[25] Petoukhov, S. V. (2021). Tensor rules in the stochastic organization of genomes and genetic stochastic resonance in algebraic biology. doi: 10.20944/preprints202110.0093.v1.

[26] Ahmed, N. and Rao, K. (1975). *Orthogonal Transforms for Digital Signal Processing.* Springer.

[27] Harmuth, H. F. (1977). *Sequency Theory: Foundations and Applications.* Academic Press.

[28] Harmuth, H. F. (1989). *Information Theory Applied to Space-time Physics.* The Catholic University of America.

[29] Karzel, H. and Kist, G. (1985). Kinematic Algebras and their Geometries. In *Rings and Geometry* (Eds. Kaya, R., Plaumann, P. and Strambach, K.), pp. 437–509. Springer.

[30] Luneburg, R. (1950). The metric of binocular visual space. *J. Opt. Soc. Am.*, **40**, 627–642.

[31] Kienle, G. (1964). Experiments Concerning the Non-Euclidean Structure of Visual Space. In *Bioastronautics*, pp. 386–400. Pergamon Press.

[32] Petoukhov, S. V. (1989). Non-Euclidean geometries and algorithms of living bodies. *Comput. Math. Appl.*, **17**(4–6), 505–534.

[33] Smolyaninov, V. V. (2000). Spatio-temporal problems of locomotion control. *Uspekhi Fizicheskikh Nauk*, **170**(10), 1063–1128.

[34] Bodnar, O. Ya. (1992). Geometry of phyllotaxis. *Reports of the Academy of Sciences of Ukraine*, (9), 9–15.

[35] Bodnar, O. Ya. (1994). *Golden Ratio and Non-Euclidean Geometry in Nature and Art.* Publishing House "Sweet".

[36] Petoukhov, S. V. (2016). The genetic code, 8-dimensional hypercomplex numbers and dyadic shifts. https://doi.org/10.48550/arXiv.1102.3596.

[37] Liberman, E. A. and Minina, S. V. (1996). Cell molecular computers and biological information as the foundation of nature's laws. *Biosystems*, **38**, 173–177.

[38] Liberman, E. A, Minina, S. V. and Shklovskiy-Kordi, N. E. (1986). The role of the electromechanical and reaction-diffusion system of intraneuronal information processing in brain function (in Russian). *Biofizika*, **31**(2), 298–303.

[39] Kalat, J. W. (2016). *Biological Psychology*, 12th edition. Cengage Learning.

[40] Fayazi, F., Fimmel, E. and Strüngmann, L. (2021). Equivalence classes of circular codes induced by permutation groups. *Theory Biosci.*, **140**(1), 107–121.

[41] Fimmel, E., Michel, Ch. J., Starman, M. and Strüngmann, L. (2018). Self-complementary circular codes in coding theory. *Theory Biosci.*, **137**(1), 51–65.

[42] Fimmel, E., Gumbel, M., Karpuzoglu, A. and Petoukhov, S. (2019). On comparing composition principles of long DNA sequences with those of random ones. *Biosystems*, **180**, 101–108.

[43] Fimmel, E., Michel, C. J., Pirot, F., Sereni, J.-S., Starman, M. and Strüngmann, L. (2020). The relation between k-circularity and circularity of codes. *Bull. Math. Biol.*, **82**, 105.

[44] Fimmel, E., Starman, M. and Strüngmann, L. (2020). Circular tessera codes in the evolution of the genetic code. *Bull. Math. Biol.*, **82**, 48.

[45] Fimmel, E., Michel, Ch. J. and Strüngmann, L. (2017). Diletter circular codes over finite alphabets. *Math Biosci.*, **294**, 120–129.

[46] Chapeville, F. and Haenni, A. L. (1974). *Biosynthese des Proteins* (in French), Hermann Collection, Paris Methodes.

[47] Bank, E. (2022). How much time does it take for a DNA molecule to replicate? https://sciencing.com/much-time-DNA molecule-replicate-21660.html.

[48] Petoukhov, S. V. (2022). The principle "Like Begets Like" in molecular and algebraic-matrix genetics. doi: 10.20944/preprints202211.0528.v2.

[49] Petoukhov, S. V. (2008). Matrix genetics, part 2: the degeneracy of the genetic code and the octave algebra with two quasi-real units (the "Yin-Yang octave algebra"). arXiv:0803.3330.

[50] Petoukhov, S. V. (2008). Matrix genetics, part 3: the evolution of the genetic code from the viewpoint of the genetic octave Yin-Yang-algebra. arXiv:0805.4692.

[51] Gusfield, D. (1997). *Algorithms on Strings, Trees, and Sequences: Computer Science and Computational Biology.* Cambridge University Press.

[52] Lehninger, A. L. (1982). *Principles of Biochemistry*, Vol. 3. Worth Publishers, Inc.

[53] Gribskov, M. and Devereux, J. (1991). *Sequence Analysis Primer.* Stockton Press.

[54] Morsella, E., Bargh, J. A. and Gollwitzer, P. M., Eds. (2009). *Oxford Handbook of Human Action.* Oxford University Press.

[55] Rizzolatti, G. and Sinigaglia, C. (2008). *Mirrors in the Brain. How We Share our Actions and Emotions.* Oxford University Press.

[56] Ferrari, P. F. and Rizzolatti, G. (2014). Mirror neuron research: the past and the future. *Philos Trans. R Soc. Lond. B Biol. Sci.*, **369**(1644), 20130169.

[57] Heyes, C. M. (2010). Where do mirror neurons come from? *Neurosci. Biobehav. Rev.*, **34**(4), 575–583.

[58] Kantor, I. L. and Solodovnikov, A. S. (1989). *Hypercomplex Numbers.* Springer.

[59] Pribram, K. (1971). *Languages of the Brain: Experimental Paradoxes and Principles in Neuropsychology.* Prentice Hall.

[60] Lashley, R. S. (1929). *Brain Mechanisms and Intelligence.* University of Chicago Press.

[61] Beloussov, L.V. (2015). *Morphomechanics of Development.* Springer.

[62] Beloussov, L. V. (1998). *The Dynamic Architecture of a Developing Organism.* Wolters Kluwer.

[63] Beloussov, L. V. (2012). Morphogenesis as a macroscopic self-organizing process. *Biosystems*, **109**, 262–279.

[64] Talbot, M. (1996). *The Holographic Universe.* HarperCollins.

[65] Karp, A. H. (1996). Bit reversal on uniprocessors. *SIAM Rev.*, **38**(1), 1–26.

[66] Lyons, R. (2010). *Understanding Digital Signal Processing*, 3rd edition. Pearson.

[67] Shishmintsev, S. (2012). Holographic properties of a bit-reversal permutation (in Russian). https://habrahabr.ru/post/155471/.

[68] Yang, Q., Ellis, J., Mamakani, K. and Ruskey, F. (2013). In-place permuting and perfect shuffling using involutions. *Inf. Process. Lett.*, **113**(10–11), 386–391.

[69] Petoukhov, S. V. (2022). Binary oppositions, algebraic holography and stochastic rules in genetic informatics. *Biosystems*, **221**, 104760.

[70] Petoukhov, S. V. (2021). Algebraic rules for the percentage composition of oligomers in genomes. doi:10.20944/preprints202101.0360.v3.

[71] Petoukhov, S. V., Petukhova, E. S. and Verevkin, V. V. (2022). Genetic Informatics and the Bit-Reversible Holography: The Problem of Inherited Holographic Features of Organisms. In *Advances in Artificial Systems for Medicine and Education V* (Hu, Z., Petoukhov, S. and He, M., Eds.), pp. 85–94. Springer.

[72] Preskill, J. (2016). Stability, topology, holography: the many facets of quantum error correction. http://theory.caltech.edu/~preskill/talks/APS-March-2016-preskill.pdf.

[73] Pastawski, F., Yoshida, B., Harlow, D. and Preskill, J. (2015). Holographic quantum error-correcting codes: toy models for the bulk/boundary correspondence. *J. High Energ. Phys.*, **2015**, 149.

[74] Pribram, K. (1998). The holographic brain. https://web.archive.org/web/20060518075852/http://homepages.ihug.co.nz/~sai/pribram.htm.

[75] Shiva, M. (2020). Could a hologram from the holographic principle bend spacetime? https://www.quora.com/Could-a-hologram-from-the-holographic-principle-bend-spacetime/answer/Shiva-Meucci.

[76] Strang, G. (1994). Wavelets. *American Scientist*, **82**(3), 250–255.

[77] Petoukhov, S. V. and Verevkin, V. V. (2020). Bisymmetric matrices and symmetrologic researches in genetic biomechanics. *Symmetry*, **31**(1), 105–112.

[78] Petoukhov, S. V. (2017). I-Ching, dyadic groups of binary numbers and the geno-logic coding in living bodies. *Prog. Biophys. Mol. Biol.*, **131**, 354–368.

[79] Petoukhov, S. V. (2019). The genetic code, algebraic codes and double numbers. doi: 10.20944/preprints201911.0301.v2.

[80] Petoukhov, S. V. (2020). Hyperbolic numbers in modeling genetic phenomena. doi: 10.20944/preprints201908.0284.v4.

[81] Petoukhov, S. V. (2020). Hyperbolic Numbers, Genetics and Musicology. In *Advances in Artificial Systems for Medicine and Education V* (Hu, Z., Petoukhov, S. and He, M., Eds.), pp. 195–207. Springer.

[82] Petoukhov, S. V. (2021). Modeling inherited physiological structures based on hyperbolic numbers. *Biosystems*, **199**, 104285.

[83] Geramita, A. V. (1979). *Orthogonal Designs: Quadratic Forms and Hadamard Matrices.* Dekker.

[84] Sklar, B. (2001). *Digital Communication: Fundamentals and Applications.* Prentice Hall.

[85] Yarlagadda, R. and Hershey, J. (1997). *Hadamard Matrix Analysis and Synthesis with Applications to Communications and Signal/Image Processing.* Wolters Kluwer.

[86] Seberry, J., Wysocki, B. J. and Wysocki, T. A. (2005). On some applications of Hadamard matrices. *Metrica*, **62**, 221–239.

[87] Petoukhov, S. V. (2005). Hadamard matrices and quint matrices in matrix presentations of molecular genetic systems. *Symmetry*, **16**(3), 247–266.

[88] Hendy, M. D. and Penny, D. (1989). A framework for the quantitative study of evolutionary trees. *Systematic Zoology*, **38**(4), 297.

[89] Hendy, M. D. and Penny, D. (1993). Spectral analysis of phylogenetic data. *J. Classif.*, **10**(1), 5–24.

[90] Székely, L. A., Erdős, P. L., Steel, M. A. and Penny, D. (1993). A Fourier inversion formula for evolutionary trees. *Appl. Math. Lett.*, **6**(2), 13–16.

[91] Chor, B., Hendy M. D., Holland, B. R. and Penny, D. (2000). Multiple maxima of likelihood in phylogenetic trees. *Mol. Biol. Evol.*, **17**(10), 1529–1541.

[92] Matsen, F. A. and Steel, M. (2007). Phylogenetic mixtures on a single tree can mimic a tree of another topology. *Syst. Biol.*, **56**(5), 767–775.

[93] Harmuth, H. F. (1970). *Transmission of Information by Orthogonal Functions.* Springer.

[94] Harmuth, H. F. (1981). *Nonsinusoidal Waves for Radar and Radio Communication.* Academic Press.

[95] Soroko, L. M. (1979). The sequency analysis in engineering and physics. *Uspekhi Fizicheskih Nauk*, **129**(10), 355–357.

[96] Soroko, L. M. (1974). "Experience" of Young-Walsh and Logical Holography. *Materials of 6th National Conference on Holography*. Leningrad Institute of Nuclear Physics.

[97] Gsponer, A. and Hurni, J.-P. (2008). Quaternions in mathematical physics (2): Analytical bibliography. http://arxiv.org/abs/math-ph/0511092.

[98] Wright, K. (2002). Times of our lives. *Scientific American*, **287**(3), 23–29.

[99] Jenuwein, Th. and Allis, C. D. (2001). Translating the histone code. *Science*, **293**, 1074–1080.

[100] Trahtman, A. M. (1972). *Introduction in Generalized Spectral Theory of Signals* (in Russian). Sovetskoye Radio.

[101] Forrest, S. and Mitchell, M. (1991). The performance of genetic algorithms on Walsh polynomials: Some anomalous results and their explanation. *Proceedings of the Fourth International Conference on Genetic Algorithms*, pp. 182–189.

[102] Geadah, Y. A. and Corinthios, M. J. (1977). Natural, dyadic and sequency order algorithms and processors for the Walsh-Hadamard transform. *IEEE Trans. Comput.*, **C-26**, 435–442.

[103] Goldberg, D. E. (1989). Genetic algorithms and Walsh functions. *Complex Syst.*, **3**(2), 129–171.

[104] Lee, M. H. and Kaveh, M. (1986). Fast Hadamard transform based on a simple matrix factorization. *IEEE Trans. Acoust. Speech Signal Process.*, **34**(6), 1666–1667.

[105] Shiozaki, A. (1980). A model of distributed type associated memory with quantized Hadamard transform. *Biol. Cybern.*, **38**(1), 19–22.

[106] Vose, M. and Wright, A. (1998). The simple genetic algorithm and the Walsh transform. *J. Evol. Comput.*, **6**(3), 253–274.

[107] Waterman, M. S., Ed. (1989). *Mathematical Methods for DNA Sequences.* CRC Press.

[108] Arquès, D. G. and Michel, C. J. (1996). A complementary circular code in the protein coding genes. *J. Theor. Biol.*, **182**, 45–58.

[109] Arques, D. and Michel, Ch. (1997). A circular code in the protein coding genes of mitochondria. *J. Theor. Biol.*, **189**, 45–58.

[110] Frey, G. and Michel, Ch. (2003). Circular codes in archaeal genomes. *J. Theor. Biol.*, **223**, 413–431.

[111] Frey, G. and Michel, Ch. (2006). Identification of circular codes in bacterial genomes. *Comput. Biol. Chem.*, **20**, 87–101.

[112] Nakagaki, T., Yamada, H. and Toth, A. (2000). Intelligence: Maze-solving by an amoeboid organism. *Nature*, **407**, 470.

[113] Goberman, O. (2011). What do mushrooms think? https://efimbog.livejournal.com/601646.html.

[114] Jabr, F. (2012). How brainless slime molds redefine intelligence. https://doi.org/10.1038/nature.2012.11811.

[115] Walecki, N. K. (2021). Can slime molds think? https://www.harvardmagazine.com/2021/11/right-now-can-slime-molds-think.

[116] Petoukhov, S. V. (2021). Algebraic harmony and probabilities in genomes. Long-range coherence in quantum code biology. *Biosystems*, **209**, 104503.

[117] Petoukhov, S. V. (2015). Music and the Modeling Approach to Genetic Systems of Biological Resonances. *Presentation at the International Conference «IS4IS Summit Vienna 2015. The Information Society at the Crossroads»*, 3–7 June 2015, Vienna, Austria. (http://sciforum.net/conference/70/paper/2812).

[118] Petoukhov, S. V. (2016). The system-resonance approach in modeling genetic structures. *Biosystems*, **139**, 1–11.

[119] Petoukhov, S. V. (2018). The concept of resonances in genetics (in Russian). *Biomashsystems*, **2**(4), 169–221.

[120] Schrödinger, E. (1952). Are there quantum jumps? Part I. *Br. J. Philos. Sci.*, **3**(10), 109–123.

[121] Schrodinger, E. (1955). *What is Life? The Physical Aspect of the Living Cell.* Cambridge University Press.

[122] Pauling, L. (1940). *The Nature of the Chemical Bond and the Structure of Molecules and Crystals: An Introduction to Modern Structural Chemistry*, 2nd edition. Oxford University Press.

[123] Cosic, I., Cosic, D. and Lazar, K. (2017). Tesla, Bioresonances and Resonant Recognition Model. *Second International Congress Nikola Tesla*, vol. 1, June 2017, Belgrade, Serbia.

[124] Blekhman, I. I. (2000). *Vibrational Mechanics*. World Scientific.

[125] Ganiev, R. F., Ganiev, S. R., Kasilov, V. P. and Pustovgar, A. P. (2015). *Wave Technology in Mechanical Engineering.* Scrivener Publishing LLC.

[126] Blekhman, I. I. (2003). Vibration "changes the laws of mechanics" (in Russian). http://vivovoco.astronet.ru/VV/JOURNAL/NATURE/11_03/VIBRATION.HTM.

[127] Jakobson, R. (1999). *Texts, Documents, Studies* (in Russian). RGGU.

[128] Gladwell, G. M. L. (2004). *Inverse Problems in Vibration.* Wolters Kluwer.

[129] Balonin, N. A. (2000). *New Course on the Theory of Motion Control* (in Russian). Saint Petersburg State University.

[130] Zharov, V. E. (2006). *Spherical Astronomy* (in Russian). Vek.

[131] Petoukhov, S. V. (2011). Matrix genetics and algebraic properties of the multi-level system of genetic alphabets. *Neuroquantology*, **9**(4), 60–81.

[132] Petoukhov, S. V. (2004). Symmetries in Biology. In *Symmetry in Art* (in Russian, Shubnikov, A. V. and Koptsik, V. A., Eds.), pp. 489–546.

[133] Petoukhov, S. V. and Petukhova, E. S. (2017). The Concept of Systemic-resonance Bioinformatics. Resonances and the Quest for Transdisciplinary. In *Information Studies and the Quest for Transdisciplinarity* (Burgin, M. and Hofkirchner, W., Eds.), pp. 467–487. World Scientific.

[134] Klein, F. (2004). *Elementary Mathematics from an Advanced Standpoint: Arithmetic, Algebra, Analysis.* Dover Publications.

[135] Shultz, E. (1916). Organism as Creativity. In *Questions of Theory and Psychology of Creativity* (in Russian), pp. 108–190.

[136] Shnoll, S. E. (1989). *Physical-Chemical Factors of Biological Evolution* (in Russian). Nauka.

[137] Darvas, G., Koblyakov, A., Petoukhov, S. and Stepanyan, I. (2012). Symmetries in molecular-genetic systems and musical harmony. *Symmetry*, **23**(3–4), 343–375.

[138] Collins, H. (1999). *Concise Dictionary*, 4th edition. Harper Collins.

[139] Shervatov, V. G. (1954). *Hyperbolic Functions.* GITTL.

[140] Stakhov, A. P. (2009). *The Mathematics of Harmony. From Euclid to Contemporary Mathematics and Computer Science.* World Scientific.

[141] Hazewinkle, M. (1994). *Double and Dual Numbers.* Wolters Kluwer.

[142] Jean, R. (2006). *Phyllotaxis: A Systemic Study in Plant Morphogenesis.* Cambridge University Press.

[143] Petoukhov, S. V. (1981). *Biomechanics, Bionics and Symmetry* (in Russian). Nauka.

[144] Coldea, R., Tennant, D. A., Wheeler, E. M., Wawrzynska, D., Prabhakaran, M., Telling, K., Habicht, P. and Smeibidl, K. K. (2010). Quantum criticality in an ising chain: experimental evidence for emergent E8 symmetry. *Science*, **327**(5962), 177–180.

[145] Petoukhov, S. V. (1999). *Biosolitons: Fundamentals of Soliton Biology.* GP Kimrskaya tipographia.

[146] Hu, Z. B., Petoukhov, S. V. and Petukhova, E. S. (2017). I-Ching, dyadic groups of binary numbers and the geno-logic coding in living bodies. *Prog. Biophys. Mol. Biol.*, **131**, 354–368.

[147] Burago, D. Yu., Burago, Yu. D. and Ivanov, S. V. (2004). Kurs metricheskoi geometrii [Metric geometry course].

[148] Chernavsky, A. V. (2010). Krivie i poverkhnosti [Curves and surfaces]. https://higeom.math.msu.su/people/chernavski/chernav-difgeom2010-3.pdf.

[149] Kagan, V. F. (1941). *Fundamentals of Surface Theory* (in Russian). OGIZ.

[150] Poznyak, E. G. and Shikin, E. V. (1990). Differential geometry: first acquaintance (in Russian). Moscow State University.

[151] Rashevsky, P. K. (1964). *Riemannian Geometry and Tensorial Analysis* (in Russian). Nauka.

[152] Sklyarenko, E. G. (2008). *Lecture Course on Classical Differential Geometry* (in Russian). Moscow State University.

[153] Haeckel, E. (1904). *Kunstformen der Natur.* Verlag.

[154] Pogorelov, V. I. (2007). *Structural Mechanics of Thin-walled Structures.* BHV-Petersburg.

[155] Petoukhov, S. V. (2020). The rules of long DNA-sequences and tetra-groups of oligonucleotides. arXiv:1709.04943v6.

[156] Thompson, D. W. (1917). *On Growth and Form.* Cambridge University Press.

[157] Petoukhov, S. V. (1988). Geometries of living nature and self-organization algorithms. In the series *Mathematics and Cybernetics*, M., Znanie, No. 6.

[158] Ramirez, J. P. (2020). The secret of the synchronized pendulums. https://physicsworld.com/a/the-secret-of-the-synchronized-pendulums/.

[159] Voeikov, V. L. (2009). The key role of a stable non-equilibrium state of water systems in bioenergetics. *Russ. Chem. J.*, **53**(6), 41–49.

[160] Pollack, G. H. and Clegg, J. (2008). Unexpected Linkage Between Unstirred Layers, Exclusion Zones, and Water. In *Phase Transitions in Cell Biology* (Pollack, G. H. and Chin, W. C., Eds.), pp. 143–152. Springer.

[161] Igamberdiev, A. U. (2012). Biomechanical and coherent phenomena in morphogenetic relaxation processes. *Biosystems*, **109**, 336–345.

[162] Rossing, T. D. (1982). Chladni's law for vibrating plates. *Am. J. Phys.*, **50**, 271–274.

[163] Jenny, H. (2002). *Cymatics: A Study of Wave Phenomena and Vibration.* Newmarket.

[164] Reid, J. S. (2021). Cymatics and cymascope. http://www.cymascope.com/index.html, https://www.dailymail.co.uk/sciencetech/article–3346588/What-dolphins-REALLY-Stunning-image-shows-diver-eyes-mammal-reveals-animals-send-images-water-other.html.

[165] Reid, J. S. (2022). Music Medicine for Mainstream Medicine. In *Comprehensive Multi-professional Approach to the Treatment of Patients with a Focus on the Use of Less Frequent Methods* (Solar, G., Ed.), pp. 293–357. The First Clinic of Acupuncture and Natural Medicine of G. Solar.

[166] Gerber, R. (1996). *Vibrational Medicine: New Choices for Healing Ourselves.* Bear & Co.

[167] Shushardzhan, S. V. and Petoukhov, S. V. (2020). Engineering in the Scientific Music Therapy and Acoustic Biotechnologies. In *Advances in Artificial Systems for Medicine and Education V* (Hu, Z., Petoukhov, S. and He, M., Eds.), pp. 273–282. Springer.

[168] Benzi, R., Sutera, A. and Vulpiani, A. (1981). The mechanism of stochastic resonance. *J. Phys. A*, **14**, 453–457.

[169] Benzi, R. and Sutera, A. (2004). Stochastic resonance in two dimensional Landau Ginzburg Equation. *J. Phys. A*, **37**, L391.

[170] Benzi, R. (2010). Stochastic resonance: from climate to biology. *Nonlin. Process. Geophys.*, **17**, 431–441.

[171] Petoukhov, S. V. (2022). Genetic code, musical harmony, stochastic resonance and the Ancient Chinese book of I-Ching. In *Comprehensive Multi-professional Approach to the Treatment of Patients with a Focus on the Use of Less Frequent Methods* (Solar, G., Ed.), pp. 160–180. The First Clinic of Acupuncture and Natural Medicine of G. Solar.

[172] Anishchenko, V. S., Neiman, A. B, Moss, F. and Shimansky-Geier, L. (1999). Stochastic resonance: noise-enhanced order. *Phys. Usp.*, **42**, 7–36.

[173] Douglass, J., Wilkens, L., Pantazelou, E. and Moss, F. (1993). Noise enhancement of information transfer in crayfish mechanoreceptors by stochastic resonance. *Nature*, **365**, 337–340.

[174] Levin, J. E. and Miller, J. P. (1996). Broadband neural encoding in the cricket cereal sensory system enhanced by stochastic resonance. *Nature*, **380**, 165–168.

[175] Pei, X., Wilkens, L. and Moss, F. (1996). Light enhances hydrodynamic signaling in the multimodal caudal photoreceptor interneurons of the crayfish. *J. Neurophysiol.*, **76**(5), 3002–3011.

[176] Krauss, P., Tziridis, K., Achim, S. A. and Schulze, H. (2018). Cross-modal stochastic resonance as a universal principle to enhance sensory processing. *Front. Neurosci.*, **12**, 578.

[177] Bezrukov, S. and Vodyanoy, I. (1997). Stochastic resonance at the single-cell level. *Nature*, **388**, 633.

[178] Moss, F. (1997). Stochastic resonance at the molecular level. *Biophys. J.*, **73**, 2249–2250.

[179] Wiesenfeld, K. and Moss, F. (1995). Stochastic resonance and benefits of noise: from ice ages to crayfish and SQUIDs. *Nature*, **373**, 33–36.

[180] Petoukhov, S. V., Svirin, V. I. and Khazina, L. V. (2015). Bionics of spiral structures. *J. Mach. Manuf. Reliab.*, **44**(3), 249–253.

[181] Cook, T. A. (1914). *The Curves of Life.* Constable and Co.

[182] Ji, S. (2016). *Molecular Theory of the Living Cell: Concepts, Molecular Mechanisms, and Biomedical Applications.* Springer.

[183] Ji, S. (2017). *The Cell Language Theory: Connecting Mind and Matter.* World Scientific.

[184] Chatterjee, P., Biswas, S. and Chakraborty, T. (2020). Hydrogen bonding effects on vibrational dynamics and photochemistry in selected binary molecular complexes. *J. Indian Inst. Sci.*, **100**, 155–165.

[185] Chargaff, E., Lipshitz, R. and Green, C. (1952). Composition of the deoxypentose nucleic acids of four genera of sea urchin. *J. Biol. Chem.*, **195**(1), 155–160.

[186] Chargaff, E. (1971). Preface to a grammar of biology: A hundred years of nucleic acid research. *Science*, **172**, 637–642.

[187] Watson, J. D. and Crick, F. H. C. (1953). A structure for deoxyribose nucleic acid. *Nature*, **171**, 737–738.

[188] Albrecht-Buehler, G. (2006). Asymptotically increasing compliance of genomes with Chargaff's second parity rules through inversions and inverted transpositions. *Proc. Natl. Acad. Sci.*, **103**(47), 17828–17833.

[189] Perez, J. C. (2010). Codon populations in single-stranded whole human genome DNA are fractal and fine-tuned by the Golden Ratio 1.618. *Interdiscip. Sci.*, **2**(3), 228–240.

[190] Shporer, S., Chor, B., Rosset, S. and Horn, D. (2016). Inversion symmetry of DNA k-mer counts: validity and deviations. *BMC Genomics*, **17**(1), 696.

[191] Prabhu, V. V. (1993). Symmetry observation in long nucleotide sequences. *Nucleic Acids Res.*, **21**, 2797–2800.

[192] Mascher, M., Schubert, I., Scholz, U. and Friedel, S. (2013). Patterns of nucleotide asymmetries in plant and animal genomes. *BioSystems*, **11**, 181–189.

[193] Nikolaou, Chr. and Almirantis, Y. (2006). Deviations from Chargaff's second parity rule in organellar DNA. Insights into the evolution of organellar genomes. *Gene*, **381**, 34–41.

[194] Rapoport, A. E. and Trifonov, E. N. (2012). Compensatory nature of Chargaff's second parity rule. *J. Biomol. Struct. Dyn.*, **31**(11), 1324–1336.

[195] Rosandic, M., Vlahovic, I., Gluncic, M. and Paar, V. (2016). Trinucleotide's quadruplet symmetries and natural symmetry law of DNA creation ensuing Chargaff's second parity rule. *J. Biomol. Struct. Dyn.*, **34**(7), 1383–1394.

[196] Yamagishi, M. E. B. (2017). *Mathematical Grammar of Biology*. Springer.

[197] Okamura, K., Wei, J. and Scherer, S. W. (2007). Evolutionary implications of inversions that have caused intra-strand parity in DNA. *BMC Genomics*, **8**, 160.

[198] Petoukhov, S. V. (2020). Hyperbolic rules of the oligomer cooperative organization of eukaryotic and prokaryotic genomes. doi:10.20944/preprints202005.0471.v2.

[199] McFadden, J. and Al-Khalili, J. (2018). The origins of quantum biology. *Proc. Royal Soc. A*, **474**(2220), 1–13.

[200] American Institute of Physics (1999). The secret nature of hydrogen bonds. www.sciencedaily.com/releases/1999/01/990121074852.htm.

[201] Shan, S. O. and Herschlag, D. (1999). Hydrogen bonding in enzymatic catalysis: analysis of energetic contributions. *Methods Enzymol.*, **308**, 246–276.

[202] Trylska, J., Grochowski, P. and McCammom, J. A. (2004). The role of hydrogen bonding in the enzymatic reaction catalyzed by HIV-1 protease. *Protein Sci.*, **13**(2), 513–528.

[203] Zhao, H., Tan, Y., Zhang, L., *et al.* (2020). Ultrafast hydrogen bond dynamics of liquid water revealed by terahertz-induced transient birefringence. *Light Sci. Appl.*, **9**, 136.

[204] Petoukhov, S. V. (2013). The genetic system and algebras of projection operators. *Presentation at the International Conference «Theoretical approaches to bioinformation systems»*, Belgrad, Serbia, 17–22 September 2013, http://www.tabis2013.ipb.ac.rs/TABIS2013_pdf_presentations/PETOUKHOV%20BELGRAD%202013%20dvtEdit.pdf.

[205] Petoukhov, S. V. (2014). Matrix genetics: algebra of projection operators, cyclic groups and inherited ensembles of biological cycles. In *Proceedings of 2nd International Conference Theoretical Approaches to BioInformation Systems*, pp. 189–204.

[206] Petoukhov, S. V. (2018). The genetic coding system and unitary matrices. doi: 10.20944/preprints201804.0131.v2.

[207] Crow, J. (1999). Hardy, Weinberg and language impediments. *Genetics*, **152**(3), 821–825.

[208] Gillespie, J. H. (2004). *Population Genetics: A Concise Guide*, 2nd edition. The Johns Hopkins University Press.

[209] Geesink, H. J. H. and Meijer, D. K. F. (2016). Quantum wave information of life revealed. *NeuroQuantology*, **14**(1), 106–125.

[210] Geesink, H. J. H. and Meijer, D. K. F. (2018). Mathematical structure for electromagnetic frequencies that may reflect pilot waves of Bohm's implicate order. *J. Mod. Phys.*, **9**, 851–897.

[211] Hofstadter, D. R. (1980). *Gödel, Escher, Bach: An Eternal Golden Braid.* Vintage Books.

[212] Nalimov, V. V. (2015). *I Am Scattering Thoughts* (in Russian). Moscow Center for Humanitarian Initiatives.

[213] Stent, G. S. (1969). *The Coming of the Golden Age.* The Natural Histore Press.

[214] Jacob, F. (1974). Le modele linguistique en biologie, Critique, Mars, tome XXX, № 322, pp. 197–205.

[215] Gurvich, A. G. (1977). *Selected Works*. Moscow Medicine.

[216] Bernstein, N. A. (1967). *The Co-ordination and Regulation of Movements*. Pergamon Press.

[217] Gastev, A. K. (1924). *Labor Installations* (in Russian). TsIT.

[218] Vetter, R. J. and Weinstein, S. (1967). The history of the phantom in congenitally absent limbs. *Neuropsychologia*, **5**, 335–338.

[219] Weinstein, S. and Sersen, E. A. (1961). Phantoms in cases of congenital absence of limbs. *Neurology*, **11**, 905–911.

[220] Weyl, H. (2016). *Symmetry.* Princeton University Press.

[221] Shubnikov, A. V. and Koptsik, V. A. (2004). *Symmetry in Science and Art*, 3rd edition. Moscow Publishing House.

[222] Engström, A. and Strandberg, B., Eds. (1968). *Symmetry and Function of Biological Systems at the Macromolecular Level. Proceedings of the Eleventh Nobel Symposium*, August 26–29, 1968 at Södergarn, Lidingö, in the County of Stockholm.

[223] Zenkevich, L. A., Ed. *The Animal Life* (in Russian). Prosvesheniye.

[224] Haeckel, E. (1898). *Naturlicheschopfungs Geschichte.* Verlag.

[225] Fyodorov, A. A., Ed. (1976). *The Vegetation Life.* Prosveshcheniye.

[226] Pertushevskaya, M. G. (1981). *Radiolaries of the Nassellaria order in the World's Oceans* (in Russian). Nauka.

[227] Bunak, V. V. (1957). Changes in the relative length of human extremity skeleton segments during the growth period. *Izv. Akad. Nauk RSFSR*, **84**, 33–45.

[228] Rokhlin, D. G. (1936). *X-ray Osteology and X-ray Anthropology* (in Russian). Biomedgiz.

[229] Klein, F. (1928). *Vorlesungen uber nicht-Euklidische Geometrie.* Springer.

[230] Petten, B. M. (1959). *The Human Embryology* (in Russian). Medgiz.

[231] Suchilin, V. A. (2017). Relativistic approach to signals and systems // Modern scientific research and innovations. https://web.snauka.ru/issues/2017/11/84761.

[232] Petoukhov, S. V. (2020). Hyperbolic rules of the cooperative organization of eukaryotic and prokaryotic genomes. *Biosystems*, **198**, 104273.

[233] Petoukhov, S. V. (2022). The stochastic organization of genomes and the doctrine of energy-information evolution based on bio-antenna arrays. *Biosystems*, **218**, 104712.

[234] Petoukhov, S. V. (2022). The stochastic organization of genomes and the doctrine of energy-information evolution based on bio-antenna arrays. doi: 10.20944/preprints202203.0100.v2.

[235] Petoukhov, S. V. (2022). Genetic intelligence, the doctrine of energy-informational evolution based on bio-antenna arrays and quantum informatics (in Russian). *Biomashsystems*, **6**(4), 60–81.

[236] Horn, R. A. and Johnson, C. R. (2012). *Matrix Analysis.* Cambridge University Press.

[237] Günther, M. and Klotz, L. (2012). Schur's theorem for a block Hadamard product. *Linear Algebra Appl.*, **437**, 948–956.

[238] Petoukhov, S. V. (2019). Nucleotide epi-chains and new nucleotide probability rules in long DNA sequences. https://www.preprints.org/manuscript/201904.0011/v2.

[239] Petoukhov, S. V. (2020). The rules of long DNA-sequences and tetra-groups of oligonucleotides. arXiv:1709.04943.

[240] Petoukhov, S. V. (1987). Higher symmetries in the biomechanics of shaping. Abstract of the dissertation for the degree of Doctor of Physical and Mathematical Sciences, Institute of Crystallography named after A.V. Shubnikov Academy of Sciences of the USSR.

[241] Petoukhov, S. V. (2022). The Universal Multilevel Relationship Between the Stochastic Organization of Genomes and the Deterministic Structure of DNA Alphabets. In *Lecture Notes on Data Engineering and Communications Technologies* (Hu, Z. B., Gavriushin S. S., Petoukhov, S. V. and He, M., Eds.), pp. 164–174. Springer.

[242] Petoukhov, S. V., Petukhova, E. S. and Svirin, V. I. (2019). Symmetries of DNA alphabets and quantum informational formalisms. *Symmetry*, **30**(2), 161–179.

[243] Slyusar, V. I. (1999). New matrix operations for digital signal processing. doi: 10.13140/RG.2.2.31620.76164/1.

[244] Minochkin, A. I., Rudakov, V. I. and Slyusar, V. I. (2011). *Fundamentals of Military-technical Research. Theory and Applications*, Vol. 2: *Synthesis of an*

Information Support Tools for Weapons and Military Equipment (in Russian). Central Research Institute of Weapons and Military Equipment of the Armed Forces of Ukraine.

[245] Planck, M. (1936). Scientific ideas, their origin, and effects. In *The Philosophy of Physics* (Planck, M., Ed.). W. W. Norton & Company.

[246] Ball, Ph. (2018). Schrödinger's cat among biology's pigeons: 75 years of *What Is Life?*. *Nature*, **560**, 548–550.

[247] Voskresensky, D. I., Gostyukhin, V. L., Maksimov, V. M. and Ponomarev, L. I. (2006). *Microwave Devices and Antennas* (in Russian). Radiotechnika.

[248] Ryapolov, A. V. and Fambulov, N. V. (2018). Anti-jamming antenna array model for satellite navigation equipment. *Journal of «Almaz — Antey» Air and Space Defence Corporation*, **3**, 20–29.

[249] Milligan, T. A. (2005). *Modern Antenna Design*, 2nd edition. John Wiley & Sons.

[250] Stutzman, W. L. and Thiele, G. A. (2012). *Antenna Theory and Design*. John Wiley & Sons.

[251] Nikandrov, V. V. (2000). Inorganic semiconductors in biological and biochemical systems: biosynthesis, properties, and photochemical activity (in Russian). *Adv. Biol. Chem.*, **40**, 357–396.

[252] Werner, D. H., Kuhirun, W. and Werner, P. L. (2003). The Peano-Gosper fractal array. *IEEE Trans. Antennas Propag.*, **51**(8), 2063–2072.

[253] Werner, D. H., Kuhirun, W. and Werner, P. L. (2004). Fractile arrays: a new class of tiled arrays with fractal boundaries. *IEEE Trans. Antennas Propag.*, **52**(8), 2008–2018.

[254] Potapov, A. A. (2002). *Fractals in Radio Physics and Radar*. Logos.

[255] Leo, V. (2015). How it works: Dolphin echolocation (in Russian). https://www.stena.ee/blog/kak-eto-rabotaet-eholokatsiya-delfinov.

[256] Albert, J. S. and Crampton, W. G. (2006). Electroreception and Electrogenesis. In *The Physiology of Fishes* (Lutz, P. L., Ed.), pp. 429–470. CRC Press.

[257] Hopkins, C. D. (1999). Design features for electric communication. *J. Exp. Biol.*, **202**(10), 1217–1228.

[258] Bullock, T. H., Hamstra, Jr. R. and Scheich, H. (1972). The jamming avoidance response of high frequency electric fish. *J. Comp. Physiol.*, **77**, 1–22.

[259] Beloussov, L. V., Voeikov, V. L. and Martynyuk, V. S. (2007). *Biophotonics and Coherent Systems in Biology*. Springer.

[260] Popp, F.-A. (1995). Modern Physical Aspects of Mitogenetic Radiation (biophotons). In *Biophotonics*, pp. 86–98. Biolnform Servicer.

[261] Voeikov, V. L. (2003). Mitogenetic Radiation, Biophotons and Non-Linear Oxidative Processes in Aqueous Media. In *Integrative Biophysics* (Popp, F.-A. and Beloussov, L., Eds.), pp. 331–360. Wolters Kluwer.

[262] Volodyaev, I. and Beloussov, L. V. (2015). Revisiting the mitogenetic effect of ultra-weak photon emission. *Front. Physiol.*, **6**, 241.

[263] Frank, G. M. (1982). *Selected Works*. Nauka.

[264] Inyushin, V. M. and Chekurov, P. R. (1975). *Biostimulation with a Laser Beam and Bioplasma*. Alma-Ata.

[265] Norman, L. J., Dodsworth, C., Foresteire, D. and Thaler, L. (2021). Human click-based echolocation: Effects of blindness and age, and real-life implications in a 10-week training program. *PloS ONE*, **16**(6), e0252330.

[266] Joannopoulos, J. D., Johnson, S. G., Winn, J. N. and Meade, R. D. (2008). *Photonic Crystals: Molding the Flow of Light*, 2nd edition. Princeton University Press.

[267] Hasan, Md. R. and Helleso, Ol. G. (2021). Dielectric optical nanoantennas. *Nanotechnology*, **32**(20), 202001.

[268] Krasnok, A. E., Maksymov, I. S., Denisyuk, A. I., Belov, P. A., Miroshnichenko, A. E., Simovskii, C. R. and Kivshar, Yu. S. (2013). Optical nanoantennas. *Phys. Usp.*, **56**, 539–564.

[269] Zaitsev, D. F. (2011). *Nanophotonics and its Applications* (in Russian). Aktion.

[270] Blau, S. (2004). Light as a feather: structural elements give peacock plumes their color. *Physics Today*, **57**(1), 18–20.

[271] Teyssier, J., Saenko, S. V., van der Marel, D. and Milinkovitch, M. C. (2015). Photonic crystals cause active colour change in chameleons. *Nat. Commun.*, **6**, 6368.

[272] Hui, W. and Ke-Qin, Zh. (2013). Photonic crystal structures with tunable structure color as colometric sensors. *Sensors*, **13**(4), 4192–4213.

[273] Herman, M. A. (1986). *Semiconductor Superlattices*. Akademie-Verlag.

[274] Shishkin, G. G. and Shishkina, A. G. (2009). *Electronics* (in Russian). Drofa.

[275] Fröhlich, H. (1988). Introduction. Theoretical Physics and Biology. In *Biological Coherence and Response to External Stimuli* (Fröhlich, H., Ed.), pp. 3–24. Springer.

[276] Sluckin, T. J., Dunmur, D. A. and Stegemeyer, H. (2004). *Crystals That Flow: Classic Papers from the History of Liquid Crystals.* Taylor & Francis.

[277] Kopp, V. I., Fan, B., Vithana, H. K. and Genack, A. Z. (1998). Low-threshold lasing at the edge of a photonic stop band in cholesteric liquid crystals. *Opt. Lett.*, **23**(21), 1707–1709.

[278] Dolgaleva, K., Wei, S. K., Lukishova, S. G., Chen, S. H., Schwertz, K. and Boyd, R. W. (2008). Enhanced laser performance of cholesteric liquid crystals doped with oligofluorene dye. *J. Opt. Soc. Am.*, **25**(9), 1496–1504.

[279] Wensink, H. H., Dunkel, J., Heidenreich, S., Drescher, K., Goldstein, R. E., Löwen, H. and Yeomans, J. M. (2012). Meso-scale turbulence in living fluids. *Proc. Natl. Acad. Sci.*, **109**(36), 14308–14313.

[280] Denisova, O. A. and Scaldin, O. A. (2013). Piezoelectric effect in liquid crystals. *Electrical and Data Processing Facilities and Systems*, **9**(4), 145–153.

[281] Yevdokimov, Yu. M., Skuridin, S. G., Salyanov, V. I., Bykov, V. A. and Palumbo, M. (2012). *Structural DNA Nanotechnology: Liquid-Crystalline Approach.* Transworld Research Network.

[282] Fujikake, H., Takizawa, K., Aida T., Negishi, T. and Kobayashi, M. (1998). Video camera system using liquid-crystal polarizing filter to reduce reflected light. *IEEE Trans. Broadcast.*, **44**(4), 419.

[283] Flack, H. D. (2009). Louis Pasteur's discovery of molecular chirality and spontaneous resolution in 1848, together with a complete review of his crystallographic and chemical work. *Acta Crystallographica A*, **65**, 371–389.

[284] Zabeo, D., Heumann, J. M., Schwartz, C. L., Suzuki-Shinjo, A., Morgan, G., Widlund, P. O. and Höög, J. L. (2018). A lumenal interrupted helix in human sperm tail microtubules. *Sci. Rep.*, **8**(1), 2727.

[285] Fraccia, T., Smith, G., Zanchetta, G., *et al.* (2015). Abiotic ligation of DNA oligomers templated by their liquid crystal ordering. *Nat. Commun.*, **6**, 6424 (2015).

[286] Feltz, B., Crommelink, M. and Goujon, Ph. (2006). *Self-organization and Emergence in Life Sciences.* Springer.

[287] Harroun, S. G., Lauzon, D., Ebert, M., Desrosiers, A., Wang, X. and Vallée-Bélisle, A. (2022). Monitoring protein conformational changes using fluorescent nanoantennas. *Nat. Methods*, **19**, 71–80.

[288] Yevdokimov, Yu. M. (2005). Nucleic acids, liquid crystals, and secrets of nanodesign (in Russian). https://www.nkj.ru/archive/articles/604/.

[289] Petoukhov, S. V. (2019). Connections Between Long Genetic and Literary Texts. The Quantum-Algorithmic Modelling. In *Advances in Computer Science for Engineering and Education II* (Hu, Z., Petoukhov, S., Dychka, I. and He, M., Eds), pp. 534–543. Springer.

[290] Barbieri, M. (2015). *Code Biology: A New Science of Life*. Springer.

[291] Drzaic, P. S. (1995). *Liquid Crystalline Dispersions*. World Scientific.

[292] Guryeva, L. L., Gerasimov, V. K., Mikhailov, Yu. M., Chalykh, A. E. and Rozenberg, B. A. (1999). Epoxyamine mixture curing with nematic liquid crystals accompanied by phase separation (in Russian). *High Molecular Weight Compounds Serie A*, **41**, 93–101.

[293] Levin, M. and Martyniuk, Ch. J. (2018). The bioelectric code: An ancient computational medium for dynamic control of growth and form. *Biosystems*, **164**, 76–93.

[294] Yuste, R. and Levin, M. (2021). New clues about the origins of biological intelligence. https://www.scientificamerican.com/article/new-clues-about-the-origins-of-biological-intelligence/.

[295] Gupta, M., Sarangi, B. R., Deschamps, J., Nematbakhsh, Y., Callan-Jones, A., Margadant, F., Mege, R. M., Lim, C. T., Voituriez, R. and Ladoux, B. (2015). Adaptive rheology and ordering of cell cytoskeleton govern matrix rigidity sensing. *Nat. Commun.*, **6**, 7525.

[296] Schakenraad, K., Ernst, J., Pomp, W., Danen, E. H. J., Merks, R. M. H., Schmidt, T. and Giomi, L. (2020). Mechanical interplay between cell shape and actin cytoskeleton organization. *Soft Matter*, **16**, 6328.

[297] Liberman, E. A., Minina, S. V., Moshkov, D. A., Santalova, I. M., Chistopolsky, I. A. and Shklovsky-Kordy, N. E. (2008). Experimental verification of the role of the cytoskeleton in solving the problems facing the animal brain by the neuron. *Biochemistry*, **73**(4), 587–591.

[298] McLaughlin, K. A. and Levin, M. (2018). Bioelectric signaling in regeneration: mechanisms of ionic controls of growth and form. *Dev. Biol.*, **433**, 177–189.

[299] Durant, F., Bischof, J., Fields, C., Morokuma, J., LaPalme, J., Hoi, A. and Levin, M. (2019). The role of early bioelectric signals in the regeneration of planarian anterior/posterior polarity. *Biophys. J.*, **116**(5), 948–961.

[300] Levin, M. (2020). The electrical blueprints that orchestrate life. https://www.youtube.com/watch?v=XheAMrS8Q1c.

[301] Pauli, W. (1948). The influence of archetypal ideas on the scientific theories of Kepler. https://docshare.tips/the-influence-of-archetypal-ideas-on-the-scientific-theories-of-kepler_5860709ab6d87fabad8b70aa.html.

[302] Anokhin, P. K. (1978). *Selected Works: Philosophical Aspects of the Theory of a Functional System* (in Russian). Nauka.

[303] Penrose, R. (1994). *Shadows of the Mind: A Search for the Missing Science of Consciousness.* Oxford University Press.

[304] Patel, A. (2001). Testing quantum dynamics in genetic information processing. *J. Genet.*, **80**(1), 39–43.

[305] Patel, A. (2001). Quantum algorithms and the genetic code. *J. Phys.*, **56**(2–3), 367–381.

[306] Patel, A. (2001). Why genetic information processing could have a quantum basis. *J. Biosci.*, **26**(2), 145–151.

[307] Abbott, D., Davies, P. C. W. and Pati, A. K., Eds. (2008). *Quantum Aspects of Life*. World Scientific.

[308] Altaisky, M. V. and Filatov, F. P. (2001). Genetic information and quantum gas.

[309] Hu, Z. B., Petoukhov, S. V. and Petukhova, E. S. (2018). On symmetries, resonances and photonic crystals in morphogenesis. *Biosystems*, **173**, 165–173.

[310] Igamberdiev, A. U. (1993). Quantum mechanical properties of biosystems: a framework for complexity, structural stability, and transformations. *Biosystems*, **31**(1), 65–73.

[311] Igamberdiev, A. I. (2004). Quantum computation, non-demolition measurements, and reflective control in living systems. *Biosystems*, **77**, 47–56.

[312] Igamberdiev, A. U. and Shklovskiy-Kordi, N. E. (2016). Computational power and generative capacity of genetic systems. *Biosystems*, **142–143**, 1–8.

[313] Igamberdiev, A. U. and Shklovskiy-Kordi, N. E. (2017). The quantum basis of spatiotemporality in perception and consciousness. *Prog. Biophys. Mol. Biol.*, **130**, 15–25.

[314] Josephson, B. D. (2018). The physics of mind and thought. https://doi.org/10.13140/RG.2.2.36516.32640/2.

[315] Khrennikov, A. (2004). *Information Dynamics in Cognitive, Psychological, Social, and Anomalous Phenomena.* Wolters Kluwer.

[316] Khrennikov, A. (2006). Quantum-like brain: Interference of minds. *Biosystems*, **84**, 225–241.

[317] Matsuno, K. (1999). Cell motility as an entangled quantum coherence. *Biosystems*, **51**, 15–19.

[318] Matsuno, K. (2003). Quantum mechanics in first, second and third person descriptions. *Biosystems*, **68**, 107–118.

[319] Matsuno, K. and Paton, R. C. (2000). Is there a biology of quantum information? *Biosystems*, **55**, 39–46.

[320] Kathuria, K., Ratan, A., McConnell, M., *et al.* (2020). Implementation of a Hamming distance — like genomic quantum classifier using inner products on ibmqx2 and ibmq_16_melbourne. *Quantum Mach. Intell.*, **2**, 7.

[321] Holmes, Z., Coble, N., Sornborger, A. T. and Subasi, Y. (2021). On nonlinear transformations in quantum computation. arXiv:2112.12307v1 [quant-ph].

[322] Guo, N., Mitarai, K. and Fujii, K. (2021). Nonlinear transformation of complex amplitudes via quantum singular value transformation. arXiv:2107.10764v1 [quant-ph].

[323] Labunets, V. G. (2003). Clifford Algebras as Unified Language for Image Processing and Pattern Recognition. In *Computational Noncommutative Algebra and Applications* (Byrnes, J. and Osteimer, G.), pp. 197–225. Wolters Kluwer.

[324] Dragovich, B., Khrennikov, A. and Misic, N. Z. (2017). Ultrametrics in the genetic code and the genome. *Appl. Math. Comput.*, **309**, 350–358.

[325] Hu, Z. B. and Petoukhov, S. V. (2017). Generalized crystallography, the genetic system and biochemical esthetics. *Struct. Chem.*, **28**(1), 239–247.

[326] Pellionisz, A. J., Graham, R., Pellionisz, P. A. and Perez, J. C. (2011). Recursive Genome Function of the Cerebellum: Geometric Unification of Neuroscience and Genomics. In *The Cerebellum and Cerebellar Disorders* (Manto, M., Ed.). Springer.

[327] Kotkovsky, S. Ya. (2020). Biosimilar properties of biquaternion zero divisors. https://all-andorra.com/biosimilar-properties-of-biquaternion-zero-divisors-by-sergey-kotkovsky/.

[328] Petoukhov, S. V. (1999). Genetic code and the ancient Chinese "Book of Changes". *Symmetry*, **10**(3–4), 211–226.

[329] Petoukhov, S. V. (2001). Genetic codes I: binary sub-alphabets, bi-symmetric matrices and golden section. *Symmetry*, **12**(3–4), 255–274.

[330] Petoukhov, S. V. (2001). Genetic codes II: numeric rules of degeneracy and a chronocyclic theory. *Symmetry*, **12**(3–4), 275–306.

[331] Petoukhov, S. V. (2001). *The Bi-periodic Table of the Genetic Code and Numbers of Protons* (in Russian). MKTS.

[332] Petoukhov, S. V. (2003). The genetic code and generalized Fibonacci matrices. *Collection of Scientific Works of Vinnytsia State Agrarian University*, **15**, 161–169.

[333] Petoukhov, S. V. (2003–2004). Attributive conception of genetic code, its bi-periodic tables and problem of unification bases of biological languages. *Symmetry*, **14–15**(1), 281–307.

[334] Petoukhov, S. V. (2004). Symmetries in Biology. In *Symmetry in Science and Art* (Shubnikov, A. V. and Koptsik, V. A., Eds.), 3rd edition, pp. 489–546. Moscow Publishing House.

[335] Petoukhov, S. V. (2005). The rules of degeneracy and segregations in genetic codes. The chronocyclic conception and parallels with Mendel's laws. In *Advances in Bioinformatics and its Applications, Proceedings of the International Conference* (He, M., Narasimhan, G. and Petoukhov, S., Eds.). World Scientific.

[336] Petoukhov, S. V. (2010). The degeneracy of the genetic code and Hadamard matrices. ARXIV:0802.3366.

[337] Petoukhov, S. V. (2012). Symmetries of the genetic code, hypercomplex numbers and genetic matrices with internal complementarities. *Symmetry*, **23**(3–4), 275–301.

[338] Petoukhov, S. V. (2016). Symmetries of the genetic code, Walsh functions and the theory of genetic logical holography. *Symmetry*, **27**(2), 95–98.

[339] Petoukhov, S. V. (2017). Genetic coding and united-hypercomplex systems in the models of algebraic biology. *Biosystems*, **158**, 31–46.

[340] Petoukhov, S. V. (2017). The genetic code, algebra of projection operators and problems of inherited biological ensembles. arXiv:1307.7882v9.

[341] Petoukhov, S. V. and Petukhova, E. S. (2017). Resonances, Walsh functions and logical holography in genetics and musicology. *Symmetry*, **28**(1), 21–40.

[342] Petoukhov, S. V. and Petukhova, E. S. (2017). Symmetries in genetic systems and the concept of geno-logical coding. *Information*, **8**(1), 2.

[343] Petoukhov, S. V., Petukhova, E. S. and Svirin, V. I. (2022). Complex and hyperbolic Fibonacci numbers and phyllotaxis. *Symmetry*, **33**(3), 209–220.

[344] Petoukhov, S. V. and Svirin, V. I. (2012). Fractal genetic nets and symmetry principles in long nucleotide sequences. *Symmetry*, **23**(3–4), 303–322.

[345] Petoukhov, S. V. and Tolokonnikov, S. V. (2020). Algebraic biology and matrix genetics systems. Presentation at the International interdisciplinary seminar "Algebraic Biology and Theory of Systems" Feb. 2020, Moscow, Russia (in Russian). https://www.youtube.com/watch?v=H2dNtvTMl1M&t=330s.

[346] Rapoport, D. R. (2016). Klein bottle logophysics, self-reference, heterarchies, genomic topologies, harmonics and evolution. *Quant. Biosyst.*, **7**(1), 1–174.

[347] Simeonov, P. L. (2013). On some recent insights in integral biomathics. *Prog. Biophys. Mol. Biol.*, **113**(1), 216–228.

[348] Tolokonnikov, G. K. and Petoukhov, S. V. (2020). From Algebraic Biology to Artificial Intelligence. In *Advances in Intelligent Systems, Computer Science and Digital Economics* (Hu, Z., Petoukhov, S. and He, M., Eds.) pp. 86–95. Springer.

[349] Tolokonnikov, G., Chernoivanov, V., Tsoi, Y. and Petoukhov, S. (2021). Systems and Algebraic-Biological Approaches to Artificial Intelligence. In *Advances in Intelligent Systems, Computer Science and Digital Economics II* (Hu, Z., Petoukhov, S. and He, M., Eds.), pp. 152–161. Springer.

[350] Fröhlich, H. (1969). Quantum Mechanical Concepts in Biology. In *From Theoretical Physics to Biology* (Marois, M., Ed.), pp. 13–22. North-Holland.

[351] Fröhlich, H. (1970). Long range coherence and the action of enzymes. *Nature*, **228**(5276), 1093.

[352] Fröhlich, H. (1978). Coherent electric vibrations in biological systems and cancer problem. *IEEE Trans. Microwave Theory Tech.*, **26**, 613–617.

[353] Fröhlich, H. (1980). The Biological Effects of Microwaves and Related Questions. In *Advances in Electronics and Electron Physics*, pp. 85–152. Academic Press.

[354] Fröhlich, H. and Kremer, F. (1983). *Coherent Excitations in Biological Systems.* Springer.

[355] Fröhlich, F. (1988). The Genetic Code as Language. In *Biological Coherence and Response to External Stimuli* (Fröhlich, H., Ed.), pp. 192–204. Springer.

[356] Vasconcellos, A. R., Vannucchi, F. S., Mascarenhas, S. and Luzzi, R. (2012). Frohlich condensate: emergence of synergetic dissipative structures in information processing biological and condensed matter systems. *Information*, **3**(4), 601–620.

[357] Hyland, G. J. (1998). Coherent GHz and THz excitations in active biosystems, and their implications. In «*The Future of Medical Diagnostics?*» Proceeding of Matra Marconi UK, Directorate of Science, Internal Report, Portsmouth, UK, 25 June 1998; pp. 14–27.

[358] Plumhof, J. D., Stoferle, Th., Mai, L., Scherf, U. and Mahrt, R. F. (2014). Room-temperature Bose-Einstein condensation of cavity exciton-polaritons in a polymer. *Nat. Mater.*, **13**(3), 247–252.

[359] Henderson, S. (2018). Bose-Einstein condensate induced at room temperature. https://www.novuslight.com/bose-einstein-condensate-induced-at-room-temperature_N8055.html.

[360] Savelev, I. and Myakishev-Rempel, M. (2020). Evidence for DNA resonance signaling via longitudinal hydrogen bonds. *Prog. Biophys. Mol. Biol.*, **156**, 14–19.

[361] Biamonte, J., Wittek, P., Pancotti, N., Rebentrost, P., Wiebe, N. and Lloud, S. (2017). Quantum machine learning. *Nature,* **549**, 195–202.

[362] Wittmann, H. G. (1961). Ansatze zur Entschlusselung des genetischen Codes. *Die Natur-wissenschaften*, **48**(24), 55.

[363] Ycas, M. (1969). *The Biological Code.* North-Holland.

[364] He, M. (2003). Genetic code, attributive mappings and stochastic matrices. *Bull. Math. Biol.*, **66**(5), 965–973.

[365] He, M., Petoukhov, S. and Ricci, P. (2004). Genetic code, Hamming distance and stochastic matrices. *Bull. Math. Biol.*, **66**, 1405–1421.

[366] He, M. and Petoukhov, S. V. (2011). *Mathematics of Bioinformatics: Theory, Practice, and Applications.* John Wiley & Sons.

[367] Prasolov, V. V. (1994). *Problems and Theorems in Linear Algebra.* American Mathematical Society.

[368] Manin, Yu. I. (1980). *Vychislimoe I nevychislimoe* [*Computable and Noncomputable*]. Sov. Radio.

[369] Petoukhov, S. V. (2020). Rules of stochastics of genomic DNAs, biological dualism "stochastics-determinism", and tensor-unitary transformations. doi: 10.20944/preprints202208.0435.v1.

[370] Bishop, R. (2017). Chaos. In *The Stanford Encyclopedia of Philosophy* (Zalta, E. N., Ed., Spring 2017 edition). Stanford University.

[371] Weinberger, D. (2019). *Everyday Chaos — Technology, Complexity, and How We're Thriving in a New World of Possibility.* Harvard Business Review Press.

[372] Graham, R. L., Knuth, D. E., and Parashnik, O. (1994). *Concrete Mathematics: A Foundations for Computer Science.* Addison-Wesley.

[373] Kappraff, J. (2006). Anne Bulckens' analysis of the proportions of the Parthenon. *Symmetry*, **17**(1–2), 91–96.

[374] Kappraff, J. (2000). The arithmetic of Nichomachus of Gerasa and its applications to systems of proportions. *Nexus Netw. J.*, **2**(4), 41.

[375] Kappraff, J. (2002). *Beyond Measure: Essays in Nature, Myth, and Number.* World Scientific.

[376] Walter, M., Gross, D. and Eisert, J. (2017). Multi-partite entanglement. arXiv:1612.02437.

[377] Horodecki, R., Horodecki, P., Horodecki, M. and Horodecki, K. (2009). Quantum entanglement. *Rev. Mod. Phys.*, **81**, 865.

[378] Gühne, O. and Tóth, G. (2009). Entanglement detection. *Phys. Rep.*, **474**, 1–75.

[379] Amico, L., Fazio, R., Osterloh, A. and Vedral, V. (2008). Entanglement in many-body systems. *Rev. Mod. Phys.*, **80**, 517.

[380] Mandelbrot, B. B. (1983). *The Fractal Geometry of Nature.* Macmillan.

[381] Jeffrey, H. J. (1990). Chaos game representation of gene structure. *Nucleic Acids Res.*, **18**(8), 2163–2170.

[382] Baish, J. W. and Jain, R. K. (2000). Fractals and cancer. *Cancer Res*, **60**, 3683–3688.

[383] Peng, C. K., Buldyrev, S. V., Goldberger, A. L., Havlin, S., Sclortino, F., Simons, M. and Stanley, H. E. (1992). Fractal landscape analysis of DNA walks. *Physica A*, **191**(1–4), 25–29.

[384] Dokukin, M. E., Guz, N. V., Woodworth, C. D. and Sokolov, I. (2015). Emergence of fractal geometry on the surface of human cervical epithelial cells during progression towards cancer. *New J. Phys.*, **17**(3), 033019.

[385] Bizzarri, M., Giuliani, A., Cucina, A., Anselmi, F. D., Soto, A. M. and Sonnenschein, C. (2011). Fractal analysis in a systems biology approach to cancer. *Semin. Cancer Biol.*, **21**(3), 175–182.

[386] Lennon, F. E., Cianci, G. C., Cipriani, N. A., Hensing, T. A., Zhang, H. J., Chen, C.-T., Murgu, S. D., Vokes, E. E., Vannier, M. W. and Salgia, R. (2015). Lung cancer — a fractal viewpoint. *Nat. Rev. Clin. Oncol.*, **12**(11), 664–675.

[387] Lieberman-Aiden, E., van Berkum, N. L., Williams, L., Imakaev, M., Ragoczy, T., Telling, A., Amit, I., Lajoie, B. R., Sabo, P. J., Dorschner, M. O., Sandstrom, R., Bernstein, B., Bender, M. A., Groudine, M., Gnirke, A., Stamatoyannopoulos, J., Mirny, L. A., Lander, E. S. and Dekker, J. (2009). Comprehensive mapping of long-range interactions reveals folding principles of the human genome. *Science*, **326**(5950), 289–293.

[388] Skiena, S. (1990). Adjacency Matrices. In *Implementing Discrete Mathematics: Combinatorics and Graph Theory with Mathematica*, pp. 81–85. Addison-Wesley.

[389] Kaczmarek, L. K. and Levitan, I. B. (1987). *Neuromodulation: The Biochemical Control of Neuronal Excitability.* Oxford University Press.

[390] Martini, F. H. (2005). *Anatomy and Physiology.* Prentice Hall.

[391] Lawler, G. F., Schramm, O. and Werner, W. (2004). Conformal invariance of planar loop-erased random walks and uniform spanning trees. *Ann. Probab.*, **32**(1B), 939–995.

[392] Kline, M. (1980). *Mathematics: The Loss of Certainty.* Oxford University Press.

[393] Annett, M. (1985). *Left, Right, Hand and Brain: The Right Shift Theory.* Lawrence Erlbaum.

[394] Annett, M. (1992). Spatial ability in subgroup of left- and right-handers. *Br. J. Psych.*, **83**, 493–962.

[395] Gazzaniga, M. S. (1995). Principles of human brain organization derived from split brain studies. *Neuron*, **14**, 217–228.

[396] Hellige, J. B. (1993). *Hemispheric Asymmetry: What's Right and What's Left.* Harvard University Press.

[397] Penrose, R. (1999). *The Emperor's New Mind: Concerning Computers, Minds, and the Laws of Physics.* Oxford University Press.

[398] Kakei, Y., Katayama, S., Lee, S., *et al.* (2022). Integration of body-mounted ultrasoft organic solar cell on cyborg insects with intact mobility. *NPJ Flex. Electron.*, **6**, 78.

[399] Arnold, V. I. (2006). *Public lecture at Mathematical Institute by V.A. Steklov* (in Russian), "Publichnaya lektsiia v Matematicheskom institute imeni V.A.Steklova", elementy.ru/lib/430178/430281.

[400] Carlevaro, C. M., Irastorza, R. M. and Vericat, F. (2016). Quaternionic representation of the genetic code. *Biosystems*, **141**, 10–19.

[401] Carlevaro, C. M., Irastorza, R. M. and Vericat, F. (2016). Chirality in a quaternionic representation of the genetic code. *Biosystems*, **150**, 99–109.

[402] Nemzer, L. R. (2017). A binary representation of the genetic code. *Biosystems*, **155**, 10–19.

[403] Seligmann, H. (2011). Two genetic codes, one genome: Frameshifted primate mitochondrial genes code for additional proteins in presence of antisense antitermination tRNAs. *Biosystems*, **105**, 271–285.

[404] Davis, P. J. (1964). Arithmetics. In *Mathematics in the Modern World*, pp. 29–45. Scientific American.

[405] Pavlov, D. G. (2004). Leading articles (in Russian). *Hypercomplex Numbers in Geometry and in Physics*, **1**(1), 4–7.

[406] Russel, B. (1967). *A History of Western Philosophy*. Simon & Schuster.

[407] Heisenberg, W. (2000). *Physics and Philosophy: The Revolution in Modern Science*. Penguin Classics.

[408] Weinberger, N. M. (2004). Music and brain. *Sci. Amer.*, **291**(5), 88–95.

[409] Voloshinov, A. V. (2000). *Mathematics and Arts* (in Russian). Prosveschenie.

[410] Hesse, H. (1975). *The Glass Bead Game*. Penguin.

[411] Feynman, R., Leighton, R. and Sands, M. (1963). *The Feynman Lectures*. Pergamon Press.

[412] Sethares, W. A. (2005). *Tuning, Timbre, Spectrum, Scale*. Springer.

[413] Apel, W. (1969). *Harvard Dictionary of Music*. Harvard University Press.

[414] Day-O'Connell, J. (2007). *Pentatonicism from the Eighteenth Century to Debussy*. University Rochester Press.

[415] Christidis, A.-F., Arapopoulou, M. and Christi, M. (2007). *A History of Ancient Greek: From the Beginnings to Late Antiquity*. Cambridge University Press.

[416] Olsen, D. A. and Sheehy, D. E., Eds. (1998). *The Garland Encyclopedia of World Music, Volume 2: South America, Mexico, Central America, and the Caribbean*. Taylor & Francis.

[417] Todd Titon, J. T. (1996). *Worlds of Music: An Introduction to the Music of the World's Peoples*, Shorter Version. Cengage Learning.

[418] Bhattacharya, S. (1968). *Ethno-musicology and India*. Indian Publications.

[419] Baines, A. (1991). *Woodwind Instruments and Their History*. Courier Dover Publications.

[420] Darvas, G. (2007). *Symmetry*. Birkhäuser.

[421] Carrasco, J., Michaelides, A., Forster, M., Haq, S., Raval, R. and Hodgson, A. (2009). A one-dimensional ice structure built from pentagons. *Nat. Mater.*, **8**, 427–431.

[422] Bandyopadhyay, A. (2014). Where does music exist? https://www.youtube.com/watch?v=N5_fhlEmJI8.

[423] Le Corbusier, Sh. (1953). *Der Modulor.* DVA.

[424] Fine, N. J. (1949). On the Walsh functions. *Trans. Am. Soc.*, **3**, 372–414.

[425] Karpovsky, M. G., Stankovic, R. S. and Astola, J. T. (2008). *Spectral Logic and its Applications for the Design of Digital Devices.* John Wiley & Sons.

[426] Derzhypolskyy, A., Melenevskyy, D. and Gnatovskyy, A. (2007). A comparative analysis of associative properties of Fourier versus Walsh digital holograms. *Acta Physica Polonica A*, **112**(5), pp. 1101–1106.

[427] Golubov, B. I., Efimov, A. V. and Skvortsov, V. A. (1987). *Series and Fourier Transforms* (in Russian). Nauka.

[428] Nie, E., Kleinermanns, K. and deVries, M. S. (2000). Pairing of isolated nucleic-acid bases in the absence of the DNA backbone. *Nature*, **408**, 949–951.

[429] Wenyon, M. (1978). *Understanding Holography.* David and Charles.

[430] Karpovsky, M. G. and Moscalev, E. S. (1973). *Spectral Methods of Analysis and Synthesis of Discrete Devices* (in Russian). Energia.

[431] Zalmanzon, L. A. (1989). *Transforms of Fourier, Walsh, Haar and their Application in Management, Communications and Other Fields* (in Russian). Nauka.

[432] Yaglom, I. M. (1980). *The Boolean Structure and its Models.* Sovetskoye Radio.

[433] Hameroff, S. R. and Penrose, R. (1996). Conscious events as orchestrated spacetime selections. *J. Conscious. Stud.*, **3**(1), 36–53.

[434] Liberman, E. A. (1972). Cell as a molecular computer. *Biofizika*, **17**, 932–943.

[435] Hameroff, S. and Penrose, R. (2014). Consciousness in the universe: A review of the 'Orch OR' theory. *Phys. Life Rev.*, **11**(1), 39–78.

[436] McClintock, B. (1950). The origin and behavior of mutable loci in maize. *Proc. Natl. Acad. Sci.*, **36**(6), 344–355.

[437] Stankovic, R. and Astola, J. (2003). *Spectral Interpretation of Decision Diagrams.* Springer.

[438] Stankovic, R. and Astola, J. (2008). *Gibbs Derivatives — the First Forty Years.* Tampere International Center for Signal Processing.

[439] Darvas, G. (2018). Petoukhov's rules on symmetries in long DNA-texts. *Symmetry*, **29**(2), 318–320.

Index

www.ingramcontent.com/pod-product-compliance
Ingram Content Group UK Ltd.
Pitfield, Milton Keynes, MK11 3LW, UK
UKHW022107170926
8624IPUK00024B/176

9 789811 278259